Annals of Mathematics Studies

Number 107

CHARACTERS OF REDUCTIVE GROUPS OVER A FINITE FIELD

BY

GEORGE LUSZTIG

PRINCETON UNIVERSITY PRESS

PRINCETON, NEW JERSEY

1984

Clothbound editions of Princeton University Press books are printed on acid-free paper, and binding materials are chosen for strength and durability. Paperbacks, while satisfactory for personal collections, are not usually suitable for library rebinding

ISBN 0-691-08350-9 (cloth)
ISBN 0-691-08351-7 (paper)

Printed in the United States of America
by Princeton University Press, 41 William Street
Princeton, New Jersey

☆

Library of Congress Cataloging in Publication data will be found on the last printed page of this book

To Miki, Irene and Tamar

TABLE OF CONTENTS

INTRODUCTION ix

1. COMPUTATION OF LOCAL INTERSECTION COHOMOLOGY OF CERTAIN LINE BUNDLES OVER A SCHUBERT VARIETY 3
2. LOCAL INTERSECTION COHOMOLOGY WITH TWISTED COEFFICIENTS OF THE CLOSURES OF THE VARIETIES X_w 30
3. GLOBAL INTERSECTION COHOMOLOGY WITH TWISTED COEFFICIENTS OF THE VARIETY $\overline{X}_w$ 58
4. REPRESENTATIONS OF WEYL GROUPS 76
5. CELLS IN WEYL GROUPS 134
6. AN INTEGRALITY THEOREM AND A DISJOINTNESS THEOREM 180
7. SOME EXCEPTIONAL GROUPS 217
8. DECOMPOSITION OF INDUCED REPRESENTATIONS 251
9. CLASSICAL GROUPS 269
10. COMPLETION OF THE PROOF OF THEOREM 4.23 296
11. EIGENVALUES OF FROBENIUS 313
12. ON THE STRUCTURE OF LEFT CELLS 324
13. RELATIONS WITH CONJUGACY CLASSES 342
14. CONCLUDING REMARKS 351

APPENDIX 358

REFERENCES 377

SUBJECT INDEX 382

NOTATION INDEX 383

INTRODUCTION

One of the aims of this book is to present a classification of the irreducible representations of the finite group $G(F_q)$, where G is a connected, reductive algebraic group with connected center defined over a finite field F_q. The representations of $G(F_q)$ will be always taken over $\overline{Q}_\ell$, an algebraic closure of the field of ℓ-adic numbers (ℓ is a fixed prime not dividing q). This is because the representations of $G(F_q)$ are studied (following [DL_1]) using the ℓ-adic cohomology with compact support of certain locally closed $G(F_q)$-stable subvarieties X_w of the flag manifold of G, with coefficients in certain locally constant ℓ-adic sheaves $\mathcal{F}_\theta$ of rank 1. (Here w is a Weyl group element, X_w is the set of Borel subgroups which are in relative position w with their transform under Frobenius and θ is a character of the group of rational points of a maximal torus T over F_q, corresponding to w.)

The main object of [DL_1] was the study of the virtual representation R_T^θ of $G(F_q)$ obtained by taking the alternating sum over i of the $G(F_q)$-modules $H_c^i(X_w, \mathcal{F}_\theta)$. The main result of the present work is an explicit formula for the multiplicities with which the various irreducible representations of $G(F_q)$ appear in R_T^θ.

It is shown in [DL_1] that the set of pairs (T, θ) can be partitioned in equivalence classes with the following properties: the equivalence classes are in 1-1 correspondence with the semisimple conjugacy classes in $G^*(F_q)$, where G^* is a reductive group over F_q, dual to G; moreover, for any irreducible representation ρ of $G(F_q)$ there is exactly one equivalence class of pairs (T, θ) such that ρ appears with nonzero multiplicity in R_T^θ for some (T, θ) in that equivalence class. If this equivalence class corresponds to the semisimple class (s) in $G^*(F_q)$, we shall write $\rho \in \mathcal{E}_{(s)}$. This gives a partition of the set of irreducible representations of $G(F_q)$ into disjoint pieces $\mathcal{E}_{(s)}$, one for each semisimple class (s) in $G^*(F_q)$. The piece $\mathcal{E}_{(1)}$ is the set of "unipotent representations" of $G(F_q)$; the corresponding equivalence class of pairs (T, θ) consists of those (T, θ) for which θ is trivial. At the other extreme, we have pieces of form $\mathcal{E}_{(s)}$ where (s) is a regular semisimple class; each of these pieces consists of a single representation of $G(F_q)$.

For a general (s), we show that $\mathcal{E}_{(s)}$ is in 1-1 correspondence with the set of unipotent representations of a smaller group $H(F_q)$, where H is the dual of the centralizer in G^* of an element $s \in (s)$. Moreover, the multiplicities of the various $\rho \in \mathcal{E}_{(s)}$ in the virtual representations R_T^θ with (T, θ) in the corresponding equivalence class, are the same (up to a sign $\varepsilon_{(s)} = \pm 1$, depending only on (s)) as the multiplicities of the unipotent representations of $H(F_q)$ in the virtual representations $R_{T'}^1$, $(T' \subset H)$, defined with respect to H; the degree of $\rho \in \mathcal{E}_{(s)}$ is obtained by multiplying the degree of the corresponding unipotent representation of $H(F_q)$ by the part prime to q of the integer $\varepsilon_{(s)} |G(F_q)| \cdot |H(F_q)|^{-1}$. This result is entirely analogous to the Jordan decomposition of an element in an algebraic group as a product of a semisimple and a unipotent element.

To state our results on unipotent representations we shall need the concept of left cells in a Weyl group. This concept has its origin in work of Robinson and Shensted, who described a 1-1 correspondence between the symmetric group $\mathfrak{S}_n$ and the set of pairs (τ, τ') of standard Young tableaux of the same shape, of size n. In this picture, a left cell of $\mathfrak{S}_n$ appears as the subset of $\mathfrak{S}_n$ corresponding to the set of pairs (τ, τ') with τ' fixed and τ of the same shape as τ'; one could also define a two-sided cell of $\mathfrak{S}_n$ as the subset of $\mathfrak{S}_n$ corresponding to the set of pairs (τ, τ') with τ, τ' of fixed shape. Thus, $\mathfrak{S}_n$ is decomposed into two-sided cells (one for each partition of n) and each two-sided cell is a disjoint union of a number of left cells equal to the dimension of the corresponding irreducible representation of $\mathfrak{S}_n$. This is the prototype of a picture which applies to any Coxeter group.

In his study of primitive ideals in the enveloping algebra of a complex semisimple Lie algebra, Joseph has defined the concept of left cells in the Weyl group W, as follows. For $w \in W$, let I_w be the annihilator of the irreducible module of the enveloping algebra with highest weight $-w\rho-\rho$, where ρ is half the sum of positive roots. Then w, w' are said to be in the same left cell precisely when $I_w = I_{w'}$. Joseph shows that each left cell can be regarded in a natural way as a basis of a representation of W; this representation is not necessarily irreducible, but Joseph [J] shows that it has a canonical "lowest" irreducible component, which appears with multiplicity one and is characterized by the property that its first occurrence in the space of harmonic polynomials on the Cartan subalgebra is earlier than for any other irreducible component.

Joseph's definition of left cells and the corresponding Weyl group representations involved some unknown quantities: the multiplicities in the Jordan-Hölder series of the Verma modules with highest weight $-w\rho \mp \rho$. In [KL_1], Kazhdan and the author have proposed an algorithm for these multiplicities and in [KL_2], [BB], [BK] it has been shown that this algorithm does indeed give the desired multiplicities. This had several consequences. It has allowed Barbasch and Vogan [BV_1], [BV_2] to determine explicitly which irreducible representations of the Weyl group can appear as "lowest" components of a left cell, (they turn out to be precisely the "special" representations defined in [L_6], [L_7]), and to determine explicitly the equivalence relation $E \underset{LR}{\sim} E'$ on irreducible representations of W generated by the relation: "E, E' appear as components of the same left cell." (It turns out that this equivalence relation is precisely that of [L_7].)

In this book we shall adopt the definition of left cells given in [KL_1] (see Chapter 5). This has the advantage that it is elementary, makes sense for arbitrary Coxeter groups and that it gives rise not only to representations of the Coxeter group, but also of the corresponding Hecke algebra.* (This last fact is crucial for applications to our group $G(F_q)$.)

---

*The Hecke algebra corresponding to a Weyl group has been first studied by N. Iwahori (On the structure of the Hecke ring of a Chevalley group over a finite field, J. Fac. Sci. Univ. Tokyo, sect 1A, 10(1964) 215-236. I think it would be most appropriate to call it the Iwahori algebra, but the name Hecke ring (or algebra) given by Iwahori himself has been in use for almost 20 years and it is probably too late to change it now.

On the other hand, as a consequence of [KL_2], [BB], [BK] and [Vo], this definition of left cells coincides with that of Joseph, in the case of Weyl groups.

We shall also need the concept of two-sided cells in W. In the language of primitive ideals, a two-sided cell could be defined as the union of all left cells with a prescribed "lowest" component. (It turns out that the number of left cells which make up a two-sided cell is equal to the dimension of that lowest component.) Again, we shall adopt the definition of two-sided cells given in [KL_1]; the definition (see Chapter 5) applies to any Coxeter group and gives rise to both left and right module structures over the Hecke algebra or the Weyl group on the vector space with basis given by the elements of the two-sided cell. The equivalence relation $E \underset{LR}{\sim} E'$ defined earlier could be also defined by the requirement that E, E' appear in the same two-sided cell, (for either the left or the right W-module structure).

There are properties of left cells which are needed in the representation theory of $G(F_q)$ for which no proof is known which is independent of the theory of primitive ideals.

We shall now given an example of such a property. Let E be an irreducible representation of W and let E(u) be the "corresponding" irreducible representation of the Hecke algebra (see 3.1) with parameter u. (Already at this stage, if one wants an explicit construction of E(u) starting from E, instead of an abstract existence statement, one must make use of the isomorphism [L_9] between the Hecke algebra and the group algebra of W over $\mathbf{Q}(u^{1/2})$, which depends on the bridge with the theory of primitive ideals.) Let $(T_x)_{x\in W}$ be the standard basis of the Hecke algebra. Let a_E be the smallest integer ≥ 0 such that the expression

$$(-1)^{\ell(x)} u^{-\frac{\ell(x)}{2}+\frac{a_E}{2}} \operatorname{Tr}(T_x, E(u))$$

is a polynomial in $u^{1/2}$, for all $x \in W$, and let $c_{x,E} \in \mathbf{Z}$ be the constant term of this polynomial. (We shall encounter $c_{x,E}$ later in this introduction.) The property we need is: "If E, E' are irreducible representations of W appearing in the same left cell then $a_E = a_{E'}$." The only known proof of this statement is by using the bridge with the theory of primitive ideals. (We deduce it from $[BV_1]$, $[BV_2]$.)

Conversely, our results on representations of $G(F_q)$ have some interesting applications to the structure of left cells in W. We show, for example (Chapter 12) that, if W is of classical type, then the number of involutions contained in a left cell is a power of 2 and this number is the same for all left cells contained in a fixed two-sided cell. For general W, we show that in any left cell Γ, there is a unique element x_0 such that $c_{x_0,E}$ is equal to the multiplicity of E in the W-module Γ, for any irreducible W-module E. (One can show that $x_0 = x_0^{-1}$, and is likely that x_0 is what in theory of primitive ideals one calls the Duflo involution in Γ.) Let Γ^* be the set of elements x in Γ such that $c_{x,E} \neq 0$ for some irreducible representation E of W. (It follows that E is necessarily a component of Γ.) These elements play an essential role in our work. We conjecture that $\Gamma^* = \Gamma \cap \Gamma^{-1}$; it will be shown elsewhere that this is true for classical groups, in which case $\Gamma \cap \Gamma^{-1}$ consists of the involutions in Γ.

With each two-sided cell c in W we associate a finite group $\mathcal{G}_c$. In the case where W is irreducible, $\mathcal{G}_c$ is either an elementary abelian 2-group or one of the symmetric groups $\mathfrak{S}_3, \mathfrak{S}_4, \mathfrak{S}_5$, and is therefore determined by its order which is equal to the denominator of the "formal dimension" (see 3.3) of the special Weyl group representation corresponding to c. Another definition (see 13.1) describes $\mathcal{G}_c$ as an explicit quotient of the group of components of the centralizer of a unipotent element u in a complex reductive group with Weyl group W, where u is associated with c by a procedure involving Springer's representations of W. The group $\mathcal{G}_c$ has also the following interpretation, (see 7.1 (ii)). There exists a left cell Γ contained in c, such that the matrix $(c_{x,E})$, (where x runs over Γ^* and E runs over the irreducible components of Γ), coincides with the character table of $\mathcal{G}_c$. Here the indices x correspond to the irreducible characters of $\mathcal{G}_c$ and the indices E correspond to the conjugacy classes of $\mathcal{G}_c$.

In the case where W is of type A_n, $\mathcal{G}_c$ is trivial. If W is of type B_n or C_n, then $\mathcal{G}_c$ is an elementary abelian 2-group of order 2^d where $d^2+d \leq n$. If W is of type D_n, then $\mathcal{G}_c$ is an elementary 2-group or order 2^d with $d^2 \leq n$. If W is of type E_8 (resp. F_4), there is a unique 2-sided cell c in W such that $\mathcal{G}_c = \mathfrak{S}_5$ (resp. $\mathfrak{S}_4$). It would be interesting to find a definition of $\mathcal{G}_c$ in the framework of primitive ideals.

Let $W^\vee(c)$ be the equivalence class corresponding to c, for the equivalence relation $E \underset{LR}{\sim} E'$ on the set of irreducible representations of W. In Chapter 4, we define (case by case) a bijection $E \leftrightarrow x_E$ between $W^\vee(c)$ and a subset of the set $\mathfrak{M}(\mathcal{G}_c)$ consisting of all pairs (x,σ) where x is an element of $\mathcal{G}_c$ (taken up to conjugacy) and σ is an irreducible representation of the centralizer of x in $\mathcal{G}_c$. (This generalizes the parametrization given in $[L_3]$ of representations of Weyl groups of classical types in terms of symbols.) If $\mathcal{G}_c = \mathfrak{S}_3, \mathfrak{S}_4$, or $\mathfrak{S}_5$, then $\mathfrak{M}(\mathcal{G}_c)$ has respectively 8, 21, 39 elements.

In the case where $\mathcal{G}_c$ is abelian, we have $\mathfrak{M}(\mathcal{G}_c) = \mathcal{G}_c \times \mathcal{G}_c^*$ where $\mathcal{G}_c^*$ is the Pontrjagin dual of $\mathcal{G}_c$, hence $\mathfrak{M}(\mathcal{G}_c)$ is a vector space of even dimension 2d over the field F_2. It has a natural pairing $\{\ ,\ \}: \mathfrak{M}(\mathcal{G}_c) \times \mathfrak{M}(\mathcal{G}_c) \to \bar{\mathbf{Q}}_\ell$ defined by $\{m,m'\} = 2^{-d}(-1)^{(m,m')}$ where $(\ ,\ )$ is the symplectic form $\mathfrak{M}(\mathcal{G}_c) \times \mathfrak{M}(\mathcal{G}_c) \to F_2$ defined by $(-1)^{(m,m')} = \sigma(y)\tau(x)$, $m = (x,\sigma)$, $m' = (y,\tau)$. The pairing $\{\ ,\ \}$ makes sense even if $\mathcal{G}_c$ is non-abelian (see Chapter 4, (4.14.3)) and using it one can define the Fourier transform of a function on $\mathfrak{M}(\mathcal{G}_c)$, just as in the abelian case (see 12.1).

Now let Γ be a left cell contained in c. As a W-module we have $\Gamma = \sum_E (E:\Gamma)E$, where E runs over the irreducible components of Γ, and $(E:\Gamma)$ is the multiplicity of E in Γ. Consider the function ϕ_Γ on $\mathcal{M}(\mathcal{G}_c)$ defined by $\phi_\Gamma(m) = (E:\Gamma)$ if $m = x_E$ for some E as above and $\phi_\Gamma(m) = 0$ for all other m. We show (Chapter 12) that the function ϕ_Γ is equal to its own Fourier transform. This gives very severe restrictions on the structure of Γ as a W-module. In particular, when W is a classical group we can deduce that ϕ_Γ is the characteristic function of a maximal isotropic subspace in $\mathcal{M}(\mathcal{G}_c)$ with respect to $(\ , \ )$ and hence the W-module Γ is multiplicity free with 2^d components, $(2^d = |\mathcal{G}_c|)$.

There is considerable evidence (which is based partly on joint work with D. Alvis on formal dimensions in type H_4) that the sets $\mathcal{M}(\mathcal{G}_c)$ with the pairing $\{ \ , \ \}$ have also a meaning for finite noncrystalographic Coxeter groups, although the groups $\mathcal{G}_c$ themselves have no meaning. It would be extremely interesting to understand the structure of left cells and two-sided cells in the case of affine Weyl groups. This would lead to an understanding of the unramified representations of split p-adic semi-simple groups.

To state our results on unipotent representations of $G(F_q)$ we shall need some definitions. For $w \in W$, we set $R_w = R^1_{T(w)}$, where $T(w)$ is a maximal torus in G, defined over F_q, obtained from a maximally split torus by twisting according to w. Let $F: W \to W$ be the automorphism of W (as a Coxeter group) induced by the Frobenius map on G.

We consider an irreducible $\mathbf{Q}[W]$-module E such that $\mathrm{Tr}(w, E) = \mathrm{Tr}(F(w), E)$ for all $w \in W$. Then there exists a $\mathbf{Q}$-linear map $\Phi: E \to E$ of finite order such that $\Phi(wv) = F(w)\Phi(v)$ for all $w \in W$, $v \in E$, (see 3.2); Φ is unique apart from the fact that it can be replaced by $-\Phi$. For such E, Φ, we define

$$R_{E,\Phi} = |W|^{-1} \sum_{w \in W} \mathrm{Tr}(\Phi w, E) R_w .$$

This is a $\mathbf{Q}$-linear combination of unipotent representations of $G(F_q)$.

Now let (E', Φ') be another pair like (E, Φ). One of our results is the following disjointness theorem.

We have $E \underset{LR}{\sim} E'$ *if and only if there exists a unipotent representation of* $G(F_q)$ *which appears with nonzero coefficient in both* $R_{E,\Phi}$ *and* $R_{E',\Phi'}$.

Since any unipotent representation appears in some $R_{E,\Phi}$ with nonzero coefficient, we can attach to it the two-sided cell of W to which E belongs; this two-sided cell is well defined by the previous statement and is necessarily stable under the automorphism $F: W \to W$. We thus get a partition of the set $\mathfrak{U}$ of unipotent representations of $G(F_q)$ into disjoint nonempty classes $\mathfrak{U}_c$, one class for each F-stable two-sided cell c of W. This reduces our problem to the following one: classify the unipotent representations of $G(F_q)$ corresponding to a fixed F-stable two-sided cell of W, and find their multiplicities in $R_{E,\Phi}$ for E appearing in that two-sided cell. If our $G(F_q)$ was a general linear group or a unitary group, each $\mathfrak{U}_c$ would consist of a single unipotent representation which is equal to $\pm R_{E,\Phi}$ where E is the unique irreducible representation of W appearing in c; the sign is easily determined from the condition that $\dim(\pm R_{E,\Phi}) > 0$. We shall now formulate the result in the case where F acts trivially on W, W arbitrary. (See Theorem 4.3 for the formulation in the general case.) We set $R_E = R_{E,\Phi}$, (Φ = identity). Then:

there exists a bijection $\rho \leftrightarrow \bar{x}_\rho$ *between the sets* $\mathfrak{U}_c$ *and* $\mathfrak{M}(\mathcal{G}_c)$ *such that for any* $E \in W^\vee(c)$ *we have*

$$(\text{multiplicity of } \rho \text{ in } R_E) = \Delta(\bar{x}_\rho)\ \{\bar{x}_\rho, x_E\}$$

where $\{\ ,\ \}$ *is the canonical pairing on* $\mathfrak{M}(\mathcal{G}_c)$ *and* $\Delta: \mathfrak{M}(\mathcal{G}_c) \to \{\pm 1\}$ *is a certain function which is identically* 1 *unless* W *has some factors of type* E_7 *or* E_8.

The question of classification of unipotent representation and the multiplicity formula for them has been already treated in $[L_8]$, $[L_{11}]$, $[L_{12}]$, $[As_3]$. (The generalization to nonunipotent representations is new.) The proof presented here has the advantage that it is independent of [L_3]; in particular, it does not depend on counting conjugacy classes, which was particularly troublesome for classical groups in characteristic 2. We have made however use of some of the easier results in [L_{11}], [L_{12}]. Unlike the proofs in [As_3], the present proofs make no use of the lifting theory of [Ka]. We have made use of the results of [L_8]; note, however, that most of [L_8] was concerned with the proof of two lemmas, which are now both trivial consequences of Theorem 6.17.

This work would have been not possible without the recent advances in étale intersection cohomology. Generalizing a construction of Goresky-MacPherson, Deligne [BBD] has defined for a singular algebraic variety over an algebraically closed field, local and global intersection cohomology groups with ℓ-adic coefficients with some very agreeable properties (such as local Poincaré duality) which one would expect if the variety was nonsingular. By results of Deligne and Gabber [D], [BBD], this theory satisfies (in the proper case) an analogue of the Weil conjectures.

It has turned out that, when applied to certain singular varieties arising from algebraic groups, intersection cohomology has representation theoretic significance. In particular, the local intersection cohomology of Schubert varieties can be interpreted in terms of multiplicities in Verma modules with regular integral highest weight for a complex semisimple Lie algebra, (see [KL_2], [BB], [BK]).

In the present work, intersection cohomology (in the étale version) is applied to the closure $\overline{X}_w$ of the variety X_w and is taken with coefficients in the locally constant ℓ-adic sheaves $\mathcal{F}_\theta$ on X_w. The global intersection cohomology groups carry representations of $G(F_q)$; their alternating sum $\overline{R}_T^\theta$ can be used as a substitute for R_T^θ. Indeed, it is a linear combination of elements $R_{T(y)}^{\theta'}$ where y runs over the elements less than w in the standard partial order on W. The coefficients in this linear combination depend on the knowledge of the local intersection cohomology groups of $\overline{X}_w$ with respect to $\mathcal{F}_\theta$. If these local groups are known, then it would follow that any information on $\bar{R}_T^\theta$ can be transferred to R_T^θ and conversely. However, the $\bar{R}_T^\theta$ have the advantage that, due to the Weil conjectures for $\bar{X}_w$ and $\mathcal{F}_\theta$, one can get hold on the individual terms whose alternating sum is $\bar{R}_T^\theta$. Thus one can work with actual representations instead of virtual ones, and this gives much additional information.

To start with, we have to compute the local intersection cohomology of $\bar{X}_w$ with coefficients in $\mathcal{F}_\theta$. When $\theta = 1$, these local groups can be identified with the corresponding local groups for a Schubert varieties and are therefore known from [KL_2]. When $\theta \neq 1$, the computation of the local groups is reduced in Chapter 2 to a problem which makes sense also in characteristic zero and involves line bundles with zero section removed over a Schubert variety. This problem (which is solved in Chapter 1) has first appeared in work of Beilinson and Bernstein on multiplicities in Verma modules with regular rational highest weight. Thus Chapter 1 has an independent interest, in that, in combination with the results of Beilinson and Bernstein, it gives formulas for the multiplicities in Verma modules with regular rational highest weight.

The third chapter, which was influenced by the papers [As_2] and [DM_2] of Asai and Digne-Michel, shows that each global intersection cohomology group of $\bar{X}_w$ with coefficients in $\mathcal{F}_\theta$, is a linear combination of $R_{T'}^{\theta'}$'s with coefficients given in terms of characters of a Hecke algebra. Given sufficient information on the characters of Hecke algebras, this provides many explicit $\mathbb{Q}$-linear combinations of $R_{T'}^{\theta'}$'s which are known to be actual representations. This is complemented by Propositions 6.9 and 6.10 in Chapter 6 which prove (assuming q is large enough) that the standard duality operator (6.8.1) applied to $(-1)^i H^i(\bar{X}_w, \mathcal{F}_\theta)$ is an actual representation of $G(F_q)$ which is again a linear combination of $R_{T'}^{\theta'}$'s, as before. The proof of this result depends on the fact that the X_w are affine, which is only known for $q \geq$ Coxeter number. (The result is later shown to be true for any q, by a rather unpleasant argument.) To be able to apply these results one needs sufficient information on the characters of Hecke algebras. The kind of information we need is the following. We need to compute as many virtual representations

$$a_x = \sum_E c_{x,E}\, E \qquad (x \in W)$$

of W as possible. (Note that $\pm c_{x,E}$ is the leading coefficient in the trace of T_x on the Hecke algebra representation $E(u)$ hence it is a piece of a character value of the Hecke algebra.) This information is developed in Chapters 4 and 5. In Chapter 5, we use the elements a_x to give a new proof of some results of Barbasch and Vogan, which are needed in the following chapters. Chapter 5 contains also a generalization (Theorem 5.7) of the usual orthogonality relations for the characters of the Hecke algebra. In Chapter 6, we prove the disjointness theorem, or rather its generalization to not necessarily unipotent representations. In the case where $\theta = 1$ and F = identity on W, the disjointness theorem is deduced from the following statement.

For any $x \in W$, $R_{a_x} \overset{\text{def.}}{=} \sum_E c_{x,E} R_E$ *is an actual representation of* $G(F_q)$. (Note that, a priori, this is only a $\mathbb{Q}$-linear combination of representations.)

The main theorem (4.23) is proved in Chapters 7, 8, 9, 10. The proof is carried out separately for classical groups and then for each exceptional group. Special arguments are needed when q is less than the Coxeter number. For example, Chapter 8 is only needed for small q; it is a preparation for the proof in Chapter 9 of the main theorem for classical groups. It contains some new results on the structure of induced representations which complement the results in $[L_2]$, $[L_3]$, [HL], $[As_1]$.

In Chapter 11, we describe (up to a power of q) all possible eigenvalues of the Frobenius map on $\bigoplus_{i,w} H_c^i(X_w, \overline{\mathbb{Q}}_\ell)$ assuming that G is split over F_q. For example, if G is of type E_8, these eigenvalues must be of the following form: $\pm\sqrt{-q}$ or a primitive root of 1 of order ≤ 6 (all these possibilities actually occur). Most of the computation centers around the case where w is an element of order 24 and length 10 in the Weyl group of type E_8, in which case the eigenvalues of the Frobenius map on $\oplus_i H^i(X_w, \overline{Q}_\ell)$ are studied by a method similar to that in $[L_2]$.

Chapter 12 contains applications of the representation theory of $G(F_q)$ to the structure of left cells in W, which have been explained already.

Chapter 13 reinterprets the parametrization of irreducible representations of $G(F_q)$ in terms of the dual group, in the spirit of Langlands; it involves the concept of special conjugacy classes in the dual group.

This chapter contains also the definition of a new orthonormal basis of the space of class function on $G(F_q)$; roughly speaking, it is obtained by applying a Fourier transform over a set $\mathfrak{M}(\mathcal{G}_c)$ (or a twisted analogue) to the standard basis consisting of irreducible characters. It is conjectured that the class functions in the new basis can be expressed in terms of certain intersection cohomology complexes on G, which have a meaning over any algebraically closed field.

Chapter 14 contains a discussion of Ree groups. It also contains a discussion of how our results extend to groups with disconnected center. It is shown that our results do extend to the case where G has finite cyclic center and is simple modulo its center.

Finally, the appendix contains several tables concerning the classification and degrees of unipotent representations.

By $[DL_1]$ our formula for multiplicities of irreducible representations in R_T^θ implies (and for large q is equivalent to) explicit formulas for the character of any irreducible representation of $G(F_q)$ on any semisimple element.

Recently there has been substantial progress (by Shoji, Beynon-Spaltenstein) in understanding the character values of R_T^θ at unipotent elements (Green functions), but this will not be discussed here. This, together with our results, implies that, at least in large characteristic, one can now write explicitly (in a form of an algorithm) the average value of the character of any irreducible representation of $G(F_q)$ on the set of F_q-rational points of any conjugacy class in G, defined over F_q.

The reader's task will be easier if he will assume throughout the book that q is larger than the Coxeter number of G. In that case he may skip 6.20, about half of Chapter 7, all of Chapter 8 and 9.8, 9.10 which are only needed for small q, when X_w are not known to be affine. (It is possible that the lifting methods of [As_3], [As_4] could provide an alternative method to reduce the case of small q to that of large q.)

My understanding of this subject has benefited very much during the last few years from my joint work with Pierre Deligne and, more recently, with David Kazhdan; I am very grateful to them.

I wish to thank the IHES and the National Science Foundation for support at various times during the preparation of this book. Much of the writing was done during a visit to the IHES in the first half of 1982. The main results were presented in a series of lectures at the Oberwolfach conference on ℓ-adic cohomology and representations, in June, 1982.

GEORGE LUSZTIG

Characters of
Reductive Groups over a
Finite Field

Chapter 1

COMPUTATION OF LOCAL INTERSECTION COHOMOLOGY OF CERTAIN LINE BUNDLES OVER A SCHUBERT VARIETY

In [KL_1], [KL_2], Kazhdan and myself have computed the local intersection cohomology (in the sense of Deligne, Goresky and MacPherson, see [BBD]) of a Schubert variety and conjectured that it describes the multiplicities in the Jordan-Hölder series of Verma modules with regular integral highest weight of the corresponding complex Lie algebra. This conjecture has been proved by Beilinson-Bernstein [BB] and by Brylinski-Kashiwara [BK].

Beilinson and Bernstein have also treated the case of Verma modules with regular rational highest weight. They have shown (unpublished) that in this case, there is again a topological interpretation for the multiplicities in the Jordan-Hölder series. The interpretation is in terms of local intersection cohomology of a line bundle (with zero section removed) over a Schubert variety with coefficients in a certain local coefficient system with monodromy of order related to the denominator of the highest weight. (Such coefficient systems were also considered by Verdier who calls them "faisceaux monodromiques".) The purpose of this chapter is to give explicit formulas for these local intersection cohomology groups. These formulas, which generalize those of [KL_2], are in terms of the polynomials $P_{y,w}$ of [KL_1] for a certain reflection subgroup of the Weyl group. The method used is an adaptation of that in [LV], but there are a number of features which didn't appear in the constant coefficient case. The proof makes use of some powerful results on intersection cohomology due to Beilinson, Bernstein, Gabber and Deligne [BBD] namely the purity theorem and the decomposition theorem.

I wish to thank Deligne for acquainting me with Beilinson and Bernstein's topological interpretation of multiplicities in Verma modules with rational highest weight. I also wish to thank Gabber for some critical remarks on this chapter.

1.0. We shall always denote by k an algebraic closure of the finite field F_p with p elements. If q is a power p^e of p, we denote F_q the subfield of k with q elements. We shall fix a prime number $\ell \neq p$.

All algebraic varieties will be over k.

Let H be an algebraic group acting on an algebraic variety X. We assume that we are given F_q-rational structures on H and X which are compatible with the action of H on X and let $F : H \to H$, $F : X \to X$ be the corresponding Frobenius maps.

Let $\mathcal{F}$ be a constructible, H-equivariant, $\overline{\mathbf{Q}}_\ell$-sheaf over X. A Φ-structure on $\mathcal{F}$ consists, by definition, in giving for each sufficiently divisible integer $m \geq 1$, an isomorphism of sheaves

$$\Phi_m : (F^m)^* \mathcal{F} \xrightarrow{\sim} \mathcal{F}$$

with the following properties.

(1.0.1) Φ_m is H-equivariant: for all $h \in H$, $x \in X$, the diagram

$$\begin{array}{ccc} \mathcal{F}_{F^m x} & \xrightarrow{\Phi_m} & \mathcal{F}_x \\ {\scriptstyle F^m(h)}\downarrow & & \downarrow{\scriptstyle h} \\ \mathcal{F}_{F^m(hx)} & \xrightarrow{\Phi_m} & \mathcal{F}_{hx} \end{array}$$

is commutative.

(1.0.2) If Φ_m is defined and m' is an integer ≥ 1, then $\Phi_{mm'}$ is defined and $\Phi_{mm'} = (\Phi_m)^{m'}$.

Two such systems $\{\Phi_m, m$ divisible by $m_0\}$, $\{\Phi'_{m'}, m'$ divisible by $m'_0\}$ are regarded as identical if there exists a multiple m''_0 of m_0 and m'_0 such that $\Phi_m = \Phi'_m$ for all m divisible by m''_0.

Given a Φ-structure on $\mathcal{F}$ and a point $x \in X$, we have $\Phi_m : \mathcal{F}_x \to \mathcal{F}_x$ for sufficiently divisible m. Hence we can define the eigenvalues of Φ on $\mathcal{F}_x$ (as elements of the group $A = \bar{\mathbf{Q}}_\ell^*/\text{roots of } 1$) as being the m-th roots of the eigenvalues of $\Phi_m : \mathcal{F}_x \to \mathcal{F}_x$, for sufficiently divisible m.

Let $\mathcal{C}_X$ be the category whose objects are the constructible, H-equivariant $\bar{\mathbf{Q}}_\ell$ sheaves over X, endowed with a Φ-structure and whose morphisms are H-equivariant maps of sheaves which commute with Φ_m for sufficiently divisible m.

If Y is another algebraic variety over k with an H-action and a fixed F_q-structure compatible with the H-action, then $\mathcal{C}_Y$ is defined in a similar way. If $f : X \to Y$ is an H-equivariant map defined over F_q, there are natural maps

$$f^* : \text{ob}\,\mathcal{C}_Y \to \text{ob}\,\mathcal{C}_X \quad \text{(inverse image)}$$
$$R^i f_* : \text{ob}\,\mathcal{C}_X \to \text{ob}\,\mathcal{C}_Y \quad \text{(higher direct image)}$$
$$R^i f_! : \text{ob}\,\mathcal{C}_X \to \text{ob}\,\mathcal{C}_Y \quad \text{(higher direct image with compact support).}$$

1.2. We shall always denote by G a connected reductive algebraic group (over k) with a fixed F_p-rational structure (with Frobenius map $F : G \to G$) such that there exists a maximal torus $T \subset G$ with the following property: $F(t) = t^p$, $\forall t \in T$. We shall fix a Borel subgroup B of G, containing T, and we shall denote by U the unipotent radical of B. We have $F(B) = B$, $F(U) = U$. Let $\mathfrak{N}(T)$ be the normalizer of T in G and let $W = \mathfrak{N}(T)/T$ be the Weyl group. Then F maps $\mathfrak{N}(T)$ into itself and induces the identity map on W. Let S be the set of simple reflection in W (with respect to B); let $\ell : W \to \mathbf{N}$ be the length function on W, regarded as a Coxeter group on the generators S. Let $\bar{n}$ denote the image in W of an element $n \in \mathfrak{N}(T)$. Let $\mathcal{B}$ be the variety of all Borel subgroups of G; it is a homogeneous space under G, which acts by conjugation. For each

$w \in W$, $\mathcal{B}_w$ denotes the set of Borel subgroups $B' \in \mathcal{B}$ such that B and B' are in relative position w, i.e. $B' = gBg^{-1}$, for some $g \in BwB$. Then $\mathcal{B}_w$ is a locally closed subvariety of $\mathcal{B}$; it is an orbit of U on $\mathcal{B}$ (a Bruhat cell). The variety $\mathcal{B}$ inherits from G an F_p-rational structure; we shall denote its Frobenius map as $F : \mathcal{B} \to \mathcal{B}$. We have $F(\mathcal{B}_w) = \mathcal{B}_w$ for all $w \in W$. For $w \in W$ and $t \in T$, we define $w(t) = \dot{w}t\dot{w}^{-1} \in T$, where $\dot{w}$ is any representative of w in $\mathfrak{N}(T)$.

1.3. Let $\mathfrak{L}$ be the set of isomorphism classes of G-equivariant line bundles L over $\mathcal{B}$, endowed with an F_p-rational structure such that the corresponding Frobenius map $F : L \to L$ satisfies $F(g\ell) = F(g)F(\ell)$, $(g \in G, \ell \in L)$ and $F(x\ell) = x^p F(\ell)$, $(x \in k^*, \ell \in L)$.

$\mathfrak{L}$ is a free abelian group of rank equal to that of G; the group structure of $\mathfrak{L}$ is defined by taking tensor products of line bundles. The Weyl group W acts on $\mathfrak{L}$ as follows. Let v be an element of W and let $O(v)$ be the set of ordered pairs $(B', B'') \in \mathcal{B} \times \mathcal{B}$ which are in relative position v. This is an orbit of G on $\mathcal{B} \times \mathcal{B}$, acting diagonally. Let $\mathfrak{L}_v$ be the set of isomorphism classes of G-equivariant line bundles over $O(v)$ endowed with an F_p-rational structure compatible with the G-action and with the scalar multiplication (as above). The two projections $pr_1, pr_2 : O(v) \to \mathcal{B}$ define maps $pr_1^* : \mathfrak{L} \to \mathfrak{L}_v$, $pr_2^* : \mathfrak{L} \to \mathfrak{L}_v$ (inverse image) and it is clear that they are both bijective. Hence, given $L \in \mathfrak{L}$, there is a unique element ${}^vL \in \mathfrak{L}$ such that there exists a G-equivariant isomorphism (over F_p):

$$\gamma : pr_1^*({}^vL) \xrightarrow{\approx} pr_2^*(L) \tag{1.3.1}$$

of line bundles over $O(v)$. (The isomorphism γ is then uniquely determined up to multiplication by an element in F_p^*.) It is easy to check that ${}^{(v'v)}L = {}^{v'}({}^vL)$ whenever $\ell(v'v) = \ell(v') + \ell(v)$ and that ${}^{v^{-1}}({}^vL) = L$ for all v. It follows that ${}^{(v'v)}L = {}^{v'}({}^vL)$ for all $v, v' \in W$, so that $(v,L) \to {}^vL$ is an action of W on $\mathfrak{L}$. It clearly preserves the group structure on $\mathfrak{L}$.

Any $s \in S$ defines a map $\check{\alpha}_s : \mathcal{L} \to \mathbf{Z}$ as follows. Let $\mathcal{P}_s$ be the class of parabolic subgroups of semisimple rank 1 corresponding to s. If $L \in \mathcal{L}$, we set $\check{\alpha}_s(L) = m$, where $m+1$ is the Euler characteristic of the (coherent) sheaf of sections of the restriction of L to any fiber ($\approx$ projective line) of the natural projection $\mathcal{B} \to \mathcal{P}_s$. It is easily checked that $\check{\alpha}_s$ is a group homomorphism. Let $L_s \in \mathcal{L}$ be the tangent bundle along the fibers of the natural projection $\mathcal{B} \to \mathcal{P}_s$. For any $L \in \mathcal{L}$, we have

$$^{s}L = L \otimes L_s^{\otimes(-m)}$$

where $m = \check{\alpha}_s(L)$.

Let R (resp. $R^\vee$) be the set of $L \in \mathcal{L}$ (resp. $\phi \in \mathrm{Hom}(\mathcal{L}, \mathbf{Z})$) such that $^{w}L = L_s$ (resp. $w(\phi) = \check{\alpha}_s$) for some $w \in W$ and some $s \in S$. Then R is a root system in $\mathcal{L}$ with Weyl group W, $R^\vee \subset \mathrm{Hom}(\mathcal{L}, \mathbf{Z})$ is the corresponding set of coroots; there is a unique W-equivariant bijection $R \leftrightarrow R^\vee$ such that $L_s \leftrightarrow \check{\alpha}_s$, $s \in S$. The $L_s (s \in S)$ form a set of simple roots for R.

Let R^+ be the corresponding set of positive roots and let $R^- = R - R^+$. We may identify $\mathcal{L}$ with the group of characters $T \to k^*$ as follows. With a character $\lambda : T \to k^*$, extended to a character $\lambda : B \to k$ (trivial on U) we associate the line bundle

(1.3.2) $$L = L_\lambda = \{(g, z) \in G \times k\}/B$$

where $b \in B$ acts by $b : (g, z) \to (gb^{-1}, \lambda(b) z)$.

The projection $L \to \mathcal{B}$ is given by $(g, z) \to gBg^{-1}$. The F_p-structure of L has Frobenius map $(g, z) \to (F(g), z^p)$. This identification is compatible with the group structures and with the action of W. (W acts on the group of characters of T by $\lambda \to {}^{v}\lambda$, $({}^{v}\lambda)(t) = \lambda(v^{-1}(t))$, $t \in T$.) Under this identification, R becomes the usual set of roots of G with respect to T, R^+ becomes the set of positive roots (with respect to B).

1.4. For any $L \in \mathcal{L}$ we shall denote by $L^{\cdot}$, the variety L minus the zero section. Let $\pi : L^{\cdot} \to \mathcal{B}$ be the natural projection. We shall take in

1.0 the variety X to be $L^{\cdot}$ and H to be $U \times k^*$. The letter n will always denote an integer such that

(1.4.1) $$n \geq 1, \ (n, p) = 1 .$$

We shall denote

$$\mu_n = \{x \in k^*, \ x^n = 1\} .$$

We define an action of H on $L^{\cdot}$, depending on n, as follows. U acts by restriction of the action of G on $L^{\cdot}$, while $x \in k^*$ acts as scalar multiplication by x^n. The variety $L^{\cdot}$ inherits an F_p-rational structure from L; H has a natural F_p-rational structure which on U is induced from that of G and on k^* has Frobenius map $x \to x^p$. We shall take in 1.0, $e = 1$, i.e. $q = p$. Let $\mathcal{C}_{L^{\cdot},n}$ be the category associated in 1.0 to $L^{\cdot}$ and to the action of H on $L^{\cdot}$ which was just defined.

The action of H on $L^{\cdot}$ has finitely many orbits: they are the inverse images $\pi^{-1}(\mathcal{B}_w)$ under $\pi : L^{\cdot} \to \mathcal{B}$ of the Bruhat cells $\mathcal{B}_w \subset \mathcal{B}$. The isotropy group in H of a point in $L^{\cdot}$ is a product $U_1 \times \mu_n \subset U \times k^*$, where U_1 is a closed connected subgroup of U.

Given $w \in W$ and a homomorphism $\psi : \mu_n \to \bar{Q}_\ell^*$, there is a unique (up to isomorphism) H-equivariant locally constant $\bar{Q}_\ell$-sheaf $\mathcal{X}_{w,\psi}^L$ of rank 1 over $\pi^{-1}(\mathcal{B}_w) \subset L^{\cdot}$ such that $x \in \mu_n$ acts on any stalk as multiplication by $\psi(x)$.

Given an element $a \in A$, (see 1.0), there is a unique Φ-structure on this sheaf extended to the whole of $L^{\cdot}$ (by zero outside $\pi^{-1}(\mathcal{B}_w)$) such that Φ has eigenvalue a on a stalk over $\pi^{-1}(\mathcal{B}_w)$. This sheaf with this Φ-structure is an object of $\mathcal{C}_{L^{\cdot},n}$ denoted $a\bar{\mathcal{X}}_{w,\psi}^L$. Clearly, the elements $a\bar{\mathcal{X}}_{w,\psi}^L$ form a **Z**-basis for the Grothendieck group $\mathcal{K}(\mathcal{C}_{L^{\cdot},n})$ of the category $\mathcal{C}_{L^{\cdot},n}$. When $a = 1$, we denote $a\bar{\mathcal{X}}_{w,\psi}^L$ simply as $\bar{\mathcal{X}}_{w,\psi}^L$. We shall regard $\mathcal{K}(\mathcal{C}_{L^{\cdot},n})$ as a free module over the group ring **Z**[A] with basis $\bar{\mathcal{X}}_{w,\psi}^L$, by defining $a \cdot \bar{\mathcal{X}}_{w,\psi}^L = a\bar{\mathcal{X}}_{w,\psi}^L$.

As in [LV], we have a Verdier duality map

$$D : \mathcal{K}(\mathcal{C}_{L^{\cdot},n}) \to \mathcal{K}(\mathcal{C}_{L^{\cdot},n}) .$$

We shall need the following two properties of D :

(1.4.2) D is Z-linear and $D(ax) = a^{-1}D(x)$, $(a \in A,\ x \in \mathcal{K}(\mathcal{C}_{L^{\cdot},n}))$.

(1.4.3) $$D(\mathfrak{X}^L_{w,\psi}) = \sum_{w' \leq w} r_{w',w,\psi}\, \mathfrak{X}^L_{w',\psi^{-1}}$$

where $r_{w',w,\psi} \in Z[A]$, $(w' \leq w)$, and $r_{w,w,\psi} = [p]^{\ell(w)-\ell(w')}$. Here, $[p]$ denotes the image of $p \in \mathbf{Q}^*_\ell$ in A ; the relation $w' \leq w$ means that we have the inclusion $\overline{\mathcal{B}}_{w'} \subset \overline{\mathcal{B}}_w$ of the Zariski closures of $\mathcal{B}_{w'}$, $\mathcal{B}_w$ in $\mathcal{B}$.

1.5. In the description (1.3.2) of L, we have

(1.5.1) $$L^{\cdot} = \{(g,z) \in G \times k^*\}/B$$

where B acts as in (1.3.2). We shall denote $\pi^{-1}(\mathcal{B}_w) \subset L^{\cdot}$ also as $L^{\cdot}|\mathcal{B}_w$. Let $\dot{w} \in \mathfrak{N}(T)$ be a representative for $w \in W$ such that $\dot{w} \in G^F$. Consider the covering space

(1.5.2) $$\begin{aligned} &(U/U \cap \dot{w}U\dot{w}^{-1}) \times k^* \to L^{\cdot}|\mathcal{B}_w \\ &(u,z) \to (u\dot{w}, z^n) . \end{aligned}$$

It is H-equivariant (where H acts on the first space by left translation) and it is the orbit map for the action of μ_n on the first space given by multiplication on the z-coordinate. The direct image of the constant sheaf $\overline{\mathbf{Q}}_\ell$ under the map (1.5.2) is an H-equivariant locally constant $\overline{\mathbf{Q}}_\ell$-sheaf over $L^{\cdot}|\mathcal{B}_w$. It has a canonical direct sum decomposition $\bigoplus_\psi \mathfrak{X}^L_{\dot{w},\psi}$, where $\mathfrak{X}^L_{\dot{w},\psi}$ is the locally constant subsheaf (of rank 1) such that the action of $x \in \mu_n \subset H$ on any stalk of $\mathfrak{X}^L_{\dot{w},\psi}$ is multiplication by $\psi(x)$. (Here ψ runs over all homomorphisms $\mu_n \to \overline{\mathbf{Q}}^*_\ell$.) Note that $\mathfrak{X}^L_{\dot{w},\psi}$ is an H-equivariant $\overline{\mathbf{Q}}_\ell$ sheaf with a natural F_{p^d}-structure (where d is the smallest integer ≥ 1 such that n divides $p^d - 1$); the corresponding Frobenius map is induced by the Frobenius map on $(U/U \cap \dot{w}U\dot{w}^{-1}) \times k^*$ (relative to F_{p^d}), given by $(u, z) \to (F^d u, z^{p^d})$. Let $\bar{\mathcal{X}}^L_{\dot{w},\psi}$ be the sheaf $\mathcal{X}^L_{\dot{w},\psi}$ extended to the whole of $L^{\cdot}$ (by zero outside $\pi^{-1}(\mathcal{B}_w)$). This is an H-equivariant constructible $\bar{Q}_\ell$-sheaf with an F_{p^d}-structure. If we take the Φ-structure associated to this F_{p^d}-structure, rather than the F_{p^d}-structure itself, then $\bar{\mathcal{X}}^L_{\dot{w},\psi}$ becomes $\bar{\mathcal{X}}^L_{w,\psi}$ (of 1.4) and, in particular, it becomes independent of the choice of representative $\dot{w}$. Thus, a choice of $\dot{w}$ gives rise to an F_{p^d}-structure, which is a refinement of the Φ-structure of $\bar{\mathcal{X}}^L_{w,\psi}$.

1.6. Let $\mathcal{F}$ be a locally constant $\bar{Q}_\ell$-sheaf of finite rank over an open dense set X_0 of an algebraic variety X such that all connected components of X are irreducible of the same dimension. Deligne, Goresky and MacPherson have associated to $\mathcal{F}$ a complex of $\bar{Q}_\ell$-sheaves $IC(X, \mathcal{F})$ (intersection cohomology complex), well defined in the derived category. We refer the reader to [BBD] for the definition and properties of this complex. We shall denote by $\mathcal{H}^i(X, \mathcal{F})$ the i-th cohomology sheaf of $IC(X, \mathcal{F})$. It is zero for $i < 0$ and $\mathcal{H}^0(X, \mathcal{F})$ restricted to X_0 is just $\mathcal{F}$. We shall denote $\mathbf{H}^i(X, \mathcal{F})$ (resp. $\mathbf{H}^i_c(X, \mathcal{F})$) the hypercohomology groups (resp. hypercohomology groups with compact support) of X with coefficients in $IC(X, \mathcal{F})$. Let us take $X = \pi^{-1}(\bar{\mathcal{B}}_w) \subset L^{\cdot}$, $X_0 = \pi^{-1}(\mathcal{B}_w) \subset L^{\cdot}$, $\mathcal{F} = \mathcal{X}^L_{\dot{w},\psi}$ ($\dot{w}$ and ψ as in 1.5). In this case, we shall denote the sheaf $\mathcal{H}^i(\pi^{-1}(\bar{\mathcal{B}}_w), \mathcal{X}^L_{\dot{w},\psi})$ extended to the whole of $L^{\cdot}$ (by zero outside $\pi^{-1}(\bar{\mathcal{B}}_w)$) as $\bar{\mathcal{H}}^i(\mathcal{X}^L_{\dot{w},\psi})$. It is in a natural way an H-equivariant, constructible $\bar{Q}_\ell$-sheaf over $L^{\cdot}$ with an F_{p^d}-structure (inherited from that of $\mathcal{X}^L_{\dot{w},\psi}$). We may regard, instead of the F_{p^d}-structure, the Φ-structure associated to it. We then get an element in $\mathcal{K}(\mathcal{C}_{L^{\cdot},n})$, which is independent of the choice of representative $\dot{w}$; we denote it $\bar{\mathcal{H}}^i(\mathcal{X}^L_{w,\psi})$. We shall set

$$\tilde{\mathfrak{X}}^L_{w,\psi} = \sum_i (-1)^i \bar{\mathcal{H}}^i(\mathfrak{X}^L_{w,\psi}) \in \mathcal{K}(\mathcal{C}_{L^\cdot,n}) .$$

We have

1.7. LEMMA. *The elements* $\tilde{\mathfrak{X}}^L_{w,\psi} \in \mathcal{K}(\mathcal{C}_{L^\cdot,n})$ *are characterized by the following two properties.*

(1.7.1) $$D(\tilde{\mathfrak{X}}^L_{w,\psi}) = [p]^{-(\ell(w)+1)}\tilde{\mathfrak{X}}^L_{w,\psi^{-1}}$$

(1.7.2) $\tilde{\mathfrak{X}}^L_{w,\psi} = \bar{\mathfrak{X}}^L_{w,\psi} + Z[A]$ – *linear combination of elements* $\bar{\mathfrak{X}}^L_{w',\psi}$,

$w' < w$, *the coefficient of* $\bar{\mathfrak{X}}^L_{w',\psi}(w' < w)$ *being an element of* $Z[A]$ *which is a* Z*-linear combination of elements* $a \in A$, *where* a *is represented by an algebraic number in* $\bar{Q}^*_\ell$ *all of whose complex conjugates have absolute value* $\leq p^{\frac{1}{2}(\ell(w)-\ell(w')-1)}$.

The fact that the elements $\tilde{\mathfrak{X}}^L_{w,\psi}$ satisfy (1.7.1), (1.7.2), follows from the definition of the intersection cohomology complex and from the purity theorem of Gabber, see [BBD].

We now show that the properties (1.7.1), (1.7.2) characterize $\tilde{\mathfrak{X}}^L_{w,\psi}$. (Compare with the proof of uniqueness in Theorem 1.1 of [KL_1].) Let $\mathfrak{A}_{w,\psi} = \bar{\mathfrak{X}}_{w,\psi} + \sum_{w' < w} a_{w',w,\psi}\bar{\mathfrak{X}}_{w',\psi} \in \mathcal{K}(\mathcal{C}_{L^\cdot n})$ be an element such that the coefficients $a_{w',w,\psi} \in Z[A]$ have the property described in (1.7.2) and such that $D(\mathfrak{A}_{w,\psi}) = [p]^{-(\ell(w)+1)}\mathfrak{A}_{w,\psi^{-1}}$. We want to show that $a_{w',w,\psi}$ are uniquely determined. It is enough to check the following statement. Let $\mathfrak{A}'_{w,\psi} = \sum_{w' < w} \beta_{w',w,\psi}\bar{\mathfrak{X}}_{w',\psi} \in \mathcal{K}(\mathcal{C}_{L^\cdot,n})$ be an element such that the coefficients $\beta_{w',w,\psi}$ have the property described in (1.7.2) and such that $D(\mathfrak{A}'_{w,\psi}) = [p]^{-(\ell(w)+1)}\mathfrak{A}'_{w,\psi^{-1}}$. Then $\beta_{w',w,\psi} = 0$ for all $w' < w$.

We have

$$[p]^{-(\ell(w)+1)} \sum_{w'<w} \beta_{w',w,\psi} \bar{\mathfrak{X}}_{w',\psi} = [p]^{-(\ell(w)+1)} \mathfrak{A}'_{w,\psi}$$
$$= D(\mathfrak{A}'_{w,\psi^{-1}})$$
$$= \sum_{w''\leq w'<w} r_{w'',w',\psi^{-1}} \bar{\beta}_{w',w,\psi^{-1}} \bar{\mathfrak{X}}_{w'',\psi} .$$

Here $\beta \to \bar{\beta}$ is the ring involution of $\mathbf{Z}[A]$ defined by $a \to a^{-1}$ for $a \in A$. (See (1.4.3).) Comparing the coefficients of $\bar{\mathfrak{X}}_{w',\psi}$, we get

$$(1.7.3) \qquad \beta_{w',w,\psi} = \sum_{\substack{w_1 \\ w'\leq w_1<w}} r_{w',w_1,\psi^{-1}} \bar{\beta}_{w_1,w,\psi^{-1}}$$

for all $w'<w$. Let us fix $w'<w$. We may assume by induction on $\ell(w)-\ell(w')$ that $\beta_{w_1,w,\psi^{-1}} = 0$ whenever $w'<w_1<w$. Then (1.7.3) becomes $\beta_{w',w,\psi} = [p]^{(\ell(w)-\ell(w'))}\bar{\beta}_{w',w,\psi^{-1}}$. Hence

$$(1.7.4) \quad [p]^{-\frac{1}{2}(\ell(w)-\ell(w'))}\beta_{w',w,\psi} = [p]^{\frac{1}{2}(\ell(w)-\ell(w'))}\bar{\beta}_{w',w,\psi^{-1}} .$$

The left (resp. right) hand side of (1.7.4) is a $\mathbf{Z}$-linear combination of elements in A represented by algebraic numbers in $\bar{\mathbf{Q}}_\ell^*$ all of whose complex conjugates have absolute value $\leq p^{-1/2}$ (resp. $\geq p^{1/2}$). Hence both sides of (1.7.4) are zero. The lemma is proved.

1.8. Let $\mathcal{L}_{\mathrm{root}}$ be the subgroup of $\mathcal{L}$ generated by R. Given $L \in \mathcal{L}$ and n as in (1.4.1), we define

$$W_{L,n} = \{w \in W \mid {}^wL = L \otimes L_1^{\otimes n} \text{ for some } L_1 \in \mathcal{L}_{\mathrm{root}}\}$$

$$R^\vee_{L,n} = \{\text{coroots in } R^\vee \text{ which take on } L \text{ a value divisible by } n\}$$

$R_{L,n}$ = set of roots corresponding to $R^{\vee}_{L,n}$

$R^{+}_{L,n} = R_{L,n} \cap R^{+}$

$R^{-}_{L,n} = R_{L,n} \cap R^{-}$

$S_{L,n}$ = {reflections in W with respect to a root in $R^{+}_{L,n}$ which is not product of two roots in $R^{+}_{L,n}$} .

It is well known that $R_{L,n}$ is a root system with Weyl group $W_{L,n}$ (which is a Coxeter group on generators $S_{L,n}$) and that $R^{+}_{L,n}$ is a set of positive roots for $R_{L,n}$. Being a Coxeter group, $W_{L,n}$ has a canonical partial order, which will be denoted $\underset{L,n}{\leq}$. It is generated by the relations $w \underset{L,n}{\leq} wr$ whenever $w \in W_{L,n}$ and r is a reflection with respect to a root $L' \in R^{+}_{L,n}$ such that ${}^{w}L' \in R^{+}_{L,n}$. (This implies that $w \leq wr$ for the canonical partial order $\leq$ of W.) Thus, the relation $w' \underset{L,n}{\leq} w$ $(w,w' \in W_{L,n})$ implies $w \leq w'$. More generally, we have

1.9. LEMMA. (i) *Any right coset* $wW_{L,n} \subset W$ $(w \in W)$ *has a unique element* w_1 *of minimal length (in* W *). It is characterized by the property that* ${}^{w_1}L' \in R^{+}$ *for any* $L' \in R^{+}_{L,n}$.

(ii) *If* $v, v' \in W_{L,n}$ *satisfy* $v' \underset{L,n}{\leq} v$ *and if* $w_1 \in W$ *has minimal length in* $w_1W_{L,n}$ *then* $w_1v' \leq w_1v$ *(for the partial order in* W *) and*

$$(1.9.1) \qquad \ell(w_1v) - \ell(w_1v') \geq \tilde{\ell}(v) - \tilde{\ell}(v')$$

where $\tilde{\ell}$ *is the length function of the Coxeter group* $(W_{L,n}, S_{L,n})$.

Proof. If $w_1 \in W$ has minimal length in $w_1 W_{L,n}$, then $\ell(w_1r) > \ell(w_1)$ for any reflection $r \in W_{L,n}$. Hence, for $L' \in R^{+}_{L,n}$, we have ${}^{w_1}L' \in R^{+}$ and for $L' \in R^{-}_{L,n}$, we have ${}^{w_1}L' \in R^{-}$. Now let $v_1 \in W_{L,n}$, $v_1 \neq e$.

Then there exists $L' \in R^+_{L,n}$ such that ${}^{v_1}L' \in R^-_{L,n}$. Hence ${}^{w_1 v_1}L' \in R^-$. Thus, by the first part of the argument, $w_1 v_1$ is not of minimal length in $w_1 W_{L,n}$ and (i) follows.

To prove (ii), we may assume that $v = v'r$, where r is the reflection with respect to $L' \in R^+_{L,n}$ and we have ${}^{v'}L' \in R^+_{L,n}$. We may, moreover, assume that $\tilde{\ell}(v) = \tilde{\ell}(v') + 1$. We have $w_1 v = (w_1 v')r$ and we have ${}^{w_1 v'}L' = {}^{w_1}({}^{v'}L') \in R^+$ by (i) hence $w_1 v' < w_1 v$. In particular, we have $\ell(w_1 v) - \ell(w_1 v') \geq 1$, as required. The lemma is proved.

1.10. Given $L \in \mathfrak{L}$, $w \in W$, n as in (1.4.1) and an injective homomorphism $\psi : \mu_n \to \bar{\mathbf{Q}}_\ell$, we define an element $\hat{\mathfrak{X}}^L_{w,\psi} \in \mathcal{K}(\mathcal{C}_{L^*,n})$ by the formula

$$(1.10.1)\quad \hat{\mathfrak{X}}^L_{w,\psi} = \sum_{\substack{v' \in W_{L,n} \\ v' \leq_{L,n} v}} [p]^{\frac{1}{2}(\ell(w_1 v) - \tilde{\ell}(w_1 v') - \tilde{\ell}(v) + \tilde{\ell}(v'))} P_{v',v}([p]) \bar{\mathfrak{X}}^L_{w_1 v',\psi}$$

where $w_1 \in W$, $v \in W_{L,n}$ are such that $w = w_1 v$ and w_1 has minimal length in $w W_{L,n}$; $P_{v',v}$ are the polynomials (with integral coefficients) attached in [KL$_1$] to the Coxeter group $(W_{L,n}, S_{L,n})$. With these notations and assumptions, we have the following.

1.11. LEMMA. $\tilde{\mathfrak{X}}^L_{w,\psi} = \hat{\mathfrak{X}}^L_{w,\psi}$.

The coefficient of $\bar{\mathfrak{X}}^L_{w_1 v',\psi}$ in $\hat{\mathfrak{X}}^L_{w,\psi}$ is a polynomial in $[p]$ with integral coefficients: indeed, $P_{v',v}[p]$ is a polynomial in $[p]$ with integral coefficients and the power of $[p]$ in front of it has as exponent an integer (since $\tilde{\ell}(v) \equiv \ell(v) \pmod 2$) which is ≥ 0 by 1.9 (ii). When $v' = v$, this polynomial is 1; when $v' \underset{L,n}{<} v$, it is a polynomial of degree $\leq \frac{1}{2}(\ell(w_1 v) - \ell(w_1 v') - 1)$, since $P_{v',v}$ has degree $\leq \frac{1}{2}(\tilde{\ell}(v) - \tilde{\ell}(v') - 1)$. Thus, $\hat{\mathfrak{X}}^L_{w,\psi}$ satisfies the property (1.7.2) of $\tilde{\mathfrak{X}}^L_{w,\psi}$. By 1.7, the lemma is thus equivalent with the following statement

1.12. LEMMA. *With the assumptions in 1.10, we have*

$$D(\hat{\mathcal{X}}^L_{w,\psi}) = [p]^{-(\ell(w)+1)}\hat{\mathcal{X}}^L_{w,\psi^{-1}}\ .$$

After some preliminaries in 1.13-1.19, the proof of 1.11 and 1.12 will be given in 1.20.

1.13. Let the notations be as in 1.10 and let $L_1 \in \mathcal{L}$. We define $L' = L \otimes L_1^{\otimes n}$. We shall now prove that the statement in 1.11 is true for (L, n, w, ψ) if and only if it is true for (L', n, w, ψ). It is clear that $W_{L,n} = W_{L',n}$, $S_{L,n} = S_{L',n}$, $R_{L,n} = R_{L',n}$.

Let M be the fiber product $L^{\cdot} \times_{\mathcal{B}} L_1^{\cdot}$. We define maps $\alpha : M \to L^{\cdot}$, $\beta : M \to (L')^{\cdot}$ by $\alpha(\ell, \ell_1) = \ell$, $\beta(\ell, \ell_1) = \ell \otimes (\ell_1^{\otimes n})$. We have natural actions of $H = U \times k^*$ on $L^{\cdot}$ and $(L')^{\cdot}$ corresponding to the integer n (see 1.4). These give rise to the categories $\mathcal{C}_{L^{\cdot},n}$, $\mathcal{C}_{(L')^{\cdot},n}$. The space M has two actions of $U \times k^* \times k^*$. For both actions, $u \in U$ acts by $u : (\ell, \ell_1) \to (u\ell, u\ell_1)$. An element $(x,y) \in k^* \times k^*$ acts by

$$(\ell, \ell_1) \to (x^n \ell, y^{-1}\ell_1), \text{ for the first action}$$

$$(\ell, \ell_1) \to (x^n y^n \ell, y^{-1}\ell_1), \text{ for the second action.}$$

These actions give rise to two categories $\mathcal{C}^{I}_M$, $\mathcal{C}^{II}_M$ (see 1.4) of $U \times k^* \times k^*$-equivariant constructible sheaves with Φ-structure on M. It is clear that α^* defines an equivalence of categories $\mathcal{C}_{L^{\cdot},n} \xrightarrow{\approx} \mathcal{C}^{I}_M$ and that β^* defines an equivalence of categories $\mathcal{C}_{(L')^{\cdot},n} \xrightarrow{\approx} \mathcal{C}^{II}_M$. On the other hand, a sheaf in $\mathcal{C}^{I}_M$ can be regarded as a sheaf in $\mathcal{C}^{II}_M$, via the change of variables $(x,y) \to (xy, y)$ in $k^* \times k^*$. (It is the same sheaf, with a different action.) This gives an equivalence of categories $\mathcal{C}^{I}_M \xrightarrow{\approx} \mathcal{C}^{II}_M$. Combining these three equivalences, we get an equivalence of categories $\mathcal{C}_{L^{\cdot},n} \to \mathcal{C}_{(L')^{\cdot},n}$, hence a canonical isomorphism $\mathcal{K}(\mathcal{C}_{L^{\cdot},n}) \to \mathcal{K}(\mathcal{C}_{(L')^{\cdot},n})$. It is clear that this isomorphism makes the elements

$\bar{\mathfrak{X}}^{L}_{w,\psi}$, $\bar{\mathfrak{X}}^{L'}_{w,\psi}$ correspond to each other. It also makes the sheaves $\bar{\mathcal{H}}^i(\mathfrak{X}^{L}_{w,\psi})$, $\bar{\mathcal{H}}^i(\mathfrak{X}^{L'}_{w,\psi})$ correspond to each other, hence it makes the elements $\tilde{\mathfrak{X}}^{L}_{w,\psi}$, $\tilde{\mathfrak{X}}^{L'}_{w,\psi}$ correspond to each other. Thus, the statement made at the beginning of this section follows.

1.14. Let $L \in \mathfrak{L}$, $s \in S$ and let n be as in (1.4.1). Let $m = \check{\alpha}_s(L)$. If $L' = {}^sL$, we have $\check{\alpha}_s(L') = -m$. We consider the pull back $\tilde{L}$ of L to $O(s)$ under $pr_2 : O(s) \to \mathcal{B}$ and the pull back $\tilde{L}'$ of L' to $O(s)$ under $pr_1 : O(s) \to \mathcal{B}$. These are line bundles over $O(s)$ and we denote by $(\tilde{L}')^{\cdot}$, $\tilde{L}^{\cdot}$ the spaces obtained from them by removing the zero sections. The group H acts naturally on both these spaces. The isomorphism γ considered in (1.3.1) defines an H-equivariant isomorphism $\gamma : (\tilde{L}')^{\cdot} \to \tilde{L}^{\cdot}$. Let $\pi_1 = (\tilde{L}')^{\cdot} \to (L')^{\cdot}$, $\pi_2 : \tilde{L}^{\cdot} \to L^{\cdot}$ be the natural projections.

Let $\mathcal{F}$ be an object of $\mathcal{C}_{L^{\cdot},n}$. Then $\gamma^* \pi_2^* \mathcal{F}$ is an H-equivariant sheaf with Φ-structure on $(\tilde{L}')^{\cdot}$. Let

$$T_s(\mathcal{F}) = \Sigma(-1)^i R^i \pi_{1,!}(\gamma^* \pi_2^* \mathcal{F}) \in \mathcal{K}(\mathcal{C}_{(L')^{\cdot},n})$$

$$T_s'(\mathcal{F}) = \Sigma(-1)^i R^i \pi_{1,*}(\gamma^* \pi_2^* \mathcal{F}) \in \mathcal{K}(\mathcal{C}_{(L')^{\cdot},n}) .$$

These extend uniquely to $Z[A]$-linear maps $T_s, T_s' : \mathcal{K}(\mathcal{C}_{L^{\cdot},n}) \to \mathcal{K}(\mathcal{C}_{(L')^{\cdot},n})$. (They are independent of the choice of γ above.)

1.15. LEMMA. *We have* $D T_s = [p]^{-1} T_s' D$ *as maps* $\mathcal{K}(\mathcal{C}_{L^{\cdot},n}) \to \mathcal{K}(\mathcal{C}_{(L')^{\cdot},n})$.

Proof. According to [Ve], we have $D\pi_{1!}(\ ) = \pi_{1*}D(\ )$, $D\pi_2^*(\ ) = (\pi_2^* D(\ )) \otimes \bar{\mathbb{Q}}_\ell(1)$, with a shift of two degrees (in the derived category), since π_2 is smooth, of relative dimension 1. Also, D commutes with γ^*. The lemma follows.

1.16. LEMMA. *If* $\psi^m \neq 1$ *on* μ_n, $(m = \check{\alpha}_s(L))$, *we have*

$$T_s(\overline{\mathfrak{X}}^L_{w,\psi}) = T'_s(\overline{\mathfrak{X}}^L_{w,\psi}) = \begin{cases} \overline{\mathfrak{X}}^{L'}_{ws,\psi}, & \text{if } ws > w \\ \\ [p]\overline{\mathfrak{X}}^{L'}_{ws,\psi} & \text{if } ws < w . \end{cases}$$

Proof. To compute $T_s(\overline{\mathfrak{X}}^L_{w,\psi})$, we must compute for any $\ell' \in (L')^{\cdot}$, the cohomology group with compact support of $\pi_1^{-1}(\ell')$ with coefficients in the restriction $\mathcal{S}_0$ of $\gamma^*\pi_2^*(\overline{\mathfrak{X}}^L_{w,\psi})$ to $\pi_1^{-1}(\ell')$. Let $y \in W$ be such that ℓ' lies over a point in $\mathcal{B}_y$. If $y \neq w$, ws, or if $y \neq ws > w$, then $\pi_1^{-1}(\ell')$ is empty. If $y = ws > w$, then $\pi_1^{-1}(\ell')$ is a single point, so that $\overline{\mathfrak{X}}^{L'}_{y,\psi}$ appears with coefficient 1 in $T_s(\overline{\mathfrak{X}}^L_{w,\psi})$. If $y = ws < w$, then $\pi_1^{-1}(\ell')$ is an affine line and $\mathcal{S}_0$ is the constant sheaf $\overline{\mathbf{Q}}_\ell$ over it, hence $\overline{\mathfrak{X}}^{L'}_{y,\psi}$ appears with coefficient $[p]$ in $T_s(\overline{\mathfrak{X}}^L_{w,\psi})$. The remaining case $y = w > ws$ is the most interesting. In this case, $\pi_1^{-1}(\ell)$ is isomorphic to k^* and $\mathcal{S}_0 \approx \mathcal{F}_{\psi^m}$ where $\mathcal{F}_{\psi^m}$ is the locally constant sheaf over k^* associated with the homomorphism $\psi^m : \mu_n \to \overline{\mathbf{Q}}_\ell^*$. (We regard μ_n as a quotient of the tame fundamental group of k^*.) By assumption, $\psi^m \not\equiv 1$ on μ_n, hence the groups $H^i_c(k^*, \mathcal{S}_0)$ vanish. Thus $\overline{\mathfrak{X}}^{L'}_{w,\psi}$ appears with coefficient 0 in $T_s(\overline{\mathfrak{X}}^L_{w,\psi})$.

It remains to show that $T_s(\overline{\mathfrak{X}}^L_{w,\psi}) = T'_s(\overline{\mathfrak{X}}^L_{w,\psi})$. Let $\hat{L}$ be the pull-back of L to $\overline{O(s)} \subset \mathcal{B} \times \mathcal{B}$ under $pr_2 : \overline{O(s)} \to \mathcal{B}$ and let $\hat{L}'$ be the pull-back of L' to $\overline{O(s)}$ under $pr_1 : \overline{O(s)} \to \mathcal{B}$. Let $\hat{L}^{\cdot}, (\hat{L}')^{\cdot}$ be obtained from $\hat{L}$, $\hat{L}'$ by removing the zero sections. Let j' be inclusion of $(\tilde{L}')^{\cdot}$ as an open subset of $(\hat{L}')^{\cdot}$.

Let $\hat{\pi}_1 : (\hat{L}')^{\cdot} \to (L')^{\cdot}$ be the natural projection. We have $\pi_1 = \hat{\pi}_1 \circ j'$.

(1.16.1)

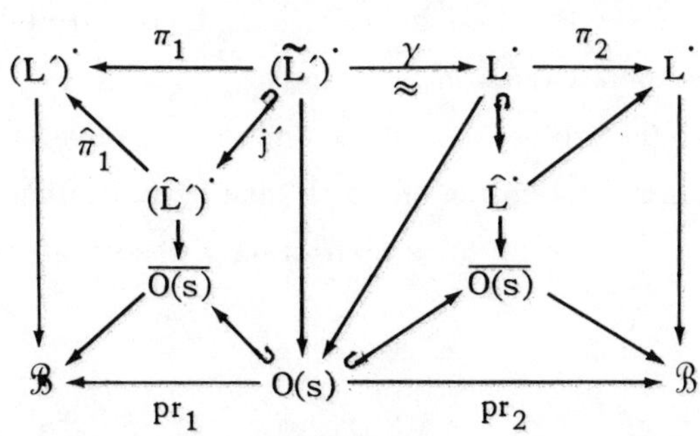

Now $\hat{\pi}_1$ is a proper map (a $\mathbf{P}^1$-bundle). Hence $R^i\pi_{1,*}(\ ) = R^i\pi_{1,!}(\ )$. Hence, in order to prove that $T_s(\bar{\mathcal{X}}^L_{w,\psi}) = T'_s(\bar{\mathcal{X}}^L_{w,\psi})$ it is enough to show that

$$R^i j'_*(\mathcal{S}') = R^i j'_!(\mathcal{S}') \tag{1.16.2}$$

where $\mathcal{S}' = \gamma^*\pi_2^*(\bar{\mathcal{X}}^L_{w,\psi})$.

Let L_Δ be the line bundle over $\overline{O(s)}$ associated to the smooth divisor $\Delta = \{(B',B'') \in \overline{O(s)} \mid B' = B''\}$ in $\overline{O(s)}$. Then L_Δ is in a natural way a G-equivariant line bundle, since Δ is G-stable in $\overline{O(s)}$. There is a unique G-equivariant isomorphism of line bundles over $\overline{O(s)} : \hat{L}' \cong \hat{L} \otimes L_\Delta^{\otimes(-m)}$. The line bundle L_Δ has a unique G-equivariant trivialization (up to a scalar) over $\overline{O(s)} - \Delta$. This gives rise to an isomorphism $\hat{L}|O(s) \cong (\hat{L} \otimes L_\Delta^{\otimes(-m)})|O(s)$, hence to an embedding $j : \tilde{L}^{\cdot} \to (\hat{L} \otimes L_\Delta^{\otimes(-m)})^{\cdot}$ of line bundles with zero section removed, compatible with the imbedding $O(s) \to \overline{O(s)}$.

Then the diagram

$$\begin{array}{ccc} (\tilde{L}')^{\cdot} & \xrightarrow{\ j'\ } & (\hat{L}')^{\cdot} \\ \gamma \downarrow \wr & & \downarrow \wr \\ \tilde{L}^{\cdot} & \xrightarrow{\ j\ } & (\hat{L} \otimes L^{\otimes(-m)})^{\cdot} \end{array}$$

is commutative (up to multiplication by an element in k^*), and we see that (1.16.2) is equivalent to

$$R^i j_*(\mathcal{S}) = R^i j_!(\mathcal{S}), \quad \text{where } \mathcal{S} = \pi_2^*(\bar{\mathcal{X}}^L_{w,\psi}) \,. \tag{1.16.3}$$

We now consider the open subset $\mathcal{U} = \{B' \in \mathcal{B} \mid B' \text{ opposed to } B_1\} \subset \mathcal{B}$ where B_1 is a fixed Borel subgroup. Then L is a trivial line bundle over $\mathcal{U}$, hence there exists an isomorphism $\delta : L^{\cdot}|\mathcal{U} \to \mathcal{U} \times k^*$ which is compatible with the projections on $\mathcal{U}$ and is k^*-equivariant (with k^* acting by scalar multiplication on $L^{\cdot}|\mathcal{U}$ and by multiplication on the second factor, on $\mathcal{U} \times k^*$). The restriction of the sheaf $\bar{\mathcal{X}}^L_{w,\psi}$ to $L^{\cdot}|\mathcal{U}$

is isomorphic to $\mathrm{pr}_{\mathfrak{U}}^*(\mathcal{G}) \boxtimes \mathrm{pr}_{k^*}^*(\mathcal{F}_\psi)$ over $\mathfrak{U} \times k^*$, where $\mathcal{G}$ is a constructible ℓ-adic sheaf over $\mathfrak{U}$. Let $\hat{\mathfrak{U}} = \{(B', B'') \in \overline{O(s)} | B' \in \mathfrak{U}, B'' \in \mathfrak{U}\}$. This is an open affine subset of $\overline{O(s)}$. We define an isomorphism $\phi : k \times \mathfrak{U} \xrightarrow{\sim} \hat{\mathfrak{U}}$ as follows. Choose a maximal torus $T_1 \subset B_1$ and let $x : k \to G$ be an isomorphism of the additive group of k onto the root subgroup corresponding to $T_1 \subset B_1$ and to minus the simple root attached to s. Then ϕ is defined by $\phi(a, B') = (x(a)B'x(a)^{-1}, B')$. When B_1 varies, the open sets $\hat{\mathfrak{U}}$ cover $O(s)$, hence, to check (1.16.3) it is enough to check that (1.16.3) is true when j is replaced by its restriction $j_{\hat{\mathfrak{U}}} : \tilde{L}^{\cdot} | \hat{\mathfrak{U}} - \Delta \to (L \otimes L_\Delta^{\otimes(-m)})^{\cdot} | \hat{\mathfrak{U}}$ and $\mathcal{S}$ is replaced by its restriction $\mathcal{S}_{\hat{\mathfrak{U}}}$ to $\tilde{L}^{\cdot} | \hat{\mathfrak{U}} - \Delta$. Using the description of $\overline{\mathcal{X}}_{w,\psi}^L$ as an external tensor product over $L^{\cdot} | \mathfrak{U} = \mathfrak{U} \times k^*$, and of $\hat{\mathfrak{U}}$ as a product $k \times \mathfrak{U}$, we get a commutative diagram

$$\begin{array}{ccc}
\tilde{L}^{\cdot} | \hat{\mathfrak{U}} - \Delta & \xrightarrow{\;j_{\hat{\mathfrak{U}}}\;} & (\hat{L} \otimes L_\Delta^{\otimes(-m)})^{\cdot} | \hat{\mathfrak{U}} \\
{\scriptstyle \alpha_1} \downarrow \wr & & \downarrow \wr \\
k^* \times \mathfrak{U} \times k^* & \xrightarrow{\;j_1\;} & k \times \mathfrak{U} \times k^* \\
{\scriptstyle \alpha_2} \downarrow \wr & & {\scriptstyle \mathrm{Id}} \downarrow \wr \\
k^* \times \mathfrak{U} \times k^* & \xrightarrow{\;j_2\;} & k \times \mathfrak{U} \times k^*
\end{array}$$

where $j_1(z, u, z') = (z, u, z^{-m}z')$, $j_2(z, u, z') = (z, u, z')$, $\alpha_2(z, u, z') = (z, u, z^{-m}z')$ and we have:

$$\mathcal{S}_{\hat{\mathfrak{U}}} = \alpha_1^*(\mathrm{pr}_{\mathfrak{U}}^*(\mathcal{G}) \boxtimes \mathrm{pr}_3^*(\mathcal{F}_\psi)) = \alpha_1^*\alpha_2^*(\mathrm{pr}_1^*(\mathcal{F}_\psi m) \boxtimes \mathrm{pr}_{\mathfrak{U}}^*(\mathcal{G}) \boxtimes \mathrm{pr}_3^*(\mathcal{F}_\psi))$$

where pr_1, pr_3 are the first and third projection of $k^* \times U \times k^*$ onto k^*. It is enough to show that $R^i(j_2)_* \; (\mathrm{pr}_1^*(\mathcal{F}_\psi m) \boxtimes \mathrm{pr}_{\mathfrak{U}}^*(\mathcal{G}) \boxtimes \mathrm{pr}_3^*(\mathcal{F}_\psi))$ is zero when restricted to $\{0\} \times \mathfrak{U} \times k^*$. But this higher direct image is

$$(R^i(j_3)_* \mathcal{F}_\psi m) \boxtimes \mathrm{pr}_3^*(\mathcal{G}) \boxtimes \mathrm{pr}_3^*(\mathcal{F}_\psi) \tag{1.16.4}$$

where j_3 is the natural inclusion $k^* \to k$. Since $\psi^m \neq 1$ on μ_n, the stalk of $R^i(j_3)_* \mathcal{F}_{\psi^m}$ at 0 is 0. Thus, (1.16.4) is zero when restricted to $\{0\} \times \mathcal{U} \times k^*$. The lemma is proved.

1.17. LEMMA. *Assume that* $ws < w$ *and that* $\psi^m \neq 1$ *on* μ_n, $(m = \check{\alpha}_s(L))$. *Then*

$$T_s \hat{\mathcal{X}}^{sL}_{ws,\psi} = \hat{\mathcal{X}}^{L}_{w,\psi} .$$

Proof. We have always $s \cdot W_{L,n} \cdot s^{-1} = W_{{}^sL,n}$ and s maps $R_{L,n}$ to $R_{{}^sL,n}$. Our assumption on ψ implies that m is not divisible by n. It follows that $s \notin W_{L,n}$, so that s maps $R^+_{L,n}$ into R^+ (since the simple root corresponding to s is not in $R^+_{L,n}$) hence s maps $R^+_{L,n}$ onto $R^+_{{}^sL,n}$. It follows that conjugation by s defines a bijection of $S_{L,n}$ onto $S_{{}^sL,n}$ hence it is an isomorphism of Coxeter groups $(W_{L,n}, S_{L,n}) \approx (W_{{}^sL,n}, S_{{}^sL,n})$. In particular, it is compatible with the partial order, length function and P-polynomials of these two Coxeter groups. Note also that, for our s, the element $w_1 \in W$ has minimal length in $w_1 W_{L,n}$ if and only if $w_1 s$ has minimal length in $(w_1 s)W_{{}^sL,n}$. (This follows from the characterization 1.9 (i).) Now the result follows immediately from the previous remarks together with Lemma 1.16.

1.18. Assume now that $L \in \mathfrak{L}$ and $s \in S$ are such that ${}^sL = L$, i.e. that $\check{\alpha}_s(L) = 0$. Then the restriction of L to any fiber ($\approx$ projective line) of the natural projection $\pi_s : \mathcal{B} \to \mathcal{P}_s$ (see 1.3) is the trivial line bundle. Hence there exists a (unique) G-equivariant line bundle $\bar{L}$ over $\mathcal{P}_s$ with a F_p-rational structure compatible with the G-action, such that $L = \pi_s^*(\bar{L})$, as G-equivariant line bundles with F_p-structure. Let $\bar{L}^{\cdot}$ denote the space obtained from $\bar{L}$ by removing the zero section, and let $\rho_s : L^{\cdot} \to \bar{L}^{\cdot}$ be the natural projection. We define an endomorphism θ_s of the Z[A]-module $\mathcal{K}(\mathcal{C}_{L^{\cdot},n})$ by setting

$$\theta_s(\mathcal{F}) = \sum_i (-1)^i \rho_s^*(R^i \rho_{s\,!}(\mathcal{F})) .$$

We can now state

1.19. LEMMA. *Assume that* ${}^{s}L = L$. *Then*

(i) $D\theta_s = [p]^{-1}\theta_s D$ *as endomorphisms of* $\mathcal{K}(\mathcal{C}_{L^{\cdot},n})$.

(ii) $$\theta_s(\mathfrak{X}^L_{w,\psi}) = \begin{cases} \mathfrak{X}^L_{w,\psi} + \mathfrak{X}^L_{ws,\psi} & , \text{ if } w < ws \\ [p](\mathfrak{X}^L_{w,\psi} + \mathfrak{X}^L_{ws,\psi}), & \text{ if } w > ws . \end{cases}$$

(iii) *If* $ws < w$, *then*

$$\theta_s(\hat{\mathfrak{X}}^L_{ws,\psi}) = \hat{\mathfrak{X}}^L_{w,\psi} + \sum_{\substack{v_1 \in W_{L,n} \\ v_1 s < v_1 < vs}} \mu(v_1,vs)[p]^{\frac{1}{2}(\ell(w)-\ell(w_1v_1))}\hat{\mathfrak{X}}_{w_1v_1,\psi}$$

where $w = w_1 \cdot v$ *is the decomposition of* w *with* $v \in W_{L,n}$ *and* w_1 *of minimal length in* $w \cdot W_{L,n}$; $\mu(v_1,vs) \in \mathbf{Z}$ *is the coefficient of*

$[p]^{\frac{1}{2}(\tilde{\ell}(vs)-\tilde{\ell}(v_1)-1)}$ *in* $P_{v_1,vs}([p])$.

Proof. (i) follows from the fact that $D_{\rho_{s!}}(\ ) = \rho_{s*}D(\ ) = \rho_{s!}D(\ )$, (in the derived category) since ρ_s is proper, and $D\rho_s^*(\ ) = (\rho_s^*D(\ ))\otimes\overline{\mathbf{Q}}_\ell(1)$, (with a shift of two degrees), since ρ_s is smooth of relative dimension 1.

(ii) follows easily from the definition.

(iii) follows from (ii) and from the property [KL$_1$, 2.2.c] of the P-polynomials.

1.20. *Proof of Lemma 1.11.* Let $h: G_1 \to G$ be a homomorphism of algebraic groups of a connected reductive group G_1 with simply connected derived subgroup, onto G, whose kernel is central in G_1 such that G_1 and h are defined over F_p and such that there exists a maximal torus in G_1 which is split over F_p. It is clear that the truth of 1.11 for G_1 implies the truth of 1.11 for G. Since we can always find such G_1 and h, we are reduced to the case where G has simply connected derived subgroup. As we have seen, 1.11 is equivalent with Lemma 1.12. We shall prove 1.12 by induction on $\ell(w)$. It is obvious when $w = e$. Assume now that 1.12 is known whenever $\ell(w) < a$ for some $a \geq 1$. We now assume that $\ell(w) = a$. Let $s \in S$ be such that $\ell(ws) < \ell(w)$. Assume first that $s \notin W_{L,n}$, i.e. that $m = \check{\alpha}_s(L)$ is not divisible by n. Since ψ is assumed to be injective, it follows that $\psi^m \neq 1$ on μ_n. By the induction hypothesis, 1.12 holds for ws and ${}^sL$:

$$D(\hat{\mathfrak{X}}^{{}^sL}_{w,\psi}) = [p]^{-\ell(w)}\hat{\mathfrak{X}}^{{}^sL}_{ws,\psi^{-1}}. \tag{1.20.1}$$

We have

$$\begin{aligned}
D(\hat{\mathfrak{X}}^{L}_{w,\psi}) &= D(T'_s(\hat{\mathfrak{X}}^{{}^sL}_{ws,\psi})) && \text{(by 1.17)}\\
&= [p]^{-1}\,T'_s(D(\hat{\mathfrak{X}}^{{}^sL}_{ws,\psi})) && \text{(by 1.15)}\\
&= [p]^{-\ell(w)-1}\,T'_s(\hat{\mathfrak{X}}^{{}^sL}_{ws,\psi^{-1}}) && \text{(by (1.20.1))}\\
&= [p]^{-\ell(w)-1}\,T_s(\hat{\mathfrak{X}}^{{}^sL}_{ws,\psi^{-1}}) && \text{(by 1.16)}\\
&= [p]^{-\ell(w)-1}\,\hat{\mathfrak{X}}^{L}_{w,\psi^{-1}} && \text{(by 1.17)}.
\end{aligned}$$

Thus, 1.12 holds for w and L.

Next, we assume that $s \in W_{L,n}$. Then ${}^sL = L \otimes L_s^{\otimes nt}$ where t is an integer. Since G is assumed to have simply-connected derived subgroup, the homomorphism $\check{\alpha}_s : \mathfrak{L} \to \mathbf{Z}$ is surjective. Hence there exists $L_1 \in \mathfrak{L}$ such that ${}^sL_1 = L_1 \otimes L_s^{\otimes(-t)}$. Then ${}^s(L \otimes L_1^{\otimes n}) = L \otimes L_1^{\otimes n}$. By the discussion in 1.13, the truth of 1.12 for (L, w) is equivalent to the truth of 1.12 for $(L \otimes L_1^{\otimes n}, w)$. Thus, we are reduced to the case where ${}^sL = L$. By the induction hypothesis, we have

$$D(\hat{\mathfrak{X}}^{L}_{w_1v_1,\psi}) = [p]^{-\ell(w_1v_1)-1}\,\hat{\mathfrak{X}}^{L}_{w_1v_1,\psi^{-1}}$$

for each term in the sum over v_1 in 1.19(iii) and

$$D(\theta_s(\hat{\mathfrak{X}}^L_{ws,\psi})) = [p]^{-\ell(w)-1}\theta_s(\hat{\mathfrak{X}}^L_{ws,\psi^{-1}})$$

(using also 1.19 (i)).

Using now the identity 1.19(iii) it follows that

$$D(\hat{\mathfrak{X}}^L_{w,\psi}) = [p]^{-\ell(w)-1}\hat{\mathfrak{X}}^L_{w,\psi^{-1}}\,.$$

Thus, 1.12 and hence 1.11, are proved.

1.21. Let $L' \in \mathfrak{L}$ and let n be as in (1.4.1). We shall define a subset $\mathfrak{N}_{L',n}$ of $\mathfrak{N}(T)$. (This subset appears also in Kilmoyer's work [Ki].) The set $\mathfrak{N}_{L',n}$ consists of all elements $n \in \mathfrak{N}(T)$ whose image $\overline{n} \in W$ belongs to $W_{L',n}$ and are such that there exists a reduced expression $\overline{n} = r_1 r_2 \cdots r_t$ in the Coxeter group $W_{L',n}$ $(r_i \in S_{L',n})$ with $n = \dot{r}_1 \dot{r}_2 \cdots \dot{r}_t$, where $\dot{r}_i$ are representatives for r_i in $\mathfrak{N}(T)$ with the following property: $\dot{r}_i$ belongs to the subgroup of $G(F_p)$ generated by the two root subgroups over F_p (in $U(F_p)$ and its T-opposite) corresponding to the reflection $r_i \in W$. Note that $\mathfrak{N}_{L',n}$ is not, in general, a subgroup.

We can now state the following.

1.22. Lemma. *Let $L' \in \mathfrak{L}$, and let n be as in (1.4.1). Let $\psi : \mu_n \to \overline{\mathbf{Q}}^*_\ell$ be an injective homomorphism. Let $w \in W$ and let w_1 be the element of minimal length in $wW_{L',n}$. Let $\dot{w}_1, \dot{w}$ be representatives for w_1, w in $\mathfrak{N}(T)^F$ such that $\dot{w}_1, \dot{w}$ belong to the subgroup of $G(F_p)$ generated by its unipotent elements and such that $\dot{w}_1^{-1}\dot{w} \in \mathfrak{N}_{L',n}$. Then the restriction of the sheaf $\mathcal{H}^i(\pi^{-1}(\overline{\mathcal{B}}_w), \mathfrak{X}^{L'}_{\dot{w},\psi})$, (see 1.6), to $\pi^{-1}(\mathcal{B}_{w'})$ $(w' \le w)$ is zero unless $w' \in w_1 W_{L',n}$ and i is even. Assume now that $w' \in w_1 W_{L',n}$ and that i is even. Let $\dot{w}'$ be a representative for w' in $\mathfrak{N}(T)^F$ such that $\dot{w}_1^{-1}\dot{w}' \in \mathfrak{N}_{L',n}$, and let $z \in F_p^*$. Then there exists an element $\varepsilon \in \overline{\mathbf{Q}}^*_\ell$,*

$\varepsilon^n = 1$ *(depending on* $L', n, \psi, z, \dot{w}_1, \dot{w}$, *but not on* $i, w', \dot{w}'$ *) with the following property. The eigenvalues of* F^d *(= Frobenius map relative to the* F_{p^d}*-structure) on the stalk of* $\mathcal{H}^i(\pi^{-1}(\overline{\mathcal{B}}_w), \mathcal{X}^L_{\dot{w},\psi})$ *at* $(\dot{w}', z) \in (L')^{\cdot}$ *(in the model (1.5.1)) are equal to* $\varepsilon p^{di/2}$.

Proof. As in 1.20, we can reduce ourselves to the case where G has simply-connected derived group. We shall prove the lemma, in this case, by induction on $\ell(w)$. The lemma is obvious for $w = e$. Assume that it is known when $\ell(w) < a$ for some $a \geq 1$; we now assume that $\ell(w) = a$. Let $s \in S$ be such that $\ell(ws) < \ell(w)$. Let $L = {}^s(L')$ and let $\lambda : T \to k^*$ be the character corresponding to L. We shall set $\mathcal{H}^i(\pi^{-1}(\overline{\mathcal{B}}_w), \mathcal{X}^{L'}_{\dot{w},\psi}) = \mathcal{H}^{i,L'}_{\dot{w}}$. We shall distinguish two cases.

Case 1: $s \notin W_{L',n}$.

Let $\dot{s}$ be any representative for s in $\mathcal{N}(T)$ such that $\dot{s}$ is a product of unipotent elements in $G(F_p)$. Let us consider the diagram (1.16.1). We shall assume (as we may) that γ in that diagram is such that $(1, z) \in$ fiber of $(L')^{\cdot}$ at B corresponds under γ to $(\dot{s}, z) \in$ fiber of $L^{\cdot}$ at $\dot{s}B\dot{s}^{-1}$; here we use the description (1.5.1) for $L^{\cdot}$ and $(L')^{\cdot}$. (Note that $(B, \dot{s}B\dot{s}^{-1}) \in O(s)$.)

We shall denote by Y the set of points in $(\widetilde{L'})^{\cdot}$ lying over the subset $\{(B', B'') \in O(s) | B'' \in \overline{\mathcal{B}}_{ws}\}$. We denote $\pi_1' : Y \to (L')^{\cdot} | \overline{\mathcal{B}}_w$, $\pi_2'\gamma : Y \to L^{\cdot} | \overline{\mathcal{B}}_{ws}$ the restriction of the maps $\pi_1, \pi_2\gamma$ in (1.16.1). Let $K_1 = IC(L^{\cdot} | \overline{\mathcal{B}}_{ws}, \mathcal{X}^L_{\dot{w}\dot{s},\psi})$, $K_2 = (\pi_2'\gamma)^* K_1$, $K_3 = (\pi_1')_! K_2$ (objects in a derived category). Since $\pi_2'\gamma$ is a locally trivial affine line bundle, we must have $K_2 = IC(Y, (\pi_2\gamma)^* \mathcal{X}^L_{\dot{w}\dot{s},\psi})$. Using (1.16.2), we see that $(\pi_1')_! K_2 = (\pi_1')_* K_2$.

Hence the Verdier dual of K_3 (is up to shift and a twist) K_3 itself, with ψ replaced by ψ^{-1}. We shall show that

$$K_3 = IC((L')^{\cdot} | \overline{\mathcal{B}}_w, \mathcal{X}^{L'}_{\dot{w},\psi}) . \tag{1.22.1}$$

By the definition of the intersection cohomology complex, it will be enough to check that the i-th cohomology sheaf $\mathcal{H}^i(K_3)$ of K_3 satisfies a suitable "perversity" condition. The stalk $\mathcal{H}^i_\ell(K_3)$ at a point $\ell \in (L')^{\cdot}|\overline{\mathcal{B}}_w$ is the i-th hypercohomology with compact support of $(\pi_1')^{-1}(\ell) \subset Y$ with coefficients in the restriction of K_2 to $(\pi_1')^{-1}(\ell)$. The fibers $(\pi')^{-1}(\ell)$ have the following structure. Let $w' \in W$ be such that ℓ lies over a point in $\mathcal{B}_{w'}$, $(w' \leq w)$. If $w' < w's \not\leq ws$ then $(\pi_1')^{-1}(\ell)$ is empty. If $w' < w's \leq ws$, then $(\pi_1')^{-1}(\ell)$ is an affine line and its i-th hypercohomology with compact support with coefficients in the restriction of K_2 is isomorphic to the stalk at a point in $\pi_1^{-1}(\ell)$ of the (i-2)th cohomology sheaf of K_2, tensored with $\overline{\mathbb{Q}}_\ell(-1)$. If $w's < w' \not\leq ws$, then $(\pi_1')^{-1}(\ell)$ is a point, hence $\mathcal{H}^i_\ell(K_3)$ is just the stalk of the i-th cohomology sheaf of K_2 at $\pi_1^{-1}(\ell)$. If $w's < w' \leq ws$, then $(\pi_1')^{-1}(\ell)$ is an affine line with a distinguished point ℓ_0. The hypercohomology with compact support of this affine line, punctured at ℓ_0, is seen to the zero, using the fact that $\psi^{\check{\alpha}_s(L)} \neq 1$ on μ_n. (Compare with the first part of the proof of 1.16.) Hence $\mathcal{H}^i_\ell(K_3)$ is just the stalk at ℓ_0 of the i-th cohomology sheaf of K_2. These computations show, in particular that, whenever $\mathcal{H}^i_\ell(K_3) \neq 0$, $i > 0$ and ℓ lies over $\mathcal{B}_{w'}$, $(w' \leq w)$, we must have $i < \ell(w) - \ell(w')$. (Indeed, the stalks of K_2 have a similar perversity property since K_2 is an intersection cohomology complex.) Thus, the perversity condition for K_3 is verified so that (1.22.1) follows.

(1.22.2) For $w' \leq w$, we choose a representative $\dot{w}'$ for w' in $\mathcal{N}(T)^F$ as in the statement of the lemma, if $w' \in w \cdot W_{L,n}$ and in an arbitrary way if $w' \notin w \cdot W_{L,n}$.

The previous computations show that the stalk of $\mathcal{H}^{i,L'}_{\dot{w}}$ at $(\dot{w}', z)$ is

– Zero, if $w' < w's \not\leq ws$.

– Canonically isomorphic to the stalk of $\mathcal{H}^{i-2,L}_{\dot{w}\dot{s}} \otimes \overline{\mathbb{Q}}_\ell(-1)$ at $(\dot{w}'\dot{s}, z)$, if $w' < w's \leq ws$.

– Canonically isomorphic to the stalk of $\mathcal{H}^{i,L}_{\dot{w}\dot{s}}$ at $(\dot{w}'\dot{s}, z)$, if $w's < w'$.

Next, we note that $w_1 s$ is an element of minimal length in $ws \cdot W_{L,n}$ and that $\dot{s}^{-1}\mathfrak{N}_{L',n}\dot{s} = \mathfrak{N}_{L,n}$ (see the proof of Lemma 1.17). Since our lemma is assumed to hold for ws, it follows that it holds also for w.

Case 2: $s \in W_{L',n}$.

As in 1.20, we may reduce ourselves to the case where $L = L'$ (using the assumption that G has simply-connected derived subgroup). In this case, we choose a representative $\dot{s}$ for s, with $\dot{s} \in \mathfrak{N}_{L,n}$.

Let $K_1 = IC(L^{\cdot}|\overline{\mathcal{B}}_{ws}, \mathcal{X}^L_{\dot{w}\dot{s},\psi})$, $K_3 = IC(L^{\cdot}|\overline{\mathcal{B}}_w, \mathcal{X}^L_{\dot{w},\psi})$. Consider the subspace $\overline{L}^{\cdot}|\pi_s(\overline{\mathcal{B}}_w)$ of $\overline{L}^{\cdot}$ (part of $\overline{L}^{\cdot}$ lying over $\pi_s(\overline{\mathcal{B}}_w) \subset \mathcal{P}_s$, with the notations in 1.18). There is a canonical locally constant $\overline{Q}_\ell$-sheaf $\mathcal{S}$ of rank 1, defined over F_p, over the open subset $\overline{L}^{\cdot}|\pi_s(\mathcal{B}_w)$ of $\overline{L}^{\cdot}|\pi_s(\overline{\mathcal{B}}_w)$ such that $\mathcal{X}^L_{\dot{w},\psi} = \rho_s^*(\mathcal{S})$, (with the notations in 1.18). Let $K_2 = IC(\overline{L}^{\cdot}|\pi_s(\overline{\mathcal{B}}_w), \mathcal{S})$.

The map $\rho_s : L^{\cdot}|\overline{\mathcal{B}}_{ws} \to \overline{L}^{\cdot}|\pi_s(\overline{\mathcal{B}}_w)$ has fibers isomorphic to a point (generically) or to $\mathbf{P}^1$. If a fiber is a projective line, then it has a distinguished point and over the complement of this point (an affine line) the restriction of the i-th cohomology sheaf of K_1 is locally constant and tame hence it is automatically constant. From this, it follows that the stalk of $\mathcal{H}^i((\rho_s)_!K_1)$ at $\rho_s(\dot{w}, z)$ (with $w' \leq w$ and $\dot{w}'$ as in (1.22.2)) is canonically isomorphic to:

– The stalk of $\mathcal{H}^{i,L}_{\dot{w}\dot{s}}$ at $(\dot{w}', z)$, if $w' < w's \not\leq ws$.

– The stalk of $\mathcal{H}^{i,L}_{\dot{w}\dot{s}}$ at $(\dot{w}'\dot{s}, z)$, if $w's < w' \not\leq ws$.

– An extension of the stalk of $\mathcal{H}^{i,L}_{\dot{w}\dot{s}}$ at $(\dot{w}', z)$ by the stalk of $\mathcal{H}^{i-2,L}_{\dot{w}\dot{s}} \otimes \overline{Q}_\ell(-1)$ at $(\dot{w}'\dot{s}, z)$, if $w' < w's \leq ws$.

– An extension of the stalk of $\mathcal{H}^{i,L}_{\dot{w}\dot{s}}$ at $(\dot{w}'\dot{s}, z)$ by the stalk of $\mathcal{H}^{i-2,L}_{\dot{w}\dot{s}} \otimes \overline{Q}_\ell(-1)$ at $(\dot{w}', z)$, if $w's < w' \leq ws$.

Since ρ_s is proper, K_2 is in a natural way a direct summand of $(\rho_s)_! K_1$. (This is a general property of intersection cohomology complexes, see [BBD].) It follows that the stalk of $\mathcal{H}^{i,L}_{\dot{w}}$ at $(\dot{w}', z)$ is canonically isomorphic to a direct summand of the stalk of $\mathcal{H}^i((\rho_s)_! K_1)$ at $\rho_s(\dot{w}', z)$; this last stalk was just described above. Since our lemma is assumed to hold for ws, it follows that it holds also for w, provided that we check the following statement.

(1.22.3) *Let* $w' \in wW_{L,n}$ *and let* $\dot{w}'$ *be a representative in* $\mathcal{N}(T)^F$ *such that* $\dot{w}_1^{-1}\dot{w}' \in \mathcal{N}_{L,n}$.

Let $s \in S_{L,n}$ *and let* $\dot{s}$ *be a representative for* s *in* $\mathcal{N}_{L,n}$. *Then there exists* $\dot{w}'' \in \mathcal{N}(T)$ *such that* $\dot{w}_1^{-1}\dot{w}'' \in \mathcal{N}_{L,n}$ *and such that* $(\dot{w}'\dot{s}, z) = (\dot{w}'', z)$ *as points in* $L^{\cdot}$.

Let $y = w_1^{-1}w'$, $\dot{y} = \dot{w}_1^{-1}\dot{w}' \in \mathcal{N}_{L,n}$. If $ys > y$, then $ys \in \mathcal{N}_{L,n}$ and we can take $w' = w_1 ys$.

Assume now that $ys < y$. Then $\dot{y}\dot{s}$ is not necessarily in $\mathcal{N}_{L,n}$. However, there exists $\dot{v} \in \mathcal{N}_{L,n}$ and a representative $\tilde{s}$ for s in $\mathcal{N}_{L,n}$ such that $\dot{y} = \dot{v}\tilde{s}$. We then have the equalities (as points in $L^{\cdot} = L^{\cdot}_{\lambda}$):

$$(\dot{w}'\dot{s}, z) = (\dot{w}_1\dot{y}\dot{s}, z) = (\dot{w}_1\dot{v}\tilde{s}\dot{s}, z) = (\dot{w}_1\dot{v}, \lambda(\tilde{s}\dot{s})z) = (\dot{w}_1\dot{v}, z) .$$

Here we have used the fact that the character λ is fixed by s, hence it is trivial on the image of the simple coroot $k^* \to T$ corresponding to s and, in particular, it is trivial on $\tilde{s}\dot{s}$. Hence, we can take $\dot{w}'' = \dot{w}_1\dot{v}$. Thus, (1.22.3) is verified and the lemma is proved.

1.23. Let $L \in \mathfrak{L}$ and let n be as in (1.4.1). Let $Z \subset W$ be a right $W_{L,n}$-coset (in particular, we have $Z \cdot W_{L,n} = Z$), and let w_1 be the element of minimal length in Z (see 1.9 (i)).

A coherent lifting of Z (to $\mathcal{N}(T)$) is a subset $Z^{\cdot}$ of $\mathcal{N}(T)^F$ mapping bijectively onto Z under $\mathcal{N}(T) \to W$ such that the following two properties are satisfied.

(a) $Z^{\cdot}$ is contained in the subgroup of $G(F_p)$ generated by the unipotent elements in $G(F_p)$.

(b) If $\dot{w}$ denotes the element of $Z^{\cdot}$ corresponding to $w \in Z$, then we have $(\dot{w}_1)^{-1}\dot{w} \in \mathfrak{N}_{L,n}$, for all $w \in Z$.

It is clear that a coherent lifting of Z (to $\mathfrak{N}(T)$) always exists.

In this set up, we can state the main result of this chapter.

1.24. THEOREM. *Let $Z^{\cdot}$ be a coherent lifting of Z, let $w \in Z$, let $\dot{w}$ be its representative in $Z^{\cdot}$ and let ψ be an injective homomorphism $\mu_n \to \overline{\mathbf{Q}}_\ell^*$. Then the restriction of the sheaf $\mathcal{H}^i(\pi^{-1}(\overline{\mathcal{B}}_w), \mathcal{X}^L_{\dot{w},\psi})$, (see 1.6), to $\pi^{-1}(\mathcal{B}_{w'})$, $(w' \leq w)$, is zero unless $w' \in Z$ and i is even. If $w' \in Z$ and i is even, then this restriction is an H-equivariant, locally constant $\overline{\mathbf{Q}}_\ell$-sheaf, defined over F_{p^d}, for which there exists a filtration by locally constant, H-equivariant sub-sheaves defined over F_{p^d}, each of whose successive quotients is isomorphic (as an H-equivariant, locally constant $\overline{\mathbf{Q}}_\ell^*$-sheaf defined over F_{p^d}) to $\mathcal{X}^i_{\dot{w}',\psi} \otimes \overline{\mathbf{Q}}_\ell(-i/2)$, ($\dot{w}'$ is the representative of w' in $Z^{\cdot}$). The number of nonzero successive quotients in this filtration is the coefficient ($\in \mathbf{Z}$) of $[p]^{i/2}\overline{\mathfrak{X}}^L_{w',\psi}$ in $\hat{\mathfrak{X}}^L_{w,\psi}$.*

Proof. The first assertion of the theorem is contained already in Lemma 1.22.

We now assume that $w' \in Z$, $w' \leq w$ and that i is even. With the notations in the proof of Lemma 1.22, we consider the restriction of $\mathcal{H}^{i,L}_{\dot{w}}$ to $\pi^{-1}(\mathcal{B}_{w'})$ and tensor it with the dual of the locally constant $\overline{\mathbf{Q}}_\ell$-sheaf $\mathcal{X}^L_{\dot{w}',\psi}$ and with $\overline{\mathbf{Q}}_\ell(i/2)$. (Here $\dot{w}'$ is the element of $Z^{\cdot}$ representing w'.) Let $\mathcal{F}$ be the resulting sheaf. It is a sheaf with a natural F_{p^d}-structure, which is U-equivariant and also k^*-equivariant for the action of k^* on $\pi^{-1}(\mathcal{B}_{w'})$ given by scalar multiplication (no n-th power involved). Hence $\mathcal{F}$ is canonically the inverse image $\pi^*\overline{\mathcal{F}}$ of an U-equivariant locally constant $\overline{\mathbf{Q}}_\ell$-sheaf $\overline{\mathcal{F}}$ on $\mathcal{B}_{w'}$, with an

F_{p^d}-structure. Now Lemma 1.22 with $z = 1$, shows that the Frobenius map F^d of $\bar{\mathfrak{F}}$ defines a unipotent transformation of the stalk of $\bar{\mathfrak{F}}$ at $\dot{w}'B(\dot{w}')^{-1}$. (Note that F^d acts as identity on the stalk of $\mathcal{X}^L_{\dot{w}',\psi}$ at $(\dot{w}',1) \in L^{\cdot}$ in the model (1.5.1) for $L^{\cdot}$.) On the other hand, $\bar{\mathfrak{F}}$ being a U-equivariant locally constant $\bar{Q}_\ell$-sheaf on $\mathcal{B}_{w'}$, must be isomorphic to a constant $\bar{Q}_\ell$-sheaf and hence it has another F_{p^d}-structure whose Frobenius map F_1^d induces F^d on $\mathcal{B}_{w'}$ and is the identity on the stalk of $\bar{\mathfrak{F}}$ at $\dot{w}'B(\dot{w}')^{-1}$. Thus, there exists an endomorphism α of the sheaf $\bar{\mathfrak{F}}$, inducing identity on $\mathcal{B}_{w'}$, whose restriction to the stalk at $\dot{w}'B(\dot{w}')^{-1}$ is unipotent, and is such that $F^d = F_1^d\alpha = \alpha F_1^d$.

We can find a filtration of $\bar{\mathfrak{F}}$ by constant $\bar{Q}_\ell$-sheaves which are α-stable (and automatically F_1^d-stable) and such that the successive quotients are of rank 1. This filtration will be stable under F^d, and it gives rise to the required filtration of the restriction of $\mathcal{H}^{i,L}_{\dot{w}}$ to $\pi^{-1}(\mathcal{B}_{w'})$. This proves the theorem except for its last statement. But the last statement follows immediately from the preceding statements together with Lemma 1.11. This completes the proof.

Note that the special case $n = 1$ of the theorem is essentially the same as Theorem 4.2 of [KL_2].

Chapter 2

LOCAL INTERSECTION COHOMOLOGY WITH TWISTED COEFFICIENTS OF THE CLOSURES OF THE VARIETIES X_w

The varieties X_w in the title of this chapter were defined by Deligne and myself in $[DL_1]$. They are locally closed subvarieties of a flag manifold and are associated with a reductive algebraic group defined over a finite field. These varieties have natural actions of the corresponding finite group and carry some natural equivariant ℓ-adic local systems $\mathcal{F}_\theta$ where θ is a character of the group of rational points of a certain maximal torus corresponding to the Weyl group element w.

In this chapter, we describe the local intersection cohomology groups of the Zariski closure $\bar{X}_w$ of X_w with coefficients in $\mathcal{F}_\theta$. It turns out that the pair $(\bar{X}_w, \mathcal{F}_\theta)$ looks locally like the objects studied in Chapter 1, and hence we may use the results of §1 to study $(\bar{X}_w, \mathcal{F}_\theta)$.

We also prove some preliminary results on the (global) intersection cohomology group $\mathbf{H}^i(\bar{X}_w, \mathcal{F}_\theta)$.

2.1. In addition to the F_p-rational structure on G, with Frobenius map $F : G \to G$ (as in 1.2), we assume given an endomorphism $F' : G \to G$ such that $F'(B) = B$, $F'(T) = T$, $F'F = FF'$ and such that some power of F' is equal to some power of F.

The map F' takes $\mathcal{N}(T)$ into itself and defines an automorphism of the Coxeter group (W, S) which will be again denoted F'. The fixed point set $G^{F'}$ is a finite subgroup of G and we shall be interested in its representations over $\bar{\mathbf{Q}}_\ell$; the map F will play only an auxiliary role. Following $[DL_1$, (1.11.1), (1.11.4)], we define for each w

$$X_w = \{gB \in G/B \mid g^{-1}F'(g) \in BwB\}$$

(a locally closed subvariety of G/B) and for each representative $\dot{w} \in \mathfrak{N}(T)^F$ for w we define

$$\tilde{X}_{\dot{w}} = \{g \in G \mid g^{-1}F'(g) \in \dot{w}U\}/U \cap \dot{w}U\dot{w}^{-1}$$

where $u_1 \in U \cap \dot{w}U\dot{w}^{-1}$ acts by $g \to gu_1^{-1}$.

Both X_w, $\tilde{X}_{\dot{w}}$ have canonical F_p-structures (inherited from (G,F)); we shall denote the corresponding Frobenius maps again by F. For $w \in W$, let $F'_w : T \to T$ be defined by $F'_w(t) = w(F'(t))$ and let $T_w = \{t \in T \mid F'_w(t) = t\}$. Then T_w is a finite abelian group of order prime to p which acts freely on $\tilde{X}_{\dot{w}}$ by $g \to gt^{-1}$, in such a way that $\tilde{X}_w/T_w \xrightarrow{\sim} X_w$, $(g \to gB)$. Note also that $G^{F'}$ acts on $\tilde{X}_{\dot{w}}$ by left multiplication. This action of $G^{F'}$ commutes with the action of T_w and induces on X_w the action of $G^{F'}$ given again by left multiplication.

Now let $\lambda : T \to k^*$ be a character and let L be the corresponding element of $\mathfrak{L}$ (see (1.3.2)). We shall also denote by λ its extension to a character of B, trivial on U. Let n be an integer as in (1.4.1). We associate to λ and n a subset $Z_{L,n}$ of W as follows.

$Z_{L,n}$ consists of the elements $w \in W$ for which there exists a character $\lambda_1 : T \to k^*$ such that

(2.1.1) $$\lambda(F'(t)) = \lambda(w^{-1}(t))\lambda_1^n(t), \text{ for all } t \in T .$$

(Note that λ_1, if it exists, is uniquely determined by w.) *We shall fix* λ, L, n *such that* $Z_{L,n}$ *is nonempty until the end of* §3. Assume now that w, λ_1 are as in (2.1.1). The restriction of λ_1 to T_w is a homomorphism

(2.1.2) $$\bar{\lambda}_1 : T_w \to \mu_n .$$

Consider the variety $(\tilde{X}_{\dot{w}} \times \mu_n)/T_w$, where $t \in T_w$ acts by

$$t : (g,\xi) \to (gt^{-1}, \bar{\lambda}_1(t)\xi), \ (\xi \in \mu_n) .$$

This is a variety with a free μ_n-action

$$\mu_n \ni \xi_0 : (g, \xi) \to (g, \xi_0^{-1}\xi)$$

and its quotient by the μ_n-action may be identified with X_w, via $(g,\xi) \to gB$. Let A_1 be the inverse image of the principal μ_n-bundle $(\tilde{X}_{\dot{w}} \times \mu_n)/T_w \to X_w$ under the map $f : G \times k^* \to G/B$, $f(g', z) = g'B$. Then A_1 is a principal μ_n-bundle over $f^{-1}(X_w)$. We have

$$\text{(2.1.3)} \qquad A_1 = \{(g, g', z, \xi) \in G \times G \times k^* \times \mu_n | g^{-1}F'(g) \in \dot{w}U, \; gB = g'B\}/(U \cap \dot{w}U\dot{w}^{-1})T_w$$

where $u_1 \in U \cap \dot{w}U\dot{w}^{-1}$ acts by

$$u_1 : (g, g', z, \xi) \to (gu_1^{-1}, g', z, \xi)$$

and $t \in T_w$ acts by

$$t : (g, g', z, \xi) \to (gt^{-1}, g', z, \overline{\lambda_1(t)}\xi) .$$

If (g, g', z, ξ) represents a point in A_1, we define $u \in U$, $b \in B$ by $g^{-1}F'(g) = \dot{w}u$, $g' = gb$. This gives an isomorphism of A_1 with

$$A_2 = \{(g', b, u, z, \xi) \in G \times B \times U \times k^* \times \mu_n | g'^{-1}F'(g') = b^{-1}\dot{w}uF'(b)\}/(U \cap \dot{w}U\dot{w}^{-1})T_w$$

where $u_1 \in U \cap \dot{w}U\dot{w}^{-1}$ acts by

$$u_1 : (g', b, u, z, \xi) \to (g', u_1 b, (\dot{w}^{-1}u_1\dot{w})\, uF'(u)^{-1}, z, \xi)$$

and $t \in T_w$ acts by

$$t : (g', b, u, z, \xi) \to (g', tb, F'(t)uF'(t)^{-1}, z, \overline{\lambda_1(t)}\xi) .$$

Now let

$$\text{(2.1.4)} \quad A_3 = \{(g', b, u, z', z) \in G \times B \times U \times k^* \times k^* | g'^{-1}F'(g') = u\dot{w}b^{-1}, \; \lambda(b) = (z/z')^n\}/U \cap \dot{w}U\dot{w}^{-1}$$

where $u_1 \in U \cap \dot{w}U\dot{w}^{-1}$ acts by

$$u_1 : (g', b, u, z', z) \to (g', b\dot{w}^{-1}u_1^{-1}\dot{w}, uu_1^{-1}, z', z) .$$

For $b \in B$, we define $t_b \in T$ and $u_b \in U$ by $b = t_b u_b$. The map

$$(g', b, u, z, \xi) \to (g', F'(b)^{-1}u^{-1}(\dot{w}^{-1}t_b\dot{w}), u_b^{-1}, \xi^{-1}\lambda_1(t_b)z, z)$$

defines an isomorphism $A_2 \xrightarrow{\sim} A_3$.

Consider the free action of $B \times k^*$ on A_1 given by

$$B \times k^* \ni (\beta, \zeta) : (g, g', z, \xi) \to (g, g'\beta^{-1}, z\zeta, \xi) .$$

Under the isomorphism $A_1 \xrightarrow{\sim} A_3$ constructed above (composition of $A_1 \xrightarrow{\sim} A_2 \xrightarrow{\sim} A_3$) this becomes the free action of $B \times k^*$ on A_3 given by

$$B \times k^* \ni (\beta, \zeta) : (g', b, u, z', z) \to (g'\beta^{-1}, F'(\beta)b(\dot{w}^{-1}t_\beta^{-1}\dot{w}),$$
$$t_\beta u_\beta u t_\beta^{-1}, \lambda_1(t_\beta)^{-1}\zeta z', \zeta z) .$$

The space of orbits of A_3 by this $B \times k^*$-action must then be isomorphic (via the isomorphism $A_1 \xrightarrow{\sim} A_3$) to the space of orbits of A_1 by its $B \times k^*$-action, i.e. to $(\tilde{X}_{\dot{w}} \times \mu_n)/T_w$. Thus we have

2.2. LEMMA. *There is a unique isomorphism* ϕ *of* $(\tilde{X}_{\dot{w}} \times \mu_n)/T_w$ *onto* $A_3/(B \times k^*)$ *such that the diagram*

$$\begin{array}{ccc} A_1 & \xrightarrow{\sim} & A_3 \\ \downarrow & & \downarrow \\ (\tilde{X}_{\dot{w}} \times \mu_n)/T_w & \xrightarrow{\phi} & A_3/(B \times k^*) \end{array}$$

is commutative. (The left vertical arrow is $(g, g', z, \xi) \to (g, \xi)$; *the right vertical arrow is the canonical orbit map for the* $B \times k^*$*-action.*

Via ϕ, the free μ_n-action on $(\tilde{X}_{\dot{w}} \times \mu_n)/T_w$ becomes the following free μ_n-action on $A_3/(B \times k^*)$:

$$\mu_n \ni \xi_0 : (g', b, u, z', z) \to (g', b, u, \xi_0 z', z) .$$

The natural $G^{F'}$-action on $(\tilde{X}_{\dot{w}} \times \mu_n)/T_w$ given by

$$G^{F'} \ni g_0 : (g, \xi) \to (g_0 g, \xi)$$

becomes, via ϕ, the following $G^{F'}$-action on $A_3/(B \times k^*)$:

$$G^{F'} \ni g_0 : (g', b, u, z', z) \to (g_0 g', b, u, z', z) .$$

Note also that $(\tilde{X}_{\dot{w}} \times \mu_n)/T_w$ and $A_3/(B \times k^*)$ have natural F_p-structures with Frobenius maps given respectively by

$$(g, \xi) \to (F(g), \xi^p)$$

$$(g', b, u, z', z) \to (F(g'), F(b), F(u), z'^p, z^p)$$

and that ϕ is compatible with these F_p-structures.

2.3. From now on, we shall fix an imbedding $\tilde{\psi} : k^* \to \bar{Q}_\ell^*$; we denote by ψ its restriction to μ_n. It is an injective homomorphism $\mu_n \to \bar{Q}_\ell^*$. Consider the direct image of the constant sheaf $\bar{Q}_\ell$ under the map $(\tilde{X}_{\dot{w}} \times \mu_n)/T_w \to X_w$ (a principal μ_n-bundle, see 2.1). This is a locally constant, $G^{F'}$-equivariant $\bar{Q}_\ell$-sheaf over X_w with a natural action of μ_n (inducing identity on X_w). Hence it can be decomposed canonically into a direct sum of $\bar{Q}_\ell$-sheaves of rank 1, according to the characters of μ_n. Let $\mathcal{F}^L_{\dot{w},n}$ be the summand corresponding to ψ; it is characterized by the fact that $x \in \mu_n$ acts on any stalk of $\mathcal{F}^L_{\dot{w},n}$ as multiplication by $\psi(x)$. (We do not indicate ψ in the notation for $\mathcal{F}^L_{\dot{w},n}$ since $\tilde{\psi}$ is fixed and ψ is determined by $\tilde{\psi}$ and n.) Note that $\mathcal{F}^L_{\dot{w},n}$ is a $G^{F'}$-equivariant locally constant $\bar{Q}_\ell$-sheaf of rank 1 over X_w. It is defined only when $w \in Z_{L,n}$ (see 2.1.1) and $\dot{w}$ is a representative for w in $\mathfrak{N}(T)^F$.

(2.3.1) Let d be the smallest integer ≥ 1 such that n divides $p^d - 1$. Then the d-th power of F acting on $(\tilde{X}_{\dot{w}} \times \mu_n)/T_w$ commutes with the action of μ_n, hence it gives rise to an F_{p^d}-rational structure on $\mathcal{F}^L_{\dot{w},n}$. We shall denote its Frobenius map by F^d. (When $\dot{w}$ is changed to another representative of w in $\mathcal{N}(T)^F$, one gets a $\overline{\mathbb{Q}}_\ell$-sheaf, $G^{F'}$-equivariantly isomorphic to $\mathcal{F}^L_{\dot{w},n}$, but with a possibly different F_{p^d}-structure.)

Note that the cohomology spaces $H^i_c(X_w, \mathcal{F}^L_{\dot{w},n})$ are $(G^{F'}, F^d)$-modules in the following sense: an element $g \in G^{F'}$ acts as $(g^{-1})^*$ and F^d acts as $(F^d)^*$; we have the identity $(F^d)^* \circ (g^{-1})^* = (F^d(g)^{-1})^* \circ (F^d)^*$.

Let $\overline{X}_w$ be the Zariski closure of X_w in G/B. This is a complete variety. The argument in [L_{11}, 4.3] shows that $\overline{X}_w$ is the union of $X_{w'}$ $(w' \leq w)$ and that:

(2.3.2) All connected components of $\overline{X}_w$ are irreducible of dimension $\ell(w)$.

We shall consider the cohomology sheaves $\mathcal{H}^i(\overline{X}_w, \mathcal{F}^L_{\dot{w},n})$ of the intersection cohomology complex $IC(\overline{X}_w, \mathcal{F}^L_{\dot{w},n})$. These sheaves are in a natural way, $G^{F'}$-equivariant constructible $\overline{\mathbb{Q}}_\ell$-sheaves on $\overline{X}_w$ with an F_{p^d}-structure (inherited from $\mathcal{F}^L_{\dot{w},n}$).

Note also that for our $w (w \in Z_{L,n})$, we have $w \cdot W_{L,n} \subset Z_{L,n}$. Indeed, if $v \in W_{L,n}$ and λ_1 is defined by (2.1.1), we have for $t \in T$:

$$\begin{aligned}
\lambda(F'(t))\lambda((wv)^{-1}(t))^{-1} &= \lambda(F'(t))\lambda(v^{-1}(w^{-1}(t))^{-1} \\
&= \lambda(F'(t))\lambda(w^{-1}(t))^{-1}\lambda_2^n(w^{-1}(t))^{-1} \\
&\quad \text{(since } v \in W_{L,n}) \\
&= \lambda_1^n(t)\lambda_2^n(w^{-1}(t))^{-1} \\
&= \lambda_3^n(t) .
\end{aligned}$$

(Here λ_2, λ_3 are characters of T.) Hence $wv \in Z_{L,n}$. We can now state

2.4. THEOREM. *Let* $w \in Z_{L,n}$, *let* $Z = w \cdot W_{L,n} \subset Z_{L,n}$ *and let* $Z^{\cdot}$ *be a coherent lifting of* Z *to* $\mathfrak{N}(T)$, *(see 1.23). Let* $\dot{w}$ *be the representative of* w *in* $Z^{\cdot}$. *Let* d *be as in (2.3.1). Then the restriction of* $\mathcal{H}^i(\overline{X}_w, \mathcal{F}^L_{\dot{w},n})$ *to* $X_{w'}$ $(w' \leq w)$ *is zero unless* $w' \in Z$ *and* i *is even.*

If $w' \in Z$ *and* i *is even, then this restriction is a* $G^{F'}$*-equivariant, locally constant* $\overline{\mathbb{Q}}_\ell$*-sheaf defined over* F_{p^d} *for which there exists a filtration by* $G^{F'}$*-equivariant, locally constant* $\overline{\mathbb{Q}}_\ell$*-subsheaves defined over* F_{p^d}, *each of whose successive quotients is isomorphic (as a* $G^{F'}$*-equivariant, locally constant,* $\overline{\mathbb{Q}}_\ell$*-sheaf defined over* F_{p^d}*) to* $\mathcal{F}^L_{\dot{w}',n} \otimes \overline{\mathbb{Q}}_\ell\left(-\frac{i}{2}\right)$, *where* $\dot{w}'$ *is the representative of* w' *in* $Z^{\cdot}$. *The number of nonzero successive quotients in this filtration is equal to* $n^i_{w',w}$, *where* $n^i_{w',w}$ *is the coefficient* $(\in \mathbb{Z})$ *of* $[p]^{i/2}\overline{\mathfrak{X}}^L_{w',\psi}$ *in* $\hat{\mathfrak{X}}^L_{w,\psi}$.

The key fact in the proof of 2.4 is given in the following:

2.5. LEMMA. *Let* $f : G \times k^* \to G/B$ *be the map* $(g, z) \to gB$ *and let* $f_1 : f^{-1}(X_w) \to X_w$, $f_2 : f^{-1}(\overline{X}_w) \to \overline{X}_w$ *be its restrictions. Let* $f' : G \times k^* \to L^{\cdot}$ *be the map given by* $f'(g,z) = (g,z)$, *(with the description (1.5.1) of* $L^{\cdot}$*) and let* $f_1' : (BwB) \times k^* \to L^{\cdot} | \mathcal{B}_w$, $f_2' : (\overline{BwB}) \times k^* \to L^{\cdot} | \overline{\mathcal{B}}_w$ *be its restrictions. Let* $f'' : G \times k^* \to G \times k^*$ *be the map given by* $f''(g,z) = (g^{-1}F'(g), z^n)$ *and let* $f_1'' : f^{-1}(X_w) \to (BwB) \times k^*$, $f_2'' : f^{-1}(\overline{X}_w) \to (\overline{BwB}) \times k^*$ *be its restrictions. Then there is a canonical* $G^{F'}$*-equivariant isomorphism of locally constant* $\overline{\mathbb{Q}}_\ell$*-sheaves over* $f^{-1}(X_w) = f''^{-1}(f'^{-1}(L^{\cdot} | \mathcal{B}_w))$ *(compatible with the* F_{p^d}*-structures)*

$$f_1^*(\mathcal{F}^L_{\dot{w},n}) \xrightarrow{\sim} f_1''^*(f_1'^*(\mathfrak{X}^L_{\dot{w},\psi})) . \tag{2.5.1}$$

Proof. In 2.1, we have constructed an isomorphism $A_1 \xrightarrow{\sim} A_3$. This is an isomorphism of principal μ_n-bundles over $f^{-1}(X_w)$. Taking the locally constant $\overline{\mathbb{Q}}_\ell$-sheaves corresponding to $\psi : \mu_n \to \overline{\mathbb{Q}}_\ell^*$ and to these principal μ_n-bundles, we find the isomorphism (2.5.1).

2.6. *Proof of Theorem 2.4.* Since f is a locally trivial fibration with smooth fibers ($\approx B \times k^*$), the sheaf $f_2^*(\mathcal{H}^i(\bar{X}_w, \mathcal{F}^L_{\dot{w},n}))$ on $f^{-1}(\bar{X}_w)$ can be canonically identified with $\mathcal{H}^i(f^{-1}(\bar{X}_w), f_1^*(\mathcal{F}^L_{\dot{w},n}))$ or, equivalently, by (2.5.1), with $\mathcal{H}^i(f^{-1}(\bar{X}_w), f_1''^* f_1'^*(\mathcal{X}^L_{\dot{w},\psi}))$.

Since f' is a locally trivial fibration with smooth fibers ($\approx B$) and f'' is an étale surjective map, the last sheaf is canonically isomorphic to $f_2''^* f_2'^* \mathcal{H}^i(L^\cdot|\bar{\mathcal{B}}_w, \mathcal{X}^L_{\dot{w},\psi})$. Using now Theorem 1.24, we see that:

(2.6.1) The restriction of the sheaf $f_2^*(\mathcal{H}^i(\bar{X}_w, \mathcal{F}^L_{\dot{w},n}))$ to $f^{-1}(X_{w'})(w' \leq w)$ is zero unless $w' \in Z$ and i is even; if $w' \in Z$ and i is even, it is locally constant and admits a filtration by locally constant, $G^{F'}$-stable $\bar{\mathbf{Q}}_\ell$-subsheaves defined over F_{p^d} each of whose successive quotients is isomorphic (as a $G^{F'}$-equivariant, $\bar{\mathbf{Q}}_\ell$-sheaf defined over F_{p^d}) to

$$f_3''^* f_3'^*(\mathcal{X}^L_{\dot{w}',\psi}) \otimes \bar{\mathbf{Q}}_\ell\left(-\frac{i}{2}\right) = f_3^*(\mathcal{F}^L_{\dot{w}',n}) \otimes \bar{\mathbf{Q}}_\ell\left(-\frac{i}{2}\right), \text{ (see 2.5.1)},$$

where $\dot{w}' \in Z^\cdot$ represents w'; the number of nonzero steps in this filtration is $n^i_{w',w}$. (Here f_3, f_3', f_3'' are defined just as f_1, f_1', f_1'' in 2.5, with w replaced by w'.)

In other words, we get the statements of Theorem 2.4, "lifted" under f^* to $G \times k^*$. The statements of 2.4 with $w' \notin Z$ or i odd follow. To get the statements of 2.4 with $w' \in Z$ and i even, before lifting under f^*, we use the following remark. Since the locally trivial fibration $f: G \times k^* \to G/B$ has connected fibers, a constructible $\bar{\mathbf{Q}}_\ell$-sheaf $\mathcal{F}$ on G/B can be reconstructed from its inverse image $f^*\mathcal{F}$: we have $\mathcal{F} = R^\circ f_*(f^*\mathcal{F})$.

Since f is a locally trivial fibration, the fact that $f_2^*(\mathcal{H}^i(\bar{X}_w, \mathcal{F}^L_{\dot{w},n}))$ is locally constant on $f^{-1}(X_{w'})$ implies that $\mathcal{H}^i(\bar{X}_w, \mathcal{F}^L_{\dot{w},n})$ is locally constant on $X_{w'}$.

Let $\hat{\mathcal{F}}$ be the restriction of $f_2^*(\mathcal{H}^i(\bar{X}_w, \mathcal{F}^L_{\dot{w},n}))$ to $f^{-1}(X_{w'})$. Let $\hat{\mathcal{F}} = \hat{\mathcal{F}}^\circ \supsetneq \hat{\mathcal{F}}^1 \supsetneq \cdots \supsetneq \hat{\mathcal{F}}^N = 0$, $(N = n^i_{w',w})$, be the filtration of $\hat{\mathcal{F}}$ whose existence is asserted in (2.6.1), and let $\mathcal{F}^j = R^\circ(f_3)_*(\hat{\mathcal{F}}^j)$ be the corresponding sheaves on $X_{w'}$. The exact sequence of sheaves

$$0 \to \hat{\mathcal{F}}^j \to \hat{\mathcal{F}}^{j+1} \to \hat{\mathcal{F}}^{j+1}/\hat{\mathcal{F}}^j \to 0$$

can be split in a $G^{F'}$-equivariant manner (but not necessarily in a way compatible with the F_{p^d}-structures). This follows from the similar property of the filtration of $\mathcal{H}^i(\pi^{-1}(\overline{\mathcal{B}}_w), \mathcal{K}^L_{\dot{w},\psi})$ restricted to $L^{\cdot}|\mathcal{B}_{w'}$, in 1.24. It follows that we have an exact sequence of sheaves

$$0 \to \mathcal{F}^j \to \mathcal{F}^{j+1} \to (R^\circ f_3)_*(\hat{\mathcal{F}}^{j+1}/\hat{\mathcal{F}}^j) \to 0$$

over $X_{w'}$. Thus, the $\mathcal{F}^j$ form a filtration of the restriction of $\mathcal{H}^i(\overline{X}_w, \mathcal{F}^L_{\dot{w},n})$ to $X_{w'}$ by locally constant, $G^{F'}$-equivariant $\overline{Q}_\ell$-sheaves defined over F_{p^d}, such that

$$\mathcal{F}^j/\mathcal{F}^{j+1} = (R^\circ f_3)_*(\hat{\mathcal{F}}^{j+1}/\hat{\mathcal{F}}^j) \approx (R^\circ f_3)_*\left(f_3^*(\mathcal{F}^L_{\dot{w}',n}) \otimes \overline{Q}_\ell\left(-\frac{i}{2}\right)\right)$$

$$\approx \mathcal{F}^L_{\dot{w}',n} \otimes \overline{Q}_\ell\left(-\frac{i}{2}\right),$$

for $j = 0, 1, \cdots, N-1$, as required. This completes the proof of the theorem.

2.7. The $\overline{Q}_\ell$-vector spaces $H^i(\overline{X}_w, \mathcal{F}^L_{\dot{w},n})$, ($w, \dot{w}$ as in 2.4), are $(G^{F'}, F^d)$-modules in a natural way (using the fact that $\mathcal{F}^L_{\dot{w},n}$ is $G^{F'}$-equivariant and defined over F_{p^d} (see 2.3)); an element $g \in G^{F'}$ acts as $(g^{-1})^*$ and F^d acts as $(F^d)^*$. Using the spectral sequence connecting hypercohomology with coefficients in a complex, with the ordinary cohomology with coefficients in the cohomology sheaves of the complex, we get an equality of virtual $(G^{F'}, F^d)$-modules

$$(2.7.1) \quad \sum_i (-1)^i H^i(\overline{X}_w, \mathcal{F}^L_{\dot{w},n}) = \sum_{i,j} (-1)^{i+j} H^i(\overline{X}_w, \mathcal{H}^j(\overline{X}_w, \mathcal{F}^L_{\dot{w},n})) .$$

Using the decomposition $\overline{X}_w = \bigcup_{w' \leq w} X_{w'}$, we can rewrite the right-hand side of (2.7.1) as

$$\sum_{w' \leq w} \sum_{i,j} (-1)^{i+j} H_c^i(X_{w'}, \mathcal{H}^j(\overline{X}_w, \mathcal{F}^L_{\dot{w},n})|X_{w'}) .$$

Using now Theorem 2.4, we get the following

2.8. COROLLARY. *In the setting of Theorem 2.4, we have an equality of* $(G^{F'}, F^d)$*-modules*

(2.8.1) $$\sum_i (-1)^i H^i(\overline{X}_w, \mathcal{F}^L_{\dot{w},n}) = \sum_i (-1)^i H^i_c(X_w, \mathcal{F}^L_{\dot{w},n}) +$$

$$+ \sum_{\substack{w' < w \\ w' \in Z}} \sum_i \sum_{\substack{j' \\ \text{even}}} (-1)^i n^{j'}_{w',w} H^i_c(X_{w'}, \mathcal{F}^L_{\dot{w}',n}) \otimes \overline{\mathbf{Q}}_\ell\left(-\frac{j'}{2}\right) .$$

(w' *are chosen as in 2.4;* $\mathbf{Q}_\ell\left(-\frac{j'}{2}\right)$ *has trivial action of* $G^{F'}$ *and* F^d *acts on it as multiplication by* $p^{j'd/2}$.)

2.9. Let r be an integer ≥ 1. We define $\mathcal{P}_{dr}$ to be the vector space of all functions $f:(G/U)^{F^{dr}} \to \overline{\mathbf{Q}}_\ell$. It is a $G^{F^{dr}}$-module:

$$(g_1 f)(gU) = f(g_1^{-1} gU), \quad g, g_1 \in G^F, \quad f \in \mathcal{P}_{dr} .$$

For each $n_1 \in \mathfrak{N}(T)^{F^{dr}}$, let $\tau_{n_1}: \mathcal{P}_{dr} \to \mathcal{P}_{dr}$ be the endomorphism

(2.9.1) $$(\tau_{n_1} f)(gU) = \sum_{\substack{g'U \in (G/U)^{F^{dr}} \\ g^{-1} g' \in U n_1 U}} f(g'U), \; (f \in \mathcal{P}_{dr}) .$$

The endomorphisms τ_{n_1} $(n_1 \in N(T)^{F^{dr}})$ form a $\overline{\mathbf{Q}}_\ell$-basis of $\operatorname{End}_{G^{F^{dr}}}(\mathcal{P}_{dr})$. For each character $\theta: T^{F^{dr}} \to \overline{\mathbf{Q}}_\ell^*$ we denote

(2.9.2) $$\mathcal{P}^{\theta} = \{f \in \mathcal{P}_{dr} | \tau_t f = \theta(t)^{-1} f, \forall t \in T^{F^{dr}}\}.$$

Then

$$\mathcal{P}_{dr} = \bigoplus_{\theta} \mathcal{P}^{\theta}$$

(sum over all characters θ of $T^{F^{dr}}$). If $n_1 \in \mathfrak{N}(T)^{F^{dr}}$, we have

(2.9.3) $$\tau_{n_1}(\mathcal{P}^{\theta}) \subset \mathcal{P}^{\theta'}$$

where θ' is defined by $\theta'(t) = \theta(n_1^{-1} t n_1)$.

It follows that

(2.9.4) The restrictions to $\mathcal{P}^{\theta}$ of the endomorphisms $\tau_{n_1} : \mathcal{P}_{dr} \to \mathcal{P}_{dr}$ (where n_1 runs over a set of representatives in $\mathfrak{N}(T)^{F^{dr}}$ for the elements in W which stabilize θ) form a $\overline{\mathbb{Q}}_{\ell}$-basis for $\mathrm{End}_{G^{F^{dr}}}(\mathcal{P}^{\theta})$.

We shall denote θ_0 the character of $T^{F^{dr}}$ defined by

(2.9.5) $$\theta_0(t) = \psi(\lambda(t)^{(p^{dr}-1)/n}), \text{ (see 2.3).}$$

If $w \in Z_{L,n}$, (see 2.1.1)) and $\dot{w}$ is a representative for w in $N(T)^F$, we have

$$\tau_{\dot{w}} : \mathcal{P}^{\theta_0} \to \mathcal{P}^{\theta_0^{F'}}$$

where, for any $i \geq 1$, $\theta_0^{(F')^i}$ is the character of $T^{F^{dr}}$ defined by

(2.9.6) $$\theta_0^{(F')^i}(t) = \theta_0((F')^i(t)).$$

Let $F' : \mathcal{P}_{dr} \to \mathcal{P}_{dr}$ be the linear map defined by

(2.9.7) $$(F'f)(gU) = f(F'^{-1}(g)U).$$

It is clear that

$$F'\tau_{n_1} = \tau_{F'(n_1)}F' : \mathcal{P}_{dr} \to \mathcal{P}_{dr}, \quad (n_1 \in \mathcal{N}(T)^{F^{dr}}). \tag{2.9.8}$$

Hence F' maps $\mathcal{P}^{\theta_0^{F'}}$ into $\mathcal{P}^{\theta_0}$ and, therefore, the composition $F'\tau_{\dot{w}}$ maps $\mathcal{P}^{\theta_0}$ into itself. It is, however, not necessarily commuting with the action of $G^{F^{dr}}$ on $\mathcal{P}^{\theta_0}$.

We now consider the following bijection, due essentially to Shintani, (see also [Ka_2], [As_2]):

$$\begin{array}{l} \{G^{F'} \text{ mod. equiv. reln. } x \sim x_0^{-1}xF^{dr}(x_0), \forall x_0 \in G^{F'}\} \\ \quad \Big\downarrow \wr \\ \{G^{F^{dr}} \text{ mod. equiv. reln. } \hat{x} \sim \hat{x}_0^{-1}\hat{x}F'(\hat{x}_0), \forall \hat{x}_0 \in G^{F^{dr}}\}. \end{array} \tag{2.9.9}$$

To $x \in G^{F'}$, there corresponds $\hat{x} \in G^{F^{dr}}$ defined by $\hat{x} = a^{-1}F'(a)$, where a is an element of G such that $x = aF^{dr}(a^{-1})$. This bijection has the following property.

(2.9.10) If $x \in G^{F'}$, $\hat{x} \in G^{F^{dr}}$ correspond to each other as above, and

$$n_x = = \{x_0 \in G^{F'} | x_0^{-1}xF^{dr}(x_0) = x\},$$

$$n_{\hat{x}} = = \{\hat{x}_0 \in G^{F'^{dr}} | \hat{x}_0^{-1}\hat{x}F^{dr}(\hat{x}_0) = \hat{x}\},$$

then $n_x = n_{\hat{x}}$.

Indeed, $n_x, n_{\hat{x}}$ are the number of points of two finite subgroups of G which are conjugate to each other (under a).

We can now state the following result which, in the special case where $n = 1$ and L is trivial, was proved by T. Asai [As_2], J. Michel-F. Digne [DM_1] and the author (unpublished).

2.10. PROPOSITION. *If* $x \in G^{F'}$, $\hat{x} \in G^{F^{dr}}$ *correspond to each other under (2.9.9) then*

$$(2.10.1) \quad \sum_i (-1)^i \mathrm{Tr}((F^{dr})^* x^*, H_c^i(X_w, \mathcal{F}^L_{\dot{w},n})) = \mathrm{Tr}(\hat{x}F'\tau_{\dot{w}} : \mathcal{P}^{\theta_0} \to \mathcal{P}^{\theta_0})$$

Proof. Let π be the left-hand side of (2.10.1). For any $t \in T_w$, the map $g \to xF^{dr}(g)t^{-1}$ of G into itself is the Frobenius map for a rational structure over $F_{p^{dr}}$. Indeed, write $x = aF^{dr}(a^{-1})$, $t = a'F^{dr}(a'^{-1})$, $a \in G$, $a' \in T$; then our map is $g \to f^{-1}(F^{dr}(f(g))$ where $f(g) = a^{-1}ga'$. Since our map leaves $\{g \in G | g^{-1}F'(g) \in \dot{w}U\}$ stable, its restriction there is again the Frobenius map for a rational structure over $F_{p^{dr}}$. We may therefore, apply to it the trace formula for Frobenius maps; hence, we have

$$\pi = |T_w|^{-1} \sum_{t \in T_w} \{g \in G | g^{-1}F'(g) \in \dot{w}U, xF^{dr}(g)t^{-1} = g\} \psi(\overline{\lambda}_1(t))^{-1} p^{-drN}$$

where $N = \dim(U \cap \dot{w}U\dot{w}^{-1})$.

We write as above $x = aF^{dr}(a^{-1})$, $t = a'F^{dr}(a'^{-1})$, $a \in G$, $a' \in T$, and set $g' = a^{-1}gb$. The equation $xF^{dr}(g)t^{-1} = g$ becomes $F^{dr}(g') = g'$; the equation $g^{-1}F'(g) \in \dot{w}U$ becomes $g'^{-1}\hat{x}F'(g') \in a'^{-1}\dot{w}F'(a')U$. Thus, π can be written as

$$\pi = |T_w|^{-1} |T^{F^{dr}}|^{-1} \sum_{\substack{a' \in T \\ a'F^{dr}(a'^{-1}) \in T_w}} \# \{g' \in G^{F^{dr}} | g'^{-1}\hat{x}F'(g') \in a'^{-1}\dot{w}F'(a')U\} \times$$

$$\times \psi(\overline{\lambda}_1(a'^{-1}F^{dr}(a')))p^{-drN} .$$

If $a' \in T$, the condition $a'F^{dr}(a'^{-1}) \in T_w$ is equivalent to the condition $a'^{-1}F'_w(a') \in T^{F^{dr}}$. Moreover, if this condition is satisfied, we have

$$\begin{aligned}\overline{\lambda}_1(a'^{-1}F^{dr}(a')) &= \lambda_1(a'^{-1}a'p^{dr}) = \lambda_1^n(a'^{(p^{dr}-1)/n}) \\ &= \lambda(F'(a'^{(p^{dr}-1)/n})\cdot w^{-1}(a'^{(p^{dr}-1)/n})^{-1}) \\ &= \lambda(\dot{w}^{-1}a'^{-1}\dot{w}F'(a'))^{(p^{dr}-1)/n} .\end{aligned}$$

Thus, π can be written as

$$\pi = |T_w|^{-1}|T^{F^{dr}}|^{-1} \sum_{\substack{a'\in T \\ a'^{-1}F'_w(a')\in T^{F^{dr}}}} \#\{g'\in G^{F^{dr}} \mid g'^{-1}\hat{x}F'(g')F'(a')^{-1}\dot{w}^{-1}a\ddot{w}\in \dot{w}U\}\times$$

$$\times\, \psi(\lambda(\dot{w}^{-1}a'^{-1}\dot{w}F'(a'))^{(p^{dr}-1)/n})p^{-drN}$$

$$= |T^{F^{dr}}|^{-1} \sum_{\hat{t}\in T^{F^{dr}}} \#\{g'\in G^{F^{dr}} \mid g'^{-1}\hat{x}F'(g')\hat{t}^{-1}\in \dot{w}U\}\psi(\lambda(\hat{t})^{(p^{dr}-1)/n})p^{-drN}$$

$$= |T^{F^{dr}}|^{-1} \sum_{\hat{t}\in T^{F^{dr}}} \#\{g'U\in (G/U)^{F^{dr}} \mid g'^{-1}\hat{x}F'(g')\hat{t}^{-1}\in U\dot{w}U\}\theta_0(\hat{t})$$

$$= \mathrm{Tr}(\hat{x}F'\tau_{\dot{w}} : \mathcal{P}^{\theta_0} \to \mathcal{P}^{\theta_0}) ,$$

and the proposition is proved.

2.11. COROLLARY. *The expression*

$$(2.11.1)\qquad \sum_{i,j} (-1)^{i+j}\mathrm{Tr}((F^{dr})^*, (H_c^i(X_w, \mathcal{F}^L_{\dot{w},n})\otimes H_c^j(X_{w'}, \mathcal{F}^{L^{-1}}_{\dot{w}',n}))^{G^{F'}})$$

is equal to the trace of the $\overline{\mathbf{Q}}_\ell$*-linear transformation*

$$(2.11.2)\qquad h \to (\tau_{(\dot{w}')^{-1}}(F')^{-1})\, h(F'\tau_{\dot{w}})$$

of $\mathrm{End}_{G^{F^{dr}}}(\mathcal{P}^{\theta_0})$ *into itself,* $(w,w'\in Z_{L,n} = Z_{L^{-1},n})$. (The upper script $G^{F'}$ in (2.11.1) means space of $G^{F'}$-invariant vectors. That space is stable under $(F^{dr})^*$, since F^{dr} maps $G^{F'}$ into itself.)

Proof. The expression (2.11.1) is equal to

$$(2.11.3)\quad \begin{aligned}&\sum_x |G^{F'}|^{-1} \sum_i (-1)^i \operatorname{Tr}((F^{dr})^* x^*, H_c^i(X_w, \mathcal{F}^L_{\dot{w},n})) \times \\ &\times \sum_j (-1)^j \operatorname{Tr}((F^{dr})^* x^*, H_c^j(X_{w'}, \mathcal{F}^{L^{-1}}_{\dot{w}',n}))\end{aligned}$$

sum over all x in $G^{F'}$.

We may replace the sum over x by a sum over a set of representatives $\{x\}$ for the orbits of $G^{F'}$ acting on $G^{F'}$ by $x \to x_0^{-1} x F^{dr}(x_0)$; we must then replace the factor $|G^{F'}|$ by n_x^{-1} (see 2.9.10). We now apply (2.10.1), (2.9.9), (2.9.10) and we find that (2.11.3) is equal to

$$(2.11.4)\quad \sum_{\hat{x}} n_{\hat{x}}^{-1} \operatorname{Tr}(\hat{x} F' \tau_{\dot{w}} : \mathcal{P}^{\theta_0} \to \mathcal{P}^{\theta_0}) \operatorname{Tr}(\hat{x} F' \tau_{\dot{w}'} : \mathcal{P}^{\theta_0^{-1}} \to \mathcal{P}^{\theta_0^{-1}}),$$

sum over a set of representatives $\{\hat{x}\}$ for the orbits of $G^{F^{dr}}$ acting on $G^{F^{dr}}$ by $\hat{x} \to \hat{x}_0^{-1} \hat{x} F'(\hat{x}_0)$. We may replace this sum by a sum over all $\hat{x}$ in $G^{F^{dr}}$, provided that we replace the factor $n_{\hat{x}}^{-1}$ by $|G^{F^{dr}}|^{-1}$. It follows that (2.11.4) is equal to

$$(2.11.5)\quad \operatorname{Tr}((F'\tau_{\dot{w}}) \otimes (F'\tau_{\dot{w}'}) : (\mathcal{P}^{\theta_0} \otimes \mathcal{P}^{\theta_0^{-1}})^{G^{F^{dr}}} \circlearrowleft).$$

Now, $\mathcal{P}^{\theta_0^{-1}}$ is in a natural way the space dual to $\mathcal{P}^{\theta_0}$. (A pairing $\mathcal{P}^{\theta_0} \times \mathcal{P}^{\theta_0^{-1}} \to \overline{\mathbf{Q}}_\ell$ can be obtained by associating to $f \in \mathcal{P}^{\theta_0}$, $f' \in \mathcal{P}^{\theta_0^{-1}}$ the sum

$$\sum_{gU \in (G/U)^{F^{dr}}} (f \cdot f')(gU) \in \overline{\mathbf{Q}}_\ell\,.)$$

Under this duality, the operator $F'\tau_{\dot{w}'}$ on $\mathcal{P}^{\theta_0^{-1}}$ corresponds to the contragredient of the operator $\tau_{(\dot{w}')^{-1}}(F')^{-1}$ on $\mathcal{P}^{\theta_0}$. It follows that (2.11.5) is equal to (2.11.2) and the corollary is proved.

2.12. We shall recall some known properties of the algebra $\mathrm{End}_{G^{F^{dr}}}(\mathcal{P}_{dr})$. It is well known that for $n_1, n_2 \in \mathfrak{N}(T)^{F^{dr}}$ we have

(2.12.1) $$\tau_{n_1}\tau_{n_2} = \tau_{n_1 n_2} : \mathcal{P}_{dr} \to \mathcal{P}_{dr}$$

whenever $\ell(\bar{n}_1\bar{n}_2) = \ell(\bar{n}_1) + \ell(\bar{n}_2)$.

We shall denote the roots of G which respect to T by Greek letters $\alpha, \beta \cdots$. They are characters of T. For each root α, let

(2.12.2) n_α be an element of $N(T) - T$ such that n_α is in the subgroup of $G^{F^{dr}}$ generated by the root subgroups (over $F_{p^{dr}}$) corresponding to $\alpha, -\alpha$.

If α is a simple root (with respect to B) then, as it is well known, we have

(2.12.3) $$\tau_{n_\alpha}\tau_{n_\alpha^{-1}} = p^{dr}\tau_e + \tau_{n_\alpha} \sum_{x \in F^*_{p^{dr}}} \tau_{\check{\alpha}(x)}.$$

(Here we regard $\check{\alpha}$ as a homomorphism $k^* \to T$.)

Given $\theta \in \mathrm{Hom}(T^{F^{dr}}, \bar{\mathbb{Q}}_\ell^*)$ we define R_θ to be the set of roots α such that $\theta(\check{\alpha}(x)) = 1$ for all $x \in F^*_{p^{dr}}$. It is a root system with $R_\theta \cap R^+$ as set of positive roots and with Weyl group W_θ, the subgroup of W generated by the reflection with respect to roots in R_θ. Let Π_θ be the set of simple roots of R_θ with respect to $R_\theta \cap R^+$ and let S_θ be the corresponding set of reflections in W_θ. Let ℓ_θ be the length function on the Coxeter group (W_θ, S_θ). When $\theta = \theta_0$ (see (2.9.5)), we have

$$R_\theta = R_{L,n}, R_\theta \cap R^+ = R^+_{L,n}, (W_\theta, S_\theta) = (W_{L,n}, S_{L,n}), \ell_\theta = \tilde{\ell}.$$

For any θ, and any simple root α, we have (cf. (2.12.3)):

$$(2.12.4) \qquad \tau_{n_\alpha}\tau_{n_\alpha^{-1}}(f) = \begin{cases} p^{dr}f & , \text{ if } \alpha \notin R_\theta \\ p^{dr}f + (p^{dr}-1)\tau_{n_\alpha}f, & \text{if } \alpha \in R_\theta \end{cases}$$

$(f \in \mathcal{P}^\theta)$. It follows that the operators $\tau_{n_\alpha} : \mathcal{P}_{dr} \to \mathcal{P}_{dr}$ are invertible. Hence the operators $\tau_{n_1} : \mathcal{P}_{dr} \to \mathcal{P}_{dr}$ are invertible for any $n_1 \in \mathcal{N}(T)^{F^{dr}}$.

We now define $\mathcal{N}_\theta$ to be the subset of $\mathcal{N}(T)^{F^{dr}}$ consisting of all elements of the form $n_{\alpha_1}n_{\alpha_2}\cdots n_{\alpha_m}$ where $\alpha_1, \cdots, \alpha_m \in \Pi_\theta$, n_{α_i} are as in (2.12.2) and $\ell_\theta(\bar{n}_{\alpha_1}\bar{n}_{\alpha_2}\cdots\bar{n}_{\alpha_m}) = m$.

When $\theta = \theta_0$, (see (2.9.5)), we have $\mathcal{N}_{L,n} \subset \mathcal{N}_\theta$, (see 1.21). The following result is due to Kilmoyer [Ki] and Howlett [H]. (The case where $\theta = 1$ goes back to Iwahori (1964).)

2.13. PROPOSITION. *Let* θ *be any character* $T^{F^{dr}} \to \bar{\mathbf{Q}}_\ell^*$ *and let* $y \in W_\theta$. *Define* $T_y : \mathcal{P}^\theta \to \mathcal{P}^\theta$ *by* $T_y = p^{-\frac{1}{2}(\ell(y)-\ell_\theta(y))dr}\tau_{\dot{y}}$ *where* $\dot{y}$ *is a representative for* y *in* $\mathcal{N}_\theta$. *Then*

(i) T_y *is independent of the choice of* $\dot{y}$.

(ii) $T_yT_{y'} = T_{yy'}$ *(on* $\mathcal{P}^\theta$*), if* $y, y' \in W_\theta$ *satisfy* $\ell_\theta(yy') = \ell_\theta(y)+\ell_\theta(y')$.

(iii) $T_{\bar{n}_\alpha}^2 = (p^{dr}-1)T_{\bar{n}_\alpha} + p^{dr}$ (on $\mathcal{P}^\theta$), *if* $\alpha \in \Pi_\theta$ *and* n_α *is as in* (2.12.2).

(iv) *Let* $n \in \mathcal{N}(T)^{F^{dr}}$, *let* $\alpha \in \Pi_\theta$ *and let* n_α *be as in (12.2.2). Assume that* $\bar{n}(\alpha) \in R^+$, *where* $\bar{n}$ *is the image of* n *in* W. *Then*

$$\tau_n\tau_{n_\alpha}(f) = p^{\frac{1}{2}(\ell(\bar{n})+\ell(\bar{n}_\alpha)-\ell(\bar{n}\,\bar{n}_\alpha))dr}\tau_{nn_\alpha}(f)$$

for all $f \in \mathcal{P}^\theta$.

The statement (iv) is proved in [Ki, (4.10)]. A completely similar proof shows that, with the assumptions of (iv), we have

(2.13.1) $$\tau^{-1}_{n^{-1}}\tau^{-1}_{n_\alpha^{-1}}(f) = p^{-\frac{1}{2}(\ell(n)+\ell(n_\alpha)-\ell(n\,n_\alpha))dr}\tau^{-1}_{(nn_\alpha)^{-1}}(f)$$

for all $f \in \mathcal{P}^\theta$.

2.14. If the center of G is connected, the stabilizer of θ in W coincide with W_θ (see [DL$_1$, 5.13]) hence from (2.9.4) we can deduce that the elements $T_y : \mathcal{P}^\theta \to \mathcal{P}^\theta$, $(y \in W_\theta)$, form a $\overline{Q}_\ell$-basis of $\mathrm{End}_{G^{F^{dr}}}(\mathcal{P}^\theta)$.

2.15. Let $w, w' \in Z_{L,n}$ be two elements such that $w \cdot W_{L,n} = w' W_{L,n}$, let w_1 be the element of minimal length in $w \cdot W_{L,n}$, let $\dot{Z}$ be a coherent lifting (see 1.23) of $Z = w_1 \cdot W_{L,n}$ to $\mathcal{N}(T)^F$ and let $\dot{w}_1, \dot{w}, \dot{w}'$ be the elements of $\dot{Z}$ corresponding to w_1, w, w'. We have $w = w_1 z$, $w' = w_1 z'$ $(z, z' \in W_{L,n})$ and $\dot{w} = \dot{w}_1\dot{z}$, $\dot{w}' = \dot{w}_1\dot{z}'$ $(\dot{z}, \dot{z}' \in \mathcal{N}_{L,n})$.

We define an automorphism γ of $W_{L,n}$ (depending on Z) by

(2.15.1) $$\gamma(y) = F'(w_1 y w_1^{-1}), (y \in W_{L,n}).$$

We have $\gamma(W_{L,n}) \subset W_{L,n}$, since $w_1 \in Z_{L,n}$. Using the fact that $w_1(R^+_{L,n}) \subset R^+$ and that $F'(B) = B$, we see that $\gamma(S_{L,n}) = S_{L,n}$, hence

(2.15.2) γ is an automorphism of the Coxeter group $(W_{L,n}, S_{L,n})$.

We can state the following.

2.16. LEMMA. *In the setting of 2.15, we have, for any* $y \in W_{L,n} = W_{\theta_0}$:

$$\tau_{(\dot{w}')^{-1}}(F')^{-1} \cdot T_y \cdot F'\tau_{\dot{w}} = p^{\frac{1}{2}(\ell(w')+\ell(w)-\tilde{\ell}(z')-\tilde{\ell}(z))dr}T_{(z')^{-1}}T_{\gamma^{-1}(y)}T_z$$

(as endomorphisms of $\mathcal{P}^{\theta_0}$*).*

Proof. We can write $z = n_{\alpha_1}n_{\alpha_2}\cdots n_{\alpha_h}$ $(\alpha_i \in \Pi_{\theta_0})$ with n_{α_i} as in (2.12.2) and with $\ell_{\theta_0}(\overline{n}_{\alpha_1}\overline{n}_{\alpha_2}\cdots\overline{n}_{\alpha_h}) = h$. We have $w_1\overline{n}_{\alpha_1}\overline{n}_{\alpha_2}\cdots\overline{n}_{\alpha_{i-1}}(\alpha_i) \in R^+$ for $i = 1, 2, \cdots, h$, since $\overline{n}_{\alpha_1}\overline{n}_{\alpha_2}\cdots\overline{n}_{\alpha_h}$ is a reduced expression in W_{θ_0} and

$w_1(R_\theta \cap R^+) \subset R^+$, (see 1.9(i)). Applying 2.13 (iv) h times we get

$$\tau_{\dot w_1}\tau_{n_{\alpha_1}}\tau_{n_{\alpha_2}}\cdots\tau_{n_{\alpha_h}} f = p^{\frac12(\ell(w_1)+\ell(\bar n_{\alpha_1})+\cdots+\ell(\bar n_{\alpha_h})-\ell(w_1\bar n_{\alpha_1}\cdots\bar n_{\alpha_h}))dr}\,\tau_{\dot w_1 n_{\alpha_1}\cdots n_{\alpha_h}} f$$

for all $f \in \mathcal{P}^{\theta_0}$. Applying again 2.13 (iv), we have

$$\tau_{n_{\alpha_1}}\tau_{n_{\alpha_2}}\cdots\tau_{n_{\alpha_h}} f = p^{\frac12(\ell(\bar n_{\alpha_1})+\cdots+\ell(\bar n_{\alpha_h})-\ell(\bar n_{\alpha_1}\cdots\bar n_{\alpha_h}))dr}\,\tau_{n_{\alpha_1}\cdots n_{\alpha_h}} f$$

for all $f \in \mathcal{P}^{\theta_0}$. Thus,

$$\tau_{\dot w_1}\tau_{\dot z} f = p^{\frac12(\ell(w_1)-\ell(w)+\ell(z))dr}\,\tau_{\dot w} f, \quad (f \in \mathcal{P}^{\theta_0}). \tag{2.16.1}$$

Similarly, using (2.13.1) instead of 2.13 (iv), we see that

$$\tau^{-1}_{\dot w_1^{-1}}\tau_{(\dot z')^{-1}} f = p^{-\frac12(\ell(w_1)-\ell(w')+\ell(z'))dr}\,\tau^{-1}_{(\dot w')^{-1}} f, \quad (f \in \mathcal{P}^{\theta_0}).$$

Setting $f' = \tau^{-1}_{\dot w_1^{-1}}\tau^{-1}_{(\dot z')^{-1}} f$, we get

$$\tau_{(\dot z')^{-1}}\tau_{\dot w_1^{-1}} f = p^{\frac12(\ell(w_1)-\ell(w')+\ell(z'))dr} f', \quad (f' \in \mathcal{P}^{\theta^{F'}_0}). \tag{2.16.2}$$

We shall prove the identity

$$\tau_{\dot w_1^{-1}(F')^{-1}}\tau_{\dot y}F'\tau_{\dot w_1}(f) = p^{\frac12(\ell(y)+2\ell(w_1)-\ell(\gamma(y)))dr}\,\tau_{\dot w_1^{-1}(F')^{-1}(\dot y)\dot w_1}(f) \tag{2.16.3}$$

($f \in \mathcal{P}^{\theta_0}$, $\dot y$ a representative for y in $\mathfrak{N}_{\theta_0}$).

The proof is based on the formula

$$\tau_{n_\alpha^{-1}}\tau_n\tau_{n_\alpha}(f) = p^{\frac12(\ell(\bar n)-\ell(\bar n_\alpha \bar n \bar n_\alpha)+2)dr}\,\tau_{n_\alpha^{-1} n n_\alpha}(f) \tag{2.16.4}$$

(α a simple root, n_α as in (2.12.2), $f \in \mathcal{P}^\theta$, $\alpha \notin R_\theta$, $n \in \mathcal{N}(T)^{F^{dr}}$, $\bar{n} \in \bar{n}_\alpha W_\theta \bar{n}_\alpha^{-1}$) which follows from (2.12.1) and (2.12.4).

We write $\dot{w}_1$ in the form $\dot{w}_1 = n_{\beta_1} n_{\beta_2} \cdots n_{\beta_m} t$ (β_i are simple roots, n_{β_i} are as in (2.12.2), $m = \ell(w_1)$, $t \in T^{F^{dr}}$). We may assume without loss of generality that $t = e$. The left-hand side of (2.16.3) is equal to

$$ {}^{\tau}{}_{n_{\beta_m}^{-1}} \cdots {}^{\tau}{}_{n_{\beta_1}^{-1}} {}^{\tau}{}_{(F')^{-1}(\dot{y})} {}^{\tau}{}_{n_{\beta_1}} \cdots {}^{\tau}{}_{n_{\beta_m}} (f) \,. $$

We wish to apply to this expression the formula (2.16.4), m times to get the right-hand side of (2.16.3). To do so, we must only check that

$$ \bar{n}_{\beta_m} \cdots \bar{n}_{\beta_{i+1}} (\beta_i) \notin R_{\theta_0} \tag{2.16.5} $$

for $i = 1, 2, \cdots, m$. Assume that (2.16.5) is violated for some i. We then have, for that i :

$$ \bar{n}_{\beta_m} \cdots \bar{n}_{\beta_{i+1}} \bar{n}_{\beta_i} \bar{n}_{\beta_{i+1}} \cdots \bar{n}_{\beta_m} \in W_{\theta_0} $$

hence

$$ \bar{n}_{\beta_i} \bar{n}_{\beta_{i+1}} \cdots \bar{n}_{\beta_m} W_{\theta_0} = \bar{n}_{\beta_{i+1}} \cdots \bar{n}_{\beta_m} W_{\theta_0} \,. $$

It follows that

$$ w_1 W_{\theta_0} = \bar{n}_{\beta_1} \bar{n}_{\beta_2} \cdots \bar{n}_{\beta_m} W_{\theta_0} = \bar{n}_{\beta_1} \bar{n}_{\beta_2} \cdots \bar{n}_{\beta_{i-1}} \bar{n}_{\beta_{i+1}} \cdots \bar{n}_{\beta_m} W_{\theta_0} \,. $$

This contradicts the assumption that w_1 has minimal length in $w_1 W_{\theta_0}$; indeed $\bar{n}_{\beta_1} \bar{n}_{\beta_2} \cdots \bar{n}_{\beta_{i-1}} \bar{n}_{\beta_{i+1}} \cdots \bar{n}_{\beta_m}$ is strictly shorter than w_1. Thus, (2.16.3) is proved.

Using (2.16.1), (2.16.2), (2.16.3), we can now compute:

$$\tau_{(\dot{w}')^{-1}(F')^{-1}} T_y F' \tau_{\dot{w}} =$$

$$= p^{-\frac{1}{2}(\ell(w_1)-\ell(w')+\ell(z'))dr} \tau_{(\dot{z}')^{-1}} \tau_{\dot{w}_1^{-1}(F')^{-1}} \times$$

$$\times\ p^{-\frac{1}{2}(\ell(y)-\tilde{\ell}(y))dr} \tau_{\dot{y}} F' p^{-1/2(\ell(w_1)-\ell(w)+\ell(z))dr} \tau_{\dot{w}_1} \tau_{\dot{z}}$$

$$= p^{\frac{1}{2}(\ell(w')+\ell(w)-\ell(z')-\ell(z)+\tilde{\ell}(y)-\ell(\gamma^{-1}(y)))dr} \tau_{(\dot{z}')^{-1}} \tau_{\dot{w}_1^{-1}(F')^{-1}\dot{y}\dot{w}_1} \tau_{\dot{z}}$$

$$= p^{\frac{1}{2}(\ell(w')+\ell(w)-\tilde{\ell}(z')-\tilde{\ell}(z)+\tilde{\ell}(y)-\tilde{\ell}(\gamma^{-1}(y)))dr} T_{(z')^{-1}} T_{\gamma^{-1}(y)} T_z\,,$$

as endomorphisms of $\mathcal{P}^{\theta_0}$.
(We have used the fact that, for $y \in \mathfrak{N}_{\theta_0}$, we have $\dot{w}_1^{-1}(F')^{-1}y\dot{w}_1 \in \mathfrak{N}_{\theta_0}$; this follows from (2.15.2).)

We see also from (2.15.2) that $\tilde{\ell}(\gamma^{-1}(y)) = \tilde{\ell}(y)$. The lemma is proved.

2.17. LEMMA. *In the setting of 2.15, and assuming that the center of* G *is connected, the expression*

$$(2.17.1)\quad \sum_{i,j} (-1)^{i+j} \mathrm{Tr}((F^{dr})^*, (H^i(\overline{X}_w, \mathcal{F}^L_{\dot{w},n}) \otimes H^j(\overline{X}_{w'}, \mathcal{F}^{L^{-1}}_{\dot{w}',n}))^{G^{F'}})$$

is equal to the trace of the $\overline{Q}_\ell$*-linear transformation of* $\mathrm{End}_{G^{F^{dr}}}(\mathcal{P}^{\theta_0})$ *given on the basis elements* T_y $(y \in W_{L,n})$ *by*

$$T_y \to p^{\frac{1}{2}(\ell(w)+\ell(w')-\tilde{\ell}(z)-\tilde{\ell}(z'))dr}\Big(\sum_{v'\in W_{L,n}} P_{v',z'}(p^{dr})T_{(v')^{-1}}\Big)\, T_{\gamma^{-1}(y)} \times$$

$$\times \Big(\sum_{v\in W_{L,n}} P_{v,z}(p^{dr})T_v\Big)$$

where $P_{v,z}$, $P_{v',z'}$ *are the polynomials (with integral coefficients) defined in* [KL$_1$] *with respect to the Coxeter group* $(W_{L,n}, S_{L,n})$.

Taking the trace of xF^{dr} (for $x \in G^{F'}$) in both sides of (2.8.1) we get the identity

$$\sum_i (-1)^i \mathrm{Tr}((F^{dr})^* x^*, \mathbf{H}^i(\overline{X}_w, \mathcal{F}^L_{\dot{w},n}))$$

$$(2.17.2) \qquad = \sum_{v \in W_{L,n}} p^{\frac{1}{2}(\ell(w_1 z) - \ell(w_1 v) - \tilde{\ell}(z) + \tilde{\ell}(v))} P_{v,z}(p^{dr}) \times$$

$$\times \sum_i (-1)^i \mathrm{Tr}((F^{dr})^* x^*, H^i_c(X_{w_1 v}, \mathcal{F}^L_{(w_1 v)^{\cdot},n})) .$$

We have a similar equality for $\Sigma(-1)^j \mathrm{Tr}((F^{dr})^* x^*, \mathbf{H}^j(\overline{X}_{w'}, \mathcal{F}^{L^{-1}}_{\dot{w}',n}))$.

The expression (2.17.1) is therefore equal to

$$|G^{F'}|^{-1} \sum_{x \in G^{F'}} \mathrm{Tr}((F^{dr})^* x^*, \mathbf{H}^*(\overline{X}_w, \mathcal{F}^L_{\dot{w},n})) \mathrm{Tr}((F^{dr})^* x^*, \mathbf{H}^*(\overline{X}_{w'}, \mathcal{F}^{L^{-1}}_{\dot{w}',n}))$$

$$= \sum_{v,v' \in W_{L,n}} p^{\frac{1}{2}(\ell(w_1 z) - \ell(w_1 v) - \tilde{\ell}(z) + \tilde{\ell}(v))} p^{\frac{1}{2}(\ell(w_1 z') - \ell(w_1 v') - \tilde{\ell}(z') + \tilde{\ell}(v'))} \times$$

$$\times P_{v,z}(p^{dr}) P_{v',z'}(p^{dr}) \sum_{i,j} (-1)^{i+j} \mathrm{Tr}((F^{dr})^*, (H^i_c(X_{w_1 v}, \mathcal{F}^L_{(w_1 v)^{\cdot},n}) \otimes$$

$$\otimes H^j_c(X_{w_1 v'}, \mathcal{F}^{L^{-1}}_{(w_1 v')^{\cdot},n}))^{G^{F'}}) .$$

We now express the last alternating sum of traces, using 2.11, as a trace of a linear transformation of $\mathrm{End}_{G^{F^{dr}}}(\mathcal{P}^{\theta}_0)$ into itself. In 2.16, we have computed the effect of that linear transformation on the elements T_y, $(y \in W_{L,n})$, which, by 2.14, form a basis of $\mathrm{End}_{G^{F^{dr}}}(\mathcal{P}^{\theta}_0)$, since the center of G is connected. From this, the lemma follows immediately.

The same proof shows at the same time that, with the assumption of the lemma, the expression

$$(2.17.3)\quad \sum_{i,j} (-1)^{i+j}\mathrm{Tr}((F^{dr})^*, (H_c^i(X_w, \mathcal{F}^L_{w,n}) \otimes H_c^j(X_{w'}, \mathcal{F}^{L^{-1}}_{w',n}))^{G^{F'}})$$

is equal to the trace of the $\overline{\mathbf{Q}}_\ell$-linear transformation of $\mathrm{End}_{G^{F^{dr}}}(\mathcal{P}^\theta \mathcal{Q})$ given on the basis elements T_y, $(y \in W_{L,n})$ by

$$(2.17.4)\qquad T_y \to p^{\frac{1}{2}(\ell(w)+\ell(w')-\widetilde{\ell}(z)-\widetilde{\ell}(z'))dr} T_{(z')^{-1}} T_{\gamma^{-1}(y)} T_z \,.$$

(Compare [L_4, 3.9].)

We can now state the following

2.18. THEOREM. *In the setting of 2.15, and assuming that the center of* G *is connected, the following holds.*

(i) *The* $\overline{\mathbf{Q}}_\ell$*-vector space*

$$(2.18.1)\qquad \mathrm{Hom}_{G^{F'}}(H^i(\overline{X}_{w'}, \mathcal{F}^L_{\dot{w}',n}), H^j(\overline{X}_w, \mathcal{F}^L_{\dot{w},n}))$$

is zero, unless $i \equiv j \pmod 2$

(ii) *If* $i \equiv j \pmod 2$, *the endomorphism* $h \to ((F^d)^*)^{-1} \cdot h \cdot (F^d)^*$ *of the vector space (2.18.1) is equal to* $p^{(j-i)d/2}$ *times a unipotent endomorphism.*

Proof. The purity theorem of Gabber [BDD] together with the results of Deligne in [D] imply that the eigenvalues of $(F^d)^*$ on $H^i(\overline{X}_w, \mathcal{F}^L_{\dot{w},n})$ are algebraic numbers all of whose complex conjugates have absolute value $p^{id/2}$. A similar result holds for the eigenvalues of $(F^d)^*$ on $H^j(\overline{X}_{w'}, \mathcal{F}^{L^{-1}}_{w',n})$. It follows that the eigenvalues of $(F^d)^*$ on the $\overline{\mathbf{Q}}_\ell$-vector space

$$(2.18.2)\qquad (H^i(\overline{X}_w, \mathcal{F}^L_{w,n}) \otimes H^j(\overline{X}_{w'}, \mathcal{F}^{L^{-1}}_{w',n}))^{G^{F'}}$$

are algebraic numbers all of whose complex conjugates have absolute value $p^{(i+j)d/2}$. From 2.17, we see that for any $r \geq 1$, the expression (2.17.1) is the value at p^{rd} of a polynomial with integral coefficients (these coefficients being independent of r). It follows that:

(2.18.3) The vector space (2.18.2) is zero unless $i \equiv j \pmod 2$. If $i \equiv j \pmod 2$, the eigenvalues of $(F^d)^*$ on (2.18.2) are all equal to $p^{(i+j)d/2}$.

By one of the main properties of intersection cohomology, we have a canonical nonsingular Poincaré duality pairing

$$H^i(\overline{X}_{w'}, \mathcal{F}^L_{\dot{w}',n}) \times H^{2\ell(w')-i}(\overline{X}_{w'}, \mathcal{F}^{L^{-1}}_{\dot{w}',n}) \to \overline{Q}_\ell(-\ell(w')) .$$

(Note that $\overline{X}_{w'}$ is complete and it satisfies (2.3.2).)

Hence the vector space (2.18.1) may be canonically identified with

$$(H^j(\overline{X}_{w'}, \mathcal{F}^L_{\dot{w},n}) \otimes H^{2\ell(w')-i}(\overline{X}_{w'}, \mathcal{F}^{L^{-1}}_{\dot{w}',n}))^{G^{F'}} \otimes \overline{Q}_\ell(\ell(w'))$$

and we see that the theorem follows from (2.18.3).

2.19. Assume that G has connected center. If $w \in Z_{L,n}$, $w' \in Z_{L,n}$, then $w^{-1}w'$ satisfies

$$\lambda((w^{-1}w')^{-1}(t)) = \lambda(t)\lambda_2^n(t), (t \in T) ,$$

for some character λ_2 of T, depending on w, w'. Since the center of G is connected it follows that $w^{-1}w' \in W_{L,n}$.

Thus, there exists a unique element w_1 of minimal length in $Z_{L,n}$ and we have $Z_{L,n} = w_1 \cdot W_{L,n}$.

We choose a coherent lifting (see 1.23) $\dot{Z}_{L,n}$ of $Z_{L,n}$ to $\mathfrak{N}(T)^F$. For any $w \in Z_{L,n}$, we denote $\dot{w}$ the corresponding element in $\dot{Z}_{L,n}$.

(2.19.1) Let b be the smallest integer ≥ 1 such that $F^{db}: G \to G$ is an integral power of F'.

(2.19.2) We shall assume chosen, once for all, a square root $p^{1/2}$ of p in $\overline{Q}_\ell$.

Let $\mathcal{E}_{L,n,w,i}$ be the set of isomorphism classes of irreducible $G^{F'}$-modules (over $\overline{\mathbf{Q}}_\ell$) appearing as components of the $G^{F'}$-module $H^i(\overline{X}_w, \mathcal{F}^L_{\dot{w},n})$, $(w \in Z_{L,n})$. This set does not depend on the choice of coherent lifting for $Z_{L,n}$. We denote by $\mathcal{E}_{L,n}$ the union (not necessarily disjoint) of the sets $\mathcal{E}_{L,n,w,i}$, where w varies in $Z_{L,n}$ and i runs through the integers ≥ 0. We shall denote $M^i_{\dot{w}}$ the $(G^{F'},F^d)$-module $H^i(X_w, \mathcal{F}^L_{\dot{w},n})$ and by $M^i_{\dot{w},\rho}$ its ρ-isotypic part $(\rho \in \mathcal{E}_{L,n,w,i})$, as a $G^{F'}$-module. Note that $M^i_{\dot{w},\rho}$ is not, a priori, F^d-stable. We have the the following

2.20. PROPOSITION. *We assume that we are in the setting of 2.19. Let $\widetilde{G}^{F'}$ be the semidirect product of $G^{F'}$ with the cyclic group of order b with generator ϕ, where ϕ acts on $G^{F'}$ by $\phi g \phi^{-1} = F^d(g)$. There exists a map $\rho \to \tilde{\rho}$ from $\mathcal{E}_{L,n}$ to the set of isomorphism classes of irreducible $G^{F'}$-modules (over $\overline{\mathbf{Q}}_\ell$), and a map $\rho \to \lambda'_\rho$ from $\mathcal{E}_{L,n}$ to the set of algebraic numbers in $\overline{\mathbf{Q}}^*_\ell$, all of whose complex conjugates have absolute value 1, such that the following hold.*

(i) *For each $\rho \in \mathcal{E}_{L,n}$, the restriction of $\tilde{\rho}$ to $G^{F'}$ is ρ itself.*

(ii) *If $\rho \in \mathcal{E}_{L,n,w,i}$ then $M^i_{\dot{w},\rho}$ is F^d-stable and moreover it admits a $(G^{F'},F^d)$ stable filtration each of whose successive quotients is isomorphic as a $G^{F'}$-module to ρ and, as a $\widetilde{G}^{F'}$-module (with ϕ acting as $(\lambda'_\rho)^{-1}p^{-id/2}F^d$) to $\tilde{\rho}$. In particular, F^{db} acts on $M^i_{\dot{w},\rho}$ as $(\lambda'_\rho)^b p^{idb/2}$ times a unipotent transformation.*

(iii) *If $x \in G^{F'}$, $r \geq 1$ is an integer such that $r \equiv 1 \pmod{b}$ and $w \in Z_{L,n}$, then*

$$\sum_i (-1)^i \mathrm{Tr}((F^{dr})^* x^*, H^i_c(\overline{X}_w, \mathcal{F}^L_{\dot{w},n})) = \sum_i (-1)^i \sum_{\rho \in \mathcal{E}_{L,n}} \langle \rho, M^i_{\dot{w}} \rangle_{G^{F'}} (\lambda'_\rho)^r p^{idr/2} \mathrm{Tr}(\phi x^{-1}, \tilde{\rho}$$

(Given two $G^{F'}$-modules M, M', of finite dimension over $\overline{\mathbf{Q}}_\ell$, we set

$$\langle M, M' \rangle_{G^{F'}} = \dim \mathrm{Hom}_{G^{F'}}(M, M') .)$$

Proof. Let $\mathfrak{A}^i_{\dot{w}}$ be the algebra $\mathrm{End}_{G^{F'}}(M^i_{\dot{w}})$ and let $\mathfrak{A}^i_{\dot{w},\rho}$, $(\rho \in \mathcal{E}_{L,n,w,i})$ be the algebra $\mathrm{End}_{G^{F'}}(M^i_{\dot{w},\rho})$. Then $\mathfrak{A}^i_{\dot{w},\rho}$ is a simple algebra and $\mathfrak{A}^i_{\dot{w}}$ is canonically the direct sum of algebras $\mathfrak{A}^i_{\dot{w}} = \oplus \mathfrak{A}^i_{\dot{w},\rho}$, (ρ runs through $\mathcal{E}_{L,n,w,i}$). Now F^d acts naturally on $M^i_{\dot{w}}$ and $h \to (F^d)^{-1} \cdot h \cdot F^d$ maps $\mathfrak{A}^i_{\dot{w}}$ into itself; it is clearly an algebra automorphism, since F^d normalizes $G^{F'}$. It follows that $h \to (F^d)^{-1} \cdot h \cdot F^d$ maps the center of $\mathfrak{A}^i_{\dot{w}}$ into itself and its restriction to that center is given by a monomial matrix (with respect to the decomposition of the center into direct sum of its intersections with the simple components of $\mathfrak{A}^i_{\dot{w}}$). On the other hand, the endomorphism $h \to (F^d)^{-1} h F^d$ of $\mathfrak{A}^i_{\dot{w}}$ is unipotent, by 2.18 (ii). Its restriction to the center of $\mathfrak{A}^i_{\dot{w}}$ is both unipotent and monomial hence it is the identity.

It follows that the endomorphism F^d of $M^i_{\dot{w}}$ maps each summand $M^i_{\dot{w},\rho}$ into itself, i.e. $h \to (F^d)^{-1} \cdot h \cdot F^d$ is an automorphism of the algebra $\mathfrak{A}^i_{\dot{w},\rho}$ for each $\rho \in \mathcal{E}_{L,n,w,i}$. This being a simple algebra, has the property that any automorphism of it, and in particular ours, is of the form $h \to A^{-1} h A$ where A is an invertible element of $\mathfrak{A}^i_{\dot{w},\rho}$, i.e. an automorphism of $M^i_{\dot{w},\rho}$ commuting with the action of $G^{F'}$. We then have $A(F^d)^{-1} h (F^d) A^{-1} = h$ for all $h \in \mathfrak{A}^i_{\dot{w},\rho}$, hence $F^d A^{-1} : M^i_{\dot{w},\rho} \to M^i_{\dot{w},\rho}$ is a $\bar{Q}_\ell$-linear combination of endomorphisms of $M^i_{\dot{w},\rho}$ defined by the various elements of $G^{F'}$. In particular, any $G^{F'}$-submodule of $M^i_{\dot{w},\rho}$ is stable under $F^d A^{-1}$. Consider a filtration f of $M^i_{\dot{w},\rho}$ by $G^{F'}$-submodules, stable under A, which cannot be refined in a nontrivial way to a filtration by $G^{F'}$-submodules, stable under A. The nonzero successive quotients of this filtration are then irreducible $G^{F'}$-modules, since A commutes with the $G^{F'}$-action. This filtration is stable under $F^d A^{-1}$ (since any $G^{F'}$-submodule is) and under A (by definition) hence also under F^d.

Now $F^{db} : M^i_{\dot{w},\rho} \to M^i_{\dot{w},\rho}$ commutes with the action of $G^{F'}$ (by the definition of b, (2.19.1)) hence it acts on each nonzero successive quotient of the filtration f as a scalar a_j times identity. Let $a_1, a_2, \cdots, a_N$ be the scalars thus associated to the nonzero successive quotients of f, ($N^2 = \dim \mathfrak{A}^i_{w,\rho}$). The eigenvalues of $h \to F^{-db} \cdot h \cdot F^{db}$, ($\mathfrak{A}^i_{w,\rho} \to \mathfrak{A}^i_{w,\rho}$), are then the scalars $a_j/a_{j'}$ $(1 \leq j, j' \leq N)$; but this linear map is unipotent by 2.18 (ii) hence $a_j/a_{j'} = 1$. Thus, we see that F^{db} acts on $M^i_{w,\rho}$ as a scalar times a unipotent transformation. By the results of Deligne and Gabber already used in the proof of 2.18, this scalar must be of the form $\lambda_\rho p^{dbi/2}$ where $\lambda_\rho \in \overline{\mathbf{Q}}_\ell^*$ is an algebraic number all of whose complex conjugates have absolute value 1. A priori, λ_ρ depends not only on λ, but also on w, and i such that $\rho \in \mathcal{E}_{L,n,w,i}$. Applying again 2.18, we see that λ_ρ is, in fact, independent of w and i. Thus to each $\rho \in \mathcal{E}_{L,n}$ we have associated an element $\lambda_\rho \in \overline{\mathbf{Q}}_\ell^*$. We select, for each $\rho \in \mathcal{E}_{L,n}$, a b-th root $\lambda'_\rho \in \overline{\mathbf{Q}}_\ell^*$ of λ_ρ.

Returning to the filtration f of $M^i_{w,\rho}$ $(\rho \in \mathcal{E}_{L,n,w,i})$ considered above, we extend the action of $G^{F'}$ on each nonzero successive quotient of f to an action of $\tilde{G}^{F'}$, by letting $\phi \in \tilde{G}^{F'}$ act as $(\lambda'_\rho)^{-1} p^{-id/2} F^d$. (It is clear from the definition of $\lambda_\rho, \lambda'_\rho$ that this is a well-defined action of $\tilde{G}^{F'}$.)

If ρ is also in $\mathcal{E}_{L,n,w',i'}$, we may consider a filtration f′ of $M^{i'}_{w',\rho}$ with properties similar to f, and we extend, as above, the action of $G^{F'}$ on each nonzero successive quotient of f′ to an action of $\tilde{G}^{F'}$

Given a nonzero successive quotient of f and a nonzero successive quotient of f′, we attach to them a root of 1 as follows. We choose a $G^{F'}$-isomorphism between them; it will take the action of ϕ on the first space to a b-th root of 1 times the action of ϕ on the second space. The collection of b-th roots of 1 thus obtained, each multiplied by $p^{(i'-i)d/2}$, is precisely the set of eigenvalues (with their multiplicities) of the linear map $h \to (F^d)^{-1} h F^d$, from $\mathrm{Hom}_{G^{F'}}(M^i_{w,\rho}, M^{i'}_{w',\rho})$ into itself. By 2.18, this linear map is equal to $p^{(i'-i)d/2}$ times a unipotent transformation, hence all the b-th roots of 1 considered above are equal to 1, hence all the $G^{F'}$-isomorphisms considered above are automatically $\tilde{G}^{F'}$-isomorphisms. This shows that the $\tilde{G}^{F'}$-module (which we call $\tilde{\rho}$) constructed above on the nonzero successive quotients of f on $M^i_{w,\rho}$ depends only on ρ, and on the function $\rho \to \lambda'_\rho$ but not on w, i. Thus, we have constructed the maps $\rho \to \lambda'_\rho$ $\rho \to \tilde{\rho}$, satisfying (i) and (ii). To check (iii), we note that F^{dr} acts on each nonzero successive quotient of the filtration of $M^i_{w,\rho}$ in (ii) as $(\lambda'_\rho p^{id/2}\phi)^r = (\lambda'_\rho)^r p^{idr/2}\phi$. (Note that $\phi^r = \phi$ since $r \equiv 1 \pmod{b}$.) Hence

$$\mathrm{Tr}((F^{dr})^* x^*, M^i_w) = \sum_\rho \langle \rho, M^i_w \rangle_{G^{F'}} (\lambda'_\rho)^r p^{idr/2} \mathrm{Tr}(\phi x^{-1}, \tilde{\rho})$$

and (iii) follows.

Chapter 3

GLOBAL INTERSECTION COHOMOLOGY WITH TWISTED COEFFICIENTS OF THE VARIETY $\overline{X}_w$

The main result of this chapter is Theorem 3.8 which connects the global intersection cohomology groups in the title with the characters of a Hecke algebra.

The fact that the characters of the Hecke algebra are closely connected with the zeta function of X_w is suggested by [L_4, 3.9]. T. Asai [As_2] and F. Digne-J. Michel [DM_2] used [L_4, 3.9] to prove a variant of Theorem 3.8 (for the part of fixed weight in the ordinary ℓ-adic cohomology with compact support of X_w) in the special case of constant coefficient sheaf.

This chapter was influenced by the papers [As_2], [DM_2] of Asai and Digne-Michel.

3.1. Let $(\mathbf{W}, S)$ be a finite Weyl group and let $\gamma : \mathbf{W} \to \mathbf{W}$ be an automorphism of the group $\mathbf{W}$, mapping S into itself. We form the semidirect product $\tilde{\mathbf{W}}$ of $\mathbf{W}$ with the infinite cyclic group with generator γ, so that in $\tilde{\mathbf{W}}$, we have the identity $\gamma \cdot w \cdot \gamma^{-1} = \gamma(w)$, $(w \in \mathbf{W})$. Let $\mathbf{W}^\vee$ be the set of isomorphism classes of irreducible $\mathbf{W}$-modules (over $\mathbf{Q}$). It is known that any irreducible $\mathbf{W}$-module over $\mathbf{Q}$ is, in fact, absolutely irreducible. Let $\mathbf{W}^\vee_{ex} \subset \mathbf{W}^\vee$ be the set consisting of those W-modules in $\mathbf{W}^\vee$ which can be extended to a $\tilde{\mathbf{W}}$-module (over an algebraic closure $\overline{\mathbf{Q}}$ of $\mathbf{Q}$). (All $\tilde{\mathbf{W}}$-modules will be tacitly assumed to be such that γ acts as a map of finite order.) We say that $\gamma : \mathbf{W} \to \mathbf{W}$ is *ordinary* if it has the following property: whenever $s \neq s'$ in S are in the same γ-orbit in S, the product ss' has order 2 or 3.

We shall need the following result.

3.2. PROPOSITION. *Assume that* γ *is ordinary and let* $E \in \mathbf{W}^{\vee}_{ex}$. *Then there exists an irreducible* $\widetilde{\mathbf{W}}$*-module over* $\mathbf{Q}$ *whose restriction to* $\mathbf{W}$ *is* E. *In fact, there are (up to isomorphism) exactly 2 such* $\widetilde{\mathbf{W}}$*-modules: one is obtained from the other by replacing the action of* γ *by* $-\gamma$.

Proof. We may assume that γ permutes transitively the irreducible components of $\mathbf{W}$. Then we may identify $\mathbf{W}$ with $\mathbf{W}_1 \times \mathbf{W}_1 \times \cdots \times \mathbf{W}_1$ and S with $S_1 \times S_1 \times \cdots \times S_1$ (t factors) where $(\mathbf{W}_1, S_1)$ is an irreducible Weyl group; the automorphism γ may be written as

$$\gamma(w_1, w_2, \cdots, w_t) = (\delta(w_t), w_1, w_2, \cdots, w_{t-1}), \quad w_i \in \mathbf{W}_1 ,$$

where δ is an automorphism of $(\mathbf{W}_1, S_1)$. The $\mathbf{W} = \mathbf{W}_1^t$-module E is of the form $E_1 \otimes \cdots \otimes E_1$ (t factors) where $E_1 \in (\mathbf{W}_1)^{\vee}_{ex}$ (with respect to δ). If we assume the result known for $\mathbf{W}_1$ and E_1, we can find $\Gamma_1 : E_1 \to E_1$ and an integer $c \geq 1$ such that

$$\Gamma_1^c = 1 , \ \Gamma_1 y_1 \Gamma_1^{-1} = \delta(y_1) : E_1 \to E_1 \quad (y_1 \in \mathbf{W}_1) .$$

We now define $\Gamma : E \to E$ by

$$\Gamma(e_1 \otimes e_2 \otimes \cdots \otimes e_t) = \Gamma_1(e_t) \otimes e_1 \otimes \cdots \otimes e_{t-1} , \quad (e_i \in E_1) .$$

It is clear that $\Gamma^{ct} = 1$ and $\Gamma y \Gamma^{-1} = \gamma(y) : E \to E \ (y \in \mathbf{W})$. This gives the required $\widetilde{\mathbf{W}}$-module. Thus, we are reduced to the case where $(\mathbf{W}, S)$ is irreducible. We may assume that the order of $\gamma : \mathbf{W} \to \mathbf{W}$ is $c \geq 2$.

Let $\widetilde{\mathbf{W}}_c$ be the semidirect product of $\mathbf{W}$ with the cyclic group of order c with generator γ, or equivalently, the quotient of $\widetilde{\mathbf{W}}$ by the subgroup generated by γ^c. Our assumption implies that E extends to a $\widetilde{\mathbf{W}}_c$-module (over $\overline{\mathbf{Q}}$).

If $(\mathbf{W}, S)$ is of type $A_n (n \geq 2)$, $D_{2n+1} (n \geq 2)$ or E_6, then $c = 2$ and we may define $\Gamma : E \to E$ to be the action of the longest element in $\mathbf{W}$. Hence we have the required $\widetilde{\mathbf{W}}$-module (over $\mathbf{Q}$). Assume now that $(\mathbf{W}, S)$

is of type $D_{2n}(n \geq 2)$, and $c = 2$. Then $\tilde{W}_c$ is isomorphic to a Weyl group of type B_{2n} and therefore, all its irreducible representations are defined over $\mathbf{Q}$. Since E is assumed to be extendable to $\tilde{W}_c$ (over $\bar{\mathbf{Q}}$) it is automatically extendable over $\mathbf{Q}$.

Finally, we assume that (W, S) is of type D_4 and $c = 3$. Consider the semidirect product W' of W with symmetric group $\mathfrak{S}_3$ acting on (W, S) in the standard way. It is well known that W' is isomorphic to a Weyl group of type F_4. If we identify the cyclic group of order 3 generated by γ with a subgroup of $\mathfrak{S}_3$, we get an imbedding of $\tilde{W}_c$ into W'. Our assumption is that E is the restriction to W of an irreducible $\tilde{W}_c$-module $\tilde{E}$ (over $\bar{\mathbf{Q}}$). But one checks that among the three possible such E, there is exactly one which is the restriction to $\tilde{W}_c$ of an irreducible W'-module. This particular $\tilde{E}$ is defined over $\mathbf{Q}$, since any irreducible representation of a Weyl group of type F_4 is defined over $\mathbf{Q}$. The last sentence in the proposition is obvious.

3.3. Let (W, S), and γ be as in 3.1, with γ ordinary. We extend the length function $\ell : W \to \mathbf{N}$ to $\tilde{W}$ by setting $\ell(\gamma^i w) = \ell(w)$, $(w \in W, i \in \mathbf{Z})$. Let H be the Hecke algebra of (W, S) over $\mathbf{Q}[u^{1/2}, u^{-1/2}]$; here $u^{1/2}$ is an indeterminate. Let $\tilde{H}$ be the algebra over $\mathbf{Q}[u^{1/2}, u^{-1/2}]$ which has a basis T_w, $(w \in \tilde{W})$, as $\mathbf{Q}[u^{1/2}, u^{-1/2}]$-module and multiplication defined by

$$T_w T_{w'} = T_{ww'} \text{ if } w, w' \in \tilde{W}, \ \ell(ww') = \ell(w) + \ell(w')$$

$$T_s^2 = uT_e + (u-1)T_s, \text{ if } s \in S.$$

Then H may be identified with the $\mathbf{Q}[u^{1/2}, u^{-1/2}]$-submodule of $\tilde{H}$ spanned by T_w, $(w \in W)$. To give an $\tilde{H}$-module is the same as to give an H-module M together with a $\mathbf{Q}[u^{1/2}, u^{-1/2}]$-linear map $T_\gamma : M \to M$ such that $T_\gamma T_w = T_{\gamma(w)} T_\gamma$ on M, $(w \in W)$.

Let E be an irreducible $\mathbf{Q}[W]$-module. We associate to it an H-module E(u), free as a $\mathbf{Q}[u^{1/2}, u^{-1/2}]$-module, by the construction in [L_{11}, 1.1, 1.2]. Assume now that E is extendable to a $\tilde{W}$-module (over an algebraic closure of $\mathbf{Q}$). By 3.2, E is also extendable to a $\widetilde{\mathbf{W}}$-module $\widetilde{E}$ over $\mathbf{Q}$; this is the same as giving a $\mathbf{Q}$-linear map $\gamma : E \to E$ such that

$$\gamma \cdot w \cdot \gamma^{-1} = \gamma(w) : E \to E\,,\ \gamma^c = 1 : E \to E\,,\ \text{ for some } c \geq 1\,.$$

Since the construction $E \to E(u)$ is canonical, and γ is a Coxeter group automorphism of $(\mathbf{W}, S)$, the map $\gamma : E \to E$ gives rise to a $\mathbf{Q}[u^{1/2}, u^{-1/2}]$ -linear map $\gamma(u) : E(u) \to E(u)$ such that

$$\gamma(u)\, T_w\, \gamma(u)^{-1} = T_{\gamma(w)} : E(u) \to E(u)\,,\ \ \gamma(u)^c = 1\,.$$

Hence the H-module E(u) becomes an $\widetilde{H}$-module, denoted $\widetilde{E}(u)$, in which T_γ acts as $\gamma(u)$. (Compare [L_{12}, 1.5].) Now, the linear map $\gamma : E \to E$ above is determined up to multiplication by ± 1. Changing γ to $-\gamma$, has the effect of changing $T_\gamma : \widetilde{E}(u) \to \widetilde{E}(u)$ to $-T_\gamma$.

Let $a \to \bar{a}$ be the involution of the ring $\mathbf{Q}[u^{1/2}, u^{-1/2}]$ such that $\overline{u^{i/2}} = u^{-i/2}$. The following properties (3.3.1)-(3.3.4) are checked as in [L_{11}, 1], [L_{12}, 1]. (In these formulas, Tr means trace over $\mathbf{Q}[u^{1/2}, u^{-1/2}]$.)

(3.3.1) $$\operatorname{Tr}(T_w, \widetilde{E}(u)) \in \mathbf{Z}[u^{1/2}],\ (w \in \widetilde{\mathbf{W}})\,.$$

(3.3.2) $$\operatorname{Tr}(T^{-1}_{w^{-1}}, \widetilde{E}(u)) = \overline{\operatorname{Tr}(T_w, \widetilde{E}(u))},\ (w \in \widetilde{\mathbf{W}})\,.$$

(3.3.3) $$\operatorname{Tr}(T_w, \widetilde{E}(u)) = \operatorname{Tr}(T_{w^{-1}}, \widetilde{E}(u)),\ (w \in \widetilde{\mathbf{W}})\,.$$

It is well known that there exists a unique polynomial $D_E(u) \in \mathbf{Q}[u]$ (the "formal dimension" of E(u)) such that

$$\sum_{w \in \mathbf{W}} u^{-\ell(w)} \operatorname{Tr}(T_w, E(u))^2 = \frac{P_{\mathbf{W}}(u)}{D_E(u)} \dim(E)$$

where

$$P_{\mathbf{W}}(u) = \sum_{w \in \mathbf{W}} u^{\ell(w)}\,.$$

If E' is another irreducible $\mathbf{Q}[\mathbf{W}]$-module which can be extended to a $\mathbf{Q}[\tilde{\mathbf{W}}]$-module $\tilde{E}'$, (see 3.1) and $\tilde{E}'(u)$ is the corresponding $\tilde{H}$-module, we have

$$(3.3.4)\quad \sum_{w\in\mathbf{W}} u^{-\ell(w)}\operatorname{Tr}(T_{\gamma w},\tilde{E}(u))\operatorname{Tr}(T_{\gamma w},\tilde{E}'(u)) = \begin{cases} \dfrac{P_W(u)}{D_E(u)}\dim(E), & \text{if } \tilde{E} \approx \tilde{E}' \\ 0 & , \text{if } E \not\approx E'. \end{cases}$$

3.4. We now return to the setting of 2.15. We assume that the center of G is connected. Then $Z = Z_{L,n}$ has a unique element of minimal length (denoted w_1) and we have $Z = w_1 \cdot W_{L,n}$. We recall that $Z^{\cdot}$ is a coherent lifting for Z in $\mathcal{N}(T)^F$, and $\dot{w}_1$ is the element of $Z^{\cdot}$ corresponding to w_1. Let H, $\tilde{H}$ be the algebras over $\mathbf{Q}[u^{1/2},u^{-1/2}]$ associated in 3.3 to $(\mathbf{W},S) = (W_{L,n},S_{L,n})$ and to its automorphism γ defined by (2.15.1).

(3.4.1) *From now on, we shall always assume, unless otherwise specified, that the map* $F':G\to G$ *in 2.1 is the Frobenius map for a rational structure on* G *over a finite subfield of* k. *It follows that* $\gamma: W_{L,n}\to W_{L,n}$ *is* ordinary (see 3.1).

Since $p^{1/2}$ has been chosen in $\bar{\mathbf{Q}}_\ell$ (2.19.2), we have a well-defined ring homomorphism $h_{dr}:\mathbf{Q}[u^{1/2},u^{-1/2}]\to\bar{\mathbf{Q}}_\ell$, taking $u^{1/2}$ to $p^{dr/2}$, (r is an integer ≥ 1). Let $H(p^{dr}) = H\otimes\bar{\mathbf{Q}}_\ell$, $\tilde{H}(p^{dr}) = \tilde{H}\otimes\bar{\mathbf{Q}}_\ell$ be the algebras over $\bar{\mathbf{Q}}_\ell$ obtained from H, $\tilde{H}$ using the homomorphism h_{dr}. We shall denote the basis elements $T_w\otimes 1$ of $\tilde{H}(p^{dr})$ again by T_w, $(w\in\tilde{W}_{L,n})$. Let θ_0 be given by (2.9.5). Recall that 2.13 and 2.14 define an isomorphism of $\bar{\mathbf{Q}}_\ell$-algebras:

$$(3.4.2)\qquad H(p^{dr}) \xrightarrow{\approx} \operatorname{End}_{G^{F^{dr}}}(\mathcal{P}^{\theta_0}).$$

For each $E\in W^{\vee}_{L,n}$ and each extension $\tilde{E}$ of E to a $\tilde{W}_{L,n}$-module (see 3.1) over $\mathbf{Q}$ (if it exists), we shall denote $E(p^{dr}) = E(u)\otimes\bar{\mathbf{Q}}_\ell$, $\tilde{E}(p^{dr}) = E(u)\otimes\bar{\mathbf{Q}}_\ell$ the corresponding modules for $H(p^{dr})$, $\tilde{H}(p^{dr})$.

3.5. Consider the endomorphism $F'\tau_{\dot{w}_1} : \mathcal{P}^{\theta_0} \to \mathcal{P}^{\theta_0}$. It has the following properties:

$$(F'\tau_{\dot{w}_1})(gf) = F'(g)((F'\tau_{\dot{w}_1})f), (g \in G^{F^{dr}}, f \in \mathcal{P}^{\theta_0}) \tag{3.5.1}$$

$$T_{\gamma(y)}((F'\tau_{\dot{w}_1})(f)) = (F'\tau_{\dot{w}_1})(T_y(f)), \ (y \in W_{L,n}, f \in \mathcal{P}^{\theta_0}). \tag{3.5.2}$$

$$(F'\tau_{\dot{w}_1})^i(f) = p^{-\frac{1}{2}(\ell(F'^{i-1}(w_1)\cdots F'(w_1)w_1) - i\ell(w_1))dr} \times \tag{3.5.3}$$

$$\times F'^i \tau_{F'^{i-1}(\dot{w}_1)\cdots F'(\dot{w}_1)\dot{w}_1}(f)$$

$(f \in \mathcal{P}^{\theta_0})$, for any $i \geq 1$.

The formula (3.5.1) is obvious. The formula (3.5.2) is proved as follows. Using (2.16.3) with $y = e$, we see that

$$\tau_{(\dot{w}_1)^{-1}} \tau_{\dot{w}_1}(f) = p^{\ell(w_1)dr} \cdot f, \ (f \in \mathcal{P}^{\theta_0}).$$

Using Lemma 2.16 with $w = w' = w_1$, and y replaced by $\gamma(y)$, we see that

$$\tau_{(\dot{w}_1)^{-1}}(F')^{-1} T_{\gamma(y)} F'\tau_{\dot{w}_1}(f) = p^{\ell(w_1)dr} T_y(f)$$

$$= \tau_{(\dot{w}_1)^{-1}} \tau_{\dot{w}_1} T_y(f), \ (f \in \mathcal{P}^{\theta_0}).$$

Cancelling $\tau_{(\dot{w}_1)^{-1}}$ (which is invertible), we get (3.5.2). We now prove (3.5.3). The left-hand side of (3.5.3) is

$$F'^i \tau_{F'^{i-1}(\dot{w}_1)} \cdots \tau_{F'(\dot{w}_1)} \tau_{\dot{w}_1}(f).$$

For $f \in \mathcal{P}^{\theta_0}$, we have

$$\tau_{\dot{w}_1}(f) \in \mathcal{P}^{\theta_0^{F'}}, \ \tau_{F'(\dot{w}_1)}\tau_{\dot{w}_1}(f) \in \mathcal{P}^{\theta_0^{F'^2}}, \quad \text{etc.};$$

moreover, we have

$$w_1(R^+_{\theta_0}) \subset R^+, \ F'(w_1)(R^+_{\theta_0^{F'}}) \subset R^+, \ F'^2(w_1)(R^+_{\theta_0^{F'^2}}) \subset R^+,$$

etc. Hence (3.5.3) is a consequence of the following result:

$$\tau_{n_1}\tau_{n_2}(f) = p^{-\frac{1}{2}(\ell(\bar{n}_1\bar{n}_2)-\ell(\bar{n}_1)-\ell(\bar{n}_2))}\tau_{n_1n_2}(f) \tag{3.5.4}$$

for $n_1, n_2 \in N(T)^{F^{dr}}$, $f \in \mathcal{P}^\theta$, $n_2(R^+_\theta) \subset R^+$.

Without loss of generality, we may assume in (3.5.4) that $n_2 = n_{\beta_1} n_{\beta_2} \cdots n_{\beta_m}$ where β_i are simple roots, n_{β_i} are as in (2.12.2) and $m = \ell(\bar{n}_2)$. The assumption $\bar{n}_2(R^+_\theta) \subset R^+$ implies that $\bar{n}_{\beta_m} \cdots \bar{n}_{\beta_{i+1}}(\beta_i) \notin R_\theta$, $(i = 1, 2, \cdots, m)$, (see (2.16.5) and the argument following it). Therefore, (3.5.4) is a consequence of [Ki, Lemma 4.9] and (3.5.3) is proved.

We shall apply (3.5.3) in the case where $i = i(r)$ is the smallest integer ≥ 1 such that we have, simultaneously:

(a) $F'^i(g) = g$ for all $g \in G^{F^{dr}}$.

(b) $F'^{(i-1)}(w_1) \cdots F'(w_1) w_1 = 1$, in W.

(c) i is divisible by c.

Such an integer i clearly exists. For this i, (3.5.3) becomes

$$(F'\tau_{\dot{w}_1})^i(f) = p^{i\ell(w_1)dr/2} F'^i \tau_t(f), \quad (f \in \mathcal{P}^{\theta_0}),$$

where $t = F'^{(i-1)}(\dot{w}_1) \cdots F'(\dot{w}_1)\dot{w}_1 \in T^F$. Note that F'^i = identity on $\mathcal{P}^{\theta_0}$. We have $\tau_t(f) = \theta_0(t)^{-1} f$, and $\theta_0(t)^n = 1$. Hence

$$(3.5.5) \qquad (F'\tau_{\dot{w}_1})^{in}(f) = p^{in\ell(w_1)dr/2}\cdot f, \quad (f \in \mathcal{P}^{\theta_0}) .$$

3.6. Let $H(p^{dr})^\vee$ be the set of isomorphism classes of irreducible representations of the $\overline{Q}_\ell$-algebra $H(p^{dr})$ and let $H(p^{dr})^\vee_{ex}$ be its subset consisting of those irreducible representations which can be extended to representations of $\widetilde{H}(p^{dr})$. Using the canonical isomorphism $\overline{Q}_\ell[W_{L,n}] \xrightarrow{\sim} H(p^{dr})$, (see $[L_9]$), which is compatible with the action of γ, we see that $E \to E(p^{dr})^\vee$ defines a bijection $W^\vee_{L,n} \xrightarrow{\sim} H(p^{dr})^\vee$, and a bijection

$$(W_{L,n})^\vee_{ex} \xrightarrow{\sim} H(p^{dr})^\vee_{ex} .$$

We now choose, in each isomorphism class of irreducible representations of $W_{L,n}$ over Q a particular representation, and we identify $W^\vee_{L,n}$ with this set of representations. Moreover, for each $E \in (W_{L,n})^\vee_{ex}$, we choose one (of the 2 possible) extensions $\widetilde{E}$ of E to a $Q[\widetilde{W}_{L,n}]$-module, (see 3.1).

For each $E \in W^\vee_{L,n}$ we denote $\mathcal{P}^{\theta_0}_E$ the $E(p^{dr})$-isotypic component of $\mathcal{P}^{\theta_0}$, regarded as an $H(p^{dr})$-module (see 3.4.2). It is always nonzero, since there exists $f \in \mathcal{P}^{\theta_0}$ which generates an $H(p^{dr})$-submodule isomorphic to the regular representation of $H(p^{dr})$. The identity (3.5.2) shows that $F'\tau_{\dot{w}_1} : \mathcal{P}^{\theta_0} \to \mathcal{P}^{\theta_0}$ permutes among them the subspaces $\mathcal{P}^{\theta_0}_E$ according to the permutation of the set $W^\vee_{L,n}$ defined by the action of γ. It follows that the isotypic components $\mathcal{P}^{\theta_0}_E$ which are mapped into themselves by $F'\tau_{\dot{w}_1}$ are precisely those for which $E \in (W_{L,n})^\vee_{ex}$. For such E, we shall denote by V_E the $G^{F^{dr}}$-module $\mathrm{Hom}_{H(p^{dr})}(E(p^{dr}), \mathcal{P}^{\theta_0}_E)$; the $G^{F^{dr}}$-action on V_E comes from the action of $G^{F^{dr}}$ on $\mathcal{P}^{\theta_0}_E$. It is clear that V_E is an irreducible $G^{F^{dr}}$-module (over $\overline{Q}_\ell$) and that we have a canonical isomorphism of $(H(p^{dr}), G^{F^{dr}})$-modules

$$E(p^{dr}) \otimes V_E \xrightarrow{\sim} \mathcal{P}_E^{\theta_0}.$$

The restriction of $F'\tau_{\dot{w}_1}$ to $\mathcal{P}_E^{\theta_0}$ becomes, via this isomorphism, an endomorphism

$$X_E : E(p^{dr}) \otimes V_E \to E(p^{dr}) \otimes V_E ,$$

and the identities (3.5.1), (3.5.2) show that there exist $\overline{Q}_\ell$-linear maps $X'_E : E(p^{dr}) \to E(p^{dr})$, $X''_E : V_E \to V_E$ such that

$$\text{(3.6.1)} \qquad \begin{cases} X_E = X'_E \otimes X_E'' \\ T_{\gamma(y)} X'_E = X'_E T_y : E(p^{dr}) \to E(p^{dr}), \ (y \in W_{L,n}) \\ X''_E \cdot g = F'(g) X''_E : V_E \to V_E, \ (g \in G^{F^{dr}}). \end{cases}$$

Moreover, (3.5.5) shows that

$$\text{(3.6.2)} \quad (X'_E \otimes X''_E)^{in} = p^{in\ell(w_1)dr/2} \cdot \text{identity, on } E(p^{dr}) \otimes V_E, (i = i(r)) .$$

In particular, X'_E, X''_E are invertible. From (3.6.1), we see that X'_E acts on $E(p^{dr})$ as a nonzero constant times T_γ (in the extension $\widetilde{E}(p^{dr})$ of $E(p^{dr})$ to an $\widetilde{H}(p^{dr})$-module). We may assume that $X'_E = T_\gamma$ on $E(p^{dr})$, by absorbing that constant into X''_E; the identities (3.6.1), (3.6.2) will be still satisfied. We now consider the $\overline{Q}_\ell$-linear endomorphism $\hat{X}''_E$ of V_E defined by $\hat{X}''_E = p^{-\frac{1}{2}\ell(w_1)dr} X''_E$. Then $\hat{X}''_E \cdot g = F'(g)\hat{X}''_E : V_E \to V_E$, $(g \in G^{F^{dr}})$ and, as a consequence of (3.6.2), we have $(X''_E)^{i(r)n} =$ identity on V_E. Hence, there is a unique extension $\widetilde{V}_E$ of the $G^{F^{dr}}$-module V_E to a module over $\widetilde{G}^{F^{dr}}$, the semidirect product of $G^{F^{dr}}$ with the cyclic group of order $i(r)n$ with generator F', and with $F' \cdot g \cdot F'^{-1} = F'(g)$, $(g \in G^{F^{dr}})$, such that $F' \in \widetilde{G}^{F^{dr}}$ acts on $\widetilde{V}_E$ as $\hat{X}''_E : V_E \to V_E$. It follows that for any element $\hat{x} \in G^{F^{dr}}$ and any element

$y \in W_{L,n}$ we have

$$(3.6.3)\quad \mathrm{Tr}(\hat{x}(F'\tau_{\dot{w}_1})T_y : \mathcal{P}^{\theta_0} \to \mathcal{P}^{\theta_0}) = \sum_{E \in (W_{L,n})^\vee_{ex}} p^{\frac{1}{2}\ell(w_1)dr} \mathrm{Tr}(\hat{x}F', \tilde{V}_E)\mathrm{Tr}(T_{\gamma y}, \tilde{E}(p^{dr})).$$

(The representations $E \in W^\vee_{L,n}$ which are not extendable do not contribute to the trace.)

3.7. Let $\mathcal{R}(G^{F'})$ be the Grothendieck group of virtual $G^{F'}$-modules of finite dimension over $\bar{\mathbf{Q}}_\ell$. For any $E \in (W_{L,n})^\vee_{ex}$, we define

$$(3.7.1)\quad R_{\tilde{E}} = |W_{L,n}|^{-1} \sum_{\substack{y \in W_{L,n}\\ i \geq 0}} (-1)^i \mathrm{Tr}(\gamma y, \tilde{E}) H^i_c(X_{w_1 y}, \mathcal{F}^L_{(w_1 y)^\cdot, n}),$$

$((w_1 y)^\cdot \in Z^\cdot)$. This is an element in $\mathcal{R}(G^{F'}) \otimes \mathbf{Q}$. It depends on the choice made in 3.6 for an extension $\tilde{E}$ of E to a $\tilde{W}_{L,n}$-module over $\mathbf{Q}$; if a different choice of $\tilde{E}$ is made, it would lead to $-R_{\tilde{E}}$. Note that $R_{\tilde{E}}$ does not depend on the choice of coherent lifting $Z^\cdot$ for $Z = Z_{L,n}$.

The pairing $\langle M, M' \rangle_{G^{F'}}$ extends to a $\mathbf{Z}$-bilinear pairing $\langle\ ,\ \rangle_{G^{F'}} : \mathcal{R}(G^{F'}) \times \mathcal{R}(G^{F'}) \to \mathbf{Z}$ and to a $\bar{\mathbf{Q}}_\ell$-bilinear pairing $\langle\ ,\ \rangle_{G^{F'}}$ on $R(G^{F'}) \otimes \bar{\mathbf{Q}}_\ell$ with values in $\bar{\mathbf{Q}}_\ell$. We can state the following.

3.8. THEOREM. *We recall that* G *is assumed to have connected center.*

(i) *For any* $\rho \in \mathcal{E}_{L,n}$, *the element* $\lambda'_\rho \in \bar{\mathbf{Q}}^*_\ell$, *considered in 2.20, is a root of* 1.

(ii) *For any* $w = w_1 z \in Z_{L,n}$, $(z \in W_{L,n})$, *we consider the element*

$$(3.8.0)\quad u^{\frac{1}{2}(\ell(w_1 z) - \tilde{\ell}(z))} \sum_E \mathrm{Tr}\Big(\sum_{v \in W_{L,n}} P_{v,z}(u) T_{\gamma v}, \tilde{E}(u)\Big) R_{\tilde{E}} \in \mathcal{R}(G^{F'}) \otimes \mathbf{Q}[u^{1/2}].$$

(*Here* E *runs over* $(W_{L,n})^\vee_{ex}$, *and the trace is taken over* $\mathbf{Q}[u^{1/2}, u^{-1/2}]$.) *Then, for any integer* $i \geq 0$, *the coefficient of* $u^{i/2}$ *in (3.8.0) (an element of* $\mathcal{R}(G^{F'})\otimes\mathbf{Q}$), *is equal to* $(-1)^i H^i(\overline{X}_{w_1z}, \mathcal{F}^L_{(w_1z)^\cdot,n})$. *Note that the expression (3.8.0) is independent of the choice (3.6) of the function* $E\to\widetilde{E}$.)

Proof. Let $w = w_1z \in Z_{L,n} = Z$, $(z\in W_{L,n})$, $\dot{w}$ its representative in $Z^\cdot$, let $x\in G^{F'}$, and let r be an integer ≥ 1 such that $r\equiv 1 \pmod b$, (see 2.19.1). Using 2.20, we can write

$$\sum_{\rho\in\mathcal{E}_{L,n}}\sum_{i\geq 0}(-1)^i\langle\rho, M^i_{(w_1z)^\cdot}\rangle_{G^{F'}}(\lambda'_\rho)^r p^{idr/2}\mathrm{Tr}(\phi x^{-1},\widetilde{\rho}) =$$

$$(3.8.1)\qquad = \sum_{i\geq 0}(-1)^i\mathrm{Tr}((F^{dr})^*x^*, H^i(\overline{X}_w, \mathcal{F}^L_{\dot{w},n}))\,.$$

Using (2.8.1), the last expression can be rewritten as

$$\sum_{\substack{v\in W_{L,n}\\ i\geq 0}} p^{\frac12(\ell(w_1z)-\ell(w_1v)-\widetilde{\ell}(z)+\widetilde{\ell}(v))dr}P_{v,z}(p^{dr})(-1)^i\times$$

$$\times\,\mathrm{Tr}((F^{dr})^*x^*, H^i_c(X_{w_1v}, \mathcal{F}^L_{(w_1v)^\cdot,n}))$$

which, by (2.10.1), is equal to

$$(3.8.2)\qquad \sum_{v\in W_{L,n}} p^{\frac12(\ell(w_1z)-\ell(w_1v)-\widetilde{\ell}(z)+\widetilde{\ell}(v))dr}P_{v,z}(p^{dr})\mathrm{Tr}(\hat{x}F'\tau_{(w_1v)^\cdot}:\mathcal{P}^{\theta_0}\to\mathcal{P}^{\theta_0}),$$

where $\hat{x}\in G^{F^{dr}}$ corresponds to $x\in G^{F'}$ under (2.9.9). By (2.16.1), we have

$$\tau_{(w_1v)^\cdot} = p^{-\frac12(\ell(w_1)-\ell(w_1v)+\widetilde{\ell}(v))dr}\tau_{\dot{w}_1}T_v,\ (\text{on } \mathcal{P}^{\theta_0})$$

hence, by (3.6.3), the expression (3.8.2) is equal to

$$(3.8.3)\qquad \sum_{v\epsilon W_{L,n}} p^{\frac12(\ell(w_1z)-\widetilde{\ell}(z))dr} P_{v,z}(p^{dr}) \sum_E \mathrm{Tr}(\widehat{x}F', \widetilde{V}_E)\mathrm{Tr}(T_{\gamma v}, \widetilde{E}(p^{dr})) .$$

(The last sum is over all $E \epsilon (W_{L,n})^{\vee}_{ex}$.)

Thus, the expressions (3.8.1) and (3.8.3) are equal.

We shall now define polynomials $Q_{z,v} \epsilon \mathbf{Z}[u]$, $(z,v \epsilon W_{L,n})$ by the equations

$$(3.8.4)\qquad \sum_{z\epsilon W_{L,n}} (-1)^{\widetilde{\ell}(v)+\widetilde{\ell}(z)} P_{v,z}(u) Q_{z,v'}(u) = \begin{cases} 1 & \text{if } v = v' \\ 0 & \text{if } v \neq v' \end{cases}$$

$(v,v' \epsilon W_{L,n})$. Note that the matrix of polynomials $(P_{v,z})$, $(v,z \epsilon W_{L,n})$ is triangular with 1 on diagonal, so that $Q_{z,v}$ are well defined.

We shall multiply (3.8.1) by

$$(3.8.5)\qquad |G^{F'}|^{-1}(-1)^{\widetilde{\ell}(z)+\widetilde{\ell}(v')} Q_{z,v'}(p^{dr}) \mathrm{Tr}(T_{\gamma v'}, \widetilde{E}'(p^{dr})) \mathrm{Tr}(x\phi^{-1}, \widetilde{\rho}') \times \\ \times \frac{D_{E'}(p^{dr}) \cdot p^{-\frac12(\ell(w_1z)-\widetilde{\ell}(z)+2\widetilde{\ell}(v'))dr}}{P_{W_{L,n}}(p^{dr}) \dim(E')}$$

where $v' \epsilon W_{L,n}$, $E' \epsilon (W_{L,n})^{\vee}_{ex}$, $\rho' \epsilon \mathcal{E}_{L,n}$, and sum over all v', $z \epsilon W_{L,n}$ and over all $x \epsilon G^{F'}$. The result is

$$(3.8.6)\qquad \frac{D_{E'}(p^{dr})}{P_{W_{L,n}}(p^{dr}) \dim(E')} \sum_{\substack{v',z\epsilon W_{L,n} \\ i\geq 0}} (-1^{\widetilde{\ell}(z)+\widetilde{\ell}(v')+i} p^{idr/2} p^{-\frac12(\ell(w_1z)-\widetilde{\ell}(z)+2\widetilde{\ell}(v'))dr} \times \\ \times Q_{z,v'}(p^{dr}) \langle \rho', M^i_{(w_1z)} \cdot \rangle_{G^{F'}} \mathrm{Tr}(T_{\gamma v'}, \widetilde{E}'(p^{dr})) (\lambda'_{\rho'})^r .$$

(Here we have used the orthogonality relation

$$(3.8.7)\qquad |G^{F'}|^{-1} \sum_{x\epsilon G^{F'}} \mathrm{Tr}(\phi x^{-1}, \widetilde{\rho}) \mathrm{Tr}(x\phi^{-1}, \widetilde{\rho}') = \begin{cases} 1, & \text{if } \rho = \rho' \\ 0, & \text{if } \rho \neq \rho' \end{cases}$$

which follows from the fact that ρ, ρ' are irreducible $G^{F'}$-modules over $\bar{Q}_\ell$.) We also multiply (3.8.3) by (3.8.5) and sum over all $v', z \in W_{L,n}$ and over all $x \in G^{F'}$. The result is

$$|G^{F'}|^{-1} \sum_{x \in G^{F'}} \mathrm{Tr}(\hat{x}F', \tilde{V}_{E'}) \mathrm{Tr}(x\phi^{-1}, \tilde{\rho}') \tag{3.8.8}$$

where $\hat{x} \in G^{F^{dr}}$ corresponds to $x \in G^{F'}$ under (2.9.9). (Here we have used the identities (3.8.4) and (3.3.4).) Thus, the expressions (3.8.6) and (3.8.8) are equal. Let $\alpha(r)$ be their common value. We regard it as a function of r ($r \geq 1$, $r \equiv 1 \pmod{b}$), for fixed $\rho' \in \mathcal{E}_{L,n}$ and $E' \in (W_{L,n})^{\vee}_{ex}$.

This function of r has the following three properties:

a) Its values are cyclotomic integers divided by $|G^{F'}|^{-1}$.

b) Its values belong to a fixed number field in $\bar{Q}_\ell$ (of finite degree over Q).

c) All complex conjugates of $\alpha(r)$ have absolute value ≤ 1.

The property a) is obvious since $|G^{F'}|\alpha(r)$ is a sum of products of character values of finite groups, see (3.8.8). By (3.3.1), the expression (3.8.6) belongs to the subfield of $\bar{Q}_\ell$ generated by $p^{1/2}$ and by $\lambda'_{\rho'}$, which is an algebraic number by 2.20. Hence $\alpha(r)$ satisfies b). To prove c), we set (for fixed r):

$$\delta = \frac{1}{|G^{F'}|} \sum_{x \in G^{F'}} \mathrm{Tr}(\hat{x}F', \tilde{V}_{E'}) \mathrm{Tr}((\hat{x}F')^{-1}, \tilde{V}_{E'})$$

$$\delta' = \frac{1}{|G^{F'}|} \sum_{x \in G^{F'}} \mathrm{Tr}(\phi x^{-1}, \tilde{\rho}') \mathrm{Tr}(x\phi^{-1}, \tilde{\rho}') .$$

By (3.8.7), we have $\delta' = 1$. Using (2.9.10) we see that

$$\delta = \frac{1}{|G^{F^{dr}}|} \sum_{\hat{x} \in G^{F^{dr}}} \mathrm{Tr}(\hat{x}F', \tilde{V}_{E'}) \mathrm{Tr}((\hat{x}F')^{-1}, \tilde{V}_{E'})$$

and this is again equal to 1, since $\tilde{V}_{E'}$ remains irreducible on restriction to $G^{F^{dr}}$.

By the Cauchy-Schwarz inequality, the absolute value squared of any complex conjugate of the sum (3.8.3) is bounded above by $\delta \cdot \delta'$ which is 1, and c) follows.

If we combine the various imbeddings of the number field in b) into $\mathbf{C}$, we get an imbedding of that number field into some $\mathbf{C}^N$. By a) and c), the set $\{\alpha(r)\}$ is carried by this imbedding into a discrete and bounded subset of $\mathbf{C}^N$. Hence the set $\{\alpha(r)\}$ is finite. This holds for any $\rho' \in \mathcal{E}_{L,n}$ and any $E' \in (W_{L,n})^\vee_{ex}$. Since the number of such (ρ', E') is a constant (independent of r), it follows that there exists an infinite subset I of the set $\{r \in \mathbf{N} | r \in 1 \pmod{b}\}$, (I independent of ρ', E') such that, for any $\rho' \in \mathcal{E}_{L,n}$, any $E' \in (W_{L,n})^\vee_{ex}$ and any $r \in I$, the expression (3.8.8) and also (3.8.6) are equal to a constant $C_{I,\rho',E'}$ (independent of $r \in I$). Let us now multiply (3.8.1) by $|G^{F'}|^{-1} \mathrm{Tr}(x\phi^{-1}, \tilde{\rho}')$ and sum over all $x \in G^{F'}$. We do the same to (3.8.3). Since the expressions (3.8.1), (3.8.3) are equal, we get the identity

$$(3.8.9) \qquad \sum_{i \geq 0} (-1)^i \langle \rho', M^i_{(w_1 z)^{\cdot}} \rangle_{G^{F'}} (\lambda'_{\rho'})^r p^{idr/2} =$$

$$= \sum_{v \in W_{L,n}} p^{\frac{1}{2}(\ell(w_1 z) - \tilde{\ell}(z))dr} P_{v,z}(p^{dr}) \sum_{E'} C_{I,\rho',E'} \mathrm{Tr}(T_{\gamma v}, \tilde{E}'(p^{dr}))$$

for any $z \in W_{L,n}$, $\rho' \in \mathcal{E}_{L,n}$, $r \in I$; the last sum is over $E' \in (W_{L,n})^\vee_{ex}$. (Here we use again (3.8.7).) Assume that $\rho' \in \mathcal{E}_{L,n}$ is such that $C_{I,\rho',E'} = 0$ for all $E' \in (W_{L,n})^\vee_{ex}$. Then the right-hand side of (3.8.9) is zero. Hence, the left-hand side is also zero. Since r runs through an infinite set, it follows that $\langle \rho', M^i_{(w_1 z)^{\cdot}} \rangle_{G^{F'}} = 0$ for all $i \geq 0$ and $z \in W_{L,n}$. But this contradicts the assumption that $\rho' \in \mathcal{E}_{L,n}$. Thus, given $\rho' \in \mathcal{E}_{L,n}$, we can find $E' \in (W_{L,n})^\vee_{ex}$, such that $C_{I,\rho',E'} \neq 0$. For such E' we consider the two identities which state that the expression (3.8.6) for $r = r_1$ and for $r = r_2$ is equal to $C_{I,\rho',E'}$, ($r_2 \neq r_1$ in I). Since $C_{I,\rho',E'} \neq 0$, we can divide term by term these two identities. We get an equality of the form

$$(\lambda'_{\rho'})^{r_1-r_2}\left(\frac{A(p^{dr_1/2})}{B(p^{dr_1/2})} : \frac{A(p^{dr_2/2})}{B(p^{dr_2/2})}\right) = 1$$

where $A, B, \in \mathbf{Q}[u]$. This identity shows that, under an isomorphism $\overline{\mathbf{Q}}_\ell \approx \mathbf{C}$, $(\lambda'_{\rho'})^{r_1-r_2}$ corresponds to a real number. On the other hand, by 2.20, $(\lambda'_{\rho'})^{r_1-r_2}$ corresponds to a complex number with absolute value 1. Being real, it must be equal to ± 1. Since $r_1 \neq r_2$, it follows that $\lambda'_{\rho'}$ must be a root of 1. Let $t \geq 1$ be an integer such that $(\lambda'_{\rho'})^t = 1$ for all $\rho' \in \mathcal{E}_{L,n}$. Since I is infinite, there exists a coset of $\mathbf{Z}$ with respect to $t\mathbf{Z}$ whose intersection I' with I is still an infinite set. The quotient $\hat{C}_{I',\rho',E'} = (\lambda'_{\rho'})^{-r} C_{I,\rho',E'}$ is then independent of r, when r runs through I'.

Consider the two polynomials in $\mathbf{Q}[u^{1/2}]$, (u is an indeterminate):

$$\sum_{i \geq 0} (-1)^i \langle \rho', M^i_{(w_1 z)^\cdot} \rangle_{G^{F'}} u^{i/2} \tag{3.8.10}$$

$$\sum_{v \in W_{L,n}} u^{\frac{1}{2}(\ell(w_1 z) - \tilde{\ell}(z))} P_{v,z}(u) \sum_{E'} \hat{C}_{I',\rho',E'} \operatorname{Tr}(T_{\gamma v}, \tilde{E}'(u)) . \tag{3.8.11}$$

(The last sum is over $E' \in (W_{L,n})^\vee_{ex}$. The trace is taken over $\mathbf{Q}[u^{1/2}, u^{1/2}]$.) By (3.8.9), these two polynomials coincide for $u = p^{dr}$, $(r \in I')$. Since I' is infinite, these two polynomials must be identically equal. In particular, they are equal for $u = 1$. Thus the value of (3.8.10) at $u = 1$, which, by (2.8.1) is the same as

$$\sum_{\substack{v \in W_{L,n} \\ i \geq 0}} (-1)^i P_{v,z}(1) \langle \rho', H^i_c(X_{w_1 v}, \mathcal{F}^L_{(w_1 v)^\cdot, n}) \rangle_{G^{F'}} \tag{3.8.12}$$

is equal to the value of (3.8.11) at $u = 1$, which is

$$\sum_{v \in W_{L,n}} P_{v,z}(1) \sum_{E'} \hat{C}_{I',\rho',E'} \operatorname{Tr}(\gamma v, \widetilde{E}') . \tag{3.8.13}$$

As we have already noted, the matrix $(P_{v,z}(u)$, $(v,z \in W_{L,n})$, is unipotent, hence invertible. Hence the matrix $(P_{v,z}(1))$ is also invertible. Hence, from the equality of (3.8.12) and (3.8.13) for all $z \in W_{L,n}$, it follows that

$$\sum_{i \geq 0} (-1)^i \langle \rho', H_c^i(X_{w_1 v}, \mathcal{F}^L_{(w_1 v)^{\cdot},n}) \rangle_{G^{F'}} = \sum_{E'} \hat{C}_{I',\rho',E'} \operatorname{Tr}(\gamma v, E') \tag{3.8.14}$$

for all $v \in W_{L,n}$.

We now multiply both sides of (3.8.14) by $|W_{L,n}|^{-1} \operatorname{Tr}(\gamma v, E)$ and sum over all $v \in W_{L,n}$. Using the identity (3.3.4) specialized for $u = 1$, we see that the right-hand side becomes $\hat{C}_{I',\rho',E}$. We obtain the equality

$$\begin{aligned} \hat{C}_{I',\rho',E} &= |W_{L,n}|^{-1} \sum_{\substack{v \in W_{L,n} \\ i \geq 0}} (-1)^i \langle \rho', H_c^i(X_{w_1 v}, \mathcal{F}^L_{(w_1 v)^{\cdot},n}) \rangle_{G^{F'}} \operatorname{Tr}(\gamma v, \widetilde{E}) \\ &= \langle \rho', R_{\widetilde{E}} \rangle_{G^{F'}} . \end{aligned} \tag{3.8.15}$$

(see (3.7.1)).

Introducing this value of $\hat{C}_{I',\rho',E}$ in (3.8.11), we can write the equality between (3.8.10) and (3.8.11) as

$$\begin{aligned} &\sum_{i>0} (-1)^i \langle \rho', M^i_{(w_1 z)^{\cdot}} \rangle_{G^{F'}} u^{i/2} = \\ &= u^{\frac{1}{2}(\ell(w_1 z) - \widetilde{\ell}(z))} \sum_{E'} \langle \rho', R_{\widetilde{E}} \rangle_{G^{F'}} \operatorname{Tr}\Big(\sum_{v \in W_{L,n}} P_{v,z}(u) T_{\gamma v}, \widetilde{E}'(u) \Big) \end{aligned} \tag{3.8.16}$$

and the theorem follows.

3.9. PROPOSITION. *Let* $E, E' \in (W_{L,n})^{\vee}_{ex}$, *and let* $\widetilde{E}, \widetilde{E}'$ *be their extensions to* $W_{L,n}$*-modules over* Q, *as in 3.6. We have*

$$\text{(i)} \quad \langle R_{\widetilde{E}}, R_{\widetilde{E}'} \rangle_{G^{F'}} = \begin{cases} 1, & \text{if } E = E' \\ 0, & \text{if } E \neq E' \end{cases}$$

(ii) *If* ρ *is an irreducible* $G^{F'}$*-module such that* $\langle \rho, R_{\widetilde{E}} \rangle_{G^{F'}} \neq 0$, *then* $\rho \in \mathcal{E}_{L,n}$.

Proof. Let $w = w_1 z$, $w' = w_1 z'$ be two elements of $Z = Z_{L,n}$, $(z, z' \in W_{L,n})$ and let $\dot{w}, \dot{w}'$ be the corresponding elements in $\dot{Z}$. We first compute the integer $N_{z,z'}$ given by

$$|G^{F'}|^{-1} \sum_{i,j} (-1)^{i+j} \sum_{x \in G^{F'}} \mathrm{Tr}(x^*, H^i_c(X_w, \mathcal{F}^L_{\dot{w},n})) \mathrm{Tr}((x^{-1})^*, H^j_c(X_{w'}, \mathcal{F}^L_{\dot{w}',n}))$$

$$= |G^{F'}|^{-1} \sum_{i,j} (-1)^{i+j} \sum_{x \in G^{F'}} \mathrm{Tr}(x^*, H^i_c(X_w, \mathcal{F}^L_{\dot{w},n})) \mathrm{Tr}(x^*, H^j_c(X_{w'}, \mathcal{F}^{L^{-1}}_{\dot{w}',n})) \text{ (using [DL}_1,$$

$$= \sum_{i,j} (-1)^{i+j} \dim (H^i_c(X_w, \mathcal{F}^L_{\dot{w},n}) \otimes H^j_c(X_{w'}, \mathcal{F}^{L^{-1}}_{\dot{w}',n}))^{G^{F'}}.$$

This is obtained by setting formally $r \to 0$ in (2.17.3), or equivalently in (2.17.4) hence it is equal to the trace of the linear transformation $v \to z'^{-1} \gamma^{-1}(v) z$ of $\overline{Q}_\ell[W_{L,n}]$ into itself, hence it is equal to $\#\{v \in W_{L,n} | v = z'^{-1} \gamma^{-1}(v) z\}$. We have

$$\langle R_{\widetilde{E}}, R_{\widetilde{E}'} \rangle_{G^{F'}} = |W_{L,n}|^{-2} \sum_{y, y' \in W_{L,n}} \mathrm{Tr}(\gamma y, \widetilde{E}) \mathrm{Tr}(\gamma y', \widetilde{E}') N_{y,y'}$$

$$= |W_{L,n}|^{-2} \sum_{y, v \in W_{L,n}} \mathrm{Tr}(\gamma y, \widetilde{E}) \mathrm{Tr}(v \gamma y v^{-1}, \widetilde{E}')$$

$$= |W_{L,n}|^{-1} \sum_{y \in W_{L,n}} \mathrm{Tr}(\gamma y, \widetilde{E}) \mathrm{Tr}(\gamma y, \widetilde{E}').$$

The identity (3.3.4) specialized for $u = 1$ states that the last expression is equal to 1 if $E = E'$ and to 0 if $E \neq E'$. Thus, (i) is proved.

Now let ρ be as in (ii). There exists $w \in Z$ such that

$$<\rho, \Sigma(-1)^i H_c^i(X_w, \mathcal{F}^L_{\dot{w},n})>_{G^{F'}} \neq 0 .$$

Take w of minimal length with this property. Then (2.8.1) shows that

$$<\rho, \Sigma(-1)^i H^i(\overline{X}_w, \mathcal{F}^L_{\dot{w},n})>_{G^{F'}} \neq 0 .$$

Hence, $\rho \in \mathcal{E}_{L,n}$. The proposition is proved.

Chapter 4

REPRESENTATIONS OF WEYL GROUPS

The purpose of this chapter is to parametrize the irreducible representations of a Weyl group. Here the concept of special representations and families (as defined in $[L_6]$, $[L_7]$, $[L_8]$) will play the main role. The parametrization is needed in the statement of the main theorem, which will be given at the end of this chapter.

From Section 4.16 until the end of this chapter, all representations will be assumed to be over $\overline{\mathbf{Q}}_\ell$ unless otherwise specified. This convention applies in particular to $\mathbf{W}^\vee$.

4.1. Let $(\mathbf{W}, S)$ be the Weyl group corresponding to a root system R in the rational vector space V (spanned by R) with a fixed set of positive roots. Then V is in a natural way an irreducible $\mathbf{Q}[W]$ module said to be the reflection representation of $\mathbf{W}$.

With the notations in 3.1, 3.3, we associate with any $E \in \mathbf{W}^\vee$ four integers $a_E \geq 0$, $b_E \geq 0$, $A_E \geq 0$, $f_E > 0$ by the requirement that

(4.1.1) $$D_E(u) = f_E^{-1} \cdot u^{a_E} + \text{higher powers of } u$$

$$= f_E^{-1} \cdot u^{A_E} + \text{lower powers of } u$$

(4.1.2) b_E = smallest integer $i \geq 0$ such that the $\mathbf{W}$-module E occurs in the i-th symmetric power of V.

We have clearly $a_E \leq A_E$. As observed in $[L_6]$, we have always

(4.1.3) $$a_E \leq b_E .$$

This justifies the following definition (cf. $[L_6]$):

(4.1.4) $E \in W^{\vee}$ is said to be *special* if $a_E = b_E$.

Now let W' be a subgroup of W generated by a subset S' of S. Then (W', S') is again a Weyl group (a standard parabolic subgroup of W). Given an irreducible $Q[W']$-module E' and an irreducible $Q[W]$-module E such that $\langle E', E \rangle_{W'} \neq 0$, (inner product as W'-modules) we have (cf. $[L_6, 4]$):

(4.1.5) $$a_{E'} \leq a_E .$$

Under the same conditions we have also $b_{E'} \leq b_E$, as one sees easily from the definition. The argument in [*loc. cit.*] gives also the following statement:

(4.1.6) If $E' \in (W')^{\vee}$ then $f_{E'}^{-1} = \sum_E \langle E', E \rangle_{W'} f_E^{-1}$

(sum over all $E \in W^{\vee}$ such that $a_{E'} = a_E$).

For any $E' \in (W')^{\vee}$, we define

(4.1.7) $$J_{W'}^{W}(E') = \sum_E \langle E', E \rangle_{W'} E$$

(sum over all $E \in W^{\vee}$ such that $a_{E'} = a_E$).

The right-hand side is an element in the Grothendieck group $\mathcal{R}(W)$ of $Q[W]$-modules. Extending by linearity, we obtain a homomorphism $J_{W'}^{W} : \mathcal{R}(W') \to \mathcal{R}(W)$. It follows immediately from (4.1.5) that the following transitivity property holds: if W'' is a subgroup of W generated by a subset of S and $W'' \subset W'$, then

(4.1.8) $$J_{W''}^{W} = J_{W'}^{W} \cdot J_{W''}^{W'} : \mathcal{R}(W'') \to \mathcal{R}(W) .$$

Similarly, for $E' \in (W')^\vee$, we define

$$(4.1.9) \qquad j_{W'}^{W}(E') = \sum_{E} \langle E', E \rangle_{W'} E$$

(sum over all $E \in W^\vee$ such that $b_{E'} = b_E$).

In the case where E' occurs with multiplicity 1 in the $b_{E'}$-th symmetric power of the reflection representation of W', we have that $j_{W'}^{W}(E')$ is again an irreducible $\mathbf{Q}[W]$-module and it occurs with multiplicity 1 in the $b_{E'}$-th symmetric power of V.

4.2. We now define, following $[L_7]$, a partition of the set $W^\vee$ into (disjoint) *families*. When W has a single element, there is only one family; it consists of the unit representation of W. Assume now that W has more than one element and that families have already been defined for any standard parabolic subgroup $W' \subsetneq W$. Then two representations $E, E' \in W^\vee$ are said to be in the same family (for W) if there exists a sequence $E = E_0, E_1, \cdots, E_m = E'$, $(E_i \in W^\vee)$ such that for each i, $(0 \le i \le m-1)$, the following condition is satisfied. There exists a standard parabolic subgroup $W_i \subsetneq W$ and representations $M_i', M_i'' \in W_i^\vee$, in the same family for W_i, such that either

$$\langle M_i', E_{i-1} \rangle_{W_i} \neq 0, \; a_{M_i'} = a_{E_{i-1}}, \qquad \langle M_i'', E_i \rangle_{W_i} \neq 0, \; a_{M_i''} = a_{E_i}$$

or

$$\langle M_i', E_{i-1} \otimes \mathrm{sign} \rangle_{W_i} \neq 0, \quad a_{M_i'} = a_{E_{i-1} \otimes \mathrm{sign}},$$

$$\langle M_i'', E_i \otimes \mathrm{sign} \rangle_{W_i} = 0, \quad a_{M_i''} = a_{E_i \otimes \mathrm{sign}}.$$

(An equivalent definition is obtained if W_i is required to be maximal, subject to the condition $W_i \subsetneq W$.)

The following property of families is clear from the definition.

If $(\mathbf{W}, \mathbf{S})$ is a product of irreducible Weyl groups $(\mathbf{W}^{(i)}, \mathbf{S}^{(i)})$, $(1 \le i \le t)$ then $E = \underset{i}{\otimes} E^{(i)}$, $F = \underset{i}{\otimes} F^{(i)}$, $(E^{(i)}, F^{(i)} \in (\mathbf{W}^{(i)})^\vee)$, are in the same family for $\mathbf{W}$ if and only if $E^{(i)}$, $F^{(i)}$ are in the same family for $\mathbf{W}^{(i)}$, for all i.

4.3. Following $[L_5]$, to any finite group $\mathcal{G}$ we associate the finite set $\mathfrak{M}(\mathcal{G})$ consisting of all pairs (x, σ) where x is an element of $\mathcal{G}$ and σ is an irreducible representation over $\overline{\mathbf{Q}}_\ell$ (up to isomorphism) of the centralizer $Z(x) = Z_{\mathcal{G}}(x)$ of x in $\mathcal{G}$, modulo the equivalence relation $(x, \sigma) \sim (gxg^{-1}, \sigma^g)$, for any $g \in \mathcal{G}$, where σ^g is the irreducible representation of $Z(gxg^{-1}) = gZ(x)g^{-1}$ defined by composing σ with conjugation by g^{-1}.

(4.3.1) If $\mathcal{G}$ is a direct product $\mathcal{G} = \mathcal{G}_1 \times \mathcal{G}_2 \times \cdots \times \mathcal{G}_t$, then $\mathfrak{M}(\mathcal{G})$ can be naturally identified with $\mathfrak{M}(\mathcal{G}_1) \times \mathfrak{M}(\mathcal{G}_2) \times \cdots \times \mathfrak{M}(\mathcal{G}_t)$; to the sequence $(x_1, \sigma_1), \cdots, (x_t, \sigma_t)$ $((x_i, \sigma_i) \in \mathfrak{M}(\mathcal{G}_i))$ there corresponds $(x, \sigma) \in \mathfrak{M}(\mathcal{G})$ given by $x = (x_1, x_2, \cdots, x_t)$, $\sigma = \sigma_1 \otimes \cdots \otimes \sigma_t$ (a representation of $Z_{\mathcal{G}}(x) = Z_{\mathcal{G}_1}(x_1) \times \cdots \times Z_{\mathcal{G}_t}(x_t)$).

Following $[L_8, 1.6]$ we shall introduce a notation for the conjugacy classes of $\mathcal{G}$ in the case where $\mathcal{G}$ is a symmetric group $\mathfrak{S}_n$, $n = 2, 3, 4$ or 5. We shall denote by 1 the neutral element of $\mathcal{G}$ and by g_i an element of $\mathcal{G}$ of order i, $(3 \le i \le 6)$. For $2 \le n \le 5$, we denote by g_2 a transposition in $\mathcal{G}$. For $n = 4$ or 5, we denote by g_2' an element of $\mathcal{G}$ which is a product of two transpositions with disjoint supports. We now introduce a notation for the irreducible representations of the centralizers of the elements in $\mathcal{G}$. In the case where the centralizer of $g \in \mathcal{G}$ is the cyclic group generated by g, we identify a one-dimensional representation of $Z(g)$ with the image of g under it (a root of 1 in $\overline{\mathbf{Q}}_\ell^*$). An exception will be the case $g = g_2 \in \mathfrak{S}_2$: the irreducible representations of $\mathfrak{S}_2$ are denoted $1, \varepsilon$ (unit, sign). The irreducible representations of

$\mathfrak{S}_3$ are denoted $1, r, \varepsilon$ (unit, reflection, sign). The irreducible representations of $\mathfrak{S}_4$ are denoted $1, \lambda^1, \lambda^2, \lambda^3, \sigma$ (unit reflection, reflection $\otimes$ sign, sign, two-dimensional). The irreducible representations of $\mathfrak{S}_5$ are denoted $1, \lambda^1, \lambda^2, \lambda^3, \lambda^4, \nu, \nu'$ (unit, reflection, its $2^{nd}, 3^{rd}, 4^{th}$ exterior power, the five-dimensional representation whose character is positive on a reflection, its tensor product with sign). The centralizer of g_2 in $\mathfrak{S}_4$ is a product of the cyclic group generated by g_2 with the cyclic group generated by another transposition τ. Let 1 be the unit representation of $Z(g_2)$, let $\varepsilon', \varepsilon''$ be the one-dimensional representations of $Z(g_2)$ such that $\varepsilon'(g_2) = -1$, $\varepsilon'(\tau) = 1$, $\varepsilon''(g_2) = 1$, $\varepsilon''(\tau) = -1$ and let $\varepsilon = \varepsilon' \otimes \varepsilon''$. The centralizer of g_2 in $\mathfrak{S}_5$ is a product of the cyclic group generated by g_2 and a symmetric group $\mathfrak{S}_3$. Its irreducible representations on which g_2 acts trivially are denoted $1, r, \varepsilon$ (corresponding to the representations of $\mathfrak{S}_3$); its irreducible representations on which g_2 acts non-trivially are denoted $-1, -r, -\varepsilon$ (corresponding to the representations of $\mathfrak{S}_3$). The centralizer of g_2' in $\mathfrak{S}_4$ may be identified with the centralizer of g_2' in $\mathfrak{S}_5$ (when $\mathfrak{S}_4$ is embedded in $\mathfrak{S}_5$ in the natural way); it is the dyhedral group of order 8. Let 1 be the unit representation of $Z(g_2')$, r its two-dimensional irreducible representation, ε' its one-dimensional character which takes the value -1 on any transposition and on any 4-cycle in $Z(g_2')$, ε'' its one-dimensional character which takes the value 1 on any transposition and the value -1 on any 4-cycle in $Z(g_2')$, and let $\varepsilon = \varepsilon' \otimes \varepsilon''$.

The centralizer of g_3 in $\mathfrak{S}_5$ is a cyclic group of order 6. We denote its characters which are trivial on the element of order 2 in $Z(g_3)$ by their value on g_3 (a root of 1 in $\overline{\mathbf{Q}}_\ell^*$); we denote its characters which are nontrivial on the element of order 2 in $Z(g_3)$ by ε times their value on g_3.

In the following sections (4.4 - 4.13) we shall describe the irreducible representations of the various irreducible Weyl groups, the way they are partitioned into families, and we shall associate to each family a finite group $\mathcal{G}$ and we shall describe an injection of that family into the set $\mathfrak{M}(\mathcal{G})$.

4.4. Assume that $(\mathbf{W}, \mathbf{S})$ is of the type A_{n-1}, hence $\mathbf{W}$ is isomorphic to the symmetric group $\mathfrak{S}_n$. The isomorphism classes of irreducible $\mathbf{Q}[\mathfrak{S}_n]$-modules are in 1-1 correspondence with the partitions of n. To the partition $\beta = (\beta_1 \leq \cdots \leq \beta_m)$ of n, $(\beta_i \geq 0,\ \sum_{i=1}^{m} \beta_i = n)$ there corresponds the irreducible $\mathbf{Q}[\mathfrak{S}_n]$-module characterized by the following two properties: its restriction to $\mathfrak{S}_{\beta_1} \times \mathfrak{S}_{\beta_2} \times \cdots \subset \mathfrak{S}_n$ contains the unit representation and its restriction to $\mathfrak{S}_{\beta_1^*} \times \mathfrak{S}_{\beta_2^*} \times \cdots \subset \mathfrak{S}_n$ (where $\beta_1^* \leq \beta_2^* \leq \cdots$ is the dual partition) contains the sign representation. We shall designate this $\mathbf{Q}[\mathfrak{S}_n]$-module as $[\lambda_1, \lambda_2, \cdots, \lambda_m]$ where $\lambda_i = \beta_i + i - 1$, $(1 \leq i \leq m)$. Here λ_i are integers ≥ 0, $\lambda_1 < \lambda_2 < \cdots < \lambda_m$ and $\sum_{i=1}^{m} \lambda_i = n + \binom{m}{2}$. We note that $[\lambda_1, \cdots, \lambda_m]$ is the same as $[0, \lambda_1+1, \lambda_2+1, \cdots, \lambda_m+1]$, and as $[0, 1, \lambda_1+2, \lambda_2+2, \cdots, \lambda_m+2]$, etc. (The number m is not uniquely determined by β: we may add as many parts equal to 0 to β as we like.)

If $E = [\lambda_1, \cdots, \lambda_m]$, we have

$$E = J_{\mathbf{W}'}^{\mathbf{W}}(\text{sign}) = j_{\mathbf{W}'}^{\mathbf{W}}(\text{sign}) \tag{4.4.1}$$

where $\mathbf{W}'$ is a standard parabolic subgroup

$$\approx \mathfrak{S}_{\beta_1^*} \times \mathfrak{S}_{\beta_2^*} \times \cdots \text{ in } \mathbf{W}.$$

We have

$$a_E = b_E = \sum_{1 \leq i < j \leq m} \inf(\lambda_i, \lambda_j) - \sum_{1 \leq i < j \leq m} \inf(i-1, j-1). \tag{4.4.2}$$

The formula for a_E follows from Steinberg's formula [St, 2.9] for the formal degree $D_E(u)$; the formula for b_E goes back to Specht. Let t be an integer such that $\lambda_i \leq t$, $(1 \leq i \leq m)$. Let $\lambda'_1 < \lambda'_2 < \cdots < \lambda'_{m'}$ be the sequence obtained from $0, 1, 2, \cdots, t$ by removing the integers $t-\lambda_1, t-\lambda_2, \cdots, t-\lambda_m$. We then have

$$[\lambda_1, \cdots, \lambda_m] \otimes \mathrm{sign} = [\lambda'_1, \lambda'_2, \cdots, \lambda'_{m'}]. \tag{4.4.3}$$

(See $[M_2$, p. 3].) For example, $[n]$ is the unit representation and $[1, 2, \cdots, n]$ is the sign representation of $\mathbf{W}$. Each representation in $\mathbf{W}^\vee$ is special (see (4.1.4)). Each family $\mathcal{F}$ in $\mathbf{W}^\vee$ consists of a single representation. The corresponding group $\mathcal{G}$ has a single element, hence the set $\mathfrak{M}(\mathcal{G})$ has again a single element. We have an obvious bijection between $\mathcal{F}$ and $\mathfrak{M}(\mathcal{G})$.

4.5. Let W_n, $(n \geq 1)$, be the group of all permutations of the set $1, 2, \cdots, n, n', \cdots, 2', 1'$ which commute with the involution $i \to i'$, $i' \to i$, $(1 \leq i \leq n)$. For each j, $1 \leq j \leq n-1$, let $s_j \in W_n$ be the permutation which interchanges j with $j+1$ and also j' with $(j+1)'$ and leaves the other elements unchanged. Let $s_n \in W_n$ be the permutation which interchanges n with n' and leaves the other elements unchanged. Let $S = \{s_1, s_2, \cdots, s_n\}$. Then (W_n, S) is a Coxeter group of type $B_n = C_n$. Let $\chi : W_n \to \{\pm 1\}$ be the homomorphism defined by the condition $\chi(s_i) = 1$, $(1 \leq i \leq n-1)$, $\chi(s_n) = -1$. A permutation in W_n defines a permutation of the n element set consisting of the unordered pairs $(1,1'), (2,2'), \cdots, (n,n')$. Thus we have a natural homomorphism of W_n onto $\mathfrak{S}_n$, the symmetric group in n letters. Let $r, \tilde{r}$ be two integers ≥ 0 such that $r + \tilde{r} = n$. The subgroup $H_{r,\tilde{r}}$ of W_n consisting of all permutations which map $\{1, 2, \cdots, r, r', \cdots, 2', 1'\}$ into itself and hence also map $\{r+1, \cdots, n, n', \cdots, (r+1)'\}$ into itself may be regarded in a natural way as a product $W_r \times W_{\tilde{r}}$. (We make the convention that W_0 has a single element.) Hence $H_{r,\tilde{r}}$ has a natural map (as above) onto the product of symmetric groups $\mathfrak{S}_r \times \mathfrak{S}_{\tilde{r}}$. Let E_1 be an irreducible $\mathbf{Q}[\mathfrak{S}_r]$-module and let E_2 be an irreducible

$\mathbf{Q}[\mathfrak{S}_{\tilde{r}}]$-module. We can regard $E_1 \otimes E_2$ as a $H_{r,\tilde{r}}$-module via the projection $H_{r,\tilde{r}} \to \mathfrak{S}_r \times \mathfrak{S}_{\tilde{r}}$. We tensor $E_1 \otimes E_2$ with the one-dimensional character of $H_{r,\tilde{r}}$ which is the identity on the B_r-factor and is the restriction of χ on the $B_{\tilde{r}}$-factor. We induce the resulting representation from $H_{r,\tilde{r}}$ to W_n. We obtain thus an irreducible $\mathbf{Q}[W_n]$-module. We shall denote it $[\Lambda]$, where Λ is the tableau

$$\Lambda = \begin{pmatrix} \lambda_1, \lambda_2, \cdots, \lambda_{m+1} \\ \\ \mu_1, \mu_2, \cdots, \mu_m \end{pmatrix} \tag{4.5.1}$$

such that, with the notations in 4.4, we have $E_1 = [\lambda_1, \lambda_2, \cdots, \lambda_{m+1}]$, $E_2 = [\mu_1, \mu_2, \cdots, \mu_m]$. (We may arrange that the number of λ's is the number of μ's, plus 1 .) Then Λ is a symbol of rank n and defect 1 in the sense of [L_3, 3.1]. In other words, $\lambda_1 < \lambda_2 < \cdots < \lambda_{m+1}$ are integers with $\lambda_1 \geq 0$, $\mu_1 < \mu_2 < \cdots < \mu_m$ are integers with $\mu_1 \geq 0$ and $\Sigma\lambda_i + \Sigma\mu_i = n + m^2$. Since m can be increased at our will, we see that $[\Lambda] = [\Lambda'] = [(\Lambda')'] = \cdots$, where

$$\Lambda' = \begin{pmatrix} 0, \lambda_1+1, \lambda_2+1, \cdots, \lambda_{m+1}+1 \\ \\ 0, \mu_1+1, \cdots, \mu_m+1 \end{pmatrix}.$$

The correspondence $\Lambda \to [\Lambda]$ is a 1-1 correspondence between the set of symbols of rank n and defect 1 (up to the equivalence relation given by the shift $\Lambda \to \Lambda'$) and the set of isomorphism classes of irreducible $\mathbf{Q}[W_n]$-modules. For example, $\begin{bmatrix} n \\ - \end{bmatrix}$ is the unit representation and $\begin{bmatrix} 0, 1, 2, \cdots, n \\ 1, 2, \cdots, n \end{bmatrix}$ is the sign representation of W_n. Let Λ be a symbol of rank n and defect 1 as in (4.5.1). We have

(4.5.2)
$$a_{[\Lambda]} = \sum_{1\le i<j\le m+1} \inf(\lambda_i,\lambda_j) + \sum_{1\le i<j\le m} \inf(\mu_i,\mu_j) + \sum_{\substack{1\le i\le m+1\\ 1\le j\le m}} \inf(\lambda_i,\mu_j) - \frac{1}{6}m(m-1)(4m+1)\,.$$

This is a consequence of Hoefsmit's formulas [Hf] for the formal degrees $D_E(u)$ for classical groups. (An alternative proof of Hoefsmit's formulas is sketched in [L_3, 9.6].) As shown in [L_6, 5] we have $a_{[\Lambda]} \le b_{[\Lambda]}$ and the equality holds if and only if

(4.5.3)
$$\lambda_1 \le \mu_1 \le \lambda_2 \le \mu_2 \le \cdots \le \mu_m \le \lambda_{m+1}\,.$$

Thus, (4.5.3) is the condition for $[\Lambda]$ to be special, see (4.1.4). Let W' be a maximal standard parabolic subgroup of W_n $(W' \ne W_n)$. Then W' may be identified with $\mathfrak{S}_r \times W_{\tilde r}$ $(r \ge 1, \tilde r \ge 0, r+\tilde r = n)$. If $\tilde r \ge 1$, we consider an irreducible representation $[\Lambda']$ of $W_{\tilde r}$ (Λ' is a symbol of rank $\tilde r$ and defect 1). If ε_r denotes the sign representation of $\mathfrak{S}_r$, we have

(4.5.4)
$$J_{W'}^{W_n}(\varepsilon_r \otimes [\Lambda']) = \begin{cases} [\Lambda] \\ \text{or } [\Lambda_1] + [\Lambda_2] \end{cases}$$

where Λ (or Λ_1, Λ_2) are obtained from Λ' by increasing the r largest entries of Λ' by 1. (We may assume, using shifts, that Λ' has at least r entries.) This can be done either in a unique way (example:

$$\Lambda' = \begin{pmatrix} 0 & & 2 \\ & 1 & \end{pmatrix},\ r = 2 \Rightarrow \Lambda = \begin{pmatrix} 0 & & 3 \\ & 2 & \end{pmatrix}),$$

leading to Λ, or in two different ways (example:

$$\Lambda' = \begin{pmatrix} 0 & & 2 \\ & 2 & \end{pmatrix},\ r = 1 \Rightarrow \Lambda_1 = \begin{pmatrix} 0 & & 3 \\ & 2 & \end{pmatrix},\ \Lambda_2 = \begin{pmatrix} 0 & & 2 \\ & 3 & \end{pmatrix}),$$

leading to Λ_1 and Λ_2. (See [L_{11}, 2.7].) This remains valid also for

$\tilde{r} = 0$, if we agree to denote the unit representation of $W_0 = \{e\}$ as

$$\begin{bmatrix} 0, 1, \cdots, m-1, m \\ 0, 1, \cdots, m-1 \end{bmatrix},$$

$(\forall m \geq 0)$, (a symbol of rank 0, defect 1).

Using (4.5.4), (4.4.1) and the transitivity formula (4.1.8), one gets an explicit description of the homomorphism $J_{W'}^{W_n}: \mathcal{R}(W') \to \mathcal{R}(W_n)$, where W' is any standard parabolic subgroup of W_n.

We shall now describe for any $[\Lambda] \in (W_n)^\vee$, with Λ as in (4.5.1), the representation $[\Lambda] \otimes \text{sign}$.

(4.5.5) Let t be an integer such that any entry of Λ is $\leq t$. Let Λ' be the symbol of rank n and defect 1 whose first row consists of the integers i, $0 \leq i \leq t$, $t-i \notin$ second row of Λ and whose second row consists of the integers j, $0 \leq j \leq t$, $t-j \notin$ first row of Λ. We then have, (see [L_{11}, 2.4]): $[\Lambda] \otimes \text{sign} = [\Lambda']$. From (4.5.4) and (4.5.5) one gets the following description of families in $(W_n)^\vee$.

(4.5.6) *Two irreducible* $\mathbf{Q}[W_n]$*-modules are in the same family if and only if they can be represented in the form* $[\Lambda]$, $[\Lambda']$ *where* Λ, Λ' *are symbols of rank* n *and defect* 1 (*with the same* m) *such that* Λ, Λ' *contain exactly the same entries (arranged perhaps differently in rows).* (For example,

$$\left\{ \begin{bmatrix} 0 & & 2 \\ & 1 & \end{bmatrix}, \begin{bmatrix} 0 & & 1 \\ & 2 & \end{bmatrix}, \begin{bmatrix} 1 & & 2 \\ & 0 & \end{bmatrix} \right\} \text{ is a family in } (W_2)^\vee.)$$

Indeed, let Λ, Λ' be two symbols of rank n and defect 1 (with the same m) such that Λ, Λ' contain exactly the same entries (perhaps permuted). We want to show that $[\Lambda]$, $[\Lambda']$ are in the same family. We may assume (by shifting) that 0 appears at most once in Λ and Λ'. Assume first that Λ and Λ' both have one of the following three properties: (a) 0 is not an entry; (b) there exist two entries $i < j$, $j - i \geq 2$ such that no integer k, $(i < k < j)$, is an entry; (c) there exist two consecutive entries i, $i+1$ which are both singles. Then by (4.5.4), we can write $[\Lambda] =$ component of $J_{W'}^{W_n}(\varepsilon_r \otimes [\Lambda_1'])$, $[\Lambda'] =$ component of $J_{W'}^{W_n}(\varepsilon_r \otimes [\Lambda_1'])$ where $W' \approx \mathfrak{G}_r \times W_{n-r}$, $r \geq 1$ and Λ_1, Λ_1' are symbols of rank $n-r$ and defect 1 (both with the same m) involving the same entries (up to permutation). We may assume by induction on n that $[\Lambda_1]$, $[\Lambda_1']$ are in the same family for W_{n-r}. From the definition of families, it follows that $[\Lambda]$, $[\Lambda']$ are in the same family. Assume next that Λ and Λ' do not have any of the properties (a), (b), (c) above. Then by (4.5.5), $[\Lambda] \otimes \text{sign} = [\Lambda_1]$, $[\Lambda'] \otimes \text{sign} = [\Lambda_1']$ where Λ_1, Λ_1' are symbols with the same entries (up to permutations) satisfying (a) or (b) or (c) above. By the first part of the argument, $[\Lambda_1]$ and $[\Lambda_1']$ are in the same family. By definition of a family, $[\Lambda]$, $[\Lambda']$ must be also in the same family. The converse follows immediately from (4.5.4) and (4.5.5).

It is clear that each family $\mathcal{F}$ contains a unique special representation. (See (4.5.3).) Let Z be a symbol of rank n and defect 1 such that $[Z] \in \mathcal{F}$; we represent Z by a definite symbol with $2m+1$ entries (not just by its equivalence class under shift). We define $\mathfrak{M}_{\mathcal{F}}$ to be the set of all arrays

$$\begin{pmatrix} \lambda_1, \lambda_2, \cdots, \lambda_b \\ \\ \mu_1, \mu_2, \cdots, \mu_{b'} \end{pmatrix}$$

where $b' < b, b+b' = 2m+1, 0 \leq \lambda_1 < \lambda_2 < \cdots < \lambda_b$, $0 \leq \mu_1 < \mu_2 < \cdots < \mu_{b'}$ and $\lambda_1, \cdots, \lambda_b, \mu_1, \cdots, \mu_{b'}$ are exactly the entries of Z, perhaps in another order (some entries may be repeated twice). In the language of $[L_3]$, the elements of $\mathfrak{M}_{\mathcal{F}}$ are symbols of rank n and odd defect. For example, if $[Z] = \begin{bmatrix} 0 & & 1 & & 3 \\ & 1 & & 2 & \end{bmatrix} \in \mathcal{F}$, then $\mathfrak{M}_{\mathcal{F}}$ consists of

$\begin{pmatrix} 0 & & 1 & & 3 \\ & 1 & & 2 & \end{pmatrix}$, $\begin{pmatrix} 0 & & 1 & & 2 \\ & 1 & & 3 & \end{pmatrix}$, $\begin{pmatrix} 1 & & 2 & & 3 \\ & 0 & & 2 & \end{pmatrix}$, $\begin{pmatrix} 0 & & 1 & 2 & 3 \\ & 1 & & & \end{pmatrix}$. (Note that if Z is replaced by an equivalent symbol, under shift, then the elements of $\mathfrak{M}_{\mathcal{F}}$ are shifted in the same way.) We may identify $\mathcal{F}$ with the subset of $\mathfrak{M}_{\mathcal{F}}$ defined by the condition that b, b' above satisfy $b = b'+1$, (using $[\Lambda] \leftrightarrow \Lambda$). This gives an injection $\mathcal{F} \hookrightarrow \mathfrak{M}_{\mathcal{F}}$. An element of $\mathfrak{M}_{\mathcal{F}}$ is of the form

$$\Lambda_M = \begin{pmatrix} Z_2 \amalg (Z_1 - M) \\ \\ Z_2 \amalg M \end{pmatrix},$$

where Z_2 is the set of elements which appear in both rows of Z, Z_1 is the set of singles of Z and M is a subset of Z_1 such that $|M| \equiv d \pmod 2$; d is defined by $|Z_1| = 2d+1$. In particular, we have $Z = \Lambda_{M_0}$ where M_0 is the set of elements of Z_1 which appear in the second row of Z. If we associate to M the set $M^{\#} \subset Z_1$ defined by $M^{\#} = (M \cup M_0) - (M \cap M_0)$ we get a bijection $\Lambda_M \leftrightarrow M^{\#}$ between $\mathfrak{M}_{\mathcal{F}}$ and the set V_{Z_1} of subsets of Z_1 of even cardinality. Note that V_{Z_1} is a vector space of dimension $2d$ over the field F_2; the sum of $M^{\#}$ and $M'^{\#}$ is defined to be their symmetric difference. The function $M^{\#}, M'^{\#} \to |M^{\#} \cap M'^{\#}| \bmod 2$ is then a nonsingular symplectic form $V_{Z_1} \times V_{Z_1} \xrightarrow{(\ ,\ )} F_2$. The symplectic vector space V_{Z_1} is a direct sum of two maximal isotropic subspaces I' and I'', defined as follows. Let us arrange the elements of Z_1 in an increasing sequence; for $1 \le i \le 2d$, let e_i be the 2-element subset of Z_1 consisting of the i-th and (i+1)-th element in this increasing sequence. Then $e_1, e_2, \cdots, e_{2d}$ is a basis of V_{Z_1} and we define I' to the subspace with basis $e_1, e_3, e_5, \cdots$ and I'' to be the subspace with basis $e_2, e_4, e_6, \cdots$. It is clear that I' and I'' are maximal isotropic subspaces of V_{Z_1} and that $V_{Z_1} = I' \oplus I''$. Assume that $(\mathbf{W}, S)$ corresponds to a root system R of type C_n (resp. B_n), see 4.1, and that $n \ge 3$. Then $(\mathbf{W}, S)$ may be canonically identified with (W_n, S) above. As promised in 4.3, we associate to a family $\mathcal{F} \subset \mathbf{W}^{\vee} = (W_n)^{\vee}$

(as above) a finite group $\mathcal{G}$ and an imbedding $\mathcal{F} \hookrightarrow \mathfrak{M}(\mathcal{G})$. If R is of type C_n, we define $\mathcal{G} = I''$; if R is of type B_n, we define $\mathcal{G} = I'$. The imbedding $\mathcal{F} \hookrightarrow \mathfrak{M}(\mathcal{G})$ is defined as the composition of the imbedding $\mathcal{F} \hookrightarrow \mathfrak{M}_{\mathcal{F}}$ (defined earlier) with the bijection $\mathfrak{M}_{\mathcal{F}} = V_{Z_1} = I' \oplus I'' \approx \mathfrak{M}(\mathcal{G})$ defined as follows. If $\mathcal{G} = I'$ we may identify I'' with the set of characters $I' \to \bar{Q}_\ell^*$ by $i'' \in I'' \leftrightarrow (i' \to (-1)^{(i',i'')})$ where (,) is the symplectic form defined earlier on V_{Z_1}; this gives an isomorphism $I' \oplus I'' \approx \mathfrak{M}(\mathcal{G})$. In the case where $G = I''$, an isomorphism $I' \oplus I'' \approx \mathfrak{M}(\mathcal{G})$ is defined similarly, interchanging the roles of I', I''.

This definition does not apply when $n = 2$, since in that case there are two distinct isomorphisms of $(\mathbf{W}, S)$ with (W_n, S). In this case, we denote $s_1 \in S$ the reflection with respect to a short simple coroot and $s_2 \in S$ the reflection with respect to a long simple coroot.

If $\mathcal{F} = \{\text{unit}\}$ or $\mathcal{F} = \{\text{sign}\}$, we take $\mathcal{G} = \{e\}$ and $\mathcal{F} \xrightarrow{\sim} \mathfrak{M}(\mathcal{G})$ to be the unique bijection. If $\mathcal{F}$ is the family containing the reflection representation, we define $G = \mathfrak{S}_2$ (symmetric group in 2 letters) and we define an imbedding $\mathcal{F} \hookrightarrow \mathfrak{M}(\mathfrak{S}_2)$ as follows: the reflection representation corresponds to $(1,1) \in \mathfrak{M}(\mathfrak{S}_2)$, the one-dimensional representation of $\mathbf{W}$ defined by $s_1 \to 1$, $s_2 \to -1$ corresponds to $(1, \varepsilon) \in \mathfrak{M}(\mathfrak{S}_2)$ and the one-dimensional representation of $\mathbf{W}$ defined by $s_1 \to -1$, $s_2 \to 1$ corresponds to $(g_2, 1) \in \mathfrak{M}(\mathfrak{S}_2)$. (We have used the notations in 4.3 for the elements of $\mathfrak{M}(\mathfrak{S}_2)$.)

4.6. Assume now that $(\mathbf{W}, S)$ is the type D_n, $(n \geq 2)$ and that we are given two commuting simple reflections s', $s'' \in S$ which are extremal points in the Coxeter graph; if $n \geq 5$, we assume that there exists a standard parabolic subgroup of type A_3 in $\mathbf{W}$ containing s', s''. (This condition determines (s', s'') uniquely, when $n \geq 5$.) We may identify $\mathbf{W}$ with the kernel of $\chi: W_n \to \{\pm 1\}$, (see 4.5), in two different ways: one such identification is characterized by the fact that $s' \in S$ corresponds to $s_{n-1} \in W_n$, $s'' \in S$ corresponds to $s_n s_{n-1} s_n$ and the other elements of S corresponds to elements in $\{s_1, s_2, \cdots, s_{n-2}\} \subset W_n$. The other identification is obtained by composing the first identification with the inner automorphism of W_n defined by s_n. Consider the irreducible $\mathbf{Q}[W_n]$-module M associated in 4.5 to (E_1, E_2) where E_1 is an irreducible $\mathbf{Q}[\mathfrak{S}_r]$-module and E_2 is an irreducible $\mathbf{Q}[\mathfrak{S}_{\tilde{r}}]$-module. Assume that $E_1 = [\lambda_1, \lambda_2, \cdots, \lambda_m]$, $E_2 = [\mu_1, \mu_2, \cdots, \mu_m]$, with the notations in 4.4. (We may arrange that the number of λ's is equal to the number of μ's .) If the sets $(\lambda_1, \cdots, \lambda_m)$, $(\mu_1, \cdots, \mu_m)$ do not coincide, the restriction of M to $\mathbf{W} \subset W_n$ (via one of the two imbeddings above) is irreducible and is denoted $[\Lambda]$, where Λ is the tableau (in which the first and second row can be interchanged among them):

$$\Lambda = \begin{pmatrix} \lambda_1, \lambda_2, \cdots, \lambda_m \\ \\ \mu_1, \mu_2, \cdots, \mu_m \end{pmatrix} = \begin{pmatrix} \mu_1, \mu_2, \cdots, \mu_m \\ \\ \lambda_1, \lambda_2, \cdots, \lambda_m \end{pmatrix}. \tag{4.6.1}$$

(Λ is a symbol of rank n and defect 0 in the sense of $[L_3, 3]$; in other words, λ_i, μ_i are integers, $0 \le \lambda_1 < \lambda_2 < \cdots < \lambda_m$, $0 \le \mu_1 < \mu_2 < \cdots < \mu_m$ and $\Sigma\lambda_i + \Sigma\mu_i = n + m^2 - m$. It is nondegenerate in the sense that it has distinct rows.)

If the sets $(\lambda_1, \cdots, \lambda_m)$, $(\mu_1, \cdots, \mu_m)$ coincide, we say that Λ given by (4.6.1) is a degenerate symbol of rank n and defect 0; in this case n is necessarily even and, the restriction of M to $\mathbf{W} \subset W_n$ (via one of the two imbeddings above) is a direct sum of two distinct irreducible $\mathbf{Q}[\mathbf{W}]$-modules, denoted $[\Lambda]_{s'}$, $[\Lambda]_{s''}$. We arrange the notation so that

$$[\Lambda]_{s'} = J_{\mathbf{W}'(s')}^{\mathbf{W}}(\text{sign}), \quad [\Lambda]_{s''} = J_{\mathbf{W}'(s'')}^{\mathbf{W}}(\text{sign}), \tag{4.6.2}$$

where $\mathbf{W}'(s') = \mathfrak{S}_{2\beta_1^*} \times \mathfrak{S}_{2\beta_2^*} \cdots \subset$ (subgroup of $\mathbf{W}$ generated by $S - \{s''\}) \approx \mathfrak{S}_n$ and $\beta_1^* \lneq \beta_2^* \le \cdots$ is the partition of $n/2$ dual to $\lambda_1 \le \lambda_2 - 1 \le \lambda_3 - 2 \le \cdots \le \lambda_m - m + 1$ (and similarly for s', s'' interchanged).

All irreducible $\mathbf{Q}[\mathbf{W}]$-modules are of the form $[\Lambda]$ with Λ nondegenerate, or of the form $[\Lambda]_{s'}$ or $[\Lambda]_{s''}$ with Λ degenerate. Moreover Λ is uniquely determined (up to a shift, as in 4.5), by the module. This parametrization of irreducible $\mathbf{Q}[\mathbf{W}]$-modules is independent of the choice of one of the two identifications of $\mathbf{W}$ with a subgroup of W_n, since the two identifications differ only by an inner automorphism of W_n. For example $\begin{bmatrix} n \\ 0 \end{bmatrix}$ is the unit representation and $\begin{bmatrix} 1, 2, \cdots, n \\ 0, 1, \cdots, n-1 \end{bmatrix}$ is the sign representation of $\mathbf{W}$.

If Λ is as in (4.6.1), we have

$$a_{[\Lambda]} = \sum_{1 \leq i < j \leq m} \inf(\lambda_i, \lambda_j) + \sum_{1 \leq i < j \leq m} \inf(\mu_i, \mu_j) + \sum_{1 \leq i, j \leq m} \inf(\lambda_i, \mu_j) - \frac{m}{6}(m-1)(4m-5) \tag{4.6.3}$$

where $[\Lambda]$ is interpreted as either $[\Lambda]_{s'}$ or $[\Lambda]_{s''}$ whenever Λ is degenerate. As shown in $[L_6, 5]$, $[\Lambda]_{s'}$ and $[\Lambda]_{s''}$ are always special; if Λ is nondegenerate, then $[\Lambda]$ is special if and only if

$$\begin{cases} \mu_1 \leq \lambda_2 \leq \mu_2 \leq \cdots \leq \lambda_m \quad \text{or if} \\ \lambda_1 \leq \mu_1 \leq \lambda_2 \leq \cdots \leq \mu_m . \end{cases} \tag{4.6.4}$$

Let $\mathbf{W}'$ be a maximal standard parabolic subgroup of type $A_{r-1} \times D_{\tilde{r}}$, $(\tilde{r} \geq 2,\ r = n - \tilde{r} \geq 1)$. The choice of s', s'' for $\mathbf{W}$ induces a similar choice for the $D_{\tilde{r}}$-factor of $\mathbf{W}'$. Consider an irreducible representation $[\Lambda']$ of the $D_{\tilde{r}}$-factor, (Λ' is a nondegenerate symbol of rank $\tilde{r}$ and defect 0), and let ε_r be the sign representation of the A_{r-1} factor. We have

$$J_{\mathbf{W}'}^{\mathbf{W}}(\varepsilon_r \otimes [\Lambda']) = \begin{cases} [\Lambda] \\ \text{or} \quad [\Lambda_1] + [\Lambda_2] \end{cases} \tag{4.6.5}$$

where Λ (or Λ_1, Λ_2) are obtained from Λ' in exactly the same way as in (4.5.4), by increasing the r largest entries of Λ' by 1; (we may assume, using shifts that, Λ' has at least r entries). Then Λ (or Λ_1, Λ_2) are necessarily nondegenerate. If $\tilde{r}$ is even and Λ' is a degenerate symbol of rank $\tilde{r}$ and defect 0, we have

$$(4.6.6)\qquad \begin{cases} J_{W'}^{W}(\varepsilon_r \otimes [\Lambda']_{s'}) = [\Lambda]_{s'} & \text{(for } r \text{ even)} \\ J_{W'}^{W}(\varepsilon_r \otimes [\Lambda']_{s'}) = [\Lambda] & \text{(for } r \text{ odd)} \end{cases}$$

where Λ is obtained from Λ' by increasing the r largest entries of Λ' by 1. The same is true with s' replaced by s''. (For example, if $[\Lambda']_{s'} = \begin{bmatrix} 2 \\ 2 \end{bmatrix}_{s'}$ and $r = 1$, then $[\Lambda] = \begin{bmatrix} 3 \\ 2 \end{bmatrix}$.)

Using (4.6.5), (4.6.6), (4.4.1) and the transitivity formula (4.1.8) we get an explicit description of the homomorphism $J_{W'}^{W} : \mathcal{R}(W') \to \mathcal{R}(W)$. To describe the homomorphisms $J_{W'(s')}^{W}$, $J_{W'(s'')}^{W}$, (see (4.6.2)), it is enough to describe their effect on the sign representation. (On irreducible representations other than sign, they can be computed by reducing to the case of W' considered above.) We have

$$(4.6.7)\qquad J_{W'(s')}^{W}(\text{sign}) = \begin{cases} \begin{bmatrix} 1, 2, \cdots, \frac{n}{2} \\ 1, 2, \cdots, \frac{n}{2} \end{bmatrix}_{s'}, & \text{if } n \text{ is even} \\ \begin{bmatrix} 1, 2, 3, \cdots, \frac{n+1}{2} \\ 0, 2, 3, \cdots, \frac{n+1}{2} \end{bmatrix}, & \text{if } n \text{ is odd} \end{cases}$$

and the same is true with s' replaced by s''.

Given a symbol Λ of rank n and defect 0, we choose t and define Λ' as in (4.5.5).

If Λ is nondegenerate, we have

(4.6.8) $$[\Lambda]\otimes\text{sign} = [\Lambda']$$

If Λ is degenerate, we have

(4.6.9) $$\{[\Lambda]_{s'}\otimes\text{sign},\ [\Lambda]_{s''}\otimes\text{sign}\} = \{[\Lambda']_{s'}, [\Lambda']_{s''}\}\,.$$

As in 4.5, we deduce the following description of families in $W^\vee$.

(4.6.10) *If* Λ *is a degenerate symbol of rank* n *and defect* 0, *then* $[\Lambda]_{s'}$ *forms a family by itself and* $[\Lambda]_{s''}$ *forms a family by itself. Let* Λ_1, Λ_2 *be two nondegenerate symbols of rank* n *and defect* 0. *Assume, as we may, by shifting, that* Λ_1, Λ_2 *have the same number of entries* 2m. *Then* $[\Lambda]$, $[\Lambda']$ *are in the same family if and only if* Λ *and* Λ' *contain exactly the same entries (arranged perhaps differently in rows).* (For example, $\left\{\begin{bmatrix}0 & 2\\ 1 & 3\end{bmatrix}, \begin{bmatrix}0 & 1\\ 2 & 3\end{bmatrix}, \begin{bmatrix}0 & 3\\ 1 & 2\end{bmatrix}\right\}$ is a family in $W^\vee$, for $n = 4$.)

It is clear that any family contains a unique special representation. (See (4.6.4).) If a family $\mathcal{F}$ is of form $\{[\Lambda]_{s'}\}$ or $\{[\Lambda]_{s''}\}$, (Λ degenerate), we associate to it the group $\mathcal{G}$ with one element and we have a unique bijection $\mathcal{F} \leftrightarrow \mathfrak{M}(\mathcal{G})$, (see 4.3).

If $\mathcal{F}$ is not of that form, then the special representation in $\mathcal{F}$ may be represented as $[Z]$, where Z is a nondegenerate symbol of rank n, defect 0 with, say 2m entries. We define $\mathfrak{M}_{\mathcal{F}}$ (resp. $\mathfrak{M}^-_{\mathcal{F}}$) to be the set of all arrays

(4.6.11) $$\begin{pmatrix}\lambda_1, \lambda_2, \cdots, \lambda_b\\ \\ \mu_1, \mu_2, \cdots, \mu_{b'}\end{pmatrix}$$

where $b - b' \equiv 0 \pmod 4$, (resp. $b - b' \equiv 2 \pmod 4$), $b + b' = 2m$, $0 \le \lambda_1 < \lambda_2 < \cdots < \lambda_b$, $0 \le \mu_1 < \mu_2 < \cdots < \mu_{b'}$ and $\{\lambda_1, \cdots, \lambda_b, \mu_1, \cdots, \mu_{b'}\}$ are exactly the entries of Z, perhaps in another order (some entries may be repeated twice); moreover the array (4.6.11) is regarded as identical with the array obtained by interchanging the two rows of (4.6.11). In the language of $[L_3]$, the elements of $\mathfrak{M}_{\mathcal{F}}$ (resp. $\mathfrak{M}_{\mathcal{F}}^-$ are symbols of rank n and defect $\equiv 0 \pmod 4$, (resp. $\equiv 2 \pmod 4$). For example, if $[Z] = \begin{bmatrix} 1 & 3 \\ 0 & 2 \end{bmatrix} \in \mathcal{F}$, then $\mathfrak{M}_{\mathcal{F}}$ consists of $\begin{pmatrix} 1 & 3 \\ 0 & 2 \end{pmatrix}, \begin{pmatrix} 0 & 1 \\ 2 & 3 \end{pmatrix}, \begin{pmatrix} 0 & 3 \\ 1 & 2 \end{pmatrix},$ $\begin{pmatrix} 0 & 1 & 2 & 3 \\ - & & & \end{pmatrix}$ and $\mathfrak{M}_{\mathcal{F}}^-$ consists of $\begin{pmatrix} 1 & 2 & 3 \\ 0 & & \end{pmatrix}, \begin{pmatrix} 0 & 2 & 3 \\ 1 & & \end{pmatrix}, \begin{pmatrix} 0 & 1 & 3 \\ 2 & & \end{pmatrix},$ $\begin{pmatrix} 0 & 1 & 2 \\ 3 & & \end{pmatrix}$. We may identify $\mathcal{F}$ with the subset of $\mathfrak{M}_{\mathcal{F}}$ defined by the condition that b, b′ above satisfy $b = b' = m$ (using $[\Lambda] \leftrightarrow \Lambda$). This gives an injection $\mathcal{F} \hookrightarrow \mathfrak{M}_{\mathcal{F}}$. An element of $\mathfrak{M}_{\mathcal{F}}$ (resp. $\mathfrak{M}_{\mathcal{F}}^-$ is of form $\Lambda_M = \begin{pmatrix} Z_2 \amalg (Z_1 - M) \\ Z_2 \amalg M \end{pmatrix}$, where Z_1, Z_2 are defined as in 4.5 and M is a subset of Z_1 such that $[M] \equiv d \pmod 2$, (resp. $|M| \equiv d+1 \pmod 2$); d is defined by $|Z_1| = 2d$. We have $\Lambda_M = \Lambda_{Z_1 - M}$. In particular, we have $Z = \Lambda_{M_0} = \Lambda_{M_0'}$ where $M_0 \amalg M_0'$ is the partition of Z_1 defined by the two rows of Z. (M_0, M_0' can be distinguished one from another by the inequality $\sum_{x \in M_0} x < \sum_{x \in M_0'} x$.) Let V_{Z_1}' be the set of all subsets of Z_1, with the F_2-vector space structure in which the sum is defined by the symmetric difference, taken modulo the line spanned by Z_1 itself. An element of V_{Z_1}' is just an unordered pair $(X, Z_1 - X)$, where $X \subset Z_1$. Let $(V_{Z_1}')^+$, (resp. $(V_{Z_1}')^-$) be the subset of V_{Z_1}' defined by the condition $|X| = $ even (resp. $|X| = $ odd). If we associate to M the set $M^\# \subset Z_1$ defined by $M^\# = (M \cup M_0) - (M \cap M_0)$, we get a bijection $\Lambda_M \leftrightarrow M^\#$ between $\mathfrak{M}_{\mathcal{F}}$ and $(V_{Z_1}')^+$ and between $\mathfrak{M}_{\mathcal{F}}^-$ and $(V_{Z_1}')^-$.

Let V_{Z_1} be the set of all subsets of even cardinality of Z_1; it is a vector space of dimension $2d-1$ over the field F_2 (under symmetric difference); note that $d > 0$ since Z is nondegenerate. Consider the nonsingular F_2-bilinear pairing $V_{Z_1}' \times V_{Z_1} \xrightarrow{(\,,\,)} F_2$ defined by $X_1, X_2 \to |X_1 \cap X_2| \bmod 2$. It defines a nonsingular symplectic form

$(V'_{Z_1})^+ \times (V'_{Z_1})^+ \xrightarrow{(,)} F_2$, since $(V'_{Z_1})^+$ is simultaneously a subspace of V'_{Z_1} and a quotient space of V_{Z_1}. The symplectic vector space $(V'_{Z_1})^+$, (of dimension $2d-2$) is a direct sum of two maximal isotropic subspaces $\bar{I}'$ and $\bar{I}''$, defined as follows. We arrange the elements of Z_1 in an increasing sequence; for $1 \le i \le 2d-1$, let e_i be the 2-element subset of Z_1 consisting of the i-th and the (i+1)-th element in this increasing sequence. Then $e_1, e_2, \cdots, e_{2d-1}$ is a basis of V_{Z_1} and their images $\bar{e}_i$ in the quotient space $(V'_{Z_1})^+$ of V_{Z_1} span that space and satisfy a single linear equation: $\bar{e}_1 + \bar{e}_3 + \bar{e}_5 + \cdots + \bar{e}_{2d-1} = 0$. Let $\bar{I}'$ be the subspace of $(V'_{Z_1})^+$ spanned by $\bar{e}_1, \bar{e}_3, \bar{e}_5, \cdots$ and let $\bar{I}''$ be the subspace of $(V'_{Z_1})^+$ spanned by $\bar{e}_2, \bar{e}_4, \cdots$. It is clear that $\bar{I}', \bar{I}''$ are maximal isotropic subspaces of $(V'_{Z_1})^+$ and that $(V'_{Z_1})^+ = \bar{I} \oplus \bar{I}''$.

We associate to $\mathcal{F}$ the finite group $\mathcal{G} = \bar{I}''$ (an F_2-vector space with basis $\bar{e}_2, \bar{e}_4, \cdots, \bar{e}_{2d-2}$). We identify $\bar{I}'$ with the set of characters $\bar{I}'' \to \bar{Q}_\ell^*$ by $i' \in \bar{I}' \leftrightarrow (i'' \to (-1)^{(i', i'')})$; this gives an isomorphism $\bar{I}'' \oplus \bar{I}' \approx \mathfrak{M}(\mathcal{G})$. Composing this with the injection (defined earlier) of $\mathcal{F}$ into $\mathfrak{M}_{\mathcal{F}} = (V'_{Z_1})^+ = \bar{I}'' \oplus \bar{I}'$, we get an injection $\mathcal{F} \hookrightarrow \mathfrak{M}(\mathcal{G})$, as stated in 4.3.

4.7. When $n \ge 5$, the choice of $s', s'' \in S$ made at the beginning of 4.6 is unique. However, when $n = 4$, that choice is not unique. In this case, the irreducible representations and the families of $\mathbf{W}$ (of type D_4) can be described without reference to this choice, as follows.

The unit representation is $\begin{bmatrix} 4 \\ 0 \end{bmatrix}$, a family by itself. The sign representation is $\begin{bmatrix} 0 & 1 & 2 & 3 \\ 1 & 2 & 3 & 4 \end{bmatrix}$, a family by itself. The reflection representation is $\begin{bmatrix} 3 \\ 1 \end{bmatrix}$, a family by itself. Its tensor product with sign is $\begin{bmatrix} 0 & 1 & 3 \\ 1 & 2 & 3 \end{bmatrix}$, a family by itself. (The notation is the one of 4.6 relative to a choice of s', s'' but it is independent of the choice.) To each of these families $\mathcal{F}$, we associate $\mathcal{G} = \{e\}$ and we have $\mathcal{F} \leftrightarrow \mathfrak{M}(\mathcal{G})$. The representations $\begin{bmatrix} 0 & 2 \\ 1 & 3 \end{bmatrix}, \begin{bmatrix} 0 & 1 \\ 2 & 3 \end{bmatrix}, \begin{bmatrix} 0 & 3 \\ 1 & 2 \end{bmatrix}$ form a family $\mathcal{F}$. (The notation is again independent of the choice of s', s''.) We associate to it $\mathcal{G} = \mathfrak{S}_2$ and we imbed $\mathcal{F} \hookrightarrow \mathfrak{M}(\mathcal{G})$ by $\begin{bmatrix} 0 & 2 \\ 1 & 3 \end{bmatrix} \to (1,1)$, $\begin{bmatrix} 0 & 1 \\ 2 & 3 \end{bmatrix} \to (g_2, 1)$, $\begin{bmatrix} 0 & 3 \\ 1 & 2 \end{bmatrix} \to (1, \varepsilon)$, with the notation for elements of $\mathfrak{M}(\mathcal{G})$, at the end of 4.5. (This coincides with the definition at the end of 4.6.)

The remaining 6 irreducible representations form families by themselves. They are: $J^{\mathbf{W}}_{<s_1,s_3>}(\text{sign})$, $J^{\mathbf{W}}_{<s_1,s_2>}(\text{sign})$, $J^{\mathbf{W}}_{<s_2,s_3>}(\text{sign})$, $J^{\mathbf{W}}_{<s_1,s_3,s_0>}(\text{sign})$, $J^{\mathbf{W}}_{<s_1,s_2,s_0>}(\text{sign})$, $J^{\mathbf{W}}_{<s_2,s_3,s_0>}(\text{sign})$ where $S = \{s_0, s_1, s_2, s_3\}$ and s_1, s_2, s_3 are the extremal points of the Coxeter diagram. We associate with each of these 6 families $\mathcal{G} = \{e\}$.

4.8. We now assume that $(\mathbf{W}, S)$ is of type G_2. Let $s_1 \in S$ be the reflection with respect to a short simple coroot and let $s_2 \in S$ be the reflection with respect to a long simple coroot. The unit representation is a family by itself; it has $a = 0$ and we set $\mathcal{G} = \{e\}$. The sign representation is a family by itself; it has $a = 6$ and we set $\mathcal{G} = \{e\}$. The remaining four irreducible representations have $a = 1$ and form a family $\mathcal{F}$. We associate to $\mathcal{F}$ the group $\mathcal{G} = \mathfrak{S}_3$ (symmetric group in 3 letters). We define an injection $\mathcal{F} \hookrightarrow \mathfrak{M}(\mathfrak{S}_3)$ as follows. The reflection representation V corresponds to $(1,1) \in \mathfrak{M}(\mathfrak{S}_3)$; the one-dimensional representation ε_1 defined by $s_1 \to 1$, $s_2 \to -1$ corresponds to $(1, r) \in \mathfrak{M}(\mathfrak{S}_3)$; the one-dimensional representation ε_2 defined by $s_1 \to -1$, $s_2 \to 1$ corresponds to $(g_3, 1) \in \mathfrak{M}(\mathfrak{S}_3)$; the 2-dimensional representation $V' = V \otimes \varepsilon_1 = V \otimes \varepsilon_2$ corresponds to $(g_2, 1) \in \mathfrak{M}(\mathfrak{S}_3)$. (We have used the notations in 4.3 for the elements in $\mathfrak{M}(\mathfrak{S}_3)$.) Each family contains exactly one special representation; the special representations are 1, sign, V. We have $J^{\mathbf{W}}_{<s_1>}(\text{sign}) = V + V' + \varepsilon_2$, $J^{\mathbf{W}}_{<s_2>}(\text{sign}) = V + V' + \varepsilon_1$, $J^{\mathbf{W}}_{<e>}(1) = 1$.

4.9. In the case where $(\mathbf{W}, S)$ is the type F_4 (resp. E_6, E_7, E_8) the character table of $\mathbf{W}$ has been determined by Kondo [Ko] (resp. by Frame

$[F_1]$, $[F_2]$). The corresponding formal degrees $D_E(u)$ have been determined by Benson-Grove-Surowski [BGS], Surowski [Su] and Benson [B]; in the process of doing this, they have worked out tables describing the decomposition of the restriction of irreducible representations of $\mathbf{W}$ to certain maximal standard parabolic subgroups. Alvis $[A_2]$ has prepared such tables for any maximal standard parabolic subgroup $\mathbf{W}'$ of $\mathbf{W}$. Using these tables and the knowledge of the a-function (4.1.1) which follows from the knowledge of the formal degrees, the function $J_{\mathbf{W}'}^{\mathbf{W}}$ (see (4.1.7)) can be determined in all cases.

4.10. Assume that $(\mathbf{W}, S)$ is of type F_4. Let s_1, s_2, s_3, s_4 be the elements of S numbered in such a way that s_1, s_4 are extremal points of the Coxeter diagram and also that s_1, s_2 correspond to short simple coroots and s_3, s_4 correspond to long simple coroots. There are precisely 25 irreducible $\mathbf{Q}[\mathbf{W}]$-modules (up to isomorphism). We shall identify them with their character. We denote δ_i the i-th character of degree δ in the list of Kondo [Ko]. (Kondo uses the notation d, a, τ, $\tau\sigma$ instead of s_4, s_3, s_2, s_1 for the elements of S.)

(4.10.1) We now list the 11 families, the values of the a-function (see (4.1.1)), which is constant on each family, and the corresponding finite groups $\mathcal{G}$.

family	a-*function*	$\mathcal{G}$
$\{1_1\}$	0	$\{e\}$
$\{1_4\}$	24	$\{e\}$
$\{4_2, 2_1, 2_3\}$	1	$\mathfrak{S}_2$
$\{4_5, 2_4, 2_2\}$	13	$\mathfrak{S}_2$
$\{9_1\}$	2	$\{e\}$
$\{9_4\}$	10	$\{e\}$
$\{8_1\}$	3	$\{e\}$
$\{8_2\}$	9	$\{e\}$
$\{8_3\}$	3	$\{e\}$
$\{8_4\}$	9	$\{e\}$
$\{12_1, 9_2, 9_3, 1_2, 1_3, 4_1, 4_3, 4_4, 6_1, 6_2, 16_1\}$	4	$\mathfrak{S}_4$.

Each family contains a unique special representation: the first one listed.

We define an imbedding of a family $\mathcal{F}$ into the corresponding $\mathfrak{M}(\mathcal{G})$, as follows. The imbedding is the obvious one when $\mathcal{G} = \{e\}$. If to $\mathcal{F}$ corresponds $\mathcal{G} = \mathfrak{G}_2$, we associate to the three elements of $\mathcal{F}$ (in the order listed), the elements $(1,1), (g_2,1), (1,\varepsilon) \in \mathfrak{M}(\mathfrak{G}_2)$, respectively. If $\mathcal{F}$ is the family containing 12_1, the imbedding $\mathcal{F} \hookrightarrow \mathfrak{M}(\mathfrak{G}_4)$ is defined by:

$$\left.\begin{array}{c} 12_1 \to (1,1),\; 9_3 \to (1,\lambda^1),\; 9_2 \to (g_2',1), \\ 1_3 \to (1,\lambda^2),\; 1_2 \to (g_2',\varepsilon'),\; 4_1 \to (g_2',\varepsilon''), \\ 4_4 \to (g_2,\varepsilon''),\; 4_3 \to (g_4,1),\; 6_1 \to (g_3,1),\; 6_2 \to (1,\sigma), \\ 16_1 \to (g_2,1)\,. \end{array}\right\} \tag{4.10.2}$$

We now describe the various irreducible representations (or characters) of $\mathbf{W}$. 1_1 is the unit character, 1_4 is the sign character, 1_2 is the one-dimensional character which is -1 on s_1, s_2 and $+1$ on $s_3, s_4, 1_3 = 1_2 \otimes \mathrm{sign}$. Moreover, 4_2 is the reflection character, $4_5 = 4_2 \otimes \mathrm{sign}$, $4_3 = 4_2 \otimes 1_2$, $4_4 = 4_2 \otimes 1_3$. The character 2_1 is that of the 2-dimensional irreducible representation on which s_1, s_2 act by reflections and s_3, s_4 act trivially; moreover, $2_2 = 2_1 \otimes \mathrm{sign}$. The character 2_3 is that of the 2-dimensional irreducible representation on which s_1, s_2 act trivially and s_3, s_4 act by reflections; moreover, $2_4 = 2_3 \otimes \mathrm{sign}$. The character 9_1 can be described as the second symmetric power of 4_2, from which we subtract the unit character; we have $9_4 = 9_1 \otimes \mathrm{sign}$, $9_2 = 9_1 \otimes 1_2$, $9_3 = 9_1 \otimes 1_3$. We have $8_1 = 4_2 \otimes 2_1$, $8_2 = 8_1 \otimes \mathrm{sign}$, $8_3 = 4_2 \otimes 2_3$, $8_4 = 8_3 \otimes \mathrm{sign}$, $4_1 = 2_1 \otimes 2_3$. The character 6_2 is that of the second exterior power of 4_2 and $6_1 = 6_2 \otimes 1_2 = 6_2 \otimes 1_3$. Finally, we have $16_1 = 4_1 \otimes 4_2$, $12_1 = 6_1 \otimes 2_1$.

(4.10.3) We have: $1_1 = J_{\langle e\rangle}^{\mathbf{W}}(1)$, $4_2 = J_{\langle s_2,s_3\rangle}^{\mathbf{W}}(A)$, $2_1 = J_{\langle s_2,s_3\rangle}^{\mathbf{W}}(B)$, $2_3 = J_{\langle s_2,s_3\rangle}^{\mathbf{W}}(C)$ where A is the reflection representation of the Weyl subgroup $\langle s_2,s_3\rangle$, (of type B_2), B is its character $s_3 \to -1$, $s_2 \to 1$

and C is its character $s_3 \to 1$, $s_2 \to -1$. We have: $9_1 = J^{\mathbf{W}}_{\langle s_1,s_4\rangle}(\text{sign})$, $8_1 = J^{\mathbf{W}}_{\langle s_1,s_2\rangle}(\text{sign})$, $8_3 = J^{\mathbf{W}}_{\langle s_3,s_4\rangle}(\text{sign})$, $8_2 = J^{\mathbf{W}}_{\langle s_2,s_3,s_4\rangle}(\text{sign})$, $8_4 = J^{\mathbf{W}}_{\langle s_1,s_2,s_3\rangle}(\text{sign})$.

(4.10.4) Let $\mathbf{W}_i$ $(1 \le i \le 4)$ be the standard parabolic subgroup of $\mathbf{W}$ generated by $S - \{s_i\}$. Then, with the notation in 4.5, we have

$$J^{\mathbf{W}}_{\mathbf{W}_1}\begin{bmatrix}0 & & 1 & & 3\\ & 1 & & 2 & \end{bmatrix} = 12_1 + 6_2 + 9_3 + 16_1$$

$$J^{\mathbf{W}}_{\mathbf{W}_1}\begin{bmatrix}1 & & 2 & & 3\\ & 0 & & 1 & \end{bmatrix} = 1_3 + 9_3 + 4_4$$

$$J^{\mathbf{W}}_{\mathbf{W}_1}\begin{bmatrix}0 & & 1 & & 2\\ & 1 & & 3 & \end{bmatrix} = 4_1 + 9_2 + 16_1$$

$$J^{\mathbf{W}}_{\mathbf{W}_4}\begin{bmatrix}0 & & 1 & & 3\\ & 1 & & 2 & \end{bmatrix} = 12_1 + 6_2 + 9_2 + 16_1$$

$$J^{\mathbf{W}}_{\mathbf{W}_4}\begin{bmatrix}1 & & 2 & & 3\\ & 0 & & 1 & \end{bmatrix} = 1_2 + 9_2 + 4_3$$

$$J^{\mathbf{W}}_{\mathbf{W}_4}\begin{bmatrix}0 & & 1 & & 2\\ & 1 & & 3 & \end{bmatrix} = 4_1 + 9_3 + 16_1$$

$$J^{\mathbf{W}}_{\mathbf{W}_3}(\text{sign}) = 12_1 + 6_1 + 9_2 + 4_3 + 16_1$$

$$J^{\mathbf{W}}_{\mathbf{W}_2}(\text{sign}) = 12_1 + 6_1 + 9_3 + 4_4 + 16_1 .$$

4.11. We now assume that $(\mathbf{W}, S)$ is of type E_6. We shall number the elements of $S = \{s_1,s_2,s_3,s_4,s_5,s_6\}$ in such a way that s_1s_3, s_3s_4, s_4s_5, s_5s_6, s_2s_4 have order 3. There are (up to isomorphism) precisely 25 irreducible $\mathbf{Q}[\mathbf{W}]$-modules. Following [F_1], $1_p, 6_p, 20_p, 30_p, 15_q, 15_p, 64_p, 60_p, 81_p, 24_p, 80_s, 60_s, 90_s, 10_s, 20_s$ denote irreducible $\mathbf{Q}[\mathbf{W}]$-modules of degree $1, 6, \cdots, 10, 20$. A prime (′) attached to such a character means that character tensored with sign. The characters $80_s, 60_s, 90_s, 10_s, 20_s$ are unchanged by tensoring with sign.

(4.11.1) We now list the 17 families, the values of the a-function (see (4.1.1)), which is constant on each family and the corresponding finite groups $\mathcal{G}$.

family	a-*function*	$\mathcal{G}$
$\{1_p\}$	0	$\{e\}$
$\{1'_p\}$	36	$\{e\}$
$\{6_p\}$	1	$\{e\}$
$\{6'_p\}$	25	$\{e\}$
$\{20_p\}$	2	$\{e\}$
$\{20'_p\}$	20	$\{e\}$
$\{30_p,15_q,15_p\}$	3	$\mathfrak{S}_2$
$\{30'_p,15'_q,15'_p\}$	15	$\mathfrak{S}_2$
$\{64_p\}$	4	$\{e\}$
$\{64'_p\}$	13	$\{e\}$
$\{60_p\}$	5	$\{e\}$
$\{60'_p\}$	11	$\{e\}$
$\{81_p\}$	6	$\{e\}$
$\{81'_p\}$	10	$\{e\}$
$\{24_p\}$	6	$\{e\}$
$\{24'_p\}$	12	$\{e\}$
$\{80_s,60_s,90_s,10_s,20_s\}$	7	$\mathfrak{S}_3$.

Each family contains a unique special representation: the first one listed. We define an imbedding of a family $\mathcal{F}$ into the corresponding $\mathfrak{M}(\mathcal{G})$ as follows. The imbedding is the obvious one when $\mathcal{G} = \{e\}$. If to $\mathcal{F}$ corresponds $\mathcal{G} = \mathfrak{S}_2$, we associate to the three elements of F (in the order listed), the elements $(1,1),(g_2,1),(1,\varepsilon) \in M(\mathfrak{S}_2)$, respectively. If $\mathcal{F}$ is the family containing 80_s, the imbedding $\mathcal{F} \hookrightarrow \mathfrak{M}(\mathfrak{S}_3)$ is defined by:

(4.11.2) $80_s \to (1,1), 60_s \to (g_2,1), 90_s \to (1,r), 10_s \to (g_3,1), 20_s \to (1,\varepsilon)$.

(4.11.3) We have:

$$1_p = J^W_{<e>}(1), \quad 6_p = J^W_{A_1}(\text{sign}), \quad 20_p = J^W_{A_1\times A_1}(\text{sign}),$$

$$20'_p = J^W_{D_5}(\text{sign}), \quad 30_p = J^W_{D_4}\begin{bmatrix}0 & 2\\ 1 & 3\end{bmatrix}, \quad 15_q = J^W_{D_4}\begin{bmatrix}0 & 1\\ 2 & 3\end{bmatrix},$$

$$15_p = J^W_{D_4}\begin{bmatrix}0 & 3\\ 1 & 2\end{bmatrix}, \quad 64_p = J^W_{A_2\times A_2}(\text{sign}), \quad 64'_p = J^W_{D_5}\begin{bmatrix}1 & 2 & 3 & 4\\ 0 & 1 & 2 & 4\end{bmatrix},$$

$$60_p = J^W_{A_2\times A_1\times A_1}(\text{sign}), \quad 60'_p = J^W_{A_4\times A_1}(\text{sign}), \quad 81_p = J^W_{A_3}(\text{sign}),$$

$$81'_p = J^W_{A_4}(\text{sign}), \quad 24_p = J^W_{A_2\times A_2}(\text{sign}), \quad 24'_p = J^W_{D_4}(\text{sign}).$$

(Here the indices $A_1, A_1\times A_1, D_5, \cdots, D_4$ denote any standard parabolic of W of the indicated type.)

(4.11.4) With a similar convention, we have

$$J^W_{D_5}\begin{bmatrix}1 & 3 & 4\\ 0 & 1 & 2\end{bmatrix} = 60_s$$

$$J^W_{D_5}\begin{bmatrix}1 & 2 & 4\\ 0 & 1 & 3\end{bmatrix} = 80_s + 90_s$$

$$J^W_{D_5}\begin{bmatrix}1 & 2 & 3\\ 0 & 1 & 4\end{bmatrix} = 90_s + 20_s$$

$$J^W_{A_2\times A_2\times A_1}(\text{sign}) = 80_s + 60_s + 10_s$$

$$J^W_{A_3\times A_1}(\text{sign}) = 80_s + 90_s + 60_s .$$

4.12. Assume now that (W, S) is of type E_7. We shall number the elements of $S = \{s_1, s_2, s_3, s_4, s_5, s_6, s_7\}$ in such a way that s_1s_3, s_3s_4, s_4s_5, s_5s_6, s_6s_7, s_2s_4 have order 3. We shall label the 60 irreducible $Q[W]$-modules (or characters) as in [BL, 4]. In particular, the longest element w_0 acts as $+1$ on representations labelled without a prime and as

−1 on those labelled with a prime. Two representations which have the same label except for the prime, are obtained one from another by tensoring with sign.

(4.12.1) We now list the 35 families, the values of the a-function (see (4.1.1)), which is constant on each family, and the corresponding finite groups $\mathcal{G}$.

family	*a-function*	$\mathcal{G}$
$\{1_a\}$	0	$\{e\}$
$\{1'_a\}$	63	$\{e\}$
$\{7'_a\}$	1	$\{e\}$
$\{7_a\}$	46	$\{e\}$
$\{27_a\}$	2	$\{e\}$
$\{27'_a\}$	37	$\{e\}$
$\{56'_a,35_b,21_a\}$	3	$\mathfrak{G}_2$
$\{56_a,35'_b,21'_a\}$	30	$\mathfrak{G}_2$
$\{21'_b\}$	3	$\{e\}$
$\{21_b\}$	36	$\{e\}$
$\{120_a,15'_a,105'_a\}$	4	$\mathfrak{G}_2$
$\{120'_a,15_a,105_a\}$	25	$\mathfrak{G}_2$
$\{189'_b\}$	5	$\{e\}$
$\{189_b\}$	22	$\{e\}$
$\{210_a\}$	6	$\{e\}$
$\{210'_a\}$	21	$\{e\}$
$\{105_b\}$	6	$\{e\}$
$\{105'_b\}$	21	$\{e\}$
$\{168_a\}$	6	$\{e\}$
$\{168'_a\}$	21	$\{e\}$
$\{189'_c\}$	7	$\{e\}$
$\{189_c\}$	20	$\{e\}$
$\{315'_a,280_b,280'_a,70'_a,35'_a\}$	7	$\mathfrak{G}_3$
$\{315_a,280'_b,280_a,70_a,35_a\}$	16	$\mathfrak{G}_3$

(4.12.1) Continued

family	*a-function*	$\mathcal{G}$
$\{405_a, 216'_a, 189_a\}$	8	$\mathfrak{S}_2$
$\{405'_a, 216_a, 189'_a\}$	15	$\mathfrak{S}_2$
$\{378'_a\}$	9	$\{e\}$
$\{378_a\}$	14	$\{e\}$
$\{210_b\}$	10	$\{e\}$
$\{210'_b\}$	13	$\{e\}$
$\{420_a, 84_a, 336'_a\}$	10	$\mathfrak{S}_2$
$\{420'_a, 84'_a, 336_a\}$	13	$\mathfrak{S}_2$
$\{105_c\}$	12	$\{e\}$
$\{105'_c\}$	15	$\{e\}$
$\{512'_a, 512_a\}$	11	$\mathfrak{S}_2$.

Each family contains a unique special representation: the first one listed.

We define an imbedding of a family $\mathcal{F}$ in the corresponding $\mathcal{M}(\mathcal{G})$ as follows. The imbedding is the obvious one when $\mathcal{G} = \{e\}$. If to $\mathcal{F}$ corresponds $\mathcal{G} = \mathfrak{S}_2$ and $\mathcal{F} \neq \{512'_a, 512_a\}$, we associate to the 3 elements of $\mathcal{F}$ (in the order listed), the elements $(1,1), (g_2,1), (1,\varepsilon) \in \mathcal{M}(\mathfrak{S}_2)$, respectively. If $\mathcal{F} = \{512'_a, 512_a\}$ the imbedding $\mathcal{F} \hookrightarrow \mathcal{M}(\mathfrak{S}_2)$ is defined by $512'_a \to (1,1)$, $512_a \to (1,\varepsilon)$. If to $\mathcal{F}$ corresponds $\mathcal{G} = \mathfrak{S}_3$, we associate to the 5 elements of $\mathcal{F}$ (in the order listed), the elements $(1,1), (g_2,1), (1,r), (g_3,1), (1,\varepsilon) \in \mathcal{M}(\mathfrak{S}_3)$, respectively.

(4.12.2) We have, with conventions as in the case of E_6, (see 4.11):

$$1_a = J_{\langle e\rangle}^{W}(1),\ 7'_a = J_{A_1}^{W}(\text{sign}),\ 27_a = J_{A_1\times A_1}^{W}(\text{sign}),$$

$$56'_a = J_{D_4}^{W}\begin{bmatrix}0 & 2\\ 1 & 3\end{bmatrix},\ 35_b = J_{D_4}^{W}\begin{bmatrix}0 & 1\\ 2 & 3\end{bmatrix},\ 21_a = J_{D_4}^{W}\begin{bmatrix}0 & 3\\ 1 & 2\end{bmatrix},$$

$$\{21'_b\} = J_{W'}^{W}(\text{sign}),\ (W' = \langle s_2, s_5, s_7\rangle),\ 21_b = J_{E_6}^{W}(\text{sign}),$$

$$120_a = J_{D_4\times A_1}^{W}\left(\begin{bmatrix}0 & 2\\ 1 & 3\end{bmatrix} \boxtimes \text{sign}\right),\ 15'_a = J_{D_4\times A_1}^{W}\left(\begin{bmatrix}0 & 1\\ 2 & 3\end{bmatrix} \boxtimes \text{sign}\right),$$

$105'_a = J^W_{D_4 \times A_1}\left(\begin{bmatrix} 0 & 3 \\ 1 & 2 \end{bmatrix} \boxtimes \text{sign}\right)$, $189'_b = J^W_{A_2 \times A_1 \times A_1}(\text{sign})$,

$210_a = J^W_{A_3}(\text{sign})$, $210'_a = J^W_{D_6}\begin{bmatrix} 1 & 2 & 3 & 4 & 5 \\ 0 & 1 & 2 & 3 & 5 \end{bmatrix}$,

$105_b = J^W_{A_2 \times A_1 \times A_1 \times A_1}(\text{sign})$, $105'_b = J^W_{A_6}(\text{sign})$,

$168_a = J^W_{A_2 \times A_2}(\text{sign})$, $168'_a = J^W_{D_5 \times A_1}(\text{sign})$, $189'_c = J^W_{W'}(\text{sign})$,

$(W' = \langle s_2, s_5, s_6, s_7 \rangle)$, $189_c = J^W_{D_5}(\text{sign})$, $315'_a = J^W_{E_6}(80_s)$, $280_b = J^W_{E_6}(60_s)$,

$280'_a = J^W_{E_6}(90_s)$, $70'_a = J^W_{E_6}(10_s)$, $35'_a = J^W_{E_6}(20_s)$,

$405_a = J^W_{D_5 \times A_1}\left(\begin{bmatrix} 1 & 2 & 4 \\ 0 & 1 & 3 \end{bmatrix} \boxtimes \text{sign}\right)$, $216'_a = J^W_{D_5 \times A_1}\left(\begin{bmatrix} 1 & 3 & 4 \\ 0 & 1 & 2 \end{bmatrix} \boxtimes \text{sign}\right)$,

$189_a = J^W_{D_5 \times A_1}\left(\begin{bmatrix} 1 & 2 & 3 \\ 0 & 1 & 4 \end{bmatrix} \boxtimes \text{sign}\right)$, $405'_a = J^W_{E_6}(30'_p)$,

$216_a = J^W_{E_6}(15'_q)$, $189'_a = J^W_{E_6}(15'_p)$, $378'_a = J^W_{A_3 \times A_2}(\text{sign})$,

$378_a = J^W_{D_5 \times A_1}\left(\begin{bmatrix} 1 & 2 & 3 & 4 \\ 0 & 1 & 2 & 4 \end{bmatrix} \boxtimes \text{sign}\right)$, $210_b = J^W_{A_3 \times A_2 \times A_1}(\text{sign})$,

$210'_b = J^W_{A_4 \times A_2}(\text{sign})$, $420_a = J^W_{D_6}\begin{bmatrix} 1 & 2 & 4 \\ 0 & 2 & 3 \end{bmatrix}$, $84_a = J^W_{D_6}\begin{bmatrix} 2 & 3 & 4 \\ 0 & 1 & 2 \end{bmatrix}$,

$336'_a = J^W_{D_6}\begin{bmatrix} 1 & 2 & 3 \\ 0 & 2 & 4 \end{bmatrix}$, $420'_a = J^W_{D_6}\begin{bmatrix} 1 & 2 & 3 & 5 \\ 0 & 1 & 2 & 4 \end{bmatrix}$,

$84'_a = J^W_{D_6}\begin{bmatrix} 1 & 2 & 4 & 5 \\ 0 & 1 & 2 & 3 \end{bmatrix}$, $336_a = J^W_{D_6}\begin{bmatrix} 1 & 2 & 3 & 4 \\ 0 & 1 & 2 & 5 \end{bmatrix}$, $105_c = J^W_{D_4}(\text{sign})$,

$105'_c = J^W_{A_5}(\text{sign})$.

We have also: $J^W_{A_4 \times A_1}(\text{sign}) = 512'_a + 512_a$.

4.13. Assume now that (W, S) is of type E_8. Let $s_1, s_2, s_3, s_4, s_5, s_6, s_7, s_8$ be the elements of S numbered so that the products $s_1 s_3$, $s_3 s_4$ $s_4 s_5$, $s_5 s_6$, $s_6 s_7$, $s_7 s_8$, $s_4 s_2$ have order 3. We shall label the 112 irreducible $\mathbf{Q}[W]$-modules (or characters) as in [F_2]. In particular, representations whose label contains a subscript x or y (resp. z or w) are acted on as +1 (resp. as −1) by w_0. Two representations whose label differ only by a prime, are obtained one from another by tensoring with sign. The representations whose label contains y or w are unchanged under tensoring with sign.

(4.13.1) We now list the 46 families, the values of the a-function, (see (4.1.1)), which is constant on each family, and the corresponding finite groups $\mathcal{G}$.

family	*a-function*	$\mathcal{G}$
$\{1_x\}$	0	$\{e\}$
$\{1'_x\}$	120	$\{e\}$
$\{8_z\}$	1	$\{e\}$
$\{8'_z\}$	91	$\{e\}$
$\{35_x\}$	2	$\{e\}$
$\{35'_x\}$	74	$\{e\}$
$\{112_z, 84_x, 28_x\}$	3	$\mathfrak{G}_2$
$\{112'_z, 84'_x, 28'_x\}$	63	$\mathfrak{G}_2$
$\{210_x, 50_x, 160_z\}$	4	$\mathfrak{G}_2$
$\{210'_x, 50'_x, 160'_z\}$	52	$\mathfrak{G}_2$
$\{560_z\}$	5	$\{e\}$
$\{560'_z\}$	47	$\{e\}$
$\{567_x\}$	6	$\{e\}$
$\{567'_x\}$	46	$\{e\}$
$\{700_x, 400_z, 300_x\}$	6	$\mathfrak{G}_2$
$\{700'_x, 400'_z, 300'_x\}$	42	$\mathfrak{G}_2$
$\{1400_z, 1344_x, 1008_z, 448_z, 56_z\}$	7	$\mathfrak{G}_3$
$\{1400'_z, 1344'_x, 1008'_z, 448'_z, 56'_z\}$	37	$\mathfrak{G}_3$
$\{1400_x, 1050_x, 1575_x, 175_x, 350_x\}$	8	$\mathfrak{G}_3$
$\{1400'_x, 1050'_x, 1575'_x, 175'_x, 350'_x\}$	32	$\mathfrak{G}_3$
$\{3240_z\}$	9	$\{e\}$
$\{3240'_z\}$	31	$\{e\}$

(4.13.1) Continued

family	*a-function*	$\mathcal{G}$
$\{2268_x, 972_x, 1296_z\}$	10	$\mathfrak{G}_2$
$\{2268'_x, 972'_x, 1296'_z\}$	30	$\mathfrak{G}_2$
$\{2240_x, 1400_{zz}, 840_z\}$	10	$\mathfrak{G}_2$
$\{2240'_x, 1400'_{zz}, 840'_z\}$	28	$\mathfrak{G}_2$
$\{4096_z, 4096_x\}$	11	$\mathfrak{G}_2$
$\{4096'_x, 4096'_z\}$	26	$\mathfrak{G}_2$
$\{525_x\}$	12	$\{e\}$
$\{525'_x\}$	36	$\{e\}$
$\{4200_x, 840_x, 3360_z\}$	12	$\mathfrak{G}_2$
$\{4200'_x, 840'_x, 3360'_z\}$	24	$\mathfrak{G}_2$
$\{2800_z, 700_{xx}, 2100_x\}$	13	$\mathfrak{G}_2$
$\{2800'_z, 700'_{xx}, 2100'_x\}$	25	$\mathfrak{G}_2$
$\{4536_z\}$	13	$\{e\}$
$\{4536'_z\}$	23	$\{e\}$
$\{2835_x\}$	14	$\{e\}$
$\{2835'_x\}$	22	$\{e\}$
$\{6075_x\}$	14	$\{e\}$
$\{6075'_x\}$	22	$\{e\}$
$\{4200_z\}$	15	$\{e\}$
$\{4200'_z\}$	21	$\{e\}$
$\{5600_z, 3200_x, 2400_z\}$	15	$\mathfrak{G}_2$
$\{5600'_z, 3200'_x, 2400'_z\}$	21	$\mathfrak{G}_2$
$\{2100_y\}$	20	$\{e\}$
$\{4480_y, 3150_y, 4200_y, 4536_y, 5670_y, 420_y, 1134_y, 1400_y, 2688_y, 1680_y, 168_y, 70_y, 7168_w, 1344_w, 2016_w, 5600_w, 448_w\}$	16	$\mathfrak{G}_5$.

Each family contains a unique special representation: the first one listed. We define an imbedding of a family $\mathfrak{F}$ in the corresponding $\mathfrak{M}(\mathcal{G})$ as follows. If $\mathcal{G} = \{e\}$, the imbedding is the obvious one. If $\mathcal{G} = \mathfrak{G}_2$ and

$\mathcal{F}$ has three elements, we associate to the three elements of $\mathcal{F}$ (in the order listed) the elements $(1,1),(g_2,1),(1,\varepsilon) \in \mathfrak{M}(\mathfrak{G}_2)$, respectively. If $\mathcal{G} = \mathfrak{G}_2$ and $\mathcal{F}$ has two elements, we associate to the two elements of $\mathcal{F}$ (in the order listed) the elements $(1,1),(1,\varepsilon) \in \mathfrak{M}(\mathfrak{G}_2)$. If to $\mathcal{F}$ corresponds $\mathcal{G} = \mathfrak{G}_3$, we associate to the 5 elements of $\mathcal{F}$ (in the order listed), the elements $(1,1),(g_2,1),(1,r),(g_3,1),(1,\varepsilon) \in \mathfrak{M}(\mathfrak{G}_3)$, respectively.

(4.13.2) If $\mathcal{F}$ is the family containing 4480_y, the imbedding $\mathcal{F} \hookrightarrow \mathfrak{M}(\mathfrak{G}_5)$ is defined by:

$$
\begin{aligned}
4480_y &\to (1,1)\\
3150_y &\to (g_3,1)\\
4200_y &\to (g_2',1)\\
4536_y &\to (1,\nu)\\
5670_y &\to (1,\lambda^1)\\
420_y &\to (g_5,1)\\
1134_y &\to (g_3,\varepsilon)\\
1400_y &\to (1,\nu')\\
2688_y &\to (g_2',\varepsilon'')\\
1680_y &\to (1,\lambda^2)\\
168_y &\to (g_2',\varepsilon')\\
70_y &\to (1,\lambda^3)\\
7168_w &\to (g_2,1)\\
1344_w &\to (g_4,1)\\
2016_w &\to (g_6,1)\\
5600_w &\to (g_2,r)\\
448_w &\to (g_2,\varepsilon)\ .
\end{aligned}
$$

(4.13.3) We have:

$$1_x = J_{<e>}^{W}(1),\ 8_z = J_{A_1}^{W}(\text{sign}),\ 35_x = J_{A_1\times A_1}^{W}(\text{sign}),$$

$$112_z = J_{D_4}^{W}\begin{bmatrix}0 & 2\\ 1 & 3\end{bmatrix},\ 84_x = J_{D_4}^{W}\begin{bmatrix}0 & 1\\ 2 & 3\end{bmatrix},\ 28_x = J_{D_4}^{W}\begin{bmatrix}0 & 3\\ 1 & 2\end{bmatrix},$$

$$210_x = J^{W}_{D_4\times A_1}\left(\begin{bmatrix}0&2\\1&3\end{bmatrix}\boxtimes \mathrm{sign}\right),\quad 50_x = J^{W}_{D_4\times A_1}\left(\begin{bmatrix}0&1\\2&3\end{bmatrix}\boxtimes \mathrm{sign}\right),$$

$$160_z = J^{W}_{D_4\times A_1}\left(\begin{bmatrix}0&3\\1&2\end{bmatrix}\boxtimes \mathrm{sign}\right),\quad 560_z = J^{W}_{A_2\times A_1\times A_1}(\mathrm{sign}),$$

$$567_x = J^{W}_{A_3}(\mathrm{sign}),\quad 567'_x = J^{W}_{E_7}(7_a),\quad 700_x = J^{W}_{D_4\times A_2}\left(\begin{bmatrix}1&3\\0&2\end{bmatrix}\boxtimes \mathrm{sign}\right),$$

$$400_z = J^{W}_{D_4\times A_2}\left(\begin{bmatrix}2&3\\0&1\end{bmatrix}\boxtimes \mathrm{sign}\right),\quad 300_x = J^{W}_{D_4\times A_2}\left(\begin{bmatrix}1&2\\0&4\end{bmatrix}\boxtimes \mathrm{sign}\right),$$

$$1400_z = J^{W}_{E_6}(80_s),\quad 1344_x = J^{W}_{E_6}(60_s),\quad 1008_z = J^{W}_{E_6}(90_s),$$

$$448_z = J^{W}_{E_6}(10_s),\quad 56_z = J^{W}_{E_6}(20_s),\quad 1400_x = J^{W}_{E_6\times A_1}(80_s\boxtimes \mathrm{sign}),$$

$$1050_x = J^{W}_{E_6\times A_1}(60_s\boxtimes \mathrm{sign}),\quad 1575_x = J^{W}_{E_6\times A_1}(90_s\boxtimes \mathrm{sign}),$$

$$175_x = J^{W}_{E_6\times A_1}(10_s\boxtimes \mathrm{sign}),\quad 350_x = J^{W}_{E_6\times A_1}(20_s\boxtimes \mathrm{sign}),$$

$$3240_z = J^{W}_{A_3\times A_2}(\mathrm{sign}),\quad 3240'_z = J^{W}_{D_4}\begin{bmatrix}1&2&3&4&5&6\\0&1&2&3&4&6\end{bmatrix},$$

$$2268_x = J^{W}_{D_6}\begin{bmatrix}1&2&3&5\\0&1&2&4\end{bmatrix},\quad 972_x = J^{W}_{D_6}\begin{bmatrix}1&2&4&5\\0&1&2&3\end{bmatrix},$$

$$1296_z = J^{W}_{D_6}\begin{bmatrix}1&2&3&4\\0&1&2&5\end{bmatrix},\quad 2268'_x = J^{W}_{E_7}(56_a),\quad 972'_x = J^{W}_{E_7}(35'_b),$$

$$1296'_z = J^{W}_{E_7}(21'_a),\quad 2240_x = J^{W}_{D_5\times A_2}\left(\begin{bmatrix}1&2&4\\0&1&3\end{bmatrix}\boxtimes \mathrm{sign}\right),$$

$$1400_{zz} = J^{W}_{D_5\times A_2}\left(\begin{bmatrix}1&3&4\\0&1&2\end{bmatrix}\boxtimes \mathrm{sign}\right),\quad 840_z = J^{W}_{D_5\times A_2}\left(\begin{bmatrix}1&2&3\\0&1&4\end{bmatrix}\boxtimes \mathrm{sign}\right),$$

$$4096_z = J^{W}_{E_7}(512'_a),\quad 4096_z = J^{W}_{E_7}(512_a),\quad 525_x = J^{W}_{D_4}(\mathrm{sign}),$$

$$525'_x = J^{W}_{E_6}(\mathrm{sign}),\quad 4200_x = J^{W}_{D_7}\begin{bmatrix}1&3&4\\0&2&3\end{bmatrix},\quad 840_x = J^{W}_{D_7}\begin{bmatrix}2&3&4\\0&3&4\end{bmatrix},$$

$$3360_z = J^{W}_{D_7}\begin{bmatrix}1&2&3\\0&3&4\end{bmatrix},\quad 2800_z = J^{W}_{D_6}\begin{bmatrix}1&2&4\\0&2&3\end{bmatrix},\quad 700_{xx} = J^{W}_{D_6}\begin{bmatrix}2&3&4\\0&1&2\end{bmatrix},$$

$2100_x = J_{D_6}^{W}\begin{bmatrix}1 & 2 & 3\\0 & 2 & 4\end{bmatrix}$, $2800'_z = J_{E_7}^{W}(120'_a)$, $700'_{xx} = J_{E_7}^{W}(15_a)$,

$2100'_x = J_{E_7}^{W}(105_a)$, $4536_z = J_{A_4\times A_2}^{W}(\text{sign})$,

$4536'_z = J_{D_5\times A_2}^{W}(\text{sign})$, $2835_x = J_{A_4\times A_2\times A_1}^{W}(\text{sign})$,

$2835'_x = J_{A_6\times A_1}^{W}(\text{sign})$, $6075_x = J_{D_5\times A_1}^{W}\left(\begin{bmatrix}1 & 2 & 3 & 4\\0 & 1 & 2 & 4\end{bmatrix}\boxtimes \text{sign}\right)$,

$6075'_x = J_{E_7}^{W}(189_b)$, $4200_z = J_{D_4\times A_2}^{W}(\text{sign})$,

$4200'_z = J_{A_6}^{W}(\text{sign})$, $5600_z = J_{E_6}^{W}(30'_p)$, $3200_x = J_{E_6}^{W}(15'_q)$,

$2400_z = J_{E_6}^{W}(15_p)$, $5600'_z = J_{D_7}^{W}\begin{bmatrix}1 & 2 & 3 & 4 & 6\\0 & 1 & 2 & 3 & 5\end{bmatrix}$,

$3200'_x = J_{D_7}^{W}\begin{bmatrix}1 & 2 & 3 & 5 & 6\\0 & 1 & 2 & 3 & 4\end{bmatrix}$, $2400'_z = J_{D_7}^{W}\begin{bmatrix}1 & 2 & 3 & 4 & 5\\0 & 1 & 2 & 3 & 6\end{bmatrix}$,

$2100_y = J_{D_5}^{W}(\text{sign})$.

(4.13.4) We have also:

$J_{E_7}^{W}(315_a) = 4480_y + 4536_y + 5670_y + 7168_w$

$J_{E_7}^{W}(280'_b) = 4200_y + 2688_y + 7168_w + 5600_w$

$J_{E_7}^{W}(280_a) = 4536_y + 5670_y + 1400_y + 1680_y + 5600_w$

$J_{E_7}^{W}(70_a) = 3150_y + 2016_w$

$J_{E_7}^{W}(35_a) = 1680_y + 70_y + 448_w$

$J_{E_6\times A_1}^{W}(30'_p\boxtimes\text{sign}) = 4480_y + 3150_y + 4536_y + 5670_y + 7168_w + 5600_w$

$J_{E_6\times A_1}^{W}(15'_q\boxtimes\text{sign}) = 4200_y + 2688_y + 7168_w + 2016_w$

$J_{E_6\times A_1}^{W}(15'_p\boxtimes\text{sign}) = 5670_y + 1134_y + 1680_y + 5600_w + 448_w$

$J_{D_7}^{W}\begin{bmatrix}1 & 2 & 3 & 5\\0 & 1 & 3 & 4\end{bmatrix} = 4480_y + 4200_y + 4536_y + 5670_y + 1400_y + 7168_w + 5600_w$

$$J_{D_7}^{W}\begin{bmatrix}1&3&4&5\\0&1&2&3\end{bmatrix} = 4200_y + 168_y + 1344_w$$

$$J_{D_7}^{W}\begin{bmatrix}1&2&3&4\\0&1&3&5\end{bmatrix} = 4536_y + 5670_y + 2688_y + 1680_y + 7168_w + 5600_w$$

$$J_{D_5\times A_2}^{W}\left(\begin{bmatrix}1&2&3&4\\0&1&2&4\end{bmatrix}\boxtimes \mathrm{sign}\right) = 4480_y + 3150_y + 4200_y + 5670_y + 1134_y + 7168_w + 1344_w + 5600_w$$

$$J_{A_4\times A_3}^{W}(\mathrm{sign}) = 4480_y + 3150_y + 4200_y + 420_y + 7168_w + 1344_w + 2016_w .$$

4.14. When (W, S) is the Weyl group corresponding to an irreducible root system R, as in 4.1, we have associated to W, S and R a decomposition of W into families, (see 4.2); to each family $\mathcal{F}$ we have associated a finite group $\mathcal{G} = \mathcal{G}_{\mathcal{F}}$ and an imbedding $\mathcal{F} \hookrightarrow \mathcal{M}(\mathcal{G}_{\mathcal{F}})$. (Type $A_n (n \geq 0)$: 4.4, Type $B_n, C_n (n \geq 2)$: 4.5, Type $D_n (n \geq 5)$: 4.6, Type D_4: 4.7, Type G_2: 4.8, Type F_4: 4.10, Type E_6: 4.11, Type E_7: 4.12, Type E_8: 4.13 .)

We define a function $\Delta : \mathcal{M}(\mathcal{G}_{\mathcal{F}}) \to \{\pm 1\}$ as follows. If $\mathcal{F}$ is the family of $512'_a$ (in E_7) or the family of 4096_z or $4096'_x$ (in E_8), we set $\Delta(1,1) = \Delta(1,\varepsilon) = 1$, $\Delta(g_2, 1) = \Delta(g_2, \varepsilon) = -1$. For all other $\mathcal{F}$, Δ is identically 1. From 4.4 - 4.13, we see that for each family $\mathcal{F} \subset W^{\vee}$, we have:

(4.14.1) *The function* $E \to a_E$, *(see 4.1.1) is constant on* $\mathcal{F}$.

(4.14.2) *If* $E \in \mathcal{F}$ *corresponds to* $(x, \sigma) \in \mathcal{M}(\mathcal{G}_{\mathcal{F}})$, *we have* $f_E = |Z_{\mathcal{G}_{\mathcal{F}}}(x)| \cdot (\dim \sigma)^{-1}$, *(see 4.1.1).* $\mathcal{F}$ *contains a unique special representation* E_0; *it corresponds to* $(1,1) \in \mathcal{M}(\mathcal{G}_{\mathcal{F}})$ *and we have* $f_{E_0} = |\mathcal{G}_{\mathcal{F}}|$.

For any finite group $\mathcal{G}$, we define, following [L_5, 4], a pairing $\{\ ,\ \} : \mathcal{M}(\mathcal{G} \times \mathcal{M}(\mathcal{G}) \to \overline{\mathbf{Q}}_\ell$ by the formula

(4.14.3)
$$\{(x,\sigma),(y,\tau)\} = \sum_{\substack{g \in \mathcal{G} \\ x \cdot gyg^{-1} = gyg^{-1} \cdot x}} |Z(x)|^{-1} \cdot |Z(y)|^{-1} \cdot \mathrm{Tr}(g^{-1}x^{-1}g, \tau)\,\mathrm{Tr}(gyg^{-1}, \sigma) .$$

Here x, y are elements of $\mathcal{G}$ and Z(x), Z(y) are their centralizers in $\mathcal{G}$. The matrix $\{(x,\sigma),(y,\tau)\}$ whose rows and columns are parametrized by $\mathfrak{M}(\mathcal{G})$ is clearly hermitian: $\{(x,\sigma),(y,\tau)\} = \overline{\{(y,\tau),(x,\sigma)\}}$ (where bar denotes an automorphis of $\overline{\mathbf{Q}}_\ell$ which sends each root of 1 to its inverse); moreover, its square is the identity matrix, [*loc. cit.*].

4.15. We shall describe the matrix of the pairing $\{\ ,\ \}: \mathfrak{M}(\mathcal{G}) \times \mathfrak{M}(\mathcal{G}) \to \overline{\mathbf{Q}}_\ell$ in the case of $\mathcal{G} = \{e\}$ or $\mathcal{G} = \mathfrak{S}_2$ and a part of that matrix (which will be of use later on) in the case of $\mathcal{G} = \mathfrak{S}_3, \mathfrak{S}_4, \mathfrak{S}_5$. This part of the matrix will have entries in $\mathbf{Q}$; the entry in the row corresponding to (y,τ) and the column corresponding to (x,σ) is the number $\{(x,\sigma),(y,\tau)\}$ defined by (4.14.3). (I wish to thank T. Asai and N. Kawanaka for spotting some misprints in the tables for $\mathfrak{S}_1$, $\mathfrak{S}_5$, in the preprint.) We shall denote by θ a fixed primitive cubic root of 1, by i a fixed primitive fourth root of 1 and by ζ a fixed primitive fifth root of 1 in $\overline{\mathbf{Q}}_\ell^*$.

Assume first that $\mathcal{G} = \{e\}$. The matrix is (1).

Assume next that $\mathcal{G} = \mathfrak{S}_2$. The matrix is:

	(1,1)	(1, ε)	$(g_2,1)$	(g_2,ε)
(1,1)	$\frac{1}{2}$	$\frac{1}{2}$	$\frac{1}{2}$	$\frac{1}{2}$
(1, ε)	$\frac{1}{2}$	$\frac{1}{2}$	$-\frac{1}{2}$	$-\frac{1}{2}$
$(g_2,1)$	$\frac{1}{2}$	$-\frac{1}{2}$	$\frac{1}{2}$	$-\frac{1}{2}$
(g_2,ε)	$\frac{1}{2}$	$-\frac{1}{2}$	$-\frac{1}{2}$	$\frac{1}{2}$

Assume next that $\mathcal{G} = \mathfrak{S}_3$. The matrix is:

	(1,1)	$(g_2,1)$	(1,r)	$(g_3,1)$	(1,ε)	(g_2,ε)	(g_3,θ)	(g_3,θ^2)
(1,1)	$\frac{1}{6}$	$\frac{1}{2}$	$\frac{1}{3}$	$\frac{1}{3}$	$\frac{1}{6}$	$\frac{1}{2}$	$\frac{1}{3}$	$\frac{1}{3}$
$(g_2,1)$	$\frac{1}{2}$	$\frac{1}{2}$	0	0	$-\frac{1}{2}$	$-\frac{1}{2}$	0	0
(1,r)	$\frac{1}{3}$	0	$\frac{2}{3}$	$-\frac{1}{3}$	$\frac{1}{3}$	0	$-\frac{1}{3}$	$-\frac{1}{3}$
$(g_3,1)$	$\frac{1}{3}$	0	$-\frac{1}{3}$	$\frac{2}{3}$	$\frac{1}{3}$	0	$-\frac{1}{3}$	$-\frac{1}{3}$
(1,ε)	$\frac{1}{6}$	$-\frac{1}{2}$	$\frac{1}{3}$	$\frac{1}{3}$	$\frac{1}{6}$	$-\frac{1}{2}$	$\frac{1}{3}$	$\frac{1}{3}$

Assume next that $\mathcal{G} = \mathfrak{G}_4$. The matrix (11×21) is:

	$(1,1)$	$(1,\lambda^1)$	$(1,\lambda^2)$	$(1,\lambda^3)$	$(1,\sigma)$	$(g_2,1)$	(g_2,ε')	(g_2,ε'')	(g_2,ε)	$(g_2',1)$	(g_2',ε'')	(g_2',ε')	(g_2',ε)	(g_2',r)	$(g_3,1)$	(g_3,θ)	(g_3,θ^2)	$(g_4,1)$	$(g_4,-1)$	(g_4,i)	$(g_4,-i)$
$(1,1)$	$\frac{1}{24}$	$\frac{1}{8}$	$\frac{1}{8}$	$\frac{1}{24}$	$\frac{1}{12}$	$\frac{1}{4}$	$\frac{1}{4}$	$\frac{1}{4}$	$\frac{1}{4}$	$\frac{1}{8}$	$\frac{1}{8}$	$\frac{1}{8}$	$\frac{1}{8}$	$\frac{1}{4}$	$\frac{1}{3}$	$\frac{1}{3}$	$\frac{1}{3}$	$\frac{1}{4}$	$\frac{1}{4}$	$\frac{1}{4}$	$\frac{1}{4}$
$(1,\lambda^1)$	$\frac{1}{8}$	$\frac{3}{8}$	$\frac{3}{8}$	$\frac{1}{8}$	$\frac{1}{4}$	$\frac{1}{4}$	$\frac{1}{4}$	$\frac{1}{4}$	$\frac{1}{4}$	$-\frac{1}{8}$	$-\frac{1}{8}$	$-\frac{1}{8}$	$-\frac{1}{8}$	$-\frac{1}{4}$	0	0	0	$-\frac{1}{4}$	$-\frac{1}{4}$	$-\frac{1}{4}$	$-\frac{1}{4}$
$(1,\lambda^2)$	$\frac{1}{8}$	$\frac{3}{8}$	$\frac{3}{8}$	$\frac{1}{8}$	$\frac{1}{4}$	$-\frac{1}{4}$	$-\frac{1}{4}$	$-\frac{1}{4}$	$-\frac{1}{4}$	$-\frac{1}{8}$	$-\frac{1}{8}$	$-\frac{1}{8}$	$-\frac{1}{8}$	$-\frac{1}{4}$	0	0	0	$\frac{1}{4}$	$\frac{1}{4}$	$\frac{1}{4}$	$\frac{1}{4}$
$(1,\sigma)$	$\frac{1}{12}$	$\frac{1}{4}$	$\frac{1}{4}$	$\frac{1}{12}$	$\frac{1}{6}$	0	0	0	0	$\frac{1}{4}$	$\frac{1}{4}$	$\frac{1}{4}$	$\frac{1}{4}$	$\frac{1}{2}$	$-\frac{1}{3}$	$-\frac{1}{3}$	$-\frac{1}{3}$	0	0	0	0
$(g_2,1)$	$\frac{1}{4}$	$\frac{1}{4}$	$-\frac{1}{4}$	$-\frac{1}{4}$	0	$\frac{1}{2}$	0	0	$-\frac{1}{2}$	$\frac{1}{4}$	$\frac{1}{4}$	$-\frac{1}{4}$	$-\frac{1}{4}$	0	0	0	0	0	0	0	0
(g_2,ε'')	$\frac{1}{4}$	$\frac{1}{4}$	$-\frac{1}{4}$	$-\frac{1}{4}$	0	0	$-\frac{1}{2}$	$\frac{1}{2}$	0	$-\frac{1}{4}$	$-\frac{1}{4}$	$\frac{1}{4}$	$\frac{1}{4}$	0	0	0	0	0	0	0	0
$(g_2',1)$	$\frac{1}{8}$	$-\frac{1}{8}$	$-\frac{1}{8}$	$\frac{1}{8}$	$\frac{1}{4}$	$\frac{1}{4}$	$-\frac{1}{4}$	$-\frac{1}{4}$	$\frac{1}{4}$	$\frac{3}{8}$	$-\frac{1}{8}$	$\frac{3}{8}$	$-\frac{1}{8}$	$-\frac{1}{4}$	0	0	0	$\frac{1}{4}$	$\frac{1}{4}$	$-\frac{1}{4}$	$-\frac{1}{4}$
(g_2',ε'')	$\frac{1}{8}$	$-\frac{1}{8}$	$-\frac{1}{8}$	$\frac{1}{8}$	$\frac{1}{4}$	$\frac{1}{4}$	$-\frac{1}{4}$	$-\frac{1}{4}$	$\frac{1}{4}$	$-\frac{1}{8}$	$\frac{3}{8}$	$-\frac{1}{8}$	$\frac{3}{8}$	$-\frac{1}{4}$	0	0	0	$-\frac{1}{4}$	$-\frac{1}{4}$	$\frac{1}{4}$	$\frac{1}{4}$
(g_2',ε')	$\frac{1}{8}$	$-\frac{1}{8}$	$-\frac{1}{8}$	$\frac{1}{8}$	$\frac{1}{4}$	$-\frac{1}{4}$	$\frac{1}{4}$	$\frac{1}{4}$	$-\frac{1}{4}$	$\frac{3}{8}$	$-\frac{1}{8}$	$\frac{3}{8}$	$-\frac{1}{8}$	$-\frac{1}{4}$	0	0	0	$-\frac{1}{4}$	$-\frac{1}{4}$	$\frac{1}{4}$	$\frac{1}{4}$
$(g_3,1)$	$\frac{1}{3}$	0	0	$\frac{1}{3}$	$-\frac{1}{3}$	0	0	0	0	0	0	0	0	0	$\frac{2}{3}$	$-\frac{1}{3}$	$-\frac{1}{3}$	0	0	0	0
$(g_4,1)$	$\frac{1}{4}$	$-\frac{1}{4}$	$\frac{1}{4}$	$-\frac{1}{4}$	0	0	0	0	0	$\frac{1}{4}$	$-\frac{1}{4}$	$-\frac{1}{4}$	$\frac{1}{4}$	0	0	0	0	$\frac{1}{2}$	$-\frac{1}{2}$	0	0

Assume next that $\mathcal{G} = \mathfrak{S}_5$. The matrix is:

	$(g_5,1)$	$(g_4,1)$	$(g_6,1)$	$(g_3,1)$	(g_3,ε)	$(g_2',1)$	(g_2',ε)	(g_2',ε')	(g_2',ε'')	$(g_2,1)$	(g_2,ε)	(g_2,r)	$(1,1)$	$(1,\lambda^1)$	$(1,\lambda^2)$
$(1,1)$	$\frac{1}{5}$	$\frac{1}{4}$	$\frac{1}{6}$	$\frac{1}{6}$	$\frac{1}{6}$	$\frac{1}{8}$	$\frac{1}{8}$	$\frac{1}{8}$	$\frac{1}{8}$	$\frac{1}{12}$	$\frac{1}{12}$	$\frac{1}{6}$	$\frac{1}{120}$	$\frac{1}{30}$	$\frac{1}{20}$
$(g_3,1)$	0	0	$\frac{1}{3}$	$\frac{1}{3}$	$\frac{1}{3}$	0	0	0	0	$\frac{1}{6}$	$\frac{1}{6}$	$-\frac{1}{6}$	$\frac{1}{6}$	$\frac{1}{6}$	0
$(g_2',1)$	0	$\frac{1}{4}$	0	0	0	$\frac{3}{8}$	$-\frac{1}{8}$	$\frac{3}{8}$	$-\frac{1}{8}$	$\frac{1}{4}$	$-\frac{1}{4}$	0	$\frac{1}{8}$	0	$-\frac{1}{4}$
$(1,\nu)$	0	$-\frac{1}{4}$	$\frac{1}{6}$	$-\frac{1}{6}$	$-\frac{1}{6}$	$\frac{1}{8}$	$\frac{1}{8}$	$\frac{1}{8}$	$\frac{1}{8}$	$\frac{1}{12}$	$\frac{1}{12}$	$\frac{1}{6}$	$\frac{1}{24}$	$\frac{1}{6}$	$\frac{1}{4}$
$(1,\lambda^1)$	$-\frac{1}{5}$	0	$-\frac{1}{6}$	$\frac{1}{6}$	$\frac{1}{6}$	0	0	0	0	$\frac{1}{6}$	$\frac{1}{6}$	$\frac{1}{3}$	$\frac{1}{30}$	$\frac{2}{15}$	$\frac{1}{5}$
$(g_5,1)$	$\frac{4}{5}$	0	0	0	0	0	0	0	0	0	0	0	$\frac{1}{5}$	$-\frac{1}{5}$	$\frac{1}{5}$
(g_3,ε)	0	0	$-\frac{1}{3}$	$\frac{1}{3}$	$\frac{1}{3}$	0	0	0	0	$-\frac{1}{6}$	$-\frac{1}{6}$	$\frac{1}{6}$	$\frac{1}{6}$	$\frac{1}{6}$	0
$(1,\nu')$	0	$\frac{1}{4}$	$-\frac{1}{6}$	$-\frac{1}{6}$	$-\frac{1}{6}$	$\frac{1}{8}$	$\frac{1}{8}$	$\frac{1}{8}$	$\frac{1}{8}$	$-\frac{1}{12}$	$-\frac{1}{12}$	$-\frac{1}{6}$	$\frac{1}{24}$	$\frac{1}{6}$	$\frac{1}{4}$
(g_2',ε'')	0	$-\frac{1}{4}$	0	0	0	$-\frac{1}{8}$	$\frac{3}{8}$	$-\frac{1}{8}$	$\frac{3}{8}$	$\frac{1}{4}$	$-\frac{1}{4}$	0	$\frac{1}{8}$	0	$-\frac{1}{4}$
$(1,\lambda^2)$	$\frac{1}{5}$	0	0	0	0	$-\frac{1}{4}$	$-\frac{1}{4}$	$-\frac{1}{4}$	$-\frac{1}{4}$	0	0	0	$\frac{1}{20}$	$\frac{1}{5}$	$\frac{3}{10}$
(g_2',ε')	0	$-\frac{1}{4}$	0	0	0	$\frac{3}{8}$	$-\frac{1}{8}$	$\frac{3}{8}$	$-\frac{1}{8}$	$-\frac{1}{4}$	$\frac{1}{4}$	0	$\frac{1}{8}$	0	$-\frac{1}{4}$
$(1,\lambda^3)$	$-\frac{1}{5}$	0	$\frac{1}{6}$	$\frac{1}{6}$	$\frac{1}{6}$	0	0	0	0	$-\frac{1}{6}$	$-\frac{1}{6}$	$-\frac{1}{3}$	$\frac{1}{30}$	$\frac{2}{15}$	$\frac{1}{5}$
$(g_2,1)$	0	0	$\frac{1}{6}$	$\frac{1}{6}$	$-\frac{1}{6}$	$\frac{1}{4}$	$-\frac{1}{4}$	$-\frac{1}{4}$	$\frac{1}{4}$	$\frac{1}{3}$	$-\frac{1}{6}$	$\frac{1}{6}$	$\frac{1}{12}$	$\frac{1}{6}$	0
$(g_4,1)$	0	$\frac{1}{2}$	0	0	0	$\frac{1}{4}$	$\frac{1}{4}$	$-\frac{1}{4}$	$-\frac{1}{4}$	0	0	0	$\frac{1}{4}$	0	0
$(g_6,1)$	0	0	$\frac{1}{3}$	$\frac{1}{3}$	$-\frac{1}{3}$	0	0	0	0	$\frac{1}{6}$	$\frac{1}{6}$	$-\frac{1}{6}$	$\frac{1}{6}$	$-\frac{1}{6}$	0
(g_2,r)	0	0	$-\frac{1}{6}$	$-\frac{1}{6}$	$\frac{1}{6}$	0	0	0	0	$\frac{1}{6}$	$\frac{1}{6}$	$\frac{1}{3}$	$\frac{1}{6}$	$\frac{1}{3}$	0
(g_2,ε)	0	0	$\frac{1}{6}$	$\frac{1}{6}$	$-\frac{1}{6}$	$-\frac{1}{4}$	$\frac{1}{4}$	$\frac{1}{4}$	$-\frac{1}{4}$	$-\frac{1}{6}$	$\frac{1}{3}$	$\frac{1}{6}$	$\frac{1}{12}$	$\frac{1}{6}$	0

$(1,\lambda^3)$	$(1,\lambda^4)$	$(1,\nu)$	$(1,\nu')$	(g_5,ζ)	(g_4,i)	$(g_4,-1)$	$(g_6,-\theta)$	(g_6,θ)	$(g_6,-1)$	(g_3,θ)	$(g_3,\varepsilon\theta)$	(g_2',r)	$(g_2,-1)$	$(g_2,-\varepsilon)$	$(g_2,-r)$
$\frac{1}{30}$	$\frac{1}{120}$	$\frac{1}{24}$	$\frac{1}{24}$	$\frac{1}{5}$	$\frac{1}{4}$	$\frac{1}{4}$	$\frac{1}{6}$	$\frac{1}{6}$	$\frac{1}{6}$	$\frac{1}{6}$	$\frac{1}{6}$	$\frac{1}{4}$	$\frac{1}{12}$	$\frac{1}{12}$	$\frac{1}{6}$
$\frac{1}{6}$	$\frac{1}{6}$	$-\frac{1}{6}$	$-\frac{1}{6}$	0	0	0	$-\frac{1}{6}$	$-\frac{1}{6}$	$\frac{1}{3}$	$-\frac{1}{6}$	$-\frac{1}{6}$	0	$\frac{1}{6}$	$\frac{1}{6}$	$-\frac{1}{6}$
0	$\frac{1}{8}$	$\frac{1}{8}$	$\frac{1}{8}$	0	$-\frac{1}{4}$	$\frac{1}{4}$	0	0	0	0	0	$-\frac{1}{4}$	$-\frac{1}{4}$	$\frac{1}{4}$	0
$\frac{1}{6}$	$\frac{1}{24}$	$\frac{5}{24}$	$\frac{5}{24}$	0	$-\frac{1}{4}$	$-\frac{1}{4}$	$\frac{1}{6}$	$\frac{1}{6}$	$\frac{1}{6}$	$-\frac{1}{6}$	$-\frac{1}{6}$	$\frac{1}{4}$	$\frac{1}{12}$	$\frac{1}{12}$	$\frac{1}{6}$
$\frac{2}{15}$	$\frac{1}{30}$	$\frac{1}{6}$	$\frac{1}{6}$	$-\frac{1}{5}$	0	0	$-\frac{1}{6}$	$-\frac{1}{6}$	$-\frac{1}{6}$	$\frac{1}{6}$	$\frac{1}{6}$	0	$\frac{1}{6}$	$\frac{1}{6}$	$\frac{1}{3}$
$-\frac{1}{5}$	$\frac{1}{5}$	0	0	$-\frac{1}{5}$	0	0	0	0	0	0	0	0	0	0	0
$\frac{1}{6}$	$\frac{1}{6}$	$-\frac{1}{6}$	$-\frac{1}{6}$	0	0	0	$\frac{1}{6}$	$\frac{1}{6}$	$-\frac{1}{3}$	$-\frac{1}{6}$	$-\frac{1}{6}$	0	$-\frac{1}{6}$	$-\frac{1}{6}$	$\frac{1}{6}$
$\frac{1}{6}$	$\frac{1}{24}$	$\frac{5}{24}$	$\frac{5}{24}$	0	$\frac{1}{4}$	$\frac{1}{4}$	$-\frac{1}{6}$	$-\frac{1}{6}$	$-\frac{1}{6}$	$-\frac{1}{6}$	$-\frac{1}{6}$	$\frac{1}{4}$	$-\frac{1}{12}$	$-\frac{1}{12}$	$-\frac{1}{6}$
0	$\frac{1}{8}$	$\frac{1}{8}$	$\frac{1}{8}$	0	$\frac{1}{4}$	$-\frac{1}{4}$	0	0	0	0	0	$-\frac{1}{4}$	$-\frac{1}{4}$	$\frac{1}{4}$	0
$\frac{1}{5}$	$\frac{1}{20}$	$\frac{1}{4}$	$\frac{1}{4}$	$\frac{1}{5}$	0	0	0	0	0	0	0	$-\frac{1}{2}$	0	0	0
0	$\frac{1}{8}$	$\frac{1}{8}$	$\frac{1}{8}$	0	$\frac{1}{4}$	$-\frac{1}{4}$	0	0	0	0	0	$-\frac{1}{4}$	$\frac{1}{4}$	$-\frac{1}{4}$	0
$\frac{2}{15}$	$\frac{1}{30}$	$\frac{1}{6}$	$\frac{1}{6}$	$-\frac{1}{5}$	0	0	$\frac{1}{6}$	$\frac{1}{6}$	$\frac{1}{6}$	$\frac{1}{6}$	$\frac{1}{6}$	0	$-\frac{1}{6}$	$-\frac{1}{6}$	$-\frac{1}{3}$
$-\frac{1}{6}$	$-\frac{1}{12}$	$\frac{1}{12}$	$-\frac{1}{12}$	0	0	0	$-\frac{1}{6}$	$\frac{1}{6}$	$-\frac{1}{6}$	$\frac{1}{6}$	$-\frac{1}{6}$	0	$\frac{1}{6}$	$-\frac{1}{3}$	$-\frac{1}{6}$
0	$-\frac{1}{4}$	$-\frac{1}{4}$	$\frac{1}{4}$	0	0	$-\frac{1}{2}$	0	0	0	0	0	0	0	0	0
$\frac{1}{6}$	$-\frac{1}{6}$	$\frac{1}{6}$	$-\frac{1}{6}$	0	0	0	$\frac{1}{6}$	$-\frac{1}{6}$	$-\frac{1}{3}$	$-\frac{1}{6}$	$\frac{1}{6}$	0	$-\frac{1}{6}$	$-\frac{1}{6}$	$\frac{1}{6}$
$-\frac{1}{3}$	$-\frac{1}{6}$	$\frac{1}{6}$	$-\frac{1}{6}$	0	0	0	$\frac{1}{6}$	$-\frac{1}{6}$	$\frac{1}{6}$	$-\frac{1}{6}$	$\frac{1}{6}$	0	$-\frac{1}{6}$	$-\frac{1}{6}$	$-\frac{1}{3}$
$-\frac{1}{6}$	$-\frac{1}{12}$	$\frac{1}{12}$	$-\frac{1}{12}$	0	0	0	$-\frac{1}{6}$	$\frac{1}{6}$	$-\frac{1}{6}$	$\frac{1}{6}$	$-\frac{1}{6}$	0	$-\frac{1}{3}$	$\frac{1}{6}$	$-\frac{1}{6}$

Here the column (g_5,ζ) stands for four identical columns (g_5,ζ^j), $j=1,2,3,4$; the column (g_4,i) stands for two identical columns: (g_4,i), $(g_4,-i)$; the columns $(g_6,-\theta)$, (g_6,θ), (g_3,θ), $(g_3,\varepsilon\theta)$ each stand for two identical columns: one being the one listed, the other being obtained by replacing θ by θ^2. Hence our matrix has 39 columns and 17 rows.

If R is of type B_n or C_n $(n\geq 3)$, the pairing $\{\ ,\ \}:\mathfrak{M}(\mathcal{G}_{\mathcal{F}})\times\mathfrak{M}(\mathcal{G}_{\mathcal{F}})\to\overline{\mathbf{Q}}_\ell$ becomes, under the identification $\mathfrak{M}(\mathcal{G}_{\mathcal{F}})=V_{Z_1}$ (see 4.5), the pairing $V_{Z_1}\times V_{Z_1}\to\mathbf{Q}$ defined by $M_1^{\#},M_2^{\#}\to\frac{1}{2^d}(-1)^{|M_1^{\#}\cap M_2^{\#}|}$ $(M_1^{\#},M_2^{\#}\in V_{Z_1}$, $|Z_1|=2d+1)$.

If R is of type D_n $(n\geq 5)$, the pairing $\{\ ,\ \}:\mathfrak{M}(\mathcal{G}_{\mathcal{F}}\times\mathfrak{M}(\mathcal{G}_{\mathcal{F}})\to\overline{\mathbf{Q}}_\ell$ becomes, under the identification $\mathfrak{M}(\mathcal{G}_{\mathcal{F}})=(V'_{Z_1})^+$, (see 4.6), the pairing $(V'_{Z_1})^+\times(V'_{Z_1})^+\to\mathbf{Q}$ defined by $M_1^{\#},M_2^{\#}\to\frac{1}{2^{d-1}}(-1)^{|M_1^{\#}\cap M_2^{\#}|}$, $(M_1^{\#},M_2^{\#}\in(V'_{Z_1})^+$, $|Z_1|=2d$, assuming $d\geq 1$).

4.16. Let $\mathcal{G}$ be a finite group; assume that we are given an imbedding of $\mathcal{G}$ into another finite group $\tilde{\mathcal{G}}$ such that $\mathcal{G}$ is normal in $\tilde{\mathcal{G}}$, $\tilde{\mathcal{G}}/\mathcal{G}$ is cyclic of order c with a distinguished generator and that there exists an element of order c in the inverse image $\mathcal{G}'\subset\tilde{\mathcal{G}}$ of that generator under $\tilde{\mathcal{G}}\to\tilde{\mathcal{G}}/\mathcal{G}$. To such a system $\mathcal{G}\subset\tilde{\mathcal{G}}$ we associate two sets $\mathfrak{M}=\mathfrak{M}(\mathcal{G}\subset\tilde{\mathcal{G}})$, $\mathfrak{M}'=\mathfrak{M}'(\mathcal{G}\subset\tilde{\mathcal{G}})$ as follows.

$\mathfrak{M}$ consists of all pairs (y,τ) where y is an element of $\mathcal{G}$ whose centralizer $Z_{\tilde{\mathcal{G}}}(y)$ in $\tilde{\mathcal{G}}$ meets $\mathcal{G}'$ and τ is an irreducible representation (up to isomorphism) of $Z_{\tilde{\mathcal{G}}}(y)$ whose restriction to $Z_{\mathcal{G}}(y)$ is still irreducible; these pairs are taken modulo the equivalence relation $(y,\tau)\sim(gyg^{-1},\tau^g)$, $\forall g\in\tilde{\mathcal{G}}$ where τ^g is the representation of $Z_{\tilde{\mathcal{G}}}(gyg^{-1})=gZ_{\tilde{\mathcal{G}}}(y)g^{-1}$ defined by composing τ with conjugation by g^{-1}.

$\mathfrak{M}'$ consists of all pairs (x,σ) where x is an element of $\mathcal{G}'$ and σ is an irreducible representation (up to isomorphism) of $Z_{\tilde{\mathcal{G}}}(x)$; these pairs are taken modulo an equivalence relation defined just as for $\mathfrak{M}$. (When $c = 1$, we may identify $\mathfrak{M}$ and $\mathfrak{M}'$.)

The sets $\mathfrak{M}$, $\mathfrak{M}'$ are contained in $\mathfrak{M}(\tilde{\mathcal{G}})$, (see 4.3). The pairing $\{\ ,\ \}: \mathfrak{M}(\tilde{\mathcal{G}}) \times \mathfrak{M}(\tilde{\mathcal{G}}) \to \bar{\mathbf{Q}}_\ell$ (see (4.14.3)), restricts to a pairing $\{\ ,\ \}: \mathfrak{M}' \times \mathfrak{M} \to \bar{\mathbf{Q}}_\ell$. Consider the operator A on the space of functions on $\mathfrak{M}(\tilde{\mathcal{G}})$ defined by

$$(Af)(x, \sigma) = \sum \{(x, \sigma), (y, \tau)\}\ f(y, \tau)$$

sum over all $(y, \tau) \in \mathfrak{M}(\tilde{\mathcal{G}})$. Let M be the group of all roots of 1 in $\bar{\mathbf{Q}}_\ell^*$ and let $M_c = \{a \in \bar{\mathbf{Q}}_\ell^* \mid a^c = 1\}$. For $a \in M_c$ let ε_a be the one-dimensional representation of $\tilde{\mathcal{G}}$ which is trivial on $\mathcal{G}$ and takes $\mathcal{G}'$ to a. The restriction of ε_a to a subgroup of $\tilde{\mathcal{G}}$ is denoted again ε_a. The maps $(y, \tau) \to (y, \tau \otimes \varepsilon_a)$ define a free M_c-action on $\mathfrak{M}$. Consider the space $\mathcal{P}_c$ of functions $\mathfrak{M}(\tilde{\mathcal{G}}) \to \bar{\mathbf{Q}}_\ell$ with support contained in $\mathfrak{M}$ and such that $f(y, \tau \otimes \varepsilon_a) = a f(y, \tau)$, for all $(y, \tau) \in \mathfrak{M}$ and all $a \in M_c$. Consider also the space $\mathcal{P}'_c$ of functions $\mathfrak{M}(\tilde{\mathcal{G}}) \to \bar{\mathbf{Q}}_\ell$ with support contained in $\mathfrak{M}'$ and such that $f(x, \sigma \otimes \varepsilon_a) = f(x, \sigma)$, for all $(x, \sigma) \in \mathfrak{M}'$ and all $a \in M_c$.

It is easy to check that $A: \mathcal{P}_c \to \mathcal{P}'_c$ and $A: \mathcal{P}'_c \to \mathcal{P}_c$. On the other hand, according to (4.14.6), we have $A^2 = 1$. It follows that:

(4.16.1) $A: \mathcal{P}_c \to \mathcal{P}'_c$ *and* $A: \mathcal{P}'_c \to \mathcal{P}_c$ *are inverse isomorphisms.*

Let us identify $\mathcal{P}'_c$ with the space $\bar{\mathcal{P}}$ of functions on the set $\bar{\mathfrak{M}} = \bar{\mathfrak{M}}(\mathcal{G} \subset \tilde{\mathcal{G}})$ consisting of all pairs $(x, \bar{\sigma})$ where x is an element of $\mathcal{G}'$ and $\bar{\sigma}$ is an irreducible representation (up to isomorphism) of $Z_{\mathcal{G}}(x)$; these pairs are taken modulo an equivalence relation defined just as for $\mathfrak{M}$. (A function on $\bar{\mathfrak{M}}$ gives rise to a function on $\mathfrak{M}'$ by composing with the natural projection $\mathfrak{M}' \to \bar{\mathfrak{M}}$, and hence to a function on $\mathfrak{M}(\tilde{\mathcal{G}})$ with support in $\mathfrak{M}'$; this gives the isomorphism $\bar{\mathcal{P}} \xrightarrow{\approx} \mathcal{P}'_c$.) We also identify the functions in $\mathcal{P}_c$ with their restrictions to $\mathfrak{M}$. For $(x, \bar{\sigma}) \in \bar{\mathfrak{M}}$ and $(y, \tau) \in \mathfrak{M}$, we set

(4.16.2) $\{(x,\bar\sigma),(y,\tau)\} \overset{\text{def.}}{=} c\{(x,\sigma),(y,\tau)\}$, where σ is any extension of $\bar\sigma$ to $Z_{\tilde{\mathcal{G}}}(x)$. Using now (4.16.1), we see that:

(4.16.3) The map $A_1 : \mathcal{P}_c \to \overline{\mathcal{P}}$ defined by $(A_1 f)(x,\bar\sigma) = \Sigma\{(x,\bar\sigma),(y,\tau)\} \times f(y,\tau), ((x,\bar\sigma) \in \overline{\mathfrak{M}})$, (sum over a set of representatives (y,τ) for the M_c-orbits on $\mathfrak{M}$) is an isomorphism; its inverse is the map $A_2 : \overline{\mathcal{P}} \to \mathcal{P}_c$ defined by $(A_2 f)(y,\tau) = \Sigma\{(x,\bar\sigma),(y,\tau)\} f(x,\bar\sigma), ((y,\tau) \in \mathfrak{M})$, (sum over all $(x,\bar\sigma) \in \overline{\mathfrak{M}}$).

4.17. Let γ be an automorphism of order c of $(\mathbf{W}, S)$, coming from an automorphism of R (assumed to be irreducible); thus, γ is ordinary, (see 3.1). Then γ acts naturally on $\mathbf{W}^\vee$. Let $\mathcal{F} \subset \mathbf{W}^\vee$ be a γ-stable family. From the explicit description of families (4.4-4.13) we see that $\mathcal{F}$ is necessarily pointwise fixed, i.e. $\mathcal{F} \subset \mathbf{W}^\vee_{ex}$. Let $\tilde{\mathbf{W}}_c$ be defined as in 3.2 (with respect to γ). Let $\tilde{\mathcal{F}}_c$ be the set of irreducible $\tilde{\mathbf{W}}_c$-modules whose restrictions to $\mathbf{W}$ are irreducible and belong to $\mathcal{F}$. The restriction to $\mathbf{W}$ defines a surjective map $\tilde{\mathcal{F}}_c \to \mathcal{F}$ whose fibers have c elements each. For $\alpha \in M_c$, we denote by E_α the one-dimensional representation of $\tilde{\mathbf{W}}_c$ on which $\mathbf{W}$ acts trivially and γ acts as multiplication by α. (Compare 4.16.) Then $E \to E \otimes E_\alpha$ defines a free action of M_c on $\tilde{\mathcal{F}}_c$ whose orbits are the fibers of $\tilde{\mathcal{F}}_c \to \mathcal{F}$. We wish to associate to $\mathcal{F}$, γ, a finite group $\tilde{\mathcal{G}}_{\mathcal{F},c}$ containing the group $\mathcal{G}_{\mathcal{F}}$ of 4.14 as a normal subgroup and a coset $\mathcal{G}'_{\mathcal{F}}$ of $\tilde{\mathcal{G}}_{\mathcal{F},c}$ with respect to $\mathcal{G}_{\mathcal{F}}$ such that $\mathcal{G}_{\mathcal{F}} \subset \tilde{\mathcal{G}}_{\mathcal{F},c} \supset \mathcal{G}'_{\mathcal{F}}$ are as in the first sentence in 4.16. We also wish to define an imbedding $\tilde{\mathcal{F}}_c \hookrightarrow \mathfrak{M}(\mathcal{G}_{\mathcal{F}} \subset \tilde{\mathcal{G}}_{\mathcal{F},c})$, (see 4.16), which is M_c-equivariant.

When $c = 1$, this has been done in 4.4-4.13. In the next 2 sections we shall do this when $c \geq 2$.

4.18. Assume that R is of type D_n $(n \geq 2)$, that $s', s'' \in S$ are chosen as in 4.6 and that γ is the automorphism (of order 2) of $(\mathbf{W}, S)$ such that $\gamma(s') = s''$, $\gamma(s'') = s'$. Let $\tilde{\mathbf{W}}_2$ be the semidirect product of $\mathbf{W}$ with the cyclic group of order 2 with generator γ such that conjugation by γ defines the automorphism of $\mathbf{W}$ which was just described. We may identify the group $\widetilde{W}_2$ with the group W_n (in 4.5) in two different ways: we identify W with a subgroup of index 2 of W_n in one of the two ways described in 4.6, and we send $\gamma \in \widetilde{W}_2$ to $s_n \in W_n$.

Consider the irreducible $Q[W_n]$-module M associated in 4.5 to E_1, E_2 where $E_1 = [\lambda_1, \lambda_2, \cdots, \lambda_m]$ is an irreducible $Q[\mathfrak{S}_r]$-module and $E_2 = [\mu_1, \mu_2, \cdots, \mu_m]$ is an irreducible $Q[\mathfrak{S}_{\tilde{r}}]$-module. We may regard M as a $\widetilde{W}_2$-module using one of the two identifications above; (the choice of identification is irrelevant since the two identifications differ only by an inner automorphism of W_n); we denote this $\widetilde{W}_2$-module as $[[\Lambda]]$, where Λ is the tableau (whose rows are *not* permutable):

$$\Lambda = \begin{pmatrix} \lambda_1, \lambda_2, \cdots, \lambda_m \\ \\ \mu_1, \mu_2, \cdots, \mu_m \end{pmatrix} \tag{4.18.1}$$

here, Λ is uniquely determined by the $\widetilde{W}_2$-module $[[\Lambda]]$, up to a shift (as in 4.5). The condition that $[[\Lambda]]$ remains irreducible on restriction to W is that the two rows of (4.18.1) are not identical.

Let $\mathfrak{F}$ be a family in $W^\vee$ (see 4.6) and assume that the special representation in $\mathfrak{F}$ is $[Z]$ where Z is a *nondegenerate* symbol of rank n, defect 0 with $2m$ entries, (see 4.6). $\widetilde{\mathfrak{F}}_2$ is the set of irreducible $Q[\widetilde{W}_2]$-modules whose restrictions to W are in $\mathfrak{F}$, (see 4.17). Let $Z_2, Z_1 = M_0 \amalg M_0'$ be defined as in 4.6. The correspondence

$M \to \left[\left[\begin{matrix} Z_2 \amalg (Z_1 - M) \\ Z_2 \amalg \quad M \end{matrix}\right]\right]$ is a 1-1 correspondence between the set $\{M | M \subset Z_1, |M| = d\}$ and the set $\widetilde{\mathfrak{F}}_2$. If we set $M^\# = M \cup M_0 - (M \cap M_0)$, we thus get (with the notation in 4.6) a bijection

$$\{M | M \subset Z_1, |M| \equiv d \,(\mathrm{mod}\ 2)\} \leftrightarrow \{M^\# | M^\# \in V_{Z_1}\}$$

and hence an injection

$$\{M | M \subset Z_1, |M| = d\} \hookrightarrow V_{Z_1}.$$

Composing with $\tilde{\mathfrak{F}}_2 \leftrightarrow \{M \mid M \subset Z_1, |M| = d\}$, we get an injection

$$\tilde{\mathfrak{F}}_2 \hookrightarrow V_{Z_1}, \tag{4.18.2}$$

such that the diagram

$$\begin{array}{ccc} \tilde{\mathfrak{F}}_2 & \hookrightarrow & V_{Z_1} \\ \downarrow & & \downarrow \\ \mathfrak{F} & \hookrightarrow & (V'_{Z_1})^+ \end{array}$$

is commutative; here, $\tilde{\mathfrak{F}}_2 \to \mathfrak{F}$ is given by restriction to $\mathbf{W}$, $V_{Z_1} \to (V'_{Z_1})^+$ is the canonical projection, $\mathfrak{F} \hookrightarrow (V'_{Z_1})^+$ is defined in 4.6.

With the notations in 4.6, we set:

$$\tilde{\mathcal{G}} = \tilde{\mathcal{G}}_{\mathfrak{F},2} = \{v \in V'_{Z_1} \mid (v, e_2) = (v, e_4) = \cdots = (v, e_{2d-2}) = 0\};$$

here $(\ ,\ )$ is the pairing $V'_{Z_1} \times V_{Z_1} \to F_2$, (see 4.6). It is clear that $\mathcal{G}$ is contained in $\tilde{\mathcal{G}}$. (Recall that $\mathcal{G} = \bar{I}''$.) Note that $\tilde{\mathcal{G}}$ is an F_2-vector space of dimension d. We take $\mathcal{G}' = \tilde{\mathcal{G}} - \mathcal{G}$. We have $V_{Z_1} = I' \oplus I''$, where I' is spanned by $e_1, e_3, e_5, \cdots$ and I'' is spanned by $e_2, e_4, e_6, \cdots$. We may identify I' with $\mathrm{Hom}(\tilde{\mathcal{G}}, \bar{\mathbf{Q}}_\ell^*)$ by $i' \leftrightarrow (v \to (-1)^{(v,i')})$; since $I'' \cong \bar{I}'' = \mathcal{G}$, we thus get an identification

$$\mathfrak{M}(\mathcal{G} \subset \tilde{\mathcal{G}}) = V_{Z_1}. \tag{4.18.3}$$

Composing this with (4.18.2), we get an imbedding

$$\tilde{\mathfrak{F}}_2 \hookrightarrow \mathfrak{M}(\mathcal{G} \subset \tilde{\mathcal{G}}). \tag{4.18.4}$$

The set $\bar{\mathfrak{M}}(\mathcal{G} \subset \tilde{\mathcal{G}})$, (see 4.16), is the set of pairs $(x, \bar{\sigma})$ where $x \in \tilde{\mathcal{G}} - \mathcal{G}$ and $\bar{\sigma} \in \mathrm{Hom}(\mathcal{G}, \bar{\mathbf{Q}}_\ell^*) = \bar{I}'$, (see 4.6). Thus, we may identify

$$\bar{\mathfrak{M}}(\mathcal{G} \subset \tilde{\mathcal{G}}) = (\tilde{\mathcal{G}} - \mathcal{G}) \times \bar{I}' = (V'_{Z_1})^-, \tag{4.18.5}$$

(see 4.6). The pairing $\{\ ,\ \}: \overline{\mathfrak{M}}(\mathcal{G} \subset \tilde{\mathcal{G}}) \times \mathfrak{M}(\mathcal{G} \subset \tilde{\mathcal{G}}) \to \overline{\mathbf{Q}}_\ell$, (see (4.16.2)), becomes under the identifications (4.18.3), (4.18.5), the pairing $(V'_{Z_1})^- \times V_{Z_1} \to \overline{\mathbf{Q}}_\ell^*$ given by

$$M_1^\#, M_2^\# \to \frac{1}{2^{d-1}} (-1)^{|M_1^\# \cap M_2^\#|}, \ (M_1^\# \in (V'_{Z_1})^-, M_2^\# \in V_{Z_1}, 2d = |Z_1| \geq 2) .$$

4.19. Assume now that R is of type A_{n-1} $(n \geq 3)$, E_6 or D_4 and that γ is an automorphism of order $c = 2$, 2 or 3 (respectively) of $(\mathbf{W}, \mathbf{S})$. If $c = 2$, any family $\mathcal{F} \subset \mathbf{W}^\vee$ is pointwise fixed by γ. If $c = 3$, there are precisely five families $\mathcal{F} \subset \mathbf{W}^\vee$ which are γ-stable: the families containing $\begin{bmatrix}4\\0\end{bmatrix}, \begin{bmatrix}0&1&2&3\\1&2&3&4\end{bmatrix}, \begin{bmatrix}3\\1\end{bmatrix}, \begin{bmatrix}0&1&3\\1&2&3\end{bmatrix}, \begin{bmatrix}0&2\\1&3\end{bmatrix}$; these families are again pointwise fixed by γ. For a γ-stable family $\mathcal{F}$, the projection $\tilde{\mathcal{F}}_c \to \mathcal{F}$ (see 4.17) has a canonical section $E \mapsto \tilde{E}$. If $c = 2$, an irreducible $\mathbf{Q}[\mathbf{W}]$-module E can be extended to a $\mathbf{Q}[\tilde{\mathbf{W}}_c]$-module $\tilde{E}$ (see 4.17) by letting γ act as the longest element in $\mathbf{W}$. If $c = 3$, an irreducible $\mathbf{Q}[\mathbf{W}]$-module E which is in $\mathbf{W}^\vee_{ex}$ can be extended uniquely to a $\mathbf{Q}[\tilde{\mathbf{W}}_c]$-module, (see 3.2).

We set $\tilde{\mathcal{G}}_{\mathcal{F},c} = \mathcal{G}_\mathcal{F} \times \mathcal{C}_c$ where $\mathcal{C}_c$ is the cyclic group of order c with generator ξ. Then $\mathcal{G}_\mathcal{F}$ is naturally a normal subgroup of $\tilde{\mathcal{G}}_{\mathcal{F},c}$; we set $\mathcal{G}'_\mathcal{F} = \mathcal{G}_\mathcal{F} \times \xi \subset \tilde{\mathcal{G}}_{\mathcal{F},c}$.

We have $\mathfrak{M} = \mathfrak{M}(\mathcal{G}_\mathcal{F} \subset \tilde{\mathcal{G}}_{\mathcal{F},c}) = \mathfrak{M}(\mathcal{G}_\mathcal{F}) \times M_c$: to $(y, \tau) \in \mathfrak{M}(\mathcal{G}_\mathcal{F} \subset \tilde{\mathcal{G}}_{\mathcal{F},c})$ we associate $((y, \overline{\tau}), a)$ where $\overline{\tau}$ is the restriction of τ to $Z_{\mathcal{G}_\mathcal{F}}(y)$ and ξ acts on τ as $a \in M_c$. We define an M_c-equivariant embedding $\tilde{\mathcal{F}}_c \hookrightarrow \mathfrak{M} = \mathfrak{M}(\mathcal{G}_\mathcal{F} \subset \tilde{\mathcal{G}}_{\mathcal{F},c})$ as follows. We require that if $E \in \mathcal{F}$ corresponds to $m \in \mathfrak{M}(\mathcal{G}_\mathcal{F})$, then $\tilde{E}$ corresponds to $(m, 1) \in \mathfrak{M}$, if $c = 3$, or if $c = 2$ and a_E is even, and to $(m, -1) \in \mathfrak{M}$ if $c = 2$ and a_E is odd. (see (4.1.1) for the definition of a_E.)

The set $\overline{\mathfrak{M}}(\mathcal{G}_\mathcal{F} \subset \tilde{\mathcal{G}}_{\mathcal{F},c})$ will be identified with $\mathfrak{M}(\mathcal{G}_\mathcal{F})$: to $(x, \overline{\sigma}) \in \overline{\mathfrak{M}}(\mathcal{G}_\mathcal{F} \subset \tilde{\mathcal{G}}_{\mathcal{F},c})$ corresponds $(x\xi^{-1}, \overline{\sigma}) \in \mathfrak{M}(\mathcal{G}_\mathcal{F})$.

The pairing $\{\ ,\ \}: \overline{\mathcal{M}}(\mathcal{G}_{\mathcal{F}} \subset \tilde{\mathcal{G}}_{\mathcal{F},c}) \times \mathcal{M}(\mathcal{G}_{\mathcal{F}} \subset \tilde{\mathcal{G}}_{\mathcal{F},c}) \to \overline{\mathbf{Q}}_\ell$, (see (4.16.2)) becomes with these identifications a pairing $\mathcal{M}(\mathcal{G}_{\mathcal{F}}) \times (\mathcal{M}(\mathcal{G}_{\mathcal{F}}) \times M_c) \to \overline{\mathbf{Q}}_\ell$ defined as follows. If $m \in \mathcal{M}(\mathcal{G}_{\mathcal{F}})$, $(m', a) \in \mathcal{M}(\mathcal{G}_{\mathcal{F}}) \times M_c$, then $\{m, (m', a)\} = a^{-1}\{m, m'\}$ where $\{m, m'\}$ is given by (4.14.3).

4.20. Now let $(\mathbf{W}, S)$ and R be as in 4.1 (not necessarily irreducible) and let γ be an automorphism of $(\mathbf{W}, S)$ coming from an automorphism of R, (so that γ is ordinary, see 3.1).

There is a unique decomposition $S = \coprod_{i \in I} S_i$ such that the subgroup $\mathbf{W}_i$ generated by S_i is an irreducible Weyl group and $\mathbf{W} = \prod_i \mathbf{W}_i$.

In this section we assume that γ permutes transitively among them the factors $\mathbf{W}_i$. We may assume that $I = Z/rZ$ and that $\gamma(\mathbf{W}_i) = \mathbf{W}_{i+1}$, $(i \in Z/rZ)$. Let γ_i be the restriction of γ^r to $\mathbf{W}_i$; it is an automorphism of $(\mathbf{W}_i, S_i)$ coming from an automorphism of the corresponding root system.

We identify $\mathbf{W}^\vee$ with $\prod_i (\mathbf{W}_i)^\vee$ using external tensor product. A family $\mathcal{F}$ in $\mathbf{W}^\vee$ is of the form $\mathcal{F} = \prod_i \mathcal{F}_i$ where $\mathcal{F}_i$ is a family in $(\mathbf{W}_i)^\vee$, $(i \in Z/rZ)$. The family $\mathcal{F}$ is γ-stable (for the natural action of γ on $\mathbf{W}^\vee$) if and only if $\gamma(\mathcal{F}_i) = \mathcal{F}_{i+1}$ for all i. (In the last equality, γ is regarded as an isomorphism $\mathbf{W}_i \xrightarrow{\sim} \mathbf{W}_{i+1}$; it gives rise to a bijection $(\mathbf{W}_i)^\vee \xrightarrow{\sim} (\mathbf{W}_{i+1})^\vee$.)

Assume now that $\mathcal{F}$ is γ-stable. Then $\mathcal{F}_i$ is pointwise fixed by γ_i for each i. Let c be the order of $\gamma_i : \mathbf{W}_i \to \mathbf{W}_i$; it is independent of i. Then $\gamma : \mathbf{W} \to \mathbf{W}$ is of order cr. Considered the group $\tilde{\mathbf{W}}_{cr}$, defined as the semidirect product of $\mathbf{W}$ (a normal subgroup) with the cyclic group or order cr with generator γ such that $\gamma w \gamma^{-1} = \gamma(w)$, $w \in \mathbf{W}$.

Let $\mathcal{F}_{cr}$ be the set of irreducible $\tilde{\mathbf{W}}_{cr}$-modules (up to isomorphism) whose restrictions to $\mathbf{W}$ are in $\mathcal{F}$.

For each i, let $(\tilde{\mathcal{F}}_i)_c$ be the set of irreducible $(\tilde{\mathbf{W}}_i)_c$-modules ($(\tilde{\mathbf{W}}_i)_c$ defined with respect to γ_i, see 4.17) whose restrictions to $\mathbf{W}_i$ are in $\mathcal{F}_i$.

We shall define a natural imbedding

(4.20.1) $(\tilde{\mathcal{F}}_1)_c \hookrightarrow \tilde{\mathcal{F}}_{cr}$ such that the diagram

$$\begin{array}{ccc} (\tilde{\mathcal{F}}_1)_c & \hookrightarrow & \tilde{\mathcal{F}}_{cr} \\ \downarrow & & \downarrow \\ \mathcal{F}_1 & \hookrightarrow & \mathcal{F} \end{array}$$

is commutative. (The vertical maps are defined by restriction to $\mathbf{W}_1$, $\mathbf{W}$ respectively. The map $\mathcal{F}_1 \to \mathcal{F}$ is defined by $E_1 \mapsto \underset{i}{\otimes} E_i \in \mathcal{F}$, $(E_1 \in \mathcal{F}_1)$, where $E_i \in \mathcal{F}_i$ corresponds to E_{i+1} under $\gamma : (\mathbf{W}_i)^\vee \xrightarrow{\sim} (\mathbf{W})^\vee_{i+1}$ for each i.) Assume that we are given an extension $\tilde{E}_1$ of E_1 to a $(\tilde{\mathbf{W}}_1)_c$-module. We can find isomorphism $\phi_i : E_i \xrightarrow{\approx} E_{i+1}$ such that $\phi_i(we_i) = \gamma(w)\,\phi_i(e_i)$, $(\forall e_i \in E_i, w \in \mathbf{W}_i)$, for all i and such that $\phi_r \cdots \phi_2\phi_1 : E_1 \to E_1$ is the map corresponding to the action of γ_1 on $\tilde{E}_1$. Then the map $\phi : \otimes E_i \to \otimes E_i$ given by $\phi(e_1 \otimes e_2 \otimes \cdots \otimes e_r) = \phi_r(e_r) \otimes \phi_1(e_1) \otimes \cdots \otimes \phi_{r-1}(e_{r-1})$ $(e_i \in E_i)$ defines a $(\tilde{\mathbf{W}})_{cr}$-module structure on $\otimes E_i$. (It is the action of γ.) Thus, we have associated to any element of $(\tilde{\mathcal{F}}_1)_c$ an element of $\tilde{\mathcal{F}}_{cr}$. As explained in 4.17, there is a natural free action of the group M_c on $(\tilde{\mathcal{F}}_1)_c$, whose orbits are just the fibers of $(\tilde{\mathcal{F}}_1)_c \to \mathcal{F}_1$. Similarly, there is a natural free action of the group M_{cr} on $\tilde{\mathcal{F}}_{cr}$, whose orbits are the nonempty fibers of $\tilde{\mathcal{F}}_{cr} \to \mathcal{F}$. The imbedding $(\tilde{\mathcal{F}}_1)_c \hookrightarrow \tilde{\mathcal{F}}_{cr}$ is M_c-equivariant $(M_c \subset M_{cr})$ and the induced map on the orbit spaces for the actions of M_c and M_{cr} is bijective. Let $\mathcal{G}_i = \mathcal{G}_{\mathcal{F}_i}$, $\tilde{\mathcal{G}}_i = \tilde{\mathcal{G}}_{\mathcal{F}_i,c}$ be the groups associated to $\mathbf{W}_i$, $\mathcal{F}_i$, $\gamma_i : \mathbf{W}_i \to \mathbf{W}_i$ in 4.4-4.13, 4.18, 4.19. We have $\mathcal{G}_i \subset \tilde{\mathcal{G}}_i$ and in $\tilde{\mathcal{G}}_i$ we have a distinguished $\mathcal{G}_i$-coset $\mathcal{G}_i'$; also $\tilde{\mathcal{G}}_i/\mathcal{G}_i$ is cyclic of order c and $\mathcal{G}_i'$ defines a generator of it, hence an isomorphism $\tilde{\mathcal{G}}_i/\mathcal{G}_i = \mathbf{Z}/c\mathbf{Z}$. Note that γ defines isomorphisms $\gamma : \mathcal{G}_i \xrightarrow{\sim} \mathcal{G}_{i+1}$, $\gamma : \tilde{\mathcal{G}}_i \xrightarrow{\sim} \tilde{\mathcal{G}}_{i+1}$, carrying $\mathcal{G}_i'$ to $\mathcal{G}_{i+1}'$. (The $\mathcal{G}_i$, $\tilde{\mathcal{G}}_i$, $\mathcal{G}_i'$ are functorial.) We associate to $\mathcal{F}$ the groups $\mathcal{G} = \mathcal{G}_{\mathcal{F}}$, $\tilde{\mathcal{G}} = \tilde{\mathcal{G}}_{\mathcal{F},cr}$ defined as follows.

$\mathcal{G}$ is the direct product $\prod_i \mathcal{G}_i$ (r factors). Consider the semidirect product $\hat{\mathcal{G}}$ of $\prod_i \tilde{\mathcal{G}}_i$ with the cyclic group of order r with generator δ, so that

$$\delta(\tilde{g}_1, \tilde{g}_2, \cdots, \tilde{g}_r)\delta^{-1} = (\gamma(\tilde{g}_r), \gamma(\tilde{g}_1), \gamma(\tilde{g}_2), \cdots)$$

$(\tilde{g}_i \in \tilde{\mathcal{G}}_i)$. There is a natural homomorphism $h : \hat{G} \to \Gamma$ where Γ is the semidirect product of $(\mathbf{Z}/c\mathbf{Z}) \times \cdots \times (\mathbf{Z}/c\mathbf{Z})$, (r factors) with the cyclic group or order r with generators, acting by

$$\delta(n_1, n_2, \cdots, n_r)\delta^{-1} = (n_r, n_1, n_2, \cdots) ,$$

$(n_i \in \mathbf{Z}/c\mathbf{Z})$; h maps δ to δ and $h(\tilde{g}_1, \tilde{g}_2, \cdots, \tilde{g}_r) = (n_1, n_2, \cdots, n_r)$ where $\tilde{g}_i \to n_i$ is the map $\tilde{\mathcal{G}}_i \to \tilde{\mathcal{G}}_i/\mathcal{G}_i = \mathbf{Z}/c\mathbf{Z}$.

The subgroup Γ_0 of Γ generated by the element $(1, 0, \cdots, 0)\delta$ is cyclic of order rc. We can now define $\tilde{\mathcal{G}}$ to be the inverse image of Γ_0 under $h : \hat{\mathcal{G}} \to \Gamma$. Clearly, $\mathcal{G}$ is a normal subgroup of $\mathcal{G}$ (it is the kernel of h) and h defines an isomorphism $\tilde{\mathcal{G}}/\mathcal{G} = \Gamma_0 = \mathbf{Z}/rc\mathbf{Z}$. Let $\mathcal{G}' = \mathcal{G}'_{\mathcal{F}}$ be the inverse image of the generator of Γ_0 considered above, under $h : \tilde{\mathcal{G}} \to \Gamma_0$. Then $\mathcal{G}'$ consists of the elements of $\hat{\mathcal{G}}$ of form $(g'_1, g_2, \cdots, g_r)\delta$ where $g'_1 \in \mathcal{G}'_1$ and $g_i \in \mathcal{G}_i$, for $i \neq 1$.

The triple $\mathcal{G}_{\mathcal{F}}$, $\mathcal{G}_{\mathcal{F},c}$, $\mathcal{G}'_{\mathcal{F}}$ just defined is as in the first sentence in 4.16. (The definition depends apparently on the choice of index $(i = 1)$; however, the choice of another index leads to a triple which is canonically isomorphic to the previous one.)

We now describe the set $\mathfrak{M}(\mathcal{G} \subset \tilde{\mathcal{G}})$, (see 4.16). There is a natural imbedding

$$\mathfrak{M}(\mathcal{G}_1 \subset \tilde{\mathcal{G}}_1) \hookrightarrow \mathfrak{M}(\mathcal{G} \subset \tilde{\mathcal{G}}) \tag{4.20.2}$$

defined as follows. Start with a pair $(g_1, \sigma_1) \in \mathfrak{M}(\mathcal{G}_1 \subset \tilde{\mathcal{G}}_1)$. Thus, g_1 is an element of $\mathcal{G}_1$ whose centralizer $Z_{\tilde{\mathcal{G}}_1}(g_1)$ meets $\mathcal{G}'_1$, and σ is an irreducible representation of $Z_{\tilde{\mathcal{G}}_1}(g_1)$ which remains irreducible on restriction to $Z_{\mathcal{G}_1}(g_1)$. Let $g = (g_1, g_1, \cdots, g_1) \in \mathcal{G}$. Then the centralizer $Z_{\tilde{\mathcal{G}}}(g)$ is the intersection $\tilde{\mathcal{G}} \cap Z_{\hat{\mathcal{G}}}(g)$. Now $Z_{\hat{\mathcal{G}}}(g)$ is the semi-direct product of $Z_{\tilde{\mathcal{G}}_1}(g_1) \times \cdots \times Z_{\tilde{\mathcal{G}}_1}(g_1)$, (r factors), with the cyclic group generated by δ ; we can construct an irreducible representation of $Z_{\hat{\mathcal{G}}}(g)$ by considering $\sigma_1 \otimes \sigma_1 \otimes \cdots \otimes \sigma_1$ as a representation of $Z_{\tilde{\mathcal{G}}_1}(g_1) \times \cdots \times Z_{\tilde{\mathcal{G}}_1}(g_1)$ and then letting δ act by cyclically permuting the factors. Let σ be the restriction of this representation to $Z_{\tilde{\mathcal{G}}}(g)$. Then (g, σ) defines an element in $\mathfrak{M}(\mathcal{G} \subset \tilde{\mathcal{G}})$. The map $(g_1, \sigma_1) \to (g, \sigma)$ gives the required imbedding (4.20.2).

Again, as explained in 4.17, the group of c-th roots of 1 acts freely on $\mathfrak{M}(\mathcal{G}_1 \subset \tilde{\mathcal{G}}_1)$; similarly, the group of cr-th roots of 1 acts freely on $\mathfrak{M}(\mathcal{G} \subset \tilde{\mathcal{G}})$. The map (4.20.2) is compatible with the two cyclic group actions and the induced map on the orbit spaces for these two group actions is bijective.

Hence, there is a unique imbedding

$$\tilde{\mathfrak{F}}_{cr} \hookrightarrow \mathfrak{M}(\mathcal{G} \subset \tilde{\mathcal{G}}) \tag{4.20.3}$$

(compatible with the actions of the group of cr-th roots of 1) such that the diagram (all of whose arrows other than (4.20.3) are already defined)

$$\begin{array}{ccc} \tilde{\mathfrak{F}}_{cr} & \hookrightarrow & \mathfrak{M}(\mathcal{G} \subset \tilde{\mathcal{G}}) \\ \uparrow & & \uparrow \\ (\tilde{\mathfrak{F}}_1)_c & \hookrightarrow & \mathfrak{M}(\mathcal{G}_1 \subset \tilde{\mathcal{G}}_1) \end{array}$$

is commutative.

We now describe the set $\bar{\mathfrak{M}}(\mathcal{G} \subset \tilde{\mathcal{G}})$, (see 4.16). There is a natural bijection

$$\bar{\mathfrak{M}}(\mathcal{G}_1 \subset \tilde{\mathcal{G}}_1) \xrightarrow{\sim} \bar{\mathfrak{M}}(\mathcal{G} \subset \tilde{\mathcal{G}}) \tag{4.20.4}$$

defined as follows. Start with a pair $(g_1', \bar{\sigma}_1') \in \bar{\mathfrak{M}}(\mathcal{G}_1 \subset \tilde{\mathcal{G}}_1)$. Thus g_1' is an element of $\mathcal{G}_1'$ and $\bar{\sigma}_1$ is an irreducible representation of $Z_{\mathcal{G}_1}(g_1')$

Let $g' = (g_1'e,e,\cdots,e)\delta \in \mathcal{G}'$. Then $Z_{\mathcal{G}}(g') = \{(z,z,\cdots,z) | z \in Z_{\mathcal{G}_1}(g_1')\}$ hence $\bar{\sigma}_1$ may be regarded as a representation of $Z_{\mathcal{G}}(g') \cong Z_{\mathcal{G}_1}(g_1')$. The map (4.20.4) is defined by $(g_1', \bar{\sigma}_1) \mapsto (g', \bar{\sigma}_1)$. It is easy to see that it is a bijection.

We define a map $\Delta : \bar{\mathfrak{M}}(\mathcal{G} \subset \tilde{\mathcal{G}}) \to \{\pm 1\}$ as follows. It is the composition of (4.20.4) and a map $\Delta : \bar{\mathfrak{M}}(\mathcal{G}_1 \subset \tilde{\mathcal{G}}_1) \to \{\pm 1\}$. The last map is the one defined in 4.14, if $c = 1$; if $c \geq 2$ it takes the constant value $(-1)^{a_{E_1} + A_{E_1}}$ where $E_1 \in \mathcal{F}_1$.

We now describe the pairing $\{\ ,\ \} : \bar{\mathfrak{M}}(\mathcal{G} \subset \tilde{\mathcal{G}}) \times \mathfrak{M}(\mathcal{G} \subset \tilde{\mathcal{G}}) \to \bar{\mathbf{Q}}_\ell$, (see (4.16.2)). Let $(x, \bar{\sigma}) \in \bar{\mathfrak{M}}(\mathcal{G} \subset \tilde{\mathcal{G}})$ and let $(x_1, \bar{\sigma}_1) \in \bar{\mathfrak{M}}(\mathcal{G}_1 \subset \tilde{\mathcal{G}}_1)$ be the element corresponding to it under (4.20.4). Let $(y, \tau) \in \mathfrak{M}(\mathcal{G} \subset \tilde{\mathcal{G}})$ and let $(y_1, \tau_1) \in \mathfrak{M}(\mathcal{G}_1 \subset \tilde{\mathcal{G}}_1)$ be an element corresponding to $(y, \tau \otimes \varepsilon^{-1})$ under (4.20.2), where ε is the one-dimensional character of $Z_{\tilde{\mathcal{G}}}(y)$ which is trivial on $Z_{\mathcal{G}}(y)$ and takes on $\mathcal{G}'$ the value α, $(\alpha^{cr} = 1)$. Then

$$\{(x, \bar{\sigma}), (y, \tau)\} = \alpha \cdot \{(x_1, \bar{\sigma}_1), (y_1, \tau_1)\} ;$$

the right-hand side is given by the pairing $\{\ ,\ \} : \bar{\mathfrak{M}}(\mathcal{G}_1 \subset \tilde{\mathcal{G}}_1) \times \mathfrak{M}(\mathcal{G}_1 \subset \tilde{\mathcal{G}}_1) \to \bar{\mathbf{Q}}_\ell$, which is already known.

4.21. Let $\mathbf{W}$, S, R be as in 4.1, (not necessarily irreducible) and let $(\mathbf{W}_i, S_i)$ be the irreducible factors of $(\mathbf{W}, S)$, $(i \in I)$. We identify $\mathbf{W}^\vee$ with $\prod_i (\mathbf{W}_i)^\vee$ using exterior tensor product. A family $\mathcal{F}$ in $\mathbf{W}^\vee$ is of the form $\prod_i \mathcal{F}_i$ where $\mathcal{F}_i$ is a family in $(\mathbf{W}_i)^\vee$.

We define $\mathcal{G}_{\mathcal{F}} = \prod_i \mathcal{G}_{\mathcal{F}_i}$, where $\mathcal{G}_{\mathcal{F}_i}$ are as in 4.4-4.13. Then $\mathfrak{M}(\mathcal{G}_{\mathcal{F}})$ may be identified with $\prod_i \mathfrak{M}(\mathcal{G}_{\mathcal{F}_i})$, (see (4.3.1)). The pairing $\{\ ,\ \} : \mathfrak{M}(\mathcal{G}_{\mathcal{F}}) \times \mathfrak{M}(\mathcal{G}_{\mathcal{F}}) \to \bar{\mathbf{Q}}_\ell$, (see (4.14.3)), is given, via this identification by

$$\{(m_i)_{i \in I}, (m_i')_{i \in I}\} = \prod_{i \in I} \{m_i, m_i'\}$$

where $m_i, m_i' \in \mathfrak{M}(\mathcal{G}_{\mathcal{F}_i})$ and $\{m_i, m_i'\}$ is given by (4.14.3) with $\mathcal{G} = \mathcal{G}_{\mathcal{F}_i}$. We define a set

$$X(\mathbf{W}) = \coprod_{\mathcal{F}} \mathfrak{M}(\mathcal{G}_{\mathcal{F}}), \tag{4.21.1}$$

disjoint union over all families $\mathcal{F}$ in $\mathbf{W}^\vee$. Actually this set depends also on S, R but these will be omitted from the notation.

We have a canonical pairing

$$\{\ ,\ \} : X(\mathbf{W}) \times X(\mathbf{W}) \to \overline{\mathbf{Q}}_\ell \tag{4.21.2}$$

defined as follows: if $m \in \mathfrak{M}(\mathcal{G}_{\mathcal{F}})$, $m' \in \mathfrak{M}(\mathcal{G}_{\mathcal{F}'})$, ($\mathcal{F}, \mathcal{F}'$ families in $\mathbf{W}$), then $\{m, m'\}$ is given by (4.14.3) with $\mathcal{G} = \mathcal{G}_{\mathcal{F}}$ if $\mathcal{F} = \mathcal{F}'$ and is zero if $\mathcal{F} \neq \mathcal{F}'$.

We have also an imbedding

$$\mathbf{W}^\vee \hookrightarrow X(\mathbf{W}) \tag{4.21.3}$$

defined by the condition that its restriction to a family $\mathcal{F}$ in $\mathbf{W}^\vee$ is the product of the imbeddings $\mathcal{F}_i \hookrightarrow \mathfrak{M}(\mathcal{G}_{\mathcal{F}_i})$ constructed in 4.4-4.13.

More generally, we consider an automorphism γ of $\mathbf{W}, S$ (coming from an automorphism of R). Grouping together the factors $\mathbf{W}_i$ in the same γ-orbit, we get a decomposition of $(\mathbf{W}, S)$ into a product of γ-stable factors $(\mathbf{W}^{(j)}, S^{(j)})$, $(j \in J)$, such that for each j, the restriction of γ to $(\mathbf{W}^{(j)}, S^{(j)})$ is of the type considered in 4.20. (It permutes transitively the irreducible components of $(\mathbf{W}^{(j)}, S^{(j)})$.) Let us identify $\mathbf{W}^\vee$ with $\prod_j (\mathbf{W}^{(j)})^\vee$ using external tensor product. A family $\mathcal{F}$ in $\mathbf{W}^\vee$ is of the form $\prod_j \mathcal{F}^{(j)}$, where $\mathcal{F}^{(j)}$ are families in $(\mathbf{W}^{(j)})^\vee, (j \in J)$; the condition that $\mathcal{F}$ is γ-stable is that each $\mathcal{F}^{(j)}$ is γ-stable. Assume that $\mathcal{F}$ is γ-stable. We associate to $\mathcal{F}$ the finite group $\mathcal{G}_{\mathcal{F}} = \prod_{j \in J} \mathcal{G}_{\mathcal{F}^{(j)}}$, where $\mathcal{G}_{\mathcal{F}^{(j)}}$ is defined in 4.20. (This is the same as the $\mathcal{G}_{\mathcal{F}}$ defined earlier in this section.)

We also associate to $\mathcal{F}$ the infinite group $\tilde{\mathcal{G}}_{\mathcal{F}}$ as follows. Let c_j be the order of $\gamma: \mathbf{W}^{(j)} \to \mathbf{W}^{(j)}$ and let $\tilde{\mathcal{G}}_{\mathcal{F}^{(j)},c_j}$ be the group associated to $\mathcal{F}^{(j)}$ and $\gamma: \mathbf{W}^{(j)} \to \mathbf{W}^{(j)}$ in 4.20. Then $\mathcal{G}_{\mathcal{F}(j)} \subset \tilde{\mathcal{G}}_{\mathcal{F}(j),c_j}$ and $\tilde{\mathcal{G}}_{\mathcal{F}(j),c_j}/\mathcal{G}_{\mathcal{F}(j)} = \mathbf{Z}/c_j\mathbf{Z}$. By definition, $\tilde{\mathcal{G}}_{\mathcal{F}}$ consists of the elements $((\tilde{g}^{(j)})_{j\in J}, m) \in \prod_j \tilde{\mathcal{G}}_{\mathcal{F}(j),c_j} \times \mathbf{Z}$ such that for each $j \in J$, the image of $\tilde{g}^{(j)}$ in $\mathbf{Z}/c_j\mathbf{Z}$ is the same as the image of m in $\mathbf{Z}/c_j\mathbf{Z}$. The group $\mathcal{G}_{\mathcal{F}}$ is naturally a normal subgroup of $\tilde{\mathcal{G}}_{\mathcal{F}}$: the element $(g^{(j)})_{j\in J} \in \prod_j \mathcal{G}_{\mathcal{F}(j)}$ is mapped to $((g^{(j)})_{j\in J}, 0)$. We have a canonical isomorphism $\tilde{\mathcal{G}}_{\mathcal{F}}/\mathcal{G}_{\mathcal{F}} = \mathbf{Z}$. Let $\mathcal{G}'_{\mathcal{F}}$ be the inverse image of the generator 1 of $\mathbf{Z}$ under $\tilde{\mathcal{G}}_{\mathcal{F}} \to \tilde{\mathcal{G}}_{\mathcal{F}}/\mathcal{G}_{\mathcal{F}} = \mathbf{Z}$. This is a coset of $\tilde{\mathcal{G}}_{\mathcal{F}}$ with respect to $\mathcal{G}_{\mathcal{F}}$. We have $\mathcal{G}'_{\mathcal{F}} = \{((\tilde{g}^{(j)})_{j\in J}, 1) \mid \tilde{g}^{(j)} \in \mathcal{G}'_{\mathcal{F}(j)}, \forall j\}$. From the triple $\mathcal{G}_{\mathcal{F}} \subset \tilde{\mathcal{G}}_{\mathcal{F}} \supset \mathcal{G}'_{\mathcal{F}}$ we define the sets $\mathfrak{M}(\mathcal{G}_{\mathcal{F}} \subset \tilde{\mathcal{G}}_{\mathcal{F}})$, $\overline{\mathfrak{M}}(\mathcal{G}_{\mathcal{F}} \subset \tilde{\mathcal{G}}_{\mathcal{F}})$ exactly as in 4.16. (In the present situation, c of 4.16 is infinite; when considering representations of an infinite group such as $Z_{\tilde{\mathcal{G}}}(x)$, $x \in \mathcal{G}'$, we tacitly assume that these representations factor through a finite quotient.)

The set $\overline{\mathfrak{M}}(\mathcal{G}_{\mathcal{F}} \subset \tilde{\mathcal{G}}_{\mathcal{F}})$ is finite; it is in a natural bijection with the set $\prod_j \overline{\mathfrak{M}}(\mathcal{G}_{\mathcal{F}(j)} \subset \tilde{\mathcal{G}}_{\mathcal{F}(j),c_j})$: to $(x_j, \bar{\sigma}_j) \in \overline{\mathfrak{M}}(\mathcal{G}_{\mathcal{F}(j)} \subset \tilde{\mathcal{G}}_{\mathcal{F}(j),c_j})$, $(j \in J)$ we associate $x = (x_j)_{j\in J} \in \mathcal{G}$ and $\bar{\sigma} = \otimes_j \bar{\sigma}_j$. Then $(x, \bar{\sigma}) \in \overline{\mathfrak{M}}(\mathcal{G}_{\mathcal{F}} \subset \tilde{\mathcal{G}}_{\mathcal{F}})$. The set $\mathfrak{M}(\mathcal{G}_{\mathcal{F}} \subset \tilde{\mathcal{G}}_{\mathcal{F}})$ is infinite. We define a map $\Delta: \overline{\mathfrak{M}}(\mathcal{G}_{\mathcal{F}} \subset \tilde{\mathcal{G}}_{\mathcal{F}}) \to \{\pm 1\}$ as follows: for $x, \bar{\sigma}$ as above, we set $\Delta(x, \bar{\sigma}) = \prod_j \Delta(x_j, \bar{\sigma}_j)$ where $\Delta(x_j, \bar{\sigma}_j)$ are as defined in 4.20. We have a natural map

$$\prod_j \mathfrak{M}(\mathcal{G}_{\mathcal{F}(j)} \subset \tilde{\mathcal{G}}_{\mathcal{F}(j),c_j}) \to \mathfrak{M}(\mathcal{G}_{\mathcal{F}} \subset \tilde{\mathcal{G}}_{\mathcal{F}}) \tag{4.21.4}$$

defined as follows. Given $(y_j, \tau_j) \in \mathfrak{M}(\mathcal{G}_{\mathcal{F}(j)}, \widetilde{\mathcal{G}}_{\mathcal{F}(j),c_j})$, $(j \in J)$, we define $y = (y_j)_{j\in J} \in \mathcal{G}$ and let τ be the restriction of the representation $(\underset{j}{\otimes}\tau_j)\otimes 1$ of $\Pi\, Z_{\widetilde{\mathcal{G}}_{\mathcal{F}(j),c_j}}(y_j)\times \mathbf{Z}$ to the subgroup $Z_{\widetilde{\mathcal{G}}}(y)$. Then $(y,\tau) \in \mathfrak{M}(\mathcal{G}_{\mathcal{F}} \subset \widetilde{\mathcal{G}}_{\mathcal{F}})$ and this gives the map (4.21.4). The two sets in (4.21.4) have natural group actions: the first set has a free action of $\prod_j M_{c_j}$ (see 4.20); the second set has a free M-action (defined just as in 4.16). The action of $(\alpha_j)_{j\in J} \in \prod_j M_{c_j}$ on the first set in (4.21.4) corresponds to the action of $\prod_j \alpha_j \in M$ on the second set in (4.21.4). The map (4.21.4) induces a bijection on the orbit spaces with respect to the two group actions. Let $\widetilde{\mathbf{W}}$ be defined as in 3.1 and let $\widetilde{\mathcal{F}}$ be the set of irreducible representations of $\widetilde{\mathbf{W}}$ whose restrictions to $\mathbf{W}$ are irreducible and are in $\mathcal{F}$. (Representations of $\widetilde{\mathbf{W}}$ are assumed to factor through a finite quotient.) We define $\widetilde{\mathcal{F}}^{(j)}_{c_j} \subset (\widetilde{\mathbf{W}}^{(j)})^{\vee}_{c_j}$ as in 4.20. We have a natural map

$$\prod_j \widetilde{\mathcal{F}}^{(j)}_{c_j} \to \widetilde{\mathcal{F}} \tag{4.21.5}$$

defined as follows. Given $\widetilde{E}^{(j)} \in \widetilde{\mathcal{F}}^{(j)}_{c_j}$ we may regard $\widetilde{E}^{(j)}$ as a $\widetilde{\mathbf{W}}$-module via the canonical homomorphism $\widetilde{\mathbf{W}} \to (\widetilde{\mathbf{W}}^{(j)})_{c_j}$. We then take the tensor product (over j) of these $\widetilde{\mathbf{W}}$-modules and we get an irreducible $\widetilde{\mathbf{W}}$-module $\widetilde{E}$. Then $(\widetilde{E}^{(j)})_{j\in J} \mapsto \widetilde{E}$ is the map (4.21.5). Again the two sets in (4.21.5) have natural free group actions: the set $\prod_j \widetilde{\mathcal{F}}^{(j)}_{c_j}$ has a free action of $\prod_j M_{c_j}$ (see 4.20), the set $\widetilde{\mathcal{F}}$ has a free M-action (defined just as in 4.17). The map (4.21.5) is compatible with these group actions in the same way as the map (4.21.4). Moreover, the map (4.21.5) induces a bijection on the orbit spaces with respect to these two group actions. It follows that there exists a unique M-equivariant imbedding

(4.21.6)
$$\tilde{\mathcal{F}} \hookrightarrow \mathcal{M}(\mathcal{G}_{\mathcal{F}} \subset \tilde{\mathcal{G}}_{\mathcal{F}})$$

such that the diagram

$$\begin{array}{ccc} \tilde{\mathcal{F}} & \hookrightarrow & \mathcal{M}(\mathcal{G}_{\mathcal{F}} \subset \tilde{\mathcal{G}}_{\mathcal{F}}) \\ \uparrow & & \uparrow \\ \prod_j \tilde{\mathcal{F}}^{(j)}_{c_j} & \longrightarrow & \prod_j \mathcal{M}(\mathcal{G}_{\mathcal{F}(j)} \subset \tilde{\mathcal{G}}_{\mathcal{F}(j),c_j}) \end{array}$$

is commutative. (The vertical maps are given by (4.21.4), (4.21.5). The lower horizontal map is a product over j of maps of type (4.20.3).) We have a pairing

$$\{\ ,\ \}: \bar{\mathcal{M}}(\mathcal{G}_{\mathcal{F}} \subset \tilde{\mathcal{G}}_{\mathcal{F}}) \times \mathcal{M}(\mathcal{G}_{\mathcal{F}} \subset \tilde{\mathcal{G}}_{\mathcal{F}}) \to \bar{\mathbf{Q}}_\ell$$

defined by

(4.21.7)
$$\{(x,\bar{\sigma}),(y,\tau)\} = \sum_{\substack{g \in \mathcal{G}_{\mathcal{F}} \\ x\,gyg^{-1} = gyg^{-1}x}} |Z_{\mathcal{G}_{\mathcal{F}}}(x)|^{-1} |Z_{\mathcal{G}_{\mathcal{F}}}(y)|^{-1} \operatorname{Tr}(g^{-1}x^{-1}g,\tau)\operatorname{Tr}(gyg^{-1},\bar{\sigma}) .$$

Let α be a root of 1 in $\bar{\mathbf{Q}}_\ell^*$ and let ε_α be the one-dimensional character of $Z_{\tilde{\mathcal{G}}}(y)$ which is trivial on $Z_{\mathcal{G}}(y)$ and takes on $\mathcal{G}' \cap Z_{\tilde{\mathcal{G}}}(y)$ the constant value α. Then

(4.21.8)
$$\{(x,\bar{\sigma}),(y,\tau\otimes\varepsilon_\alpha)\} = \alpha^{-1}\{(x,\bar{\sigma}),(y,\tau)\} .$$

Assume that $(x,\bar{\sigma})$ above corresponds to $((x^{(j)},\bar{\sigma}^{(j)}))_{j\in J} \in \prod_j \bar{\mathcal{M}}(\mathcal{G}_{\mathcal{F}(j)} \subset \tilde{\mathcal{G}}_{\mathcal{F}(j),c_j})$ and that $(y,\tau\otimes\varepsilon_\alpha)$ is the image under (4.21.4) of

$$((y^{(j)},\tau^{(j)}))_{j\in J} \in \prod_j \mathcal{M}(\mathcal{G}_{\mathcal{F}(j)} \subset \tilde{\mathcal{G}}_{\mathcal{F}(j),c_j})) .$$

Then

(4.21.9) $$\{(x,\bar\sigma),(y,\tau)\} = \alpha \prod_j \{(x^{(j)},\bar\sigma^{(j)}),(y^{(j)},\tau^{(j)})\}$$

where the right-hand side is computed by (4.16.2).

As in 4.16, we introduce the $\bar{\mathbf{Q}}_\ell$-vector spaces $\mathcal{P},\bar{\mathcal{P}}$ as follows. $\mathcal{P}$ is the space of all functions $f:\mathfrak{M}(\mathcal{G}\subset\tilde{\mathcal{G}})\to\bar{\mathbf{Q}}_\ell$ such that $f(y,\tau\otimes\varepsilon_\alpha) = \alpha f(y,\tau)$ for any $(y,\tau)\in\mathfrak{M}(\mathcal{G}\subset\tilde{\mathcal{G}})$ and any $\alpha\in M$. We define $\bar{\mathcal{P}}$ to be the space of all functions $f:\bar{\mathfrak{M}}(\mathcal{G}\subset\tilde{\mathcal{G}})\to\bar{\mathbf{Q}}_\ell$. We have a linear map $A_1:\mathcal{P}\to\bar{\mathcal{P}}$ defined by

$$(A_1f)(x,\bar\sigma) = \sum \{(x,\bar\sigma),(y,\tau)\}f(y,\tau), \qquad (x,\bar\sigma)\in\bar{\mathfrak{M}}(\mathcal{G}\subset\tilde{\mathcal{G}}),$$

sum over a set of representatives (y,τ) for the orbits of M on $\mathfrak{M}(\mathcal{G}\subset\tilde{\mathcal{G}})$.

We have also a linear map $A_2:\bar{\mathcal{P}}\to\mathcal{P}$ defined by

$$(A_2f)(y,\tau) = \sum \overline{\{(x,\bar\sigma),(y,\tau)\}}f(x,\bar\sigma), \qquad (y,\tau)\in\mathfrak{M}(\mathcal{G}\subset\tilde{\mathcal{G}}),$$

sum over all $(x,\bar\sigma)\in\bar{\mathfrak{M}}(\mathcal{G}\subset\tilde{\mathcal{G}})$.

As in (4.16.3) we have:

(4.21.10) *The linear maps* A_1, A_2 *are isomorphisms inverse to each other.*

We now define two sets

(4.21.11) $$X(\mathbf{W},\gamma) = \coprod_{\mathcal{F}} \mathfrak{M}(\mathcal{G}_{\mathcal{F}}\subset\tilde{\mathcal{G}}_{\mathcal{F}})$$

and

(4.21.12) $$\bar{X}(\mathbf{W},\gamma) = \coprod_{\mathcal{F}} \bar{\mathfrak{M}}(\mathcal{G}_{\mathcal{F}}\subset\tilde{\mathcal{G}}_{\mathcal{F}});$$

in both cases we take disjoint union over all γ-stable families $\mathcal{F}$ in $\mathbf{W}^\vee$. The set (4.21.11) has a natural free M-action given by $\alpha:(y,\tau)\to(y,\tau\otimes\varepsilon_\alpha)$.

(The set of orbits for this action is finite and has the same cardinal as the set (4.21.12), cf. (4.21.10).) When γ = identity, we have $\bar{X}(\mathbf{W},\gamma) = X(\mathbf{W})$, $X(\mathbf{W},\gamma) = X(\mathbf{W}) \times M$. We have a canonical pairing

$$\{\ ,\ \} : \bar{X}(\mathbf{W},\gamma) \times X(\mathbf{W},\gamma) \to \bar{\mathbf{Q}}_\ell \tag{4.21.13}$$

defined as follows. If $m \in \bar{\mathfrak{M}}(\mathcal{G}_{\mathcal{F}} \subset \tilde{\mathcal{G}}_{\mathcal{F}})$, $m' \in \mathfrak{M}(\mathcal{G}_{\mathcal{F}'} \subset \tilde{\mathcal{G}}_{\mathcal{F}'})$ then $\{m, m'\}$ is given by (4.21.7) if $\mathcal{F} = \mathcal{F}'$ and is zero if $\mathcal{F} \neq \mathcal{F}'$.

Let $(\tilde{\mathbf{W}})^\vee_{ex}$ be the set of isomorphism classes of irreducible representations of $\tilde{\mathbf{W}}$, which factor through a finite quotient and which remain irreducible on restriction to $\mathbf{W}$. It has a natural free M-action defined by $\alpha : \tilde{E} \mapsto \tilde{E} \otimes E_\alpha$ where E_α is the one dimensional representation of $\tilde{\mathbf{W}}$ on which $\mathbf{W}$ acts trivially and γ acts as multiplication by α, $(\alpha \in M)$. We have an M-equivariant imbedding

$$(\tilde{\mathbf{W}})^\vee_{ex} \hookrightarrow X(\mathbf{W},\gamma) \tag{4.21.14}$$

defined by the condition that its restriction to $\tilde{\mathcal{F}}$, for a γ-stable family $\mathcal{F}$ in $\mathbf{W}^\vee$ is the imbedding (4.21.6).

4.22. Let $F' : G \to G$ be as in (3.4.1). We assume that G has connected center. Let $L \in \mathfrak{L}$ (see 1.3) and let n be as in (1.4.1). Assume that the set $Z_{L,n}$ (see (2.1.1)) is nonempty. Let $(W_{L,n}, S_{L,n})$, $R_{L,n}$ be as in 1.8, and let γ be the automorphism of the Coxeter group $(W_{L,n}, S_{L,n})$ defined by (2.15.1). (It comes from an automorphism of $R_{L,n}$.) Let $\mathcal{E}_{L,n}$ be the set of irreducible $G^{F'}$-modules defined in 2.19.

Let $\tilde{W}_{L,n}$ be defined as in 3.1 from $W_{L,n}$ and $\gamma : W_{L,n} \to W_{L,n}$. Recall from the previous section that $(\tilde{W}_{L,n})^\vee_{ex}$ is the set of irreducible $\tilde{W}_{L,n}$-modules (over $\bar{\mathbf{Q}}_\ell$) which factor through a finite quotient and which remain irreducible on restriction to $W_{L,n}$. The set $(\tilde{W}_{L,n})^\vee_{ex}$ has a natural free M-action, (see 4.20), and the orbit space of this action can be identified (via restriction to $W_{L,n}$) with the set $(W_{L,n})^\vee_{ex}$, (see 3.1). Let $(\tilde{W}_{L,n})^\vee_{\mathbf{Q}}$ be the subset of $(\tilde{W}_{L,n})^\vee_{ex}$ consisting of those representations which are defined over $\mathbf{Q}$. Then each M-orbit on $(\widetilde{W}_{L,n})^{\vee}_{ex}$ contains exactly 2 representations in $(\widetilde{W}_{L,n})^{\vee}_{\mathbf{Q}}$. (See 3.2.) For each $\widetilde{E} \in (\widetilde{W}_{L,n})^{\vee}_{\mathbf{Q}}$, let $R_{\widetilde{E}} \in \mathcal{R}(G^{F'}) \otimes \mathbf{Q}$ be defined by (3.7.1), and let $x_{\widetilde{E}} \in X(W_{L,n}, \gamma)$ be the image of $\widetilde{E}$ under the imbedding (4.21.14), (with $\mathbb{W} = W_{L,n}$). With these notations, we can now state:

4.23. MAIN THEOREM. *There exists a bijection*

$$\mathcal{E}_{L,n} \leftrightarrow \bar{X}(W_{L,n}, \gamma), \quad (\rho \leftrightarrow \bar{x}_\rho),$$

such that for any $\rho \in \mathcal{E}_{L,n}$ *and any* $\widetilde{E} \in (\widetilde{W}_{L,n})^{\vee}_{\mathbf{Q}}$, *we have*

$$\langle \rho, R_{\widetilde{E}} \rangle_{G^{F'}} = (-1)^{\ell(w_1)} \Delta(\bar{x}_\rho)\{\bar{x}_\rho, x_{\widetilde{E}}\}$$

where $\{\ ,\ \}$ *is the pairing (4.22.13), with* $\mathbb{W} = W_{L,n}$ *and* w_1 *is the element of minimal length in* $Z_{L,n}$. *Here,* $\Delta(\bar{x}_\rho) = \pm 1$ *is as defined in 4.21.*

The proof will be given in the following chapters. In the special case where $n = 1$ (corresponding to "unipotent" representations) and for large enough ground field, a proof can be found in [L_{11}], [L_{12}], [L_8]. (See also [As_3]

Using now 3.9 (ii) we find the following multiplicity formula for $R_{\widetilde{E}}$:

4.24. COROLLARY. $R_{\widetilde{E}} = (-1)^{\ell(w_1)} \cdot \sum_{\rho \in \mathcal{E}_{L,n}} \{\bar{x}_\rho, x_{\widetilde{E}}\}\Delta(\bar{x}_\rho)\rho$.

This suggests defining for any element $x \in X(W_{L,n}, \gamma)$

$$R_x = (-1)^{\ell(w_1)} \sum_{\rho \in \mathcal{E}_{L,n}} \{\bar{x}_\rho, x\}\Delta(\bar{x}_\rho)\rho \in \mathcal{R}(G^{F'}) \otimes \bar{\mathbf{Q}}_\ell . \tag{4.24.1}$$

This reduces to $R_{\widetilde{E}}$ when x is the image of $\widetilde{E}$ under the imbedding (4.21.14). Up to a root of 1, R_x depends only on the M-orbit of $x \in X(W_{L,n}, \gamma)$. (See (4.21.8).)

4.25. COROLLARY. *Let* X_0 *be a set of representatives for the* M*-orbits on* $X(W_{L,n}, \gamma)$. *Then, for any* $\rho \in \mathcal{E}_{L,n}$ *we have*

$$(-1)^{\ell(w_1)} \Delta(\bar{x}_\rho)\rho = \sum_{x \in X_0} \{\overline{\bar{x}_\rho, x}\} R_x .$$

Moreover, if $x, x' \in X_0$ *we have* $\langle R_x, R_{x'} \rangle_{G^{F'}} = \begin{cases} 1, & \text{if } x = x' \\ 0, & \text{if } x \neq x' \end{cases}$.

This follows immediately from 4.24 and (4.21.10).

4.26. If $g \in G^{F'}$ is a semisimple element, the character of $R_{\tilde{E}}$ at g is known explicitly for any $\tilde{E} \in (\tilde{W}_{L,n})^{\vee}_{\mathbf{Q}}$; indeed, the trace of g on $\Sigma(-1)^i H^i_c(X_w, \mathcal{F}^L_{\dot{w},n})$, $(w \in Z_{L,n})$ can be computed by [DL_1, 7.2]. On the other hand, if a class function f on $G^{F'}$ is orthogonal to all $R_{\tilde{E}}$ (for all (L,n)) then f must vanish at g. (This follows from [DL_1, 7.5].) In particular, $\mathrm{Tr}(g, R_x) = 0$, (see (4.24.1)), whenever $x \in X(W_{L,n}, \gamma)$ is not in the image of (4.21.14). (See 4.25.) Hence, from 4.25 we see that for any $\rho \in \mathcal{E}_{L,n}$ we have (assuming the truth of 4.23):

$$\text{(4.26.1)} \quad \mathrm{Tr}(g,\rho) = (-1)^{\ell(w_1)} \Delta(\bar{x}_\rho) \sum_{E \in (W^{\vee}_{L,n})_{ex}} \{\bar{x}_\rho, x_{\tilde{E}}\} \mathrm{Tr}(g, R_{\tilde{E}})$$

(where $\tilde{E}$ are as in 3.6), hence $\mathrm{Tr}(g,\rho)$ can be computed explicitly. This applies in particular, in the case where $g = e$ is the identity element of $G^{F'}$. Using [L_4, (3.19.1)] it is easy to see that

$$\dim(R_{\tilde{E}}) = c_0 q^{b_E} + c_1 q^{b_E + 1} + \cdots$$

where $c_0, c_1, \cdots$ are integers and q is the power of p such that $F' : G \to G$ is a Frobenius map relative to an F_q-rational structure. Moreover, the coefficient c_0 is ± 1 whenever E is a special representation. Let $\mathcal{F}$ be the family corresponding to $\bar{x}_\rho$, let E_1 be the special representation in $\mathfrak{F}$. Then $\{\overline{x}_\rho, x_{\widetilde{E}_1}\} = \pm\frac{1}{d}$, (where d is an integer ≥ 1), $\{\overline{x}_\rho, x_{\widetilde{E}}\} \in \frac{1}{d}\mathbf{Z}$ if $E \in (W_{L,n})^\vee_{ex}$, $E \in \mathfrak{F}$ (see the tables in 4.15) and $\{\overline{x}_\rho, x_{\widetilde{E}}\} = 0$ if $E \in (W_{L,n})^\vee_{ex}$, $E \notin \mathfrak{F}$. Also $b_{E_1} = a_{E_1}$ and $b_E > a_{E_1}$ if $E \in \mathfrak{F}$, $E \neq E_1$ (by the definition of a special representation). Hence from the formula

$$(4.26.2) \qquad \dim(\rho) = (-1)^{\ell(w_1)} \Delta(\overline{x}_\rho) \sum_{\substack{E \in (W_{L,n})^\vee_{ex} \\ E \in \mathfrak{F}}} \{\overline{x}_\rho, x_{\widetilde{E}}\} \dim R_{\widetilde{E}}$$

we get

$$\pm\, d \cdot \dim(\rho) = q^{a_{E_1}} + \text{integral combination of powers of } q \text{ with exponent } > a_{E_1}.$$

In particular, we have (again assuming the truth of 4.23):

$$(4.26.3) \qquad \dim(\rho) = \pm\{\overline{x}_\rho, x_{\widetilde{E}_1}\} q^{a_{E_1}} \cdot N$$

where N is an integer satisfying $N \equiv 1 \pmod q$.

Chapter 5
CELLS IN WEYL GROUPS

This chapter is concerned with certain virtual representations a_x of a Weyl group $\mathbf{W}$, defined for any element $x \in \mathbf{W}$. Their definition involves considering the leading coefficients of the trace of the basis element T_x of the Hecke algebra H in the various irreducible representations of H. We establish various properties of the elements a_x and show how they are related to the cells of $\mathbf{W}$ (in the sense of $[KL_1]$). This leads to a new proof of part of the recent results of Barbasch-Vogan $[BV_1]$, $[BV_2]$.

5.1. We place ourselves again in the situation of 3.1 and 3.3. Thus, we are given a Weyl group $(\mathbf{W}, S)$ and an (ordinary) automorphism γ of it. We define $\widetilde{\mathbf{W}}$, ℓ, $\mathbf{W}^\vee$, $\mathbf{W}^\vee_{ex}$, H, $\widetilde{H}$ as in *loc. cit.* Given any $x \in \mathbf{W}$, there is a canonical element $C_x \in H$ defined uniquely by the following property (see $[KL_1, 1.1]$):

(5.1.1)
$$\begin{aligned} C_x &= \sum_{y \leq x} (-1)^{\ell(x)-\ell(y)} P_{y,x}(u^{-1})\, u^{(\ell(x)/2)-\ell(y)} T_y \\ &= \sum_{y \leq x} (-1)^{\ell(x)-\ell(y)} P_{y,x}(u)\, u^{-(\ell(x)/2)+\ell(y)} T^{-1}_{y^{-1}} \end{aligned}$$

where $P_{y,x}(u)$ are polynomials in u with integral coefficients of degree $\leq \frac{1}{2}(\ell(x)-\ell(y)-1)$ if $y < x$ and $P_{x,x}(u) = 1$. Here $\leq$ is the standard partial order on $\mathbf{W}$; compare (1.4.3). (The polynomials $P_{y,x}$ have been used already in (1.10.1), for example.) As in $[KL_1, (1.10)]$, we shall denote

(5.1.2) $$C'_x = \sum_{y \leq x} P_{y,x}(u) u^{-\ell(x)/2} T_y = \sum_{y \leq x} P_{y,x}(u^{-1}) u^{\ell(x)/2} T^{-1}_{y^{-1}} .$$

Following [KL_2, (2.1.6)] we define, for any $y \leq x$ in $\mathbf{W}$ polynomials $Q_{y,x} \in \mathbf{Z}[u]$ by

$$(5.1.3) \qquad \sum_{\substack{y \in \mathbf{W} \\ z \leq y \leq x}} (-1)^{\ell(x)-\ell(y)} P_{z,y} Q_{y,x} = \begin{cases} 1 & \text{if } z = x \\ 0 & \text{if } z < x . \end{cases}$$

(5.1.4) $Q_{y,x}(u)$ has degree $\leq \frac{1}{2}(\ell(x)-\ell(y)-1)$ if $y < x$ and $Q_{x,x} = 1$.

This follows by induction on $\ell(x) - \ell(y)$ from (5.1.3) or, alternatively, from the formula

$$(5.1.5) \qquad Q_{y,x} = P_{w_0x,w_0y} \qquad (y \leq x)$$

(see [KL_1, 3.1]) where w_0 denotes the longest element of $\mathbf{W}$.

For any $x \in \mathbf{W}$, we set

$$(5.1.6) \qquad D_x = \sum_{\substack{y \\ x \leq y}} Q_{x,y}(u^{-1}) u^{-(\ell(x)/2)} T_y$$

$$(5.1.7) \qquad D'_x = \sum_{\substack{y \\ x \leq y}} (-1)^{\ell(y)-\ell(x)} Q_{x,y}(u) u^{(\ell(x)/2)-\ell(y)} T_y .$$

Thus, we have 5 basis (T_x), (C_x), (C'_x), (D_x), (D'_x) of H as a $\mathbf{Q}[u^{1/2}, u^{-1/2}]$-module. It is easy to see, using (5.1.5), that

$$(5.1.8) \qquad D_x = u^{-\nu/2} C'_{xw_0} T_{w_0} , \quad D'_x = u^{-\nu/2} C_{xw_0} T_{w_0}$$

where $\nu = \ell(w_0)$.

Let $\tau : H \to \mathbf{Q}[u^{1/2}, u^{-1/2}]$ be the $\mathbf{Q}$-linear map such that $\tau(T_w) = 0$ if $w \neq e$, $\tau(T_e) = 1$. It is well known that

$$(5.1.9) \qquad \tau(T_x T_{y^{-1}}) = \begin{cases} u^{\ell(x)} , & \text{if } x = y \\ 0 , & \text{if } x \neq y . \end{cases}$$

Using (5.1.3), and the identity $Q_{x,y} = Q_{x^{-1},y^{-1}}$, it follows that

$$(5.1.10) \qquad \tau(C_x D_{y^{-1}}) = \tau(C'_x D'_{y^{-1}}) = \begin{cases} 1, & \text{if } x = y \\ 0, & \text{if } x \neq y . \end{cases}$$

We note also that τ has the following property which follows immediately from (5.1.9):

$$(5.1.11) \qquad \begin{cases} \tau(hh') = \tau(h'h) & , (h, h' \in H) \\ \tau(hh') = 0 \quad \text{for all } h' \in H \Rightarrow h = 0 . \end{cases}$$

For any $y < x$ in W such that $\ell(x) \neq \ell(y) \pmod 2$, we define $\mu(y,x)$ to be the coefficient of $u^{\frac{1}{2}(\ell(x)-\ell(y)-1)}$ in $P_{x,y}(u)$; (cf. [KL_1]). For $s \in S$ and $x \in W$, we have:

$$(5.1.12) \quad \begin{cases} C_s \cdot C_x = \begin{cases} -(u^{1/2}+u^{-1/2})C_x & , \text{ if } sx < x \\ C_{sx} + \sum\limits_{\substack{y<x \\ \ell(y)\neq\ell(x)(2) \\ sy<y}} \mu(y,x)C_y , & \text{ if } sx > x \end{cases} \\ \\ C_x \cdot C_s = \begin{cases} -(u^{1/2}+u^{-1/2})C_x & , \text{ if } xs < x \\ C_{xs} + \sum\limits_{\substack{y<x \\ \ell(y)\neq\ell(x)(2) \\ ys<y}} \mu(y,x)C_y , & \text{ if } xs > x \end{cases} \end{cases}$$

(see [KL_1, 1.2]). Let $\underset{L}{\leq}$ be the preorder relation on W generated by the relation " $x' \underset{L}{\leq} x$ if there exists $s \in S$ such that $C_{x'}$ appears with non-zero coefficient in $C_s \cdot C_x$ (expressed in the C_w-basis)." The equivalence relation associated to $\underset{L}{\leq}$ is denoted $\underset{L}{\sim}$ and the corresponding equivalence classes on $\mathbf{W}$ are called the left cells of $\mathbf{W}$, (cf. [KL_1, §1]). We define $x \underset{R}{\leq} y$ to be the same as $x^{-1} \underset{L}{\leq} y^{-1}$. The equivalence relation associated to the preorder $\underset{R}{\leq}$ is denoted $\underset{R}{\sim}$ and the corresponding equivalence classes on $\mathbf{W}$ are called the right cells of $\mathbf{W}$ (cf. [KL_1, §1]). Given $x, y \in W$, we say that $x \underset{LR}{\leq} y$ if there exists a sequence $x = x_0, x_1, \cdots, x_n = y$ of elements in $\mathbf{W}$ such that, for $i = 0, 1, \cdots, n-1$, we have either $x_i \underset{L}{\leq} x_{i+1}$ or $x_i \underset{R}{\leq} x_{i+1}$. The equivalence relation on $\mathbf{W}$ corresponding to the preorder $\underset{LR}{\leq}$ is denoted $\underset{LR}{\sim}$ and the corresponding equivalence classes on $\mathbf{W}$ are called the two-sided cells of $\mathbf{W}$, (cf. [KL_1, §1]).

We shall need the following result.

(5.1.13) If $y \underset{L}{\leq} x$ and $y \underset{LR}{\sim} x$ then $y \underset{L}{\sim} x$.

No elementary proof of this result is known; in [L_9, 4], it is deduced from properties of primitive ideals in an enveloping algebra. For any $x \in X$, we denote I_x^L (resp. $\hat{I}_x^L$) the $\mathbf{Q}[u^{1/2}, u^{-1/2}]$ submodule of H spanned by the elements C_y $(y \underset{L}{\leq} x)$ (resp. by elements C_y $(y \underset{L}{\leq} x, y \underset{L}{\not\sim} x)$). We define similarly I_x^{LR} and $\hat{I}_x^{LR}$ by replacing $\underset{L}{\leq}$ and $\underset{L}{\not\sim}$ by $\underset{LR}{\leq}$ and $\underset{LR}{\not\sim}$ in the previous definition. It is clear from (5.1.12) that I_x^L, $\hat{I}_x^L$ are left ideals of H and that I_x^{LR}, $\hat{I}_x^{LR}$ are two-sided ideals of H.

We shall also need the following characterization of the relations $y \underset{L}{\leq} x$, $y \underset{R}{\leq} x$:

$$(5.1.14) \qquad \begin{cases} \text{We have } y \underset{L}{\leq} x \text{ if and only if } C_x D_{y^{-1}} \neq 0 \\ \text{We have } y \underset{R}{\leq} x \text{ if and only if } D_{y^{-1}} C_x \neq 0 \end{cases}$$

We shall only prove the first statement; the second statement is proved in a similar way. Assume first that $y \underset{L}{\leq} x$.

Since the μ function in (5.1.12) has ≥ 0 values [KL$_2$, 4.8], there exists $s_1, \cdots, s_p \in S$ such that $C_{s_1} C_{s_2} \cdots C_{s_p} C_x = \sum_{z \underset{L}{\leq} x} a_z C_z$ ($a_z \in \mathbf{Q}[u^{1/2}, u^{-1/2}]$), with $a_y \neq 0$. (All a_z in the previous sum are in fact polynomials with ≥ 0 integral coefficients in $-(u^{1/2}+u^{-1/2})$.) Using (5.1.10) we have

$$a_y = \tau(C_{s_1} C_{s_2} \cdots C_{s_p} C_x D_{y^{-1}})$$

hence $C_x D_{y^{-1}} \neq 0$. Conversely, assume that $C_x D_{y^{-1}} \neq 0$. From (5.1.11) we see that there exists $z \in \mathbf{W}$ such that $\tau(T_z C_x D_{y^{-1}}) \neq 0$. We have $T_z C_x = \sum_{x' \underset{L}{\leq} x} \beta_{x'} C_{x'}$, $\beta_{x'} \in \mathbf{Q}[u^{1/2}, u^{-1/2}]$ (since I_x^L is a left ideal of H). By (5.1.10), we have $\tau(T_x C_z D_{y^{-1}}) = \sum_{x' \underset{L}{\leq} x} \beta_{x'} \tau(C_{x'} D_{y^{-1}}) = \begin{cases} \beta_y, & \text{if } y \underset{L}{\leq} x \\ 0, & \text{otherwise} \end{cases}$. Since $\tau(T_x C_z D_{y^{-1}})$ is known to be $\neq 0$, it follows that $y \underset{L}{\leq} x$ and the first statement of (5.1.14) is proved.

Let $\Phi: H \to H$ be the ring homomorphism defined by $\Phi(u^{1/2}) = -u^{1/2}$, $\Phi(T_x) = (-u)^{\ell(x)} T_{x^{-1}}^{-1}$, $(x \in \mathbf{W})$. Then $\Phi^2 = 1$ and it is clear that $\Phi(C_x) = C'_x$, $(x \in \mathbf{W})$. Applying Φ to (5.1.12) we get for $s \in S$, $x \in \mathbf{W}$:

$$(5.1.15) \quad C'_s \cdot C'_x = \begin{cases} (u^{1/2}+u^{-1/2}) C'_x & , \text{ if } sx < x \\ C'_{sx} + \sum\limits_{\substack{y<x \\ \ell(y) \not\equiv \ell(x) (2) \\ sy<y}} \mu(y,x) C'_y & , \text{ if } sx > x \end{cases}$$

and a similar formula for $C'_x \cdot C'_s$.

Replacing in (5.1.15) x by xw_0 and multiplying both sides by $u^{-\nu/2}T_{w_0}$, we get

$$(5.1.16)\quad \begin{cases} C'_sD_x = \begin{cases} (u^{1/2}+u^{-1/2})D_x & , \text{ if } sx > x \\ D_{sx} + \sum\limits_{\substack{y>x \\ \ell(y)\not\equiv\ell(x)(2) \\ sy>y}} \mu(x,y)D_y & , \text{ if } sx < x \end{cases} \\ D_xC'_s = \begin{cases} (u^{1/2}+u^{-1/2})D_x & , \text{ if } xs > x \\ D_{xs} + \sum\limits_{\substack{y>x \\ \ell(y)\not\equiv\ell(x)(2) \\ ys>y}} \mu(x,y)D_y & , \text{ if } xs < x . \end{cases} \end{cases}$$

For each $x \in \mathbf{W}$, we denote by J_x^{LR} the $\mathbf{Q}[u^{1/2}, u^{-1/2}]$-submodule of H spanned by the elements D_y, ($y \underset{LR}{\geq} x$). It follows from (5.1.16) that J_x^{LR} is a two-sided ideal of H.

Let $\mathcal{K}$ be the field of formal power series $\mathbf{Q}((u^{1/2}))$. We shall denote $H_{\mathcal{K}} = H \otimes \mathcal{K}$ (a semisimple $\mathcal{K}$-algebra), $I^{LR}_{x,\mathcal{K}} = I^{LR}_x \otimes \mathcal{K}$, $J^{LR}_{x,\mathcal{K}} = J^{LR}_x \otimes \mathcal{K}$. We define similarly $\hat{I}^{LR}_{x,\mathcal{K}}$, $I^{L}_{x,\mathcal{K}}$, $\hat{I}^{L}_{x,\mathcal{K}}$.

For each irreducible $\mathbf{Q}[\mathbf{W}]$-module E, the H-module $E(u)$ (see 3.3) gives rise to an $H_{\mathcal{K}}$-module $E(u)_{\mathcal{K}} = E(u) \otimes \mathcal{K}$. This gives a 1-1 correspondence between the set of isomorphism classes of irreducible $\mathbf{Q}[\mathbf{W}]$-modules and the set of isomorphism classes of (absolutely) irreducible $H_{\mathcal{K}}$-modules.

We shall denote $I_{E,\mathcal{K}}$ the minimal 2-sided ideal (simple component) of $H_{\mathcal{K}}$ corresponding to the irreducible $H_{\mathcal{K}}$-module $E(u)_{\mathcal{K}}$.

We shall say that $E \underset{LR}{\leq} x$, ($x \in \mathbf{W}$), if $I_{E,\mathcal{K}} \subset I^{LR}_{x,\mathcal{K}}$. We shall say that $E \underset{LR}{<} x$ if $I_{E,\mathcal{K}} \subset \hat{I}^{LR}_{x,\mathcal{K}}$. We shall say that $E \underset{LR}{\sim} x$ if $E \underset{LR}{\leq} x$ but

$E \underset{LR}{\not\leq} x$. It is then clear that

$$I^{LR}_{z,\mathcal{K}} = \bigoplus_{\substack{E \\ E \underset{LR}{\leq} x}} I_{E,\mathcal{K}} \,.$$

Consider the $\mathcal{K}$-bilinear form $(\ ,\ ): H_{\mathcal{K}} \times H_{\mathcal{K}} \to \mathcal{K}$ defined by $(h,h') = \tau(hh')$. It is symmetric and nonsingular (5.1.11). Since $I_{E,\mathcal{K}} \cdot I_{E',\mathcal{K}} = 0$ for $E \not\approx E'$, the decomposition $H_{\mathcal{K}} = \oplus_E I_{E,\mathcal{K}}$ is an orthogonal decomposition for $(\ ,\ )$. Taking the orthogonals with respect to $(\ ,\ )$ of the two sides in (5.1.17) we get (cf. (5.1.10)):

(5.1.18) $$\bigoplus_{\substack{E \\ E \underset{LR}{\not\leq} x}} I_{E,\mathcal{K}} = \mathcal{K}\text{-subspace of } H_{\mathcal{K}} \text{ spanned by the } D_{y^{-1}}\ (y \underset{LR}{\not\leq} x)\,.$$

We shall say that $E \underset{LR}{\geq} x'$ if there exists $x'' \in W$ with $E \underset{LR}{\sim} x''$, $x'' \underset{LR}{\geq} x'$. Assume that $E \underset{LR}{\geq} x'$ and let $x'' \in W$ be such that $E \underset{LR}{\sim} x''$, $x'' \underset{LR}{\geq} x'$. From (5.1.18) we see that $I_{E,\mathcal{K}}$ is contained in the $\mathcal{K}$-subspace

$$\bigcap_{\substack{x \\ x'' \underset{LR}{\not\leq} x}} \{\mathcal{K}\text{-subspace of } H_{\mathcal{K}} \text{ spanned by the } D_{y^{-1}}, y \underset{LR}{\not\leq} x\}$$

$$= \mathcal{K}\text{-subspace of } H_{\mathcal{K}} \text{ spanned by the } D_{y^{-1}}, (y \underset{LR}{\geq} x'')\,,$$

hence it is contained in $J^{LR}_{x'^{-1},\mathcal{K}}$. It follows that

$$\bigoplus_{\substack{E \\ E \underset{LR}{\geq} x'}} I_{E,\mathcal{K}} \subset J^{LR}_{x'^{-1},\mathcal{K}} \,.$$

Conversely, let $y \in W$ be such that $y \underset{LR}{\geq} x'$. From (5.1.18) we see that

$D_{y^{-1}}$ is contained in $\bigcap\limits_{\substack{x \\ y \not\leq_{LR} x}} \Big(\bigoplus\limits_{\substack{E \\ E \not\leq_{LR} x}} I_{E,\mathcal{K}} \Big) = \bigoplus\limits_{\substack{E \\ E \geq_{LR} y}} I_{E,\mathcal{K}}$ hence $J^{LR}_{x'^{-1},\mathcal{K}} \subset \bigoplus\limits_{\substack{E \\ E \geq_{LR} x'}} I_{E,\mathcal{K}}$, and hence

(5.1.19)
$$J^{LR}_{x'^{-1},\mathcal{K}} = \bigoplus_{\substack{E \\ E \geq_{LR} x'}} I_{E,\mathcal{K}} .$$

(5.1.20) If $E \in \mathbf{W}^{\vee}_{ex}$ and $\tilde{E}$ is one of the two extensions of E to an irreducible $\mathbf{Q}[\tilde{\mathbf{W}}]$-module, we shall denote $\tilde{E}(u)_{\mathcal{K}} = \tilde{E}(u) \otimes \mathcal{K}$ (see (3.3); it is an irreducible module for $\tilde{H}_{\mathcal{K}} = \tilde{H} \otimes \mathcal{K}$.

Using (3.3.1), (3.3.4), we see that for any $x \in \mathbf{W}$, there is a unique element $c_{\gamma x,\tilde{E}} \in \mathbf{Z}$ such that

(5.1.21) $\mathrm{Tr}(u^{-\ell(x)/2} T_{\gamma x}, \tilde{E}(u)) = (-1)^{\ell(x)} c_{\gamma x,\tilde{E}} u^{-a_E/2} +$higher powers of $u^{1/2}$.

(The sum over $x \in \mathbf{W}$ of squares of the expressions (5.1.21) must be equal, by (3.3.4), to $\dim(E) u^{-a_E}$ + higher powers of u ; a_E is defined in (4.1.1).)

(5.1.22) When γ = identity, and if $\gamma \in \tilde{\mathbf{W}}$ acts trivially on $\tilde{E}$, we set $c_{\gamma x,\tilde{E}} = c_{x,E}$ where $E = \tilde{E}|\mathbf{W}$.

Similarly, for any $x \in W$, there is a unique element $c'_{\gamma x,\tilde{E}} \in \mathbf{Z}$ such that

(5.1.23) $\mathrm{Tr}(u^{-\ell(x)/2} T_{\gamma x}, \tilde{E}(u)) = c'_{\gamma x,\tilde{E}} u^{(\nu - A_E)/2} +$lower powers of $u^{1/2}$.

(Recall that $\nu = \ell(w_0)$; A_E is defined in (4.1.1).)

We have the following result:

5.2. LEMMA. (i) *If* $E, \tilde{E}$ *are given as above, then there exists at least one* $x \in \mathbf{W}$ *such that* $c_{\gamma x,\tilde{E}} \neq 0$.

(ii) *If* $c_{\gamma x,\tilde{E}} \neq 0$, *then* $E \underset{LR}{\sim} x$.

(iii) *For any* $x \in W$, *we have* $x \underset{LR}{\sim} x^{-1}$.

Proof. From the remark following (5.1.21) we see that $\sum_{x \in W} c^2_{\gamma x,\tilde{E}} = \dim(E)$, and (i) follows. We now prove (ii).

We have

(5.2.1) $\operatorname{Tr}(T_\gamma C_x, \tilde{E}(u)) = (-1)^{\ell(x)} c_{\gamma x,\tilde{E}} u^{-a_E/2} + \text{higher powers of } u^{1/2}$

(5.2.2) $\operatorname{Tr}(T_\gamma D_x, \tilde{E}(u)) = (-1)^{\ell(x)} c_{\gamma x,\tilde{E}} u^{-a_E/2} + \text{higher powers of } u^{1/2}$.

Indeed,

$$\operatorname{Tr}(T_\gamma C_x, \tilde{E}(u)) = \operatorname{Tr}(u^{-\ell(x)/2} T_{\gamma x}, \tilde{E}(u)) +$$

$$+ \sum_{y<x} (-1)^{\ell(x)-\ell(y)} P_{y,x}(u^{-1}) u^{(\ell(x)-\ell(y))/2} \operatorname{Tr}(u^{-\ell(y)/2} T_{\gamma y}, \tilde{E}(u)).$$

From the bounds (5.1.1) satisfied by the degree of $P_{y,x}$ and from (5.1.21) we see that the terms in the sum $\sum_{y<x}$ contribute only powers of $u^{1/2}$ with exponent $> -a_E$ and (5.2.1) follows. The proof of (5.2.2) is similar; it is based on the bounds (5.1.4) for the degree of $Q_{y,x}$.

Assume now that $c_{\gamma x,\tilde{E}} \neq 0$. Then $\operatorname{Tr}(T_\gamma C_x, \tilde{E}(u)) \neq 0$ hence $C_x \cdot I_{E,\mathcal{K}} \neq 0$.

The $\mathcal{K}$-linear map $\tilde{H}_{\mathcal{K}} \to \tilde{H}_{\mathcal{K}}$ defined by $T_w \to T_{w^{-1}}$ ($w \in \tilde{W}$) is an algebra anti-automorphism. It carries $T_\gamma D_x$ ($x \in W$) to $D_{x^{-1}} T_{\gamma^{-1}}$. Using (3.3.3) we see then that

(5.2.3) $$\operatorname{Tr}(T_\gamma D_x, \tilde{E}(u)) = \operatorname{Tr}(D_{x^{-1}} T_{\gamma^{-1}}, \tilde{E}(u)).$$

This is nonzero, by (5.2.2) and by the assumption $c_{\gamma x,\tilde{E}} \neq 0$. It follows that $D_{x^{-1}} : \tilde{E}(u) \to \tilde{E}(u)$ is nonzero, hence $D_{x^{-1}} \cdot I_{E,\mathcal{K}} \neq 0$.

Since $I^{LR}_{x,\mathcal{K}}$, $I_{E,\mathcal{K}}$ are two-sided ideals, $C_x \cdot I_{E,\mathcal{K}} \neq 0$ implies that $C_x \cdot I_{E,\mathcal{K}} \subset I^{LR}_{x,\mathcal{K}} \cap I_{E,\mathcal{K}}$ hence $I_{E,\mathcal{K}} \subset I^{LR}_{x,\mathcal{K}}$. Thus, $E \underset{LR}{\leq} x$. Assume now that $E \underset{LR}{<} x$. Then there exists x', $x' \underset{LR}{<} x$, such that $I_{E,\mathcal{K}} \subset I^{LR}_{x',\mathcal{K}}$. From $D_{x^{-1}} \cdot I_{E,\mathcal{K}} \neq 0$, it follows that $D_{x^{-1}} \cdot I^{LR}_{x',\mathcal{K}} \neq 0$ hence $D_{x^{-1}} \cdot C_{x''} \neq 0$ for some x'', $x'' \underset{LR}{<} x$. From (5.1.14) we see then that $x \underset{R}{\leq} x''$. But this contradicts $x'' \underset{LR}{<} x$. Hence, we have $E \underset{LR}{\sim} x$, and (ii) is proved.

For the proof of (iii), we may assume that $\gamma : \mathbf{W} \to \mathbf{W}$ is the identity. Let $x \in \mathbf{W}$, and let E be an irreducible representation of $\mathbf{W}$ such that $E \underset{LR}{\sim} x$. (Such E exists since $I^{LR}_{x,\mathcal{K}} / \hat{I}^{LR}_{x,\mathcal{K}} \neq 0$.) By (i), there exists $x' \in \mathbf{W}$ such that $c_{x',E} = 0$. By (ii) we have $E \underset{LR}{\sim} x'$. From (3.3.3) and (5.1.21) with $\gamma = 1$, we have $c_{x',E} = c_{x'^{-1},E}$. Using again (ii), we have $E \underset{LR}{\sim} x'^{-1}$. This, together with $E \underset{LR}{\sim} x$, $E \underset{LR}{\sim} x'$ shows that $x \underset{LR}{\sim} x'$, $x' \underset{LR}{\sim} x'^{-1}$. From $x \underset{LR}{\sim} x'$ it follows that $x^{-1} \underset{LR}{\sim} x'^{-1}$. We deduce that $x \underset{LR}{\sim} x^{-1}$. The lemma is proved.

5.3. From (5.1.17) we have, for any $x \in \mathbf{W}$:

$$C_x \in \bigoplus_{\substack{E \\ E \underset{LR}{\leq} x}} I_{E,\mathcal{K}} = \Big(\bigoplus_{\substack{E \\ E \underset{LR}{<} x}} I_{E,\mathcal{K}}\Big) \oplus \Big(\bigoplus_{\substack{E \\ E \underset{LR}{\sim} x}} I_{E,\mathcal{K}}\Big) \subset \Big(\bigoplus_{\substack{E \\ E \underset{LR}{<} x}} I_{E,\mathcal{K}}\Big) \oplus \Big(\bigoplus_{\substack{E \\ E \underset{LR}{\geq} x}} I_{E,\mathcal{K}}\Big).$$

The components of C_x in the two summands in the last direct sum decomposition can be written (cf. (5.1.17), (5.1.19)):

$$\sum_{y \underset{LR}{<} x} a_{y,x} C_y \,, \qquad \sum_{y \underset{LR}{\geq} x} b_{y,x} D_{y^{-1}} \,.$$

We can state:

(5.3.1) *For any* $x \in \mathbf{W}$, *there exist uniquely defined elements* $a_{y,x}$ $(y \underset{LR}{<} x)$ *and* $b_{y,x}$ $(y \underset{LR}{\geq} x)$ *in* $\mathcal{K}$ *such that*

$$C_x - \sum_{\substack{y \underset{LR}{<} x}} a_{y,x} C_y = \sum_{\substack{y \underset{LR}{\geq} x}} b_{y,x} D_{y^{-1}} \in \bigoplus_{\substack{E \\ E \underset{LR}{\sim} x}} I_{E,\mathcal{K}} \, .$$

Moreover, we must have $a_{y,x} \in \mathbf{Z}[[u^{1/2}]]$, $(y \underset{LR}{<} x)$, $b_{y,x} \in \mathbf{Z}[[u^{1/2}]]$, $(y \underset{LR}{\geq} x)$ *and* $a_{y,x}(0) = 0$ $(y \underset{LR}{<} x)$.

The existence and uniqueness of the $a_{y,x}$ and $b_{y,x}$ (in $\mathcal{K}$) has been already proved. We shall now prove the remaining statements in (5.3.1).

We invert (5.1.6) using (5.1.3); we get

$$T_y = \sum_{\substack{z \\ z \geq y}} (-1)^{\ell(x)-\ell(y)} P_{y,z}(u^{-1}) u^{\ell(z)/2} D_z \, .$$

Introducing this in (5.1.1) we get

$$C_x = \sum_{z} \sum_{\substack{y \\ y \leq x \\ y \leq z}} (-1)^{\ell(x)+\ell(z)} P_{y,x}(u^{-1}) P_{y,z}(u^{-1}) u^{\frac{\ell(x)+\ell(z)}{2} - \ell(y)} D_z \, .$$

The coefficient of D_z is equal by (5.1.10) to $\tau(C_{z^{-1}} C_x)$. Hence

(5.3.2) $$\tau(C_{z^{-1}} C_x) = \sum_{\substack{y \\ y \leq x \\ y \leq z}} (-1)^{\ell(x)+\ell(z)} P_{y,x}(u^{-1}) P_{y,z}(u^{-1}) u^{\frac{\ell(x)+\ell(z)}{2} - \ell(y)} \, .$$

From the bounds (5.1.1) for the degrees of the polynomials $P_{y,x}$, $P_{y,z}$, we see immediately that

(5.3.3) $\tau(C_{z^{-1}} C_x) \in \mathbf{Z}[u^{1/2}]$ and its constant term is zero, if $x \neq z$ and 1 if $x = z$.

We now express C_x, C_y in terms of D_z in (5.3.1). We get

$$\text{(5.3.4)}\quad \sum_z \tau(C_{z^{-1}}C_x)D_z - \sum_{\substack{y \\ y \underset{LR}{<} x}} \left(a_{y,x}\sum_z \tau(C_{z^{-1}}C_y)D_z\right) = \sum_{\substack{z \\ z \underset{LR}{\geq} x}} b_{z^{-1},x} D_z$$

(here we use the fact that $z \underset{LR}{\geq} x$ if and only if $z^{-1} \underset{LR}{\geq} x$, which follows from 5.2 (iii)). Comparing the coefficient of D_z, $(z \underset{LR}{<} x)$ in the two sides of the previous equality, we get

$$\text{(5.3.5)}\quad \tau(C_{z^{-1}}C_x) - \sum_{\substack{y \\ y \underset{LR}{<} x}} \tau(C_{z^{-1}}C_y)a_{y,x} = 0,\ (\forall z,\ z \underset{LR}{<} x)\,.$$

We shall regard this as a system of linear equations for the unknowns $a_{y,x}$ (where x is fixed). This system has coefficients in $\mathbf{Z}[u^{1/2}]$, (see (5.3.3)). Its matrix, $(\tau(C_{z^{-1}}C_y))$, where $y,z \underset{LR}{<} x$, is square. (We may assume that its index set is nonempty; otherwise, there is nothing to prove.) This matrix reduces to the identity matrix when we specialize $u^{1/2}$ to 0 (see (5.3.3)). Hence its determinant is a polynomial in $\mathbf{Z}[u^{1/2}]$ with constant term 1. It follows that the inverse of this determinant is in $\mathbf{Z}[[u^{1/2}]]$ and has constant term 1. Therefore, from (5.3.4) we can deduce that $a_{y,x} \in \mathbf{Z}[[u^{1/2}]]$ for all y, $y \underset{LR}{<} x$, and that the constant term of $a_{y,x}$ is the same as the constant term of $\tau(C_{y^{-1}}C_x)$, hence it is zero (by (5.3.3)). Once the $a_{y,x}$ are known to be in $\mathbf{Z}[[u^{1/2}]]$, the fact that $b_{y,x}$ $(y \underset{LR}{\geq} x)$ are in $\mathbf{Z}[[u^{1/2}]]$ follows immediately from (5.3.3) and (5.3.4). Thus, (5.3.1) is proved.

Note that $u^{(\ell(x)+\ell(z))/2}\tau(C_{z^{-1}}C_x) \in \mathbf{Z}[u]$. Therefore, from (5.3.5) we deduce

$$\text{(5.3.6)}\qquad a_{y,x}u^{(\ell(x)+\ell(y))/2} \in \mathbf{Z}[[u]]\,.$$

5.4. LEMMA. *Let* Γ *be a left cell of* W. *Let* $[\Gamma]$ *be the left* $H_{\mathcal{K}}$*-module* $I^L_{x,\mathcal{K}}/\hat{I}^L_{x,\mathcal{K}}$ *where* $x \in \Gamma$. *(This is independent of the choice of* x *in* Γ.*) Let* I_Γ *be the* $\mathcal{K}$*-submodule of* $H_{\mathcal{K}}$ *spanned by the products* $C_x D_{y^{-1}}$, $(x \in \Gamma, y \in \Gamma)$. *Then* I_Γ *is a two-sided ideal of* $H_{\mathcal{K}}$. *If* $E \in W^\vee$ *and* $E(u)_{\mathcal{K}}$ *is isomorphic to an* $H_{\mathcal{K}}$*-submodule of* $[\Gamma]$, *then* $I_{E,\mathcal{K}} \subset I_\Gamma$.

Proof. Let $x \in \Gamma$, $y \in \Gamma$, $s \in S$. Then by (5.1.12), $C_s C_x$ is a $\mathcal{K}$-linear combination of elements C_z $(z \underset{L}{\leq} x)$, hence $C_s C_x D_{y^{-1}}$ is a $\mathcal{K}$-linear combination of elements $C_z D_{y^{-1}}$, $(z \underset{L}{\leq} x)$. By (5.1.14), we may assume that z satisfies in addition $y \underset{L}{\leq} z$, hence $y \underset{L}{\leq} z \underset{L}{\leq} x$, hence $z \in \Gamma$. Thus $C_s C_x D_{y^{-1}} \in I_\Gamma$. By (5.1.16), $D_{y^{-1}} C'_s$ is a $\mathcal{K}$-linear combination of elements $D_{z^{-1}}$, $(z^{-1} \underset{R}{\geq} y^{-1})$, hence $C_x D_{y^{-1}} C'_s$ is a $\mathcal{K}$-linear combination of elements $C_x D_{z^{-1}}$, $(z \underset{L}{\geq} y)$. By (5.1.14), we may assume that z satisfies in addition $z \underset{L}{\leq} x$, hence $y \underset{L}{\leq} z \underset{L}{\leq} x$, hence $z \in \Gamma$. Thus $C_x D_{y^{-1}} C'_s \in I_\Gamma$. Since (C_s), (C'_s) form two systems of algebra generators of $H_{\mathcal{K}}$, it follows that I_Γ is a two-sided ideal of $H_{\mathcal{K}}$.

For any $y \in \Gamma$, the map $\phi_y : [\Gamma] \to I_\Gamma$ defined by $\phi_y(h) = h \cdot D_{y^{-1}}$ is well defined (by (5.1.14)). It is a homomorphism of left $H_{\mathcal{K}}$-modules. The homomorphism $\phi : [\Gamma] \to \bigoplus_{y \in \Gamma} I_\Gamma$, $\phi(h) = (\phi_y(h))$ is injective: if $h \in I^L_{x,\mathcal{K}}$, $(x \in \Gamma)$, satisfies $hD_{y^{-1}} = 0$ for all $y \in \Gamma$, then $\tau(hD_{y^{-1}}) = 0$ for all $y \in \Gamma$ hence, by (5.1.10), the coefficient of C_y in h is zero for any $y \in \Gamma$; hence $h \in \hat{I}^L_{x,\mathcal{K}}$. It follows that, if $E \in W^\vee$ is such that $E(u)_{\mathcal{K}}$ is isomorphic to an $H_{\mathcal{K}}$-submodule of $[\Gamma]$, then $E(u)_{\mathcal{K}}$ is also isomorphic to an $H_{\mathcal{K}}$-submodule of I_Γ, hence $I_{E,\mathcal{K}} \subset I_\Gamma$. The lemma is proved.

5.5. LEMMA. *Let* $\mathfrak{M}, \mathfrak{M}'$ *be two* $H_{\mathcal{K}}$*-modules which are isomorphic to* $H_{\mathcal{K}}$*-submodules of* I_Γ. *Let* $f : \mathfrak{M} \to \mathfrak{M}'$ *be a* $\mathcal{K}$*-linear map. Define* $\bar{f} : \mathfrak{M} \to \mathfrak{M}'$ *by* $\bar{f}(v) = \sum_{x\in\Gamma} C_x f(D_{x^{-1}} v)$, $(v \in \mathfrak{M})$. *Then* $\bar{f}$ *is* $H_{\mathcal{K}}$*-linear.*

Proof. For two elements $x, y \in W$, we shall write $x - y$ whenever the following conditions are satisfied: $\ell(x) \not\equiv \ell(y) \pmod 2$, and either $x < y$, $\mu(x,y) \neq 0$ or $y < x$, $\mu(y,x) \neq 0$. If $x - y$, $x < y$, we define $\mu(y,x)$ to be $\mu(x,y)$.

Let $v \in \mathfrak{M}$ and let $s \in S$. Using (5.1.12), we have

$$C_s(\bar{f}(v)) = -\sum_{\substack{x\in\Gamma \\ sx<x}} (u^{1/2}+u^{-1/2}) C_x f(D_{x^{-1}} v) + \sum_{\substack{x\in\Gamma \\ sx>x}} \sum_{\substack{y \leq_L x \\ sy<y \\ y-x}} \mu(y,x) C_y f(D_{x^{-1}} v) . \tag{5.5.1}$$

We now show that for y in the last sum, such that $C_y f(D_{x^{-1}} v) \neq 0$, we have necessarily $y \in \Gamma$. We identify $\mathfrak{M}'$ with an $H_{\mathcal{K}}$-submodule of I_Γ. We then have $C_y C_a D_{b^{-1}} \neq 0$ for some $a \in \Gamma$, $b \in \Gamma$ and $y \leq_L a$. Now $C_y C_a$ is a $\mathcal{K}$-linear combination of elements C_z, $(z \leq_R y)$, hence $C_y C_a D_{b^{-1}}$ is a $\mathcal{K}$-linear combination of elements $C_z D_{b^{-1}}$, $(z \leq_R y)$. Hence there exists $z \in W$, $(z \leq_R y)$, such that $C_z D_{b^{-1}} \neq 0$, hence (by (5.1.14)) such that $b \leq_L z \leq_R y$. Since $a \sim_L b$, it follows that $a \leq_{LR} y$. This, together with $y \leq_L a$ implies, by (5.1.13), that $y \sim_L a$, i.e. $y \in \Gamma$, as required.

Hence, in the last sum in (5.5.1) we may replace " $y \leq_L x$ " by " $y \in \Gamma$." Next, we compute, using (5.1.16):

$$\overline{f}(C_s v) = \overline{f}(C'_s v - (u^{1/2}+u^{-1/2})v)$$

$$= -(u^{1/2}+u^{-1/2}) \sum_{x\epsilon\Gamma} C_x f(D_{x^{-1}} v) + \sum_{x\epsilon\Gamma} C_x f(D_{x^{-1}} C'_s v)$$

$$= -(u^{1/2}+u^{-1/2}) \sum_{x\epsilon\Gamma} C_x f(D_{x^{-1}} v) + (u^{1/2}+u^{-1/2}) \sum_{\substack{x\epsilon\Gamma \\ x^{-1}s>x^{-1}}} C_x f(D_{x^{-1}} v)$$

$$+ \sum_{\substack{x\epsilon\Gamma \\ x^{-1}s<x^{-1}}} \sum_{\substack{y \\ y^{-1} \geq_R x^{-1} \\ y^{-1}s>y^{-1} \\ y-x}} \mu(x^{-1},y^{-1}) C_x f(D_{y^{-1}} v)$$

$$= -(u^{1/2}+u^{-1/2}) \sum_{\substack{x\epsilon\Gamma \\ sx<x}} C_x f(D_{x^{-1}} v) + \sum_{\substack{x\epsilon\Gamma \\ sx<x}} \sum_{\substack{y \geq_L x \\ sy>y \\ y-x}} \mu(x,y) C_x f(D_{y^{-1}} v) .$$

We now show that for y in the last sum such that $C_x f(D_{y^{-1}} v) \neq 0$, we have necessarily $y \in \Gamma$. We identify $\mathfrak{M}$ with an $H_{\mathcal{K}}$-submodule of I_Γ. We then have $D_{y^{-1}} C_a D_{b^{-1}} \neq 0$ for some $a \in \Gamma$, $b \in \Gamma$, and $y \geq_L a$. In particular, we have $D_{y^{-1}} C_a \neq 0$, hence, by (5.1.14) we have $y \leq_R a$. This, together with $y \geq_L a$ shows that $y \underset{LR}{\sim} a$. Using again $y \geq_L a$ and (5.1.13), we see that $y \underset{L}{\sim} a$, i.e. $y \in \Gamma$, as required.

We now see that

$$C_s(\overline{f}(v)) = \overline{f}(C_s v)$$

$$= -(u^{1/2}+u^{-1/2}) \sum_{\substack{x\epsilon\Gamma \\ sx<x}} C_x f(D_{x^{-1}} v) + \sum_{\substack{x\epsilon\Gamma, y\epsilon\Gamma \\ sx>x,\, sy<y \\ y-x}} \mu(y,x) C_y f(D_{x^{-1}} v) .$$

Since $H_{\mathcal{K}}$ is generated by the elements C_s, $(s \in S)$ as an algebra, $\bar{f}$ must be $H_{\mathcal{K}}$-linear. The lemma is proved.

5.6. LEMMA. (i) *For any* $h \in I_\Gamma$, *the element* $\sum_{x\in\Gamma} C_x h D_{x^{-1}}$ *is in the center of* $H_{\mathcal{K}}$.

(ii) $\sum_{x\in\Gamma} C_x D_{x^{-1}}$ *is in the center of* $H_{\mathcal{K}}$.

Proof. We apply Lemma 5.5 to the $\mathcal{K}$-linear map $f: I_\Gamma \to I_\Gamma$ given by $f(a) = ha$ where $h \in I_\Gamma$ or $h = 1$. We deduce that the map $\bar{f}: I_\Gamma \to I_\Gamma$ given by $f(a) = \sum_{x\in\Gamma} C_x h D_{x^{-1}} a$ is $H_{\mathcal{K}}$-linear. Hence, for any $\beta \in H_{\mathcal{K}}$ and any $a \in I_\Gamma$ we have $\sum_{x\in\Gamma} C_x h D_{x^{-1}} \beta a = \sum_{x\in\Gamma} \beta C_x h D_{x^{-1}} a$. We take a to be the unit element of the algebra I_Γ. It follows that $\sum_{x\in\Gamma} C_x h D_{x^{-1}} \beta = \sum_{x\in\Gamma} \beta C_x h D_{x^{-1}}$ for all $\beta \in H_{\mathcal{K}}$. (If $h \in I_\Gamma$, this follows from the fact that $C_x h D_{x^{-1}} \beta$ and $\beta C_x h D_{x^{-1}}$ are in I_Γ; if $h = 1$, this follows from the fact that $\sum_{x\in\Gamma} C_x D_{x^{-1}}$ is in I_Γ.) The lemma is proved.

5.7. THEOREM. *Let* $E, E' \in W^\vee_{ex}$ *and let* $\tilde{E}$, $\tilde{E}'$ *be extensions of* E, E' *to* $\mathbf{Q}[\tilde{W}]$*-modules. Assume that the multiplicities* $(E(u)_{\mathcal{K}} : [\Gamma])$, $(E'(u)_{\mathcal{K}} : [\Gamma])$ *are strictly positive. (We set* $(E(u)_{\mathcal{K}} : [\Gamma]) = \dim_{\mathcal{K}} \mathrm{Hom}_{H_{\mathcal{K}}}(E(u)_{\mathcal{K}}, [\Gamma])$.) *Then*:

(i) *The sum*

$$\sum_{x\in\Gamma} \mathrm{Tr}(C_x T_\gamma, \tilde{E}'(u)_{\mathcal{K}})\, \mathrm{Tr}(T_{\gamma^{-1}} D_{x^{-1}}, \tilde{E}(u)_{\mathcal{K}}) \tag{5.7.1}$$

is zero, if $E(u)_{\mathcal{K}}$, $E'(u)_{\mathcal{K}}$ *are not-isomorphic as* $H_{\mathcal{K}}$*-modules and is equal to* $(E(u)_{\mathcal{K}} : [\Gamma])\ P_{\mathbf{W}}(u) D_E(u)^{-1}$ *(see 3.3) and also to the scalar by which the central element* $\sum_{x\in\Gamma} C_x D_{x^{-1}}$ *(see 5.6 (ii)) acts on* $E(u)_{\mathcal{K}}$, *if* $\tilde{E} = \tilde{E}'$.

(ii) *We have*

$$\sum_{x\in\Gamma} C_x D_{x^{-1}} = \sum_{x\in W} \mathrm{Tr}(C_x, [\Gamma])D_{x^{-1}} .$$

Proof. Assume first that $E(u)_{\mathcal{K}}$, $E'(u)_{\mathcal{K}}$ are not isomorphic as $H_{\mathcal{K}}$-modules.

Let $\xi_1, \cdots, \xi_d$ be a basis of the $\mathcal{K}$-vector space $\tilde{E}(u)_{\mathcal{K}}$ and let $\xi'_1, \cdots, \xi'_{d'}$ be a basis of the $\mathcal{K}$-vector space $\tilde{E}'(u)_{\mathcal{K}}$. Let $f_{ij} : \tilde{E}(u)_{\mathcal{K}} \to \tilde{E}'(u)_{\mathcal{K}}$ be the $\mathcal{K}$-linear map

$$f_{ij}(\xi_k) = \xi'_j \delta_{ik}, \ (1 \le i \le d, 1 \le j \le d'), \text{ where } \delta_{ij} = \begin{cases} 1, & \text{if } i = k \\ 0, & \text{if } i \ne k \end{cases} .$$

Let $f'_{ij} : \tilde{E}(u)_{\mathcal{K}} \to \tilde{E}'(u)_{\mathcal{K}}$ be the $\mathcal{K}$-linear map $v \to T_\gamma f_{ij}(T_\gamma^{-1} v)$. Applying Lemma 5.5 to f'_{ij} we obtain an $H_{\mathcal{K}}$-linear map $\bar{f}'_{ij}$ from $\tilde{E}(u)_{\mathcal{K}}$ to $\tilde{E}'(u)_{\mathcal{K}}$ which, by our assumption, must be zero. Thus, we have

$$\sum_{x\in\Gamma} C_x T_\gamma (f_{ij}(T_\gamma^{-1} D_{x^{-1}} v)) = 0$$

for all $v \in \tilde{E}(u)_{\mathcal{K}}$, all i, $(1 \le i \le d)$, and all j, $(1 \le j \le d')$. Let $T_\gamma^{-1} D_{x^{-1}}(\xi_j) = \sum_p q^j_{x,p} \xi_p$, $C_x T_\gamma(\xi'_j) = \sum_{\tilde{p}} \tilde{q}^j_{x,\tilde{p}} \xi'_{\tilde{p}}$, $(q^j_{x,p} \in \mathcal{K}, \ \tilde{q}^j_{x,\tilde{p}} \in \mathcal{K})$. Take $v = \xi_r$, $(1 \le r \le d)$. We have

$$0 = \sum_{\substack{x\in\Gamma \\ p}} C_x T_\gamma f_{ij}(q^r_{x,p} \xi_p) = \sum_{\substack{x\in\Gamma \\ p}} C_x T_\gamma (q^r_{x,i} \xi'_j) = \sum_{\substack{x\in\Gamma \\ \tilde{p}}} q^r_{x,i} \tilde{q}^j_{x,\tilde{p}} \xi'_{\tilde{p}} .$$

Thus, we have $\sum_{x\in\Gamma} q^r_{x,i} \tilde{q}^j_{x,\tilde{p}} = 0$ for all $i, j, r, \tilde{p}$. Now take $i = r$, $\tilde{p} = j$ and sum over all $1 \le i \le d$, $1 \le j \le d'$. We get

$$0 = \sum_{x\in\Gamma} \sum_{i,j} q^i_{x,i} \tilde{q}^j_{x,j} = \sum_{x\in\Gamma} \mathrm{Tr}(C_x T_\gamma, \tilde{E}'(u)_{\mathcal{K}}) \mathrm{Tr}(T_\gamma^{-1} D_{x^{-1}}, \tilde{E}(u)_{\mathcal{K}})$$

as required.

Next, we assume that $\tilde{E} = \tilde{E}'$. Let $\lambda \in \mathcal{K}$ be the scalar by which the central element $\sum_{x\in\Gamma} C_x D_{x^{-1}} \in H_{\mathcal{K}}$ acts on $E(u)_{\mathcal{K}}$.

Let $\Omega : I_{E,\mathcal{K}} \to I_{E,\mathcal{K}}$ be defined by $\Omega(a) = \sum_{x\in\Gamma} C_x T_\gamma a T_{\gamma^{-1}} D_{x^{-1}}$. (This is well defined since $T_\gamma I_{E,\mathcal{K}} T_\gamma^{-1} = I_{E,\mathcal{K}}$, as $E(u)_{\mathcal{K}}$ is extendable to an $\tilde{H}_{\mathcal{K}}$-module.)

The sum (5.7.1) is equal (for $\tilde{E} = \tilde{E}'$) to $\mathrm{Tr}(\Omega : I_{E,\mathcal{K}} \to I_{E,\mathcal{K}})$. By Lemma 5.6, applied to $h = T_\gamma a T_\gamma^{-1}$, we have $\Omega(I_{E,K}) \subset \text{center } (H_{\mathcal{K}}) \cap I_{E,\mathcal{K}}$.

This last vector space is one-dimensional. Let a_0 be a basis element for it.

The trace of Ω is the scalar $\lambda' \in \mathcal{K}$ such that $\Omega a_0 = \lambda' a_0$. We have $T_\gamma a_0 T_\gamma^{-1} = a_0$ since $E(u)_{\mathcal{K}}$ is extendable to an $\tilde{H}_{\mathcal{K}}$-module. Hence $\Omega a_0 = \sum_{x\in\Gamma} C_x a_0 D_{x^{-1}} = \sum_{x\in\Gamma} C_x D_{x^{-1}} a_0 = \lambda a_0$ (using that a_0 is central). It follows that $\lambda = \lambda' =$ expression (5.7.1).

Assuming that (ii) is proved already, we have

$$
\begin{aligned}
(5.7.2)\quad \lambda &= (\dim E)^{-1}\mathrm{Tr}\Big(\sum_{x\in\Gamma} C_x D_{x^{-1}}, E(u)_{\mathcal{K}}\Big) \\
&= (\dim E)^{-1} \sum_{x\in W} \mathrm{Tr}(C_x,[\Gamma])\,\mathrm{Tr}(D_{x^{-1}},E(u)_{\mathcal{K}}) \\
&= (\dim E)^{-1} \sum_{E''\in W^\vee} (E''(u)_{\mathcal{K}} : [\Gamma]) \sum_{x\in W} \mathrm{Tr}(C_x,E''(u)_{\mathcal{K}})\,\mathrm{Tr}(D_{x^{-1}},E(u)_{\mathcal{K}}).
\end{aligned}
$$

We now compute (using (3.3.4) with $\gamma = 1$, (5.1.1), (5.1.6) and (5.1.3)) the last sum:

$$(5.7.3)\quad \sum_{x\in W} \mathrm{Tr}(C_x, E''(u)_{\mathcal{K}})\,\mathrm{Tr}(D_{x^{-1}}, E(u)_{\mathcal{K}}) =$$

$$= \sum_{\substack{x,y,y' \\ y\le x\le y'}} (-1)^{\ell(x)-\ell(y)} u^{-\ell(y)} P_{y,x}(u^{-1}) Q_{x,y'}(u^{-1}) \mathrm{Tr}(T_{y'}, E''(u)_{\mathcal{K}}) \mathrm{Tr}(T_{y'^{-1}}, E(u)_{\mathcal{K}})$$

$$= \sum_{y\in W} u^{-\ell(y)} \mathrm{Tr}(T_y, E''(u)_{\mathcal{K}}) \mathrm{Tr}(T_{y^{-1}}, E(u)_{\mathcal{K}})$$

$$= \sum_{y\in W} u^{-\ell(y)} \mathrm{Tr}(T_y, E''(u)_{\mathcal{K}}) \mathrm{Tr}(T_y, E(u)_{\mathcal{K}}) = \begin{cases} P_W(u) D_E(u)^{-1} \dim E, & \text{if } E \approx E'' \\ 0, & \text{if } E \not\approx E'' \end{cases}.$$

Introducing this into (5.7.2) we find

$$\lambda = (E(u)_{\mathcal{K}} : [\Gamma])\, P_W(u)\, D_E(u)^{-1}$$

as required.

It remains to prove (ii). For $y \in W$, $x \in \Gamma$ let $n_x^y \in \mathcal{K}$ be defined by

$$(5.7.4)\qquad C_y C_x = n_x^y C_x + \text{combination of } C_{x'},\quad x' \neq x .$$

We have $\mathrm{Tr}(C_y, [\Gamma]) = \sum_{x\in\Gamma} n_x^y$.

Multiply (5.7.4) on the right by $D_{x^{-1}}$ and apply τ to both sides. Using (4.1.10), we see that $n_x^y = \tau(C_y C_x D_{x^{-1}})$. Thus

$$\mathrm{Tr}(C_y, [\Gamma]) = \sum_{x\in\Gamma} \tau(C_y C_x D_{x^{-1}}) = \tau(C_y \sum_{x\in\Gamma} C_x D_{x^{-1}}) .$$

Let us now write $\sum_{x\in\Gamma} C_x D_{x^{-1}}$ as a $\mathcal{K}$-linear combination $\sum_{y\in W} m_y D_{y^{-1}}$. Using again (5.1.10), we see that $m_y = \tau(C_y \sum_{x\in\Gamma} C_x D_{x^{-1}})$. Thus, $m_y = \mathrm{Tr}(C_y, [\Gamma])$, $(y \in W)$ and (ii) is proved.

5.8. COROLLARY. *With the assumptions of Theorem 5.7, we have*

$$(5.8.1)\qquad \sum_{x\in\Gamma} c_{\gamma x,\tilde{E}}\cdot c_{\gamma x,\tilde{E}'} = \begin{cases} (E(u)_{\mathcal{K}}:[\Gamma])f_E, & \text{if } \tilde{E} = \tilde{E}' \\ 0 & \text{, if } E \not\approx E' \end{cases}.$$

(We recall that f_E *is defined in (4.1.1).) In particular, if* $(E(u)_{\mathcal{K}}:[\Gamma]) > 0$, *there exists at least one* $x \in \Gamma$ *such that* $c_{\gamma x,\tilde{E}} \neq 0$.

Proof. Theorem 5.7 (i) gives a formula for (5.7.1) in terms of $D_E(u)$. Taking the coefficients of $u^{-(1/2)(a_E+a_{E'})}$ in both sides of this formula and taking into account (5.2.1), (5.2.2) and (5.2.3), we obtain (5.8.1).

5.9. Let $(\mathbf{W}', S')$ be a γ-stable standard parabolic subgroup of $\mathbf{W}$. Define $\tilde{\mathbf{W}}'$ from $\mathbf{W}'$, $\gamma: \mathbf{W}' \to \mathbf{W}'$ in the same way that $\tilde{\mathbf{W}}$ was defined from $\mathbf{W}$, $\gamma: \mathbf{W} \to \mathbf{W}$. Define the Hecke algebras H', $\tilde{H}'$, $H'_{\mathcal{K}}$, $\tilde{H}'_{\mathcal{K}}$ associated with $\mathbf{W}'$, $\tilde{\mathbf{W}}'$ in the same way as H, $\tilde{H}$, $H_{\mathcal{K}}$, $\tilde{H}_{\mathcal{K}}$ were defined for $\mathbf{W}$, $\tilde{\mathbf{W}}$.

Let $\tilde{E}$ be an irreducible $\mathbf{Q}[\tilde{\mathbf{W}}]$-module whose restriction E to $\mathbf{W}$ is still irreducible. If we assume either that $\gamma^2 = 1$ on $\mathbf{W}$ or that $(\mathbf{W}, S)$ is of type D_4 then we have for any $x \in \mathbf{W}'$:

$$(5.9.1)\qquad c_{\gamma x,\tilde{E}} = \sum_{\tilde{E}'} \langle \tilde{E}', \tilde{E} \rangle_{\tilde{\mathbf{W}}'} c_{\gamma x,\tilde{E}'}$$

(sum over all irreducible $\mathbf{Q}[\tilde{\mathbf{W}}']$-modules $\tilde{E}'$ with the following property: the restriction E' of $\tilde{E}'$ to $\mathbf{W}'$ is irreducible and $a_E = a_{E'}$. Recall that we are considering $\mathbf{W}'$-modules which factor through a finite quotient).

This is proved as follows. Assume first that $\gamma^2 = 1$ on $\mathbf{W}$. By definition, $(-1)^{\ell(x)} c_{\gamma x,\tilde{E}}$ is the coefficient of $u^{-a_E/2}$ in $\mathrm{Tr}(u^{-\ell(x)/2} T_{\gamma x}, \tilde{E}(u)_{\mathcal{K}}) \in \mathbf{Z}[u^{1/2}, u^{-1/2}]$. The restriction of $\tilde{E}(u)_{\mathcal{K}}$ to the subalgebra $\tilde{H}'_{\mathcal{K}}$ of $\tilde{H}_{\mathcal{K}}$ is a direct sum of irreducible submodules of two kinds: some which become reducible on restriction to $H'_{\mathcal{K}}$ (they clearly do not contribute to the trace of $u^{-\ell(x)/2}T_{\gamma x}$) and some of the form $\tilde{E}'(u)_{\mathcal{K}}$ where $\tilde{E}'$ is an irreducible $Q[\tilde{W}']$-module which remains irreducible on restriction to W'. (Here we use the rationality result 3.2 and the assumption that $\gamma^2 = 1$.) Given such an $\tilde{E}'(u)_{\mathcal{K}}$, the coefficient of $u^{-a_E/2}$ in $\mathrm{Tr}(u^{-\ell(x)/2}T_{\gamma x}, \tilde{E}'(u)_{\mathcal{K}})$ is $(-1)^{\ell(x)}c_{\gamma x,\tilde{E}}$, if $a_E = a_{E'}$ and it is zero if $a_E > a_{E'}$. The case $a_E < a_{E'}$ cannot arise, by (4.1.5). From this, (5.9.1) follows. The proof in the case where (W, S) is of type D_4 and $\gamma^3 = 1$ is left to the reader.

5.10. We define $\mathcal{R}(W, \gamma)$ to be the group of formal linear combinations with coefficients in $\frac{1}{2}\mathbf{Z}$ of irreducible $Q[\tilde{W}]$-modules up to isomorphism (see 3.1), whose restrictions to W are still irreducible, and such that if $\tilde{E}_1, \tilde{E}_2$ are two such $Q[\tilde{W}]$-modules, and if $\tilde{E}_1 \not\approx \tilde{E}_2$, $\tilde{E}_1|W \approx \tilde{E}_2|W$, then the coefficient ($\epsilon\frac{1}{2}\mathbf{Z}$) of $\tilde{E}_1$ is minus the coefficient of $\tilde{E}_2$.

(5.10.1) If γ = identity, we may identify $\mathcal{R}(W, \gamma)$ with the Grothendieck group $\mathcal{R}(W)$ of all $Q[W]$-modules by attaching to an irreducible $Q[W]$-module E, the difference $\frac{1}{2}(\hat{E} - \hat{E}') \in \mathcal{R}(W, \gamma)$ where $\hat{E}$ is E with γ acting as 1 and $\hat{E}'$ is E with γ acting as -1.

We now define, for any $x \in W$, an element $a_{\gamma x} \in \mathcal{R}(W, \gamma)$ by the formula

(5.10.2)
$$a_{\gamma x} = (1/2) \sum_{\tilde{E}} c_{\gamma x, \tilde{E}} \tilde{E}$$

(sum over all irreducible $Q[\tilde{W}]$-modules whose restrictions to W are irreducible). If γ = identity, this corresponds under (5.10.1) to the element $a_x \in \mathcal{R}(W)$ defined by

(5.10.3)
$$a_x = \sum_{E} c_{x,E} E ,$$

(sum over all irreducible $Q[W]$-modules; see (5.1.22) for the definition of

$c_{x,E}$). Now let $(\mathbf{W}', S')$ be as in 5.9. In the following definition we assume that either $\gamma^2 = 1$ on $\mathbf{W}$ or that $\mathbf{W}$ is of type D_4.

(5.10.4) DEFINITION. For any irreducible $\mathbf{Q}[\widetilde{\mathbf{W}}']$-module $\widetilde{E}'$ whose restriction to $\mathbf{W}'$ is irreducible we define $J_{\widetilde{\mathbf{W}}'}^{\widetilde{\mathbf{W}}}(\widetilde{E}')$ to be the formal $\mathbf{Z}$-linear combination $\Sigma \langle \widetilde{E}', \widetilde{E} \rangle_{\widetilde{\mathbf{W}}'} \widetilde{E}$ (sum over all irreducible $\mathbf{Q}[\widetilde{\mathbf{W}}]$-modules $\widetilde{E}$ which remain irreducible on restriction to $\mathbf{W}$ and which satisfy $a_{\widetilde{E}|\mathbf{W}} = a_{\widetilde{E}'|\mathbf{W}'}$). This extends to a $\mathbf{Z}$-linear map $J_{\widetilde{\mathbf{W}}'}^{\widetilde{\mathbf{W}}} : \mathcal{R}(\mathbf{W}', \gamma) \to \mathcal{R}(\mathbf{W}, \gamma)$.

(If γ acts trivially on $\mathbf{W}$, the map $J_{\widetilde{\mathbf{W}}'}^{\widetilde{\mathbf{W}}}$ becomes, under the identification (5.10.1) for $\mathbf{W}$ and $\mathbf{W}'$, the map $J_{\mathbf{W}'}^{\mathbf{W}}$ (see (4.1.7)).)

The following property is a restatement of (5.9.1); it holds under the same conditions as (5.9.1):

(5.10.5) If $x \in \mathbf{W}'$, then $a_{\gamma x}^{\mathbf{W}} = J_{\widetilde{\mathbf{W}}'}^{\widetilde{\mathbf{W}}}(a_{\gamma x}^{\mathbf{W}'}) \in \mathcal{R}(\mathbf{W}, \gamma)$.

Here $a_{\gamma x}^{\mathbf{W}}$ means $a_{\gamma x}$ relative to $\mathbf{W}$; $a_{\gamma x}^{\mathbf{W}'}$ means $a_{\gamma x}$ relative to $\mathbf{W}'$.

5.11. If $E \in \mathbf{W}_{ex}^{\vee}$ and $\widetilde{E}$ is an extension of E to an irreducible $\mathbf{Q}[\widetilde{\mathbf{W}}]$-module we have defined for any $x \in \mathbf{W}$ an integer $c'_{\gamma x, \widetilde{E}}$ by (5.1.23).

An equivalent definition is

(5.11.1) $\operatorname{Tr}(T_\gamma C'_x, \widetilde{E}(u)) = c'_{\gamma x, \widetilde{E}} u^{(\nu - A_E)/2} +$ lower powers of $u^{1/2}$.

(Recall that C'_x is given by (5.1.2).) Indeed

$$\operatorname{Tr}(T_\gamma C'_x, \widetilde{E}(u)) =$$

$$= \operatorname{Tr}(u^{-\ell(x)/2} T_{\gamma x}, \widetilde{E}(u)) + \sum_{y<x} P_{y,x}(u) u^{\frac{-\ell(x)+\ell(y)}{2}} \operatorname{Tr}(u^{-\frac{\ell(y)}{2}} T_{\gamma y}, \widetilde{E}(u)).$$

From the bounds (5.1.1) satisfied by the degree of $P_{y,x}$ and from (5.1.22) we see that the terms in the sum $\sum_{y<x}$ contribute only powers of $u^{1/2}$ with exponent $< \nu - A(E)$ and the equivalence of (5.1.23) and (5.11.1) follows.

(5.11.2) Let $\widetilde{\text{sign}}$ be the extension of the sign representation of $\mathbf{W}$ to $\widetilde{\mathbf{W}}$, on which γ acts trivially.

We have

$$c'_{\gamma x,\widetilde{E}} = c_{\gamma x,\widetilde{E}\otimes\widetilde{\text{sign}}} . \tag{5.11.3}$$

This follows from the known identity

$$\mathrm{Tr}\,(T_{\gamma x},(\widetilde{E}\otimes\widetilde{\text{sign}})\,(u)) = (-u)^{\ell(x)}\,\overline{\mathrm{Tr}\,(T_{\gamma x},\widetilde{E}(u))}\ , \tag{5.11.4}$$

(where the bar denotes the involution of $\mathbf{Q}[u^{1/2},u^{-1/2}]$ considered in 3.3) and from its consequence

$$a_{E\otimes\widetilde{\text{sign}}} = \nu - A_E\ , \tag{5.11.5}$$

by considering the coefficient of $u^{(\ell(x)/2)-((a_{E\otimes\widetilde{\text{sign}}})/2)}$ in the two sides of (5.11.4). We now define, for any $x \in \mathbf{W}$, an element $\mathcal{A}_{\gamma x} \in \mathcal{R}(\mathbf{W},\gamma)$ by the formula

$$\mathcal{A}_{\gamma x} = (1/2)\sum_{\widetilde{E}} c'_{\gamma x,\widetilde{E}}\widetilde{E}\ , \tag{5.11.6}$$

sum over the same set as in (5.10.2). (When γ = identity, we shall denote $\mathcal{A}_x$ the element of $\mathcal{R}(\mathbf{W})$ corresponding to $\mathcal{A}_{\gamma x}$ under (5.10.1).)

Using (5.11.3), we see that for any $x \in \mathbf{W}$, we have

$$\mathcal{A}_{\gamma x} = a_{\gamma x} \otimes \widetilde{\text{sign}}\ . \tag{5.11.7}$$

5.12. Let $\sigma: \mathbf{W} \to \mathbf{W}$ be the automorphism of $(\mathbf{W}, S)$ given by $\sigma(w) = w_0 w w_0$ where w_0 is the longest element of $\mathbf{W}$. It is clear that σ commutes with $\gamma: \mathbf{W} \to \mathbf{W}$.

For any $x \in \mathbf{W}$, we have

$$T_{w_0} T^{-1}_{x^{-1}} = T_{w_0 x} T_{x^{-1}} T^{-1}_{x^{-1}} = T_{w_0 x}$$

$$T^{-1}_{\sigma(x)^{-1}} T_{w_0} = T^{-1}_{\sigma(x)^{-1}} T_{\sigma(x)^{-1}} T_{\sigma(x) w_0} = T_{w_0 x}$$

hence

$$T_y T_{w_0} = T_{w_0} T_{\sigma(y)} \quad \text{for all} \quad y \in \mathbf{W}.$$

Let E be an irreducible $\mathbf{Q}[\mathbf{W}]$-module. By the discussion in 3.3 applied to $\sigma: \mathbf{W} \to \mathbf{W}$, there is an involution $\sigma(u)$ of the $\mathbf{Q}[u^{1/2}, u^{-1/2}]$ module $E(u)$ such that $\sigma(u) T_w \sigma(u)^{-1} = T_{\sigma(w)}: E(u) \to E(u)$ for all $w \in \mathbf{W}$. This involution of $E(u)$ is uniquely determined up to multiplication by (± 1). The composition $\sigma(u) T_{w_0}: E(u) \to E(u)$ satisfies $T_y(\sigma(u) T_{w_0}) = (\sigma(u) T_{w_0}) T_y$ $(y \in \mathbf{W})$; since $E(u)_{\mathcal{K}}$ is an irreducible $H_{\mathcal{K}}$-module, it follows that $\sigma(u) T_{w_0}$ acts on $E(u)$ as a scalar $\lambda \in \mathbf{Q}[u^{1/2}, u^{-1/2}]$. We have

$$\text{(5.12.1)} \quad \operatorname{Tr}(T_{w_0 x}, E(u)) = \operatorname{Tr}(T_{w_0} T^{-1}_{x^{-1}}, E(u)) = \lambda \operatorname{Tr}(\sigma(u) T^{-1}_{x^{-1}}, E(u))$$
$$= \lambda \overline{\operatorname{Tr}(\sigma(u) T_x, E(u))}.$$

(The last equality follows from (3.3.2), with γ replaced by σ.) It follows that

$$\sum_x u^{-\ell(x)} \operatorname{Tr}(T_x, E(u))^2 = \sum_x u^{-\ell(w_0 x)} \operatorname{Tr}(T_{w_0 x}, E(u))^2$$
$$= \sum_x u^{-\nu + \ell(x)} \lambda^2 \overline{\operatorname{Tr}(\sigma(u) T_x, E(u))^2}$$
$$= \lambda^2 u^{-\nu} \overline{\sum_x u^{-\ell(x)} \operatorname{Tr}(\sigma(u) T_x, E(u))^2}.$$

Using (3.3.4), once with $\gamma = 1$ and once with $\gamma = \sigma$, we get

$$P_{\mathbf{W}}(u) \cdot D_E(u)^{-1} = \lambda^2 u^{-\nu} P_{\mathbf{W}}(u^{-1}) D_E(u^{-1})^{-1} .$$

We now use the identity

$$D_E(u^{-1}) = u^{-a_E - A_E} D_E(u)$$

(see [L_{11}, (1.8.2)]) and the identity $P_{\mathbf{W}}(u^{-1}) = u^{-\nu} P_{\mathbf{W}}(u)$. It follows that

$$1 = u^{-2\nu} \lambda^2 u^{a_E + A_E}$$

hence

$$\lambda = \pm u^{\nu - ((a_E + A_E)/2)} .$$

We have therefore proved:

(5.12.2) *For any irreducible* $\mathbf{Q}[\mathbf{W}]$*-module* E, *there is a unique* $\mathbf{Q}[u^{1/2}, u^{-1/2}]$*-linear involution* $\sigma(u) : E(u) \to E(u)$ *such that*

$$T_{w_0} = u^{\nu - ((a_E + A_E)/2)} \sigma(u) : E(u) \to E(u) .$$

It satisfies $\sigma(u) T_w = T_{\sigma(w)} \sigma(u) : E(u) \to E(u)$ *for all* $w \in \mathbf{W}$.

(Compare [L_{11}, 1.11].)

Recall that we are given an automorphism γ of $(\mathbf{W}, S)$. Let $\gamma' = \gamma\sigma = \sigma\gamma : \mathbf{W} \to \mathbf{W}$. There is a canonical 1-1 correspondence between irreducible $\mathbf{Q}[\tilde{\mathbf{W}}]$-modules which remain irreducible on restriction to $\mathbf{W}$ (where $\tilde{\mathbf{W}}$ is defined with respect to γ) and irreducible $\mathbf{Q}[\tilde{\mathbf{W}}]$-modules which remain irreducible on restriction to $\mathbf{W}$ (where $\tilde{\mathbf{W}}$ is defined with respect to γ'): we replace the action by T_γ by $T_{\gamma'} \overset{\text{def}}{=} \sigma(u) \cdot T_\gamma = T_\gamma \cdot \sigma(u)$, where $\sigma(u)$ is defined by (5.12.2). This extends to an isomorphism

(5.12.3) $$\zeta : \mathcal{R}(\mathbf{W}, \gamma) \approx \mathcal{R}(\mathbf{W}, \gamma') .$$

(5.12.4) *Under the isomorphism (5.12.3),* $\mathfrak{A}_{\gamma w_0 x} \in \mathcal{R}(\mathbf{W}, \gamma)$ *corresponds to* $(-1)^{\ell(x)} a_{\gamma' x} \in \mathcal{R}(\mathbf{W}, \gamma')$ *for any* $x \in \mathbf{W}$.

Indeed, the computation (5.12.1) together with (5.12.2) show that, for an irreducible $\mathbf{Q}[\widetilde{\mathbf{W}}]$-module $\widetilde{E}$ which remains irreducible on restriction to $\mathbf{W}$ (where $\widetilde{\mathbf{W}}$ is defined with respect to γ or γ'), we have

$$\text{(5.12.5)} \qquad \operatorname{Tr}(T_\gamma T_{w_0 x}, \widetilde{E}(u)) = u^{\nu-((a_E+A_E)/2)} \overline{\operatorname{Tr}(T_{\gamma'} T_x, \widetilde{E}(u))}$$

and (5.12.4) follows by taking the coefficient of $u^{(\ell(w_0 x)+\nu-A_E)/2}$ in the two sides of (5.12.5).

(5.12.6) In the special case where w_0 is in the center of $\mathbf{W}$, we have $\gamma = \gamma'$ and ζ in (5.12.3) takes $\sum_{\widetilde{E}} c_{\widetilde{E}} \widetilde{E} \in \mathcal{R}(\mathbf{W}, \gamma)$ to $\sum_{\widetilde{E}} c_{\widetilde{E}} \varepsilon_{\widetilde{E}} \widetilde{E}$ where $\varepsilon_{\widetilde{E}} \in \{1, -1\}$ is defined by the condition that $w_0 = \varepsilon_{\widetilde{E}} \times$ identity on $\widetilde{E}$. (This follows immediately from (5.12.2).)

5.13. LEMMA. (i) *For any* $x \in \mathbf{W}$, *the elements* $a_{\gamma x}, \mathfrak{A}_{\gamma w_0 x} \in \mathcal{R}(\mathbf{W}, \gamma)$ *are linear combinations of irreducible* $\mathbf{Q}[\widetilde{\mathbf{W}}]$*-modules* $\widetilde{E}$ *with the following property: the restriction* E *of* $\widetilde{E}$ *to* $\mathbf{W}$ *is irreducible and* $E \underset{LR}{\sim} x$.

(ii) *Conversely, let* $E \in (\mathbf{W})^{\vee}_{ex}$ *and let* Γ *be a left cell of* $\mathbf{W}$ *such that* $(E(u)_{\mathcal{K}} : [\Gamma]) > 0$, *(see 5.7). We denote by* $\widetilde{E}, \widetilde{E}'$ *the two (distinct) extensions of* E *to* $\mathbf{Q}[\widetilde{\mathbf{W}}]$*-modules. We have*

$$\sum_{x \in \Gamma} c'_{\gamma w_0 x, \widetilde{E}} \mathfrak{A}_{\gamma w_0 x} = \sum_{x \in \Gamma} c_{\gamma x, \widetilde{E}} a_{\gamma x} = (E(u)_{\mathcal{K}} : [\Gamma]) f_E (1/2)(\widetilde{E} - \widetilde{E}') .$$

Proof. (i) and (ii) for the elements $a_{\gamma x}$ follow from 5.2 (ii) and 5.8 respectively. This implies the corresponding statements for the elements $\mathfrak{A}_{\gamma w_0 x}$ in view of (5.11.3), (5.11.7) and the following result.

5.14. LEMMA. *Let* Γ *be a left of* $\mathbf{W}$. *Then*

(i) $w_0\Gamma$ *is again a left cell.*

(ii) *If* E *is an irreducible* $\mathbf{Q}[\mathbf{W}]$*-module, we have*

$$(E(u)_{\mathcal{K}} : [\Gamma]) = ((E \otimes \mathrm{sign})(u)_{\mathcal{K}} : [w_0\Gamma]) .$$

(iii) *Given* E *as in (ii) and* $x \in \mathbf{W}$, *the conditions* $E \underset{LR}{\sim} x$ and $E \otimes \mathrm{sign} \underset{LR}{\sim} w_0 x$ *are equivalent.*

(iv) *If* γ = *identity and* $c = 1$, *then* $a_{w_0} = \mathrm{sign} \in \mathcal{R}(\mathbf{W})$, *(see (5.10.3)).*

The statement (i) follows from $[KL_1, 3.2]$. The statements (ii) and (iii) can also be deduced from $[KL_1, 3.2]$; see $[BV_2]$ for details. Statement (iv) is proved in $[L_{11}, 1.22]$.

5.15. The algebra $H_{\mathcal{K}}$ is on the one hand filtered by the two sided ideals $I^{LR}_{x,\mathcal{K}}$ (see 5.1) and on the other hand it has a decomposition $\oplus I_{E,\mathcal{K}}$ into minimal two-sided ideals. This decomposition is clearly compatible with the filtration. It follows that there is a well-defined preorder $\underset{LR}{\leq}$ on the set (of isomorphism classes) of irreducible $\mathbf{Q}[\mathbf{W}]$-modules such that $E \underset{LR}{\leq} E'$ if and only if there exists $x \in \mathbf{W}$ with $I_{E,\mathcal{K}}, I_{E',\mathcal{K}} \subset I^{LR}_{x,\mathcal{K}}$, $I_{E',K} \not\subset \hat{I}^{LR}_{x,\mathcal{K}}$, or, equivalently, if there exists $x \in \mathbf{W}$ such that $E' \underset{LR}{\sim} x$, $E \underset{LR}{\leq} x$, (see 4.1). The equivalence relation associated to this preorder will be denoted $\underset{LR}{\sim}$; thus, we have $E \underset{LR}{\sim} E'$ if and only if $E \underset{LR}{\leq} E' \underset{LR}{\leq} E$, or, equivalently, if there exist $x \in \mathbf{W}$ such that $E \underset{LR}{\sim} x$, $E' \underset{LR}{\sim} x$. If $\mathbf{W}$ is a product of Weyl groups $\mathbf{W}_1 \times \mathbf{W}_2$ and E_1, E_1' (resp. E_2, E_2') are irreducible $\mathbf{Q}[\mathbf{W}_1]$ (resp. $\mathbf{Q}[\mathbf{W}_2]$) modules, we have $E_1 \otimes E_2 \underset{LR}{\sim} E_1' \otimes E_2'$ if and only if $E_1 \underset{LR}{\sim} E_2$ and $E_1' \underset{LR}{\sim} E_2'$. In Chapter 4 we have defined a decomposition of the set of irreducible $\mathbf{Q}[\mathbf{W}]$-modules into families, and in [L_7] we conjectured that E, E' are in the same family if and only if $E \underset{LR}{\sim} E'$. This conjecture was recently proved by Barbasch and Vogan [BV_1], [BV_2].

5.16. Assume that $(\mathbf{W}, S)$ is of type A_{n-1}. Using [KL_1, 1.4], we see that for any left cell $\Gamma \subset \mathbf{W}$, the $H_{\mathcal{K}}$-module $[\Gamma]$ is irreducible hence it is of form $E(u)_{\mathcal{K}}$ for a well-defined irreducible $\mathbf{Q}[\mathbf{W}]$-module E, which is necessarily of form $J_{\mathbf{W}'}^{\mathbf{W}}$ (sign) for some standard parabolic subgroup $\mathbf{W}' \subset \mathbf{W}$. (See (4.4.1).) Assume now that $\gamma : \mathbf{W} \to \mathbf{W}$ is the identity. If x is the longest element in $\mathbf{W}'$ then

$$(5.16.1) \qquad a_x = E \in \mathcal{R}(\mathbf{W}) \quad \text{and} \quad \ell(x) = a_E .$$

This follows from (4.4.1), (4.4.2), (5.10.5), 5.15.

Since $f_E = 1$ (cf. (4.1.6) and (4.4.1)), we see from (5.8.1) that $\sum_{x \in \Gamma} c_{x,E}^2 = 1$ hence there is a unique element $x \in \Gamma$ such that $c_{x,E} = \pm 1$; for all other elements $x' \in \Gamma$ we have $c_{x',E} = 0$.

Hence the following holds:

(5.16.2) For any $x \in \mathbf{W}$ we have either $a_x = 0$ or $a_x = \pm$ irreducible $\mathbf{Q}[\mathbf{W}]$-module.

The second alternative occurs for a unique x in each left cell.

Assume now that $n \geq 3$ and that γ is the unique automorphism of order 2 of $(\mathbf{W}, S)$. Then $\gamma(w) = w_0 w w_0$ for all $w \in \mathbf{W}$.

Using (5.12.4) and (5.11.7) we deduce the following statement.

(5.16.3) Let E and x be as in (5.16.1). Let $\hat{E}$ (resp. $\hat{E}'$) be the extension of E to a $\mathbf{Q}[\widetilde{\mathbf{W}}]$-module on which γ acts as w_0 (resp. as $-w_0$). Then

$$\mathcal{A}_{\gamma w_0 x} = (-1)^{\ell(x)}(1/2)(\hat{E} - \hat{E}'),$$

$$a_{\gamma w_0 x} = (-1)^{\ell(x)}(1/2)(\hat{E} \otimes \widetilde{\text{sign}} - \hat{E}' \otimes \widetilde{\text{sign}})$$

in $\mathcal{R}(\mathbf{W}, \gamma)$.

5.17. Assume that $(\mathbf{W}, \mathbf{S})$ is a Weyl group of type $B_2 = C_2$. With the notations at the end of 4.5 we have

$$(5.17.1)\quad \begin{cases} a_e = \mathcal{A}_{w_0} & = 1 \\ a_{s_1} = \mathcal{A}_{s_2} & = V + \varepsilon_2 \\ a_{s_2} = \mathcal{A}_{s_1} & = V + \varepsilon_1 \\ a_{s_1 s_2 s_1} = \mathcal{A}_{s_2 s_1 s_2} & = V - \varepsilon_2 \\ a_{s_2 s_1 s_2} = \mathcal{A}_{s_1 s_2 s_1} & = V - \varepsilon_1 \\ a_{w_0} = \mathcal{A}_e & = \text{sign} \\ a_x = \mathcal{A}_x & = 0, \text{ whenever } x \in \mathbf{W} \text{ is not an involution.} \end{cases}$$

Assume now that $(\mathbf{W}, \mathbf{S})$ is a Weyl group of type B_n or C_n, $(n \geq 3)$. Let Z be a symbol of rank n and defect 1 such that $[Z]$ is special (see (4.5.3)), with $2m+1$ entries (not just its equivalence class under shift). As in 4.5, Z_1 denotes the set of "singles" in Z; let $2d+1 = |Z_1|$.

We shall consider arrangements Φ of the set Z_1 into d disjoint pairs, (one element of Z_1 will remain outside Φ) such that each pair in Φ contains one element of Z_1 from the first row of Z and one from the second row of Z.

We want to define, by induction on d, what it means for Φ to be an admissible arrangement for Z, (cf. [L_{11}, 2.11]). If $d = 0$, there is a unique arrangement: the empty one; it is, by definition, admissible. Assume now that $d > 0$. Then Φ is admissible for Z if and only if Φ

contains a pair of singles $\xi < \xi'$ in different rows of Z such that there are no singles $z \in Z_1$ with $\xi < z < \xi'$ and if the corresponding arrangement for the symbol obtained from Z by removing ξ, ξ' is admissible.

For example, the special symbol $\begin{pmatrix} 0 & & 2 & & 4 \\ & 1 & & 3 & \end{pmatrix}$ has five admissible arrangements: $\Phi_1 : (0,1), (2,3)$; $\Phi_2 : (1,2), (3,4)$; $\Phi_3 : (0,3), (1,2)$; $\Phi_4 : (0,1), (3,4)$; $\Phi_5 : (1,4), (2,3)$.

If Ψ is a set of pairs in Φ, we denote by Ψ^* (resp. Ψ_*) the set of elements of Z_1 in the first (resp. second) row of Z which appear in a pair of Ψ. Let Z_0^* (resp. $(Z_0)_*$) be the set of elements in the first (resp. second) row of Z which do not appear in any pair of Φ. In particular any entry of Z which is not a single is in Z_0^* or in $(Z_0)_*$.

Let Φ be an admissible arrangement for the symbol Z of rank n and defect 1 such that $[Z]$ is special and let $\hat{\Phi} \subset \Phi$ be a subset of Φ. We define (with notations in 4.5):

$$\text{(5.17.2)} \qquad c(Z, \Phi, \hat{\Phi}) = \sum_{\Psi \subset \Phi} (-1)^{e(\Psi)} \begin{bmatrix} (Z_0)^* \amalg \Psi_* \amalg (\Phi - \Psi)^* \\ \\ (Z_0)_* \amalg \Psi^* \amalg (\Phi - \Psi)_* \end{bmatrix}$$

(an element of $\mathcal{R}(W)$) where $e(\Psi) = |\hat{\Phi} \cap \Psi|$ and the rows are arranged in increasing order. This is a sum of 2^d terms, each of which is $\pm$ an irreducible $\mathbf{Q}[W]$-module (corresponding to a symbol of rank r and defect 1), in the same family as $[Z]$, which is the term corresponding to $\Psi = \emptyset$.

Using (4.5.4), (4.5.5), (5.10.5), (5.12.4) and (5.11.7) it is easy to prove the following result (see $[L_{11}, 2.20]$ for details):

(5.17.3) *For any* $Z, \Phi, \hat{\Phi}$ *as above, there exists an involution* $x \in W$ *such that* $c(Z, \Phi, \hat{\Phi}) = a_x \in \mathcal{R}(W)$ *and* $\ell(x) \equiv a_{[Z]} \pmod 2$.

If $[\Lambda]$ is any irreducible representation of W in the same family as $[Z]$ then there exists an admissible arrangement Φ for Z such that for any $\hat{\Phi} \subset \Phi$, $\pm[\Lambda]$ is one of the terms in the sum (5.17.2) defining $c(Z, \Phi, \hat{\Phi})$. (See $[L_{11}, 2.22]$.)

Using now (5.17.2) and 5.2 (ii) we see that the following holds:

(5.17.4) *If* $[\Lambda]$ *is an irreducible representation of* $\mathbf{W}$ *in the same family as the special representation* $[Z]$, *then there exists an involution* $x \in \mathbf{W}$ *such that* $[\Lambda] \underset{LR}{\sim} x$, $[Z] \underset{LR}{\sim} x$.

5.18. Assume that $(\mathbf{W}, S)$ is a Weyl group of type D_n, $(n \geq 4)$, with given $s', s'' \in S$ as in the first sentence in 4.6 and let γ be the automorphism of $(\mathbf{W}, S)$ interchanging s', s''.

Let Z be a nondegenerate symbol of rank n and defect 0 with $2m$ entries (not just its equivalence class under shift) such that $[Z]$ is special, (see (4.6.4)). We write the two rows in a definite order, in such a way that the sum of entries in the first row is strictly bigger than the sum of entries in the second row. As in 4.6, Z_1 denotes the set of singles in Z. It has $2d > 0$ elements. We shall consider arrangements of Φ of the set Z_1 into d disjoint pairs such that each pair in Φ contains one element of Z_1 in the first row of Z and one from the second row of Z.

We define, by induction on d, what it means for Φ to be an admissible arrangement for Z, in exactly the same way as in 5.17, except that here the induction starts with $d = 1$, in which case the unique arrangement is, by definition, admissible.

For example the symbol $\begin{pmatrix} 1 & 3 & 5 \\ 0 & 2 & 4 \end{pmatrix}$ has five admissible arrangements: $\Phi_1 : (0,1), (2,3), (4,5)$; $\Phi_2 : (0,3), (1,4), (2,5)$; $\Phi_3 : (0,3), (1,2), (4,5)$; $\Phi_4 : (0,1), (2,5), (3,4)$; $\Phi_5 : (0,5), (1,4), (2,3)$.

If Ψ is a set of pairs in Φ, we define Ψ^*, Ψ_*, $(Z_0)^*$, $(Z_0)_*$ as in 4.18. If $\hat{\Phi} \subset \Phi$ is a subset of Φ, we define $c(Z, \Phi, \hat{\Phi})$ to be $1/2$ of the right-hand side of (5.17.2); this sum is now interpreted as follows. If $|\hat{\Phi}|$ is even, then $(-1)^{e(\Psi)} = (-1)^{e(\Phi - \Psi)}$ and the term corresponding to Ψ is an irreducible $\mathbf{Q}[\mathbf{W}]$-module (the same for Ψ as for $\Phi - \Psi$). Thus, our sum is an element of $\mathcal{R}(\mathbf{W})$. If $|\hat{\Phi}|$ is odd, then we replace the brackets [ ] in (5.17.2) by brackets [[ ]]; thus (5.17.2) is now a linear combination of representations of $\tilde{\mathbf{W}}_2$ (see 4.18), hence it may be also regarded as a linear combination of representations over $\mathbf{Q}$ of $\tilde{\mathbf{W}}$ (defined with respect to γ). Since in this case $(-1)^{e(\Psi)} = -(-1)^{e(\Phi-\Psi)}$, our sum (with 2^d terms) is an element of $\mathcal{R}(\mathbf{W},\gamma)$.

The following result (analogous to (5.17.3)) appears in [L_{12}, 1.16, 1.17].

(5.18.1) *Let* $Z, \Phi, \hat{\Phi}$ *be as above* (Z *nondegenerate*). *If* $|\hat{\Phi}|$ *is even, then there exists an involution* $x \in \mathbf{W}$ *such that* $c(Z,\Phi,\hat{\Phi}) = a_x \in \mathcal{R}(\mathbf{W})$ *and* $\ell(x) \equiv a_{[Z]} \pmod 2$. *If* $|\hat{\Phi}|$ *is odd, then there exists an element* $x \in \mathbf{W}$ *such that the image of* $\gamma x \in \tilde{\mathbf{W}}$ *in* $\tilde{\mathbf{W}}_2$ *is an involution,* $c(Z,\Phi,\hat{\Phi}) = a_{\gamma x} \in \mathcal{R}(\mathbf{W},\gamma)$ *and* $\ell(x) \equiv a_{[Z]} \pmod 2$.

If $|\hat{\Phi}|$ is even, all irreducible $\mathbf{Q}[\mathbf{W}]$-modules appearing in the sum $c(Z,\Phi,\hat{\Phi})$ are in the same family as $[Z]$. Conversely, if $[\Lambda]$ is any irreducible $\mathbf{Q}[\mathbf{W}]$-module in the same family as $[Z]$, then there exists an admissible arrangement Φ of Z such that for any $\hat{\Phi} \subset \Phi$ with $|\hat{\Phi}|$ even, $\pm[\Lambda]$ is one of the terms in the sum defining $c(Z,\Phi,\hat{\Phi})$. (See [L_{12}, 1.1.9].) Using now (5.18.1) and 5.2 (ii), we see that the following holds:

(5.18.2) *If* $[\Lambda]$ *is an irreducible representation of* $\mathbf{W}$ *in the same family as the special representation* $[Z]$, (Z *nondegenerate*), *then there exists an involution* $x \in \mathbf{W}$ *such that* $[\Lambda] \underset{LR}{\sim} x$, $[Z] \underset{LR}{\sim} x$.

The following statement follows from (4.6.6) and (5.16.1):

(5.18.3) *If* Λ *is a degenerate symbol of rank* n *and defect* 0, $(n \geq 4)$, *then there exists involutions* $x', x'' \in \mathbf{W}$ *such that* $[\Lambda]_{s'} = a_{x'}$, $[\Lambda]_{s''} = a_{x''}$ *(with notations in 4.6) and* $\ell(x') = \ell(x'') = a_{[\Lambda]_{s'}} = a_{[\Lambda]_{s''}}$.

We now give an example in D_4. With the notations in 4.7, we have

(5.18.4)
$$\begin{cases} \begin{bmatrix}1&3\\0&2\end{bmatrix} + \begin{bmatrix}2&3\\0&1\end{bmatrix} = a_{s_1s_2s_3} \\ \begin{bmatrix}1&3\\0&2\end{bmatrix} - \begin{bmatrix}2&3\\0&1\end{bmatrix} = a_{w_0s_1s_2s_3} \\ \begin{bmatrix}1&3\\0&2\end{bmatrix} + \begin{bmatrix}1&2\\0&3\end{bmatrix} = a_{s_0s_1s_0} \\ \begin{bmatrix}1&3\\0&2\end{bmatrix} - \begin{bmatrix}1&2\\0&3\end{bmatrix} = a_{w_0s_0s_1s_0} . \end{cases}$$

5.19. Assume now that $(\mathbf{W}, \mathbf{S})$ is of type G_2. With the notations in 4.8, we have

$$\begin{aligned} a_e &= \mathfrak{a}_{w_0} &&= 1 \\ a_{s_1} &= \mathfrak{a}_{s_2} &&= V + V' + \varepsilon_2 \\ a_{s_2} &= \mathfrak{a}_{s_1} &&= V + V' + \varepsilon_1 \\ a_{s_1s_2s_1} &= \mathfrak{a}_{s_2s_1s_2} &&= 2V - \varepsilon_2 \\ a_{s_2s_1s_2} &= \mathfrak{a}_{s_1s_2s_1} &&= 2V - \varepsilon_1 \\ a_{w_0s_2} &= \mathfrak{a}_{w_0s_1} &&= V - V' + \varepsilon_2 \\ a_{w_0s_1} &= \mathfrak{a}_{w_0s_2} &&= V - V' + \varepsilon_1 \\ a_{w_0} &= \mathfrak{a}_e &&= \text{sign} \\ a_x &= \mathfrak{a}_x &&= 0, \text{ whenever } x \in \mathbf{W} \text{ is not an involution.} \end{aligned}$$

These formulas are proved using the simple explicit matrix description of the irreducible representation of the Hecke algebra of type G_2, (see [CIK]), from which one gets easily the traces of T_x for all $x \in \mathbf{W}$.

5.20. Let $(\mathbf{W}, \mathbf{S})$ be a Weyl group of type F_4.

(5.20.1) *With the notations in 4.10, the following elements of* $\mathcal{R}(\mathbf{W})$ *are of the form* a_x *for some involution* $x \in \mathbf{W}$ *such that* $\ell(x) \equiv a_E \pmod 2$, *where* E *is the unique special component of* a_x :

$1_1, 1_4, 4_2 + 2_1, 4_2 - 2_1, 4_2 + 2_3, 4_2 - 2_3, 4_5 + 2_2,$

$4_5 - 2_2, 4_5 + 2_4, 4_5 - 2_4, 9_1, 9_4, 8_1, 8_2, 8_3, 8_4,$

$$Y_{1,1} = 12_1 + 16_1 + 9_2 + 6_1 + 4_3$$

$$Y_{2,1} = 12_1 + 16_1 + 9_3 + 6_1 + 4_4$$

$$Y_{3,1} = 12_1 + 16_1 + 2.9_2 + 6_2 + 4_3 + 1_2$$

$$Y_{4,1} = 12_1 + 16_1 + 2.9_3 + 6_2 + 4_4 + 1_3$$

$$Y_{5,1} = 12_1 + 2.16_1 + 9_2 + 9_3 + 6_2 + 4_1 \ .$$

This is proved as follows. We use notations in 4.10. It is clear that $1_1 = \alpha_e$ and, by 5.14 (iv), we have $1_4 = \alpha_{w_0}$. We have $4_2 \pm 2_1 = J^{\mathbf{W}}_{<s_2,s_3>}$ $(A \pm B)$, (see (4.10.3)). By (5.17.1), $A \pm B$ are of form α_x (with respect to $<s_2, s_3>$) where x is an involution of odd length in $<s_2, s_3>$. Using (5.10.5), it follows that $4_2 \pm 2_1$ are of form α_x where x is an involution of odd length in $\mathbf{W}$. The cases of $4_2 \pm 2_3$, 9_1, 8_1, 8_2, 8_3, 8_4 are dealt with in the same way, using (4.10.3) and (5.10.5).

We now apply (5.12.4) and (5.12.6) to the equation $4_2 + 2_1 = \alpha_x$ $(x \in \mathbf{W}, \ \ell(x) = \text{odd}, \ x^2 = 1)$. Using the fact that w_0 acts as -1 on 4_2 and as $+1$ on 2_1, we get $\mathcal{A}_{w_0 x} = -\zeta(\alpha_x) = 4_2 - 2_1 = \alpha_{w_0 x} \otimes \text{sign}$. It follows that $\alpha_{w_0 x} = (4_2 - 2_1) \otimes \text{sign} = 4_5 - 2_2$. Hence $4_5 - 2_2$ is of the required form, since $\ell(w_0)$ is even. Similarly, from $4_2 - 2_1 = \alpha_x$ we get $\alpha_{w_0 x} = 4_5 + 2_1$. A similar argument shows that $4_5 \pm 2_4$ and 9_4 are of the required form.

Using now (4.10.4) we see that $Y_{5,1} = J^{\mathbf{W}}_{<s_2,s_3>}$ (sign) and that

$$Y_{4,1} = J^{\mathbf{W}}_{\mathbf{W}_1}\left(\begin{bmatrix} 0 & 1 & 3 \\ & 1 & 2 & \end{bmatrix} + \begin{bmatrix} 1 & 2 & 3 \\ & 0 & 1 & \end{bmatrix}\right).$$

Using now (5.10.5), 5.14 (ii) for $<s_2, s_3>$ and (5.17.3) for $\mathbf{W}_1$, we see that $Y_{4,1}, Y_{5,1}$ have the required form. Similarly, replacing in this argument $\mathbf{W}_1$ by $\mathbf{W}_4$, we see that $Y_{3,1}$ has the required form. By (4.10.4) we have $Y_{1,1} = J^{\mathbf{W}}_{\mathbf{W}_3}$ (sign). Using 5.14 (iv) and (5.10.5), it follows that $Y_{1,1} = \alpha_x$ where x is the longest element in $\mathbf{W}_3$. (It has even length.) Replacing in this argument $\mathbf{W}_3$ by $\mathbf{W}_2$ we see that $Y_{2,1}$ has the required form. This completes the verification of (5.20.1).

5.21. Let $(\mathbf{W}, \mathbf{S})$ be a Weyl group of type E_6, and let γ be its unique automorphism of order 2.

(5.21.1) *With the notations in 4.11, the following elements of* $\mathcal{R}(\mathbf{W})$ *are of the form* α_x *for some involution* $x \in \mathbf{W}$ *such that* $\ell(x) \equiv a_E$ (mod 2) *where* E *is the unique special component of* α_x :

$$1_p, 1'_p, 6_p, 6'_p, 20_p, 20'_p, 30_p + 15_q, 30_p - 15_q,$$

$$30_p + 15_p, 30_p - 15_p, 30'_p + 15'_q, 30'_p - 15'_q,$$

$$30'_p + 15'_p, 30'_p - 15'_p, 64_p, 64'_p, 60_p, 60'_p, 81_p, 81'_p,$$

$$24_p, 24'_p, 80_s + 2.90_s + 20_s, 80_s + 60_s + 10_s,$$

$$80_s + 60_s + 90_s, 80_s - 60_s + 90_s .$$

This is proved as follows. It is clear that $1_p = \alpha_e$ and, by 5.14 (iv), we have $1'_p = \alpha_{w_0}$. We have $6_p = J^{\mathbf{W}}_{<s_2>}(\text{sign}) = \alpha_{s_2}$ (by (4.11.3) and (5.10.5)), $30_p + 15_q = J^{\mathbf{W}}_{D_4}\left(\begin{bmatrix} 0 & 2 \\ 1 & 3 \end{bmatrix} + \begin{bmatrix} 0 & 1 \\ 2 & 3 \end{bmatrix}\right) = \alpha_{s_2 s_3 s_5}$ (by (4.11.3), (5.18.4), (5.10.5)), hence 6_p, $30_p + 15_q$ are of the required form.

The cases of $30_p, 20'_p, 30_p - 15_q, 30_p \pm 15_p, 64_p, 64'_p, 60_p, 60'_p, 81_p, 81'_p, 24_p, 24'_p$ are dealt with in the same way, using (4.11.3) and (5.10.5).

We now show that $6'_p$ is of the required form. Let $\tilde{6}_p$ (resp. $\hat{6}_p$) be the $\mathbf{Q}[\tilde{\mathbf{W}}]$-module ($\tilde{\mathbf{W}}$ defined in terms of $\gamma: \mathbf{W} \to \mathbf{W}$) which extends 6_p and is such that γ acts on it as $w_0 : 6_p \to 6_p$ (resp. as $-w_0$). The restriction of $\tilde{6}_p$ to the subgroup generated by s_2 and γ has a unique line on which s_2 acts as -1; γ acts on that line also as -1. It follows that $\hat{6}_p - \tilde{6}_p = J^{\tilde{\mathbf{W}}}_{<s_2,\gamma>}(\widetilde{\text{sign}} - \widehat{\text{sign}})$, where $\widetilde{\text{sign}}$ (resp. $\widehat{\text{sign}}$)

is the one dimensional representation of $\langle s_2, \gamma \rangle$ on which s_2 acts as -1 and γ acts as 1 (resp. as -1). Using (5.10.5), it follows that $(1/2)(\hat{6}_p - \tilde{6}_p) = a_{\gamma s_2} \in \mathcal{R}(\tilde{W}, \gamma)$. Using (5.12.4), we have $\mathcal{A}_{w_0 s_2} = -\zeta(a_{\gamma s_2}) = \zeta((1/2)(\tilde{6}_p - \hat{6}_p)) = 6_p$ (where we have identified $\mathcal{R}(W)$ with $\mathcal{R}(W, \mathrm{id})$ as in (5.10.1)), hence $6'_p = \mathcal{A}_{w_0 s_2} \otimes \mathrm{sign} = a_{w_0 s_2}$. Thus, $6'_p$ has the required form.

Let W_{16} be the parabolic subgroup generated by $S - \{s_1, s_6\}$. It is of type D_4.

Let $\widetilde{30}_p, \widetilde{15}_q, \widetilde{15}_p$ (resp. $\widehat{30}_p, \widehat{15}_q, \widehat{15}_p$) be the $Q[\tilde{W}]$-modules extending $30_p, 15_q, 15_p$, on which γ acts as w_0 (resp. as $-w_0$). Let $\tilde{W}_{16}$ be the subgroup of $\tilde{W}$ generated by W_{16} and γ.

There exists $\varepsilon \in \{\pm 1\}$ such that

$$\widehat{30}_p - \widetilde{30}_p + \widehat{15}_q - \widetilde{15}_q = J_{\tilde{W}_{16}}^{\tilde{W}}\left(\begin{bmatrix}1 & 3\\0 & 2\end{bmatrix} - \begin{bmatrix}0 & 2\\1 & 3\end{bmatrix} + \varepsilon\begin{bmatrix}2 & 3\\0 & 1\end{bmatrix} - \varepsilon\begin{bmatrix}0 & 1\\2 & 3\end{bmatrix}\right)$$

$$= 2a_{\gamma x} \quad \text{(by (5.10.5) and (5.18.1) for } (W_{16}, \gamma))$$

$$\widehat{30}_p - \widetilde{30}_p - (\widehat{15}_q - \widetilde{15}_q) = J_{\tilde{W}_{16}}^{\tilde{W}}\left(\begin{bmatrix}1 & 3\\0 & 2\end{bmatrix} - \begin{bmatrix}0 & 2\\1 & 3\end{bmatrix} - \varepsilon\begin{bmatrix}2 & 3\\0 & 1\end{bmatrix} + \varepsilon\begin{bmatrix}0 & 1\\2 & 3\end{bmatrix}\right)$$

$$= 2a_{\gamma x'} \quad \text{(by (5.10.5) and (5.18.1))}$$

where $x, x' \in W_{16}$ have odd length and $\gamma(x)x = \gamma(x')x' = 1$.

Using (5.12.4), it follows that

$$\mathcal{A}_{w_0 x} = -\zeta(a_{\gamma x}) = 30_p + 15_q$$

$$\mathcal{A}_{w_0 x'} = -\zeta(a_{\gamma x'}) = 30_p - 15_q$$

hence $30'_p + 15'_q = \mathcal{A}_{w_0 x} \otimes \mathrm{sign} = a_{w_0 x}$, $30'_p - 15'_q = \mathcal{A}_{w_0 x'} \otimes \mathrm{sign} = a_{w_0 x'}$ so that $30'_p + 15'_q$, $30'_p - 15'_q$ have the required form. A similar argument shows that $30'_p + 15'_p$, $30'_p - 15'_p$ have the required form.

Using now (4.11.4), we have: $80_s + 60_s + 10_s = J^{W}_{A_2 \times A_2 \times A_1}(\mathrm{sign}) = a_x$, where x is the longest element of $A_2 \times A_2 \times A_1 \subset W$ (by (5.10.5) and 5.14 (iv)), as required. We have $80_s + 60_s + 90_s = J^{W}_{A_3 \times A_1}(\mathrm{sign}) = a_x$, where x is the longest element of $A_3 \times A_1 \subset W$ (by (5.10.5) and 5.14 (iv)), as required.

We have

$$80_s + 2.90_s + 20_s = J^{W}_{D_5}\left(\begin{pmatrix}1 & 2 & 4\\ 0 & 1 & 3\end{pmatrix} + \begin{pmatrix}1 & 2 & 3\\ 0 & 1 & 4\end{pmatrix}\right)$$

$$80_s - 60_s + 90_s = J^{W}_{D_5}\left(\begin{pmatrix}1 & 2 & 4\\ 0 & 1 & 3\end{pmatrix} - \begin{pmatrix}1 & 3 & 4\\ 0 & 1 & 2\end{pmatrix}\right)$$

(see (4.11.4)). Using (5.10.5) and (5.18.1) for D_5, we see that the last three expressions are of the form a_x for certain involutions x of odd length.

This completes the verification of (5.21.1).

5.22. Let (W, S) be a Weyl group of type E_7.

(5.22.1) *With the notations in 4.12, the following elements of* $\mathcal{R}(W)$ *are of the form* a_x *for some involution* $x \in W$ *such that* $\ell(x) \equiv a_E \pmod 2$ *where* E *is the unique special component of* a_x :

(a) *Any irreducible* $Q[W]$*-module which forms a family by itself.*

(b) $E+E_1$, $E-E_1$, $E+E_2$, $E-E_2$ *for any family of form* $\{E, E_1, E_2\}$, *with* E *special.*

(c) $E+2E_2+E_4$, $E+E_1+E_3$, $E+E_1+E_2$, $E-E_1+E_2$ *for any family of form* $\{E, E_1, E_2, E_3, E_4\}$, *listed in this order.*

(d) $512'_0 + 512_0$.

Using the formulas (4.12.2) and the statement (5.10.5) we can deduce that several of the elements listed above have the required form, since similar results are already known for proper parabolic subgroups. In this way, we see that the following elements of $\mathcal{R}(W)$ have the required form:

1_a, $7'_a$, 27_a, $56'_a \pm 35_b$, $56'_a \pm 21_a$, $21'_b$, 21_b, $120_a \pm 15'_a$,

$120_a \pm 105'_a$, $189'_b$, 210_a, $210'_a$, 105_b, $105'_b$, 168_a, $168'_a$,

$189'_c$, 189_c, $315'_a + 2.280'_a + 35'_a$, $315'_a + 280_b + 70'_a$,

$315'_a + 280_b + 280'_a$, $315'_a - 280_b + 280'_a$, $405_a \pm 216'_a$,

$405_a \pm 189_a$, $405'_a \pm 216_a$, $405'_a \pm 189'_a$, $378'_a$, 378_a, 210_b,

$210'_b$, $420_a \pm 84_a$, $420_a \pm 336'_a$, $420'_a \pm 84'_a$, $420'_a \pm 336_a$,

105_c, $105'_c$, $512'_a + 512_a$.

Note also that $315_a + 280'_b + 70_a = J^{\mathbf{W}}_{A_5 \times A_1}$ (sign) is of the required form. To deal with the remaining elements in (5.22.1) we apply (5.12.4) to the elements already dealt with; note that (5.12.4) is particularly simple in our case, since w_0 is central, (see (5.12.6)). This completes the verification of (5.22.1). We now prove:

(5.22.2) *The virtual representation* $512'_a - 512_a \in \mathcal{R}(\mathbf{W})$ *is of the form* $-a_x$ *for some involution* $x \in \mathbf{W}$ *such that* $\ell(x) \equiv a_E + 1 \pmod 2$, *where* $E = 512'_a$ *is the special component of* a_x.

Let $y \in \mathbf{W}$ be an involution of odd length such that $512' + 512 = a_y$. (See (5.22.1).) Using (5.12.4) and (5.12.6) we have

$$512'_a - 512_a = -\zeta(512'_a + 512_a) = -\zeta(a_y) = \mathfrak{A}_{w_0 y} = a_{w_0 y} \otimes \text{sign}$$

hence $512'_a - 512_a = -a_{w_0 y}$ and $w_0 y$ is an involution of even length, as required.

5.23. Let $(\mathbf{W}, S)$ be a Weyl group of type E_8.

(5.23.1) *Each of the following elements of* $\mathcal{R}(\mathbf{W})$ *is of the form* a_x *for some involution* $x \in W$ *such that* $\ell(x) \equiv a_E \pmod 2$ *where* E *is the unique special component of* a_x.

(a) *Any irreducible* $\mathbf{Q}[W]$*-module which forms a family by itself.*

(b) $E+E_1$, $E-E_1$, $E+E_2$, $E-E_2$, *for any family of the form* $\{E,E_1,E_2\}$ *with* E *special.*

(c) $E+2E_2+E_4$, $E+E_1+E_3$, $E+E_1+E_2$, $E-E_1+E_2$, *for any family of form* $\{E,E_1,E_2,E_3,E_4\}$, *listed in this order.*

(d) 4096_z+4096_x, $4096'_x+4096'_z$.

(e)
$$X_{1,1}=4480_y+4200_y+2.4536_y+2.5670_y+1400_y+2688_y+1680_y+2.7168_w+2.5600_w$$
$$X_{2,1}=4480_y+3150_y+4200_y+4536_y+5670_y+2688_y+2.7168_w+2016_w+5600_w$$
$$X_{3,1}=4480_y+3.4536_y+3.5670_y+2.1400_y+3.1680_y+70_y+7168_w+2.5600_w+448_w$$
$$X_{4,1}=4480_y+3150_y+4536_y+2.5670_y+1134_y+1680_y+7168_w+2.5600_w+448_w$$
$$X_{5,1}=4480_y+2.4200_y+4536_y+5670_y+1400_y+168_y+7168_w+1344_w+5600_w$$
$$X_{6,1}=4480_y+3150_y+4200_y+5670_y+1134_y+7168_w+1344_w+5600_w$$
$$X_{7,1}=4480_y+3150_y+4200_y+420_y+7168_w+1344_w+2016_w.$$

(5.23.2) *The following elements of* $\mathcal{R}(\mathbf{W})$ *are of the form* $-a_x$ *for some involution* $x\in\mathbf{W}$ *such that* $\ell(x)\equiv a_E+1 \pmod 2$ *where* E *is the unique special component of* a_x :

$$4096_z-4096_x,\quad 4096'_x-4096'_z.$$

The fact that the elements of type (a), (b) are of the required form is proved just as for E_7, using (4.13.3), (5.10.5), (5.12.4) and (5.12.6).

The same method applies to the elements of type (c) except for $1400'_z+1344'_x+448'_z$. But this element is equal to $J^{\mathbf{W}}_{E_6\times A_1}$ (sign) and we may then use (5.10.5) again to see that it is of the required form.

For elements of type (d), we use (4.13.4) and (5.10.5):

$$X_{1,1} = J_{D_6}^{W}\left(\begin{bmatrix}1 & 2 & 3 & 4\\ 0 & 1 & 3 & 4\end{bmatrix}\right)$$

$$X_{2,1} = J_{A_5\times A_1}^{W}(\text{sign})$$

$$X_{3,1} = J_{E_7}^{W}(315_a + 2.280_a + 35_a)$$

$$X_{4,1} = J_{E_6\times A_1}^{W}(30_p' \boxtimes \varepsilon + 15_p' \boxtimes \varepsilon), \quad \varepsilon = \text{sign on } A_1$$

$$X_{5,1} = J_{D_7}^{W}\left(\begin{bmatrix}1 & 2 & 3 & 5\\ 0 & 1 & 3 & 4\end{bmatrix} + \begin{bmatrix}1 & 3 & 4 & 5\\ 0 & 1 & 2 & 3\end{bmatrix}\right)$$

$$X_{6,1} = J_{D_5\times A_2}^{W}\left(\begin{bmatrix}1 & 2 & 3 & 4\\ 0 & 1 & 2 & 4\end{bmatrix} \boxtimes \varepsilon\right), \quad \varepsilon = \text{sign on } A_2$$

$$X_{7,1} = J_{A_4\times A_3}^{W}(\text{sign})$$

hence $X_{i,1}$ $(1 \le i \le 7)$ have the required form.

Using the results in 5.22 and $J_{E_7}^{W}$, we see that

$$4096_z + 4096_x = a_r, \quad 4096_z - 4096_x = -a_{r'}$$

where r is an involution of odd length and r' is an involution of even length in W. Using (5.12.4) and (5.12.6), we have

$$4096_x + 4096_z = \zeta(4096_x - 4096_z) = \zeta(a_{r'}) = \mathfrak{a}_{w_0 r'} = a_{w_0 r'} \otimes \text{sign}$$

hence $a_{w_0 r'} = 4096_x \otimes \text{sign} + 4096_z \otimes \text{sign} = 4096_x' + 4096_z'$ and

$4096_z - 4096_x = -\zeta(4096_z + 4096_x) = -\zeta(a_r) = \mathfrak{a}_{w_0 r} = a_{w_0 r} \otimes \text{sign}$ hence

$-a_{w_0 r} = 4096_x \otimes \text{sign} - 4096_z \otimes \text{sign} = 4096_x' - 4096_z'$.

This completes the verification of (5.23.1) and (5.23.2).

5.24. We note the following property of an irreducible $\mathbf{Q}[W]$-module E, where (W, S) is any Weyl group.

(5.24.1) *If* E_0 *is the special representation in the same family as* E*, then there exists an involution* $x \in \mathbf{W}$ *such that* E *and* E_0 *both appear with nonzero coefficient in* $a_x \in \mathcal{R}(\mathbf{W})$ *and, hence (5.2 (ii)) we have* $E \underset{LR}{\sim} x$, $E_0 \underset{LR}{\sim} x$.

We may assume that $(\mathbf{W}, S)$ is irreducible. If $(\mathbf{W}, S)$ is of type A, B, C, or D, then (5.24.1) has been noted already. (See 5.16, (5.17.1), (5.17.4), (5.18.2), (5.18.3).) For exceptional Weyl groups (5.24.1) follows from the results in 5.19-5.23.

Since $E \underset{LR}{\sim} x$, $E_0 \underset{LR}{\sim} x$ implies $E \underset{LR}{\sim} E_0$, (cf. 5.15), we see that we have proved the implication (i) $\Rightarrow$ (ii) in the following result of Barbasch and Vogan.

5.25 THEOREM ([BV_1], [BV_2]). *Let* E, E′ *be two irreducible* $\mathbf{Q}[\mathbf{W}]$*-modules. The following conditions are equivalent*:

(i) E, E′ *are in the same family, (see 4.2).*

(ii) $E \underset{LR}{\sim} E'$, *(see 5.15).*

It remains to prove the implication (ii) $\Rightarrow$ (i). (I do not know a proof entirely in the framework of Hecke algebras.) Using the dictionary [BB], [BK], [KL_2], [Vo], between left cells and two-sided cells in Hecke algebras and the corresponding notions in the theory of primitive ideals in enveloping algebras, one can draw the following conclusions ((5.25.1)-(5.25.3)) from the work of Joseph [J] on Goldie rank representations.

(5.25.1) The b-function (4.1.2) restricted to any equivalence class (under $\underset{LR}{\sim}$) of irreducible $\mathbf{Q}[\mathbf{W}]$-modules, reaches its minimum at a unique irreducible representation (called J-special) in that class.

(5.25.2) Let $\Gamma \subset \mathbf{W}$ be a left cell, let $[\Gamma]$ be the corresponding left $H_{\mathcal{K}}$-module (see 5.4) and let $[\Gamma]_1$ be the corresponding $\mathbf{Q}[\mathbf{W}]$-module.

Then there is a unique J-special representation E of $\mathbb{W}$ such that $<E, [\Gamma]_1>_{\mathbb{W}} \neq 0$.

(5.25.3) If $\mathbb{W}'$ is a standard parabolic subgroup of $\mathbb{W}$ and E' is a J-special representation of $\mathbb{W}'$, then $j_{\mathbb{W}'}^{\mathbb{W}}(E')$, (see (4.1.9)), is a J-special representation of $\mathbb{W}$.

Since the implication (i) $\Rightarrow$ (ii) of the theorem is already proved, we see from (5.25.1), (4.14.1), (4.14.2) and 4.13 that any J-special representation of $\mathbb{W}$ must be special. The implication (ii) $\Rightarrow$ (i) is a consequence of the following.

5.26 LEMMA. *Any special representation of* $\mathbb{W}$ *must be* J*-special.*

We shall sketch a proof of this lemma, on the lines of [BV_2]; this will differ from the proof given in [BV_1] in the case of classical groups. If $\Gamma \subset W$ is a disjoint union of left cells $\Gamma_1, \Gamma_2, \cdots, \Gamma_m$ of W, it gives rise to a $\mathbb{Q}[W]$-module $[\Gamma]_1 = \bigoplus_{i=1}^{m} [\Gamma_i]_1$, (see (5.25.2)).

The following statement is proved in [BV_2].

(5.26.1) Let $\mathbb{W}'$ be a standard parabolic subgroup of $\mathbb{W}$, let M be the set of all $w \in \mathbb{W}$ such that w has minimal length in $w \cdot \mathbb{W}'$ and let $\Gamma' \subset \mathbb{W}'$ be a left cell of $\mathbb{W}'$. Then $M \cdot \Gamma'$ is a union of left cells of $\mathbb{W}$ and the $\mathbb{Q}[\mathbb{W}]$-module $[M \cdot \Gamma']_1$ is $\mathrm{Ind}_{\mathbb{W}'}^{\mathbb{W}}([\Gamma']_1)$.

To prove the lemma, we may assume that $\mathbb{W}$ is irreducible, $\neq \{e\}$ and that the lemma is already proved for all proper standard parabolic subgroups of $\mathbb{W}$.

Let $\mathcal{E}$ be the set of special representations of $\mathbb{W}$ which are not of the form $j_{\mathbb{W}'}^{\mathbb{W}}(E')$, for $\mathbb{W}'$ a proper standard parabolic subgroup of $\mathbb{W}$ and E' a special representation of $\mathbb{W}'$.

Consider the following statement.

(5.26.2) For any $E \in \mathcal{E}$ there exist a proper standard parabolic subgroup $W' \subset W$ and a left cell $\Gamma' \subset W'$ such that

a) the $Q[W']$-modules $E|W'$ and $[\Gamma']_1$ are not disjoint.

b) For any $E_1 \in \mathcal{E}$, the $Q[W']$-modules $(E_1 \otimes \text{sign})|W'$ and $[\Gamma']_1$ are disjoint.

c) If W is of type E_7 (resp. E_8) then for any irreducible $Q[W]$-module E^0 occurring in $\text{Ind}_{W'}^{W}([\Gamma']_1)$ we have $a_{E^0} > 11$ (resp. $a_{E^0} > 26$).

We now show how (5.26.2) implies the lemma. If E is a special representation of W such that $E \notin \mathcal{E}$, then the induction hypothesis together with (5.25.3) show that E is J-special. Assume now that $E \in \mathcal{E}$. To E we can attach W' and Γ' as in (5.26.2). Let $M \cdot \Gamma' \subset W$ be defined as in (5.26.1). Then E is isomorphic to a $Q[W]$-submodule of $\text{Ind}_{W'}^{W}([\Gamma']_1) = [M \cdot \Gamma']_1$. Since $M \cdot \Gamma'$ is a union of left cells, there exists a left cell $\Gamma \subset M \cdot \Gamma'$ such that E occurs in the $Q[W]$-module $[\Gamma]_1$.

Let E^0 be the unique J-special component of $[\Gamma]_1$, (see (5.25.2)). Then E, E^0 are special and satisfy $a_E > 11$, $a_{E^0} > 11$ (if $W = E_7$) and $a_E > 26$, $a_{E^0} > 26$ (if $W = E_8$), (cf. (5.26.2)(c)), hence $E \otimes \text{sign}$, $E^0 \otimes \text{sign}$ are again special (cf. [L_6]). Moreover, $E \otimes \text{sign} \notin \mathcal{E}$, $E^0 \otimes \text{sign} \notin \mathcal{E}$ (cf. (5.26.2) (b)). As we have seen earlier, any special representation which is not in $\mathcal{E}$ is J-special. Thus, $E \otimes \text{sign}$ and $E^0 \otimes \text{sign}$ are J-special. They are both components of the $Q[W]$-module $[w_0\Gamma]_1$, (see 5.14 (ii)). But $w_0\Gamma$ is a left cell (see 5.14 (i)) and hence, by (5.25.2), $[w_0\Gamma]_1$ can only have one J-special component. It follows that $E \otimes \text{sign} = E^0 \otimes \text{sign}$, hence $E = E^0$. Thus, E is J-special, as desired.

It remains to prove (5.26.2).

If W is of type A_n, $(n \geq 1)$, we have $\mathcal{E} = \{\text{sign}\}$ and we can take W' of type A_{n-1} and $\Gamma' = \{\text{longest element in } W'\}$, so that $[\Gamma']_1 = \text{sign}$ representation of W'.

Assume now that W is of type B_n or C_n $(n \geq 2)$ hence it is isomorphic to W_n (see 4.5). Then $\mathcal{E}$ consists of the representations $[\Lambda]$ where Λ is a symbol of rank n and defect 1, with the following properties: $[\Lambda]$ is special; if t is the largest entry in Λ then each number in $\{0,1,\cdots,t\}$ appears in Λ at least once, 0 appears exactly once and two consecutive numbers i, $i+1$ cannot be both singles. The set $\mathcal{E}' = \{[\Lambda]\otimes \text{sign} \mid [\Lambda] \in \mathcal{E}\}$ consists of the representations $[\Lambda]$ where Λ is a symbol of rank n and defect 1 with the following properties: $[\Lambda]$ is special; all entries of Λ are singles and two consecutive numbers i, $i+1$ cannot both appear in Λ. Take $W' \subset W$ of type B_{n-1} (or C_{n-1}). One can check easily that if an irreducible representation E_1 (resp. E_2) of W' occurs in the restriction to W' of a representation in $\mathcal{E}$ (resp. in $\mathcal{E}'$), then E_1, E_2 cannot be in the same family of W'. If $E \in \mathcal{E}$ is given, we take $\Gamma' \subset W'$ to be any left cell such that $E|W'$ and $[\Gamma']_1$ are not disjoint. Let E_1 be an irreducible component of $E|W'$ occurring in $[\Gamma']_1$. Assume that $E' \in \mathcal{E}'$ is such that $E'|W'$ has an irreducible component E_2 occurring in $[\Gamma']_1$. Then, as we have seen, E_1 and E_2 cannot be in the same family of W'; as 5.25 is known to hold for W', E_1, E_2 cannot be both components of $[\Gamma']_1$. This contradiction shows that for our choice of W', Γ', (5.26.2) is satisfied.

The proof of (5.26.2) for W of type D_n is completely similar to that for B_n or C_n.

If W is of type G_2, we have $\mathcal{E} = \{\text{sign}\}$ and we can take W' of type A_1 and $\Gamma' = \{\text{longest element of } W'\}$.

If W is of type F_4, we have $\mathcal{E} = \{1_4, 4_5, 9_4\}$ and we can take W' of type B_3, $\Gamma' = \{\text{longest element of } W'\}$ and (5.26.2) can be easily verified.

If W is of type E_n, $(n = 6,7,8)$, we have

$$\mathcal{E} = \{1_p', 6_p'\} \text{ if } n=6,\ \mathcal{E} = \{1_a', 7_a, 27_a', 189_b\} \text{ if } n=7 \text{ and}$$

$$\mathcal{E} = \{1_x', 8_z', 35_x', 210_x', 560_z', 1400_x'\} \text{ if } n=8;$$

we can take W' of type A_{n-1}, $\Gamma' = \{\text{longest element of } W'\}$ and (5.26.2) can be easily verified.

This completes the proof of the lemma and hence that of Theorem 5.25.

5.27. COROLLARY. *If* E, E' *are irreducible* $\mathbf{Q}[\mathbf{W}]$*-modules such that* $E \underset{LR}{\sim} E'$, *then*

(i) $a_E = a_{E'}$

(ii) $A_E = A_{E'}$.

Proof. (i) follows immediately from 5.25 and (4.14.1). From $E \underset{LR}{\sim} E'$, it follows that $E \otimes \text{sign} \underset{LR}{\sim} E' \otimes \text{sign}$ (see 5.14 (iii)). By (i) we have $a_{E \otimes \text{sign}} = a_{E' \otimes \text{sign}}$, and (ii) follows from (5.11.5).

From this corollary, it follows that there are well-defined integer valued functions on $\mathbf{W}: x \mapsto a(x)$, $x \mapsto A(x)$ such that

$$a(x) = a_E, \quad A(x) = A_E \tag{5.27.1}$$

for any irreducible $\mathbf{Q}[\mathbf{W}]$-module E such that $E \underset{LR}{\sim} x$.

5.28. Let $\mathbf{W}'$ be a standard parabolic subgroup of a Weyl group $(\mathbf{W}, S)$ and let $\Gamma' \subset \mathbf{W}'$ be a left cell of $\mathbf{W}'$. The following result appears in $[BV_2]$:

(5.28.1) *There exists a left cell* $\Gamma \subset \mathbf{W}$ *such that* $J_{\mathbf{W}'}^{\mathbf{W}}([\Gamma']_1) = [\Gamma]_1$.

This can be proved as follows. Consider the set $M \cdot \Gamma' \subset \mathbf{W}$ defined in (5.26.1) and write $M \cdot \Gamma' = \Gamma_1 \amalg \Gamma_2 \amalg \cdots \amalg \Gamma_m$ (a disjoint union of left cells in $\mathbf{W}$). Let $a_{\Gamma'}$ be defined by $a_{\Gamma'} = a(x')$, relative to $\mathbf{W}'$, (see 5.27), where x' is any element of Γ'. Define similarly a_{Γ_i} (relative to $\mathbf{W}$), for $i = 1, \cdots, m$. By (5.26.1), $\sum_i [\Gamma_i]_1 = \mathrm{Ind}_{\mathbf{W}'}^{\mathbf{W}}([\Gamma']_1)$, and it follows from (4.1.5) that $a_{\Gamma_i} \geq a_{\Gamma'}$ for all i. Moreover, by definition of $J_{\mathbf{W}'}^{\mathbf{W}}$, we have $J_{\mathbf{W}'}^{\mathbf{W}}([\Gamma']_1) = \Sigma [\Gamma_i]_1$, sum over all i such that $a_{\Gamma_i} = a_{\Gamma'}$. It remains to show that there cannot exist $i \neq j$ with $a_{\Gamma_i} = a_{\Gamma_j} = a_{\Gamma'}$. Assume that such i, j exist. Let E_i, E_j be the special representations of $\mathbf{W}$

appearing in $[\Gamma_i]_1$, $[\Gamma_j]_1$ respectively and let E' be the special representation of W' appearing in $[\Gamma']_1$. Then $b_{E_i} = b_{E_j} = b_{E'}$, (see (4.1.2)), hence $j_{W'}^{W}(E)$ contains $E_i \oplus E_j$ as a submodule. This contradicts the sentence following (4.1.9), since E' occurs with multiplicity 1 in the $b_{E'}$-th symmetric power of the reflection representation of W'. (See $[L_6, 2]$.)

5.29. A representation ρ of W (over $\mathbf{Q}$) is said to be *constructible* if it satisfies:

(a) If $W = \{e\}$, then ρ = unit representation.

(b) If $W \neq \{e\}$, then there exists a standard parabolic subgroup $W' \subsetneq W$ and a constructible representation ρ' of W' such that either $\rho = J_{W'}^{W}(\rho')$ or $\rho \otimes \text{sign} = J_{W'}^{W}(\rho')$.

(This is an inductive definition, in which we assume that the concept of constructible representation has been already defined for groups $W' \subsetneq W$, as above.)

It follows from (5.28.1) and 5.14 that for any constructible representation ρ of W there exists a left cell $\Gamma \subset W$ such that $[\Gamma]_1 = \rho$. For W irreducible, the constructible representations are described explicitly in $[L_7]$ where it is conjectured that for any left cell $\Gamma \subset W$, $[\Gamma]_1$ is constructible.

For example, using the definition, we see that the following representations of W are constructible: $Y_{1,1}, Y_{2,1}, Y_{3,1}, Y_{4,1}, Y_{5,1}$ (type F_4, see 5.20), $80_s + 2.90_s + 20_s$, $80_s + 60_s + 10_s$, $80_s + 60_s + 90_s$ (type E_6, see 5.21), $X_{1,1}, X_{2,1}, X_{3,1}, X_{4,1}, X_{5,1}, X_{6,1}, X_{7,1}$ (type E_8, see 5.23). Hence each of these representations is of the form $[\Gamma]_1$, for a suitable left cell Γ in W.

5.30. By direct inspection, we see that any irreducible $\mathbf{Q}[W]$-module ρ appears with multiplicity 1 in some constructible representation of W. It follows that ρ appears with multiplicity 1 in the W-module $[\Gamma]_1$ for some left cell $\Gamma \subset W$.

Chapter 6

AN INTEGRALITY THEOREM AND A DISJOINTNESS THEOREM

The main results of this chapter are the integrality theorem 6.14, which states that certain elements $R_{\alpha_{\gamma x}}$, $R_{\mathfrak{A}_{\gamma x}}$ of $\mathcal{R}(G^{F'}) \otimes \mathbb{Q}$ actually belong to $\mathcal{R}(G^{F'})$, and the disjointness theorem 6.17 which, under certain assumptions (automatically verified for large q), states that $\pm R_{\alpha_{\gamma x}}$, $\pm R_{\mathfrak{A}_{\gamma x}}$ (for some choice of signs) are actual representations of $G^{F'}$, not only virtual representations. The assumptions will be removed in later chapters.

6.1. We now return to the setting in 3.4. However, in 6.1-6.4, the center of G need not be connected. Given $L \in \mathfrak{L}$, $n \geq 1$, $(n,p) = 1$, and $w \in Z_{L,n}$, we have, for any integer $n' \geq 1$, $(n',p) = 1$, a canonical isomorphism $\mathcal{F}^{L}_{\dot{w},n} \cong \mathcal{F}^{L^{\otimes n'}}_{\dot{w},nn'}$, of $G^{F'}$-equivariant $\overline{\mathbb{Q}}_\ell$-sheaves over X_w. ($\dot{w}$ represents w in $N(T)^F$). A pair (L,n) is said to be *indivisible* if there is no $L_1 \in \mathfrak{L}$, $n_1 \geq 1$, $(n_1,p) = 1$ and $n_1' > 1$, $(n_1',p) = 1$ such that $L = L_1^{\otimes n_1'}$, $n = n_1 n_1'$.

Note also that $\mathcal{F}^{L}_{\dot{w},n} \cong \mathcal{F}^{L \otimes L_1^n}_{\dot{w},n}$ (isomorphism of $G^{F'}$-equivariant $\overline{\mathbb{Q}}_\ell$-sheaves over X_w).

6.2. LEMMA. (i) *If* $w \in Z_{L,n}$ *and* $x \in W$, *then* $w' = F'^{-1}(x) w x^{-1}$ *is in* $Z_{{}^xL,n}$ *and*

$$\sum (-1)^i H^i_c(X_w, \mathcal{F}^L_{\dot{w},n}) = \sum (-1)^i H^i_c(X_{w'}, \mathcal{F}^{{}^xL}_{(\dot{w}'),n}) \in \mathcal{R}(G^{F'}) .$$

(ii) *Let* $L \in \mathfrak{L}$, $L' \in \mathfrak{L}$, $n \geq 1$, $n' \geq 1$, $(n,p) = 1$, $(n',p) = 1$. *Assume that* (L,n), (L',n') *are indivisible. Let* $w \in Z_{L,n}$, $w' \in Z_{L',n'}$. *Then*

$$< \sum (-1)^i H^i_c(X_w, \mathcal{F}^L_{\dot{w},n}), \sum (-1)^i H^i_c(X_{w'}, \mathcal{F}^{L'}_{(w')^{\cdot},n'}) >_{G^{F'}} = 0$$

unless $n = n'$ *and there exists* $x \in W$ *such that* $w' = F'^{-1}(x) w x^{-1}$ *and* $L = {}^x L \otimes L_0^{\otimes n}$ *for some* $L_0 \in \mathfrak{L}$.

(iii) *For any irreducible representation* ρ *of* $G^{F'}$ *(over* $\overline{Q}_\ell$*) there exist* $L \in \mathfrak{L}$, $n \geq 1$, $(n,p) = 1$, *(*(L,n) *indivisible) such that*

$$<\rho, \sum (-1)^i H^i_c(X_w, \mathcal{F}^L_{\dot{w},n})>_{G^{F'}} \neq 0 .$$

Proof. Let $T_w \subset T$ be as in 2.1 and let $\theta : T_w \to \overline{Q}_\ell^*$ be a homomorphism. The T_w-covering $\widetilde{X}_{\dot{w}} \to X_w$ and θ give rise to a locally constant, $G^{F'}$-equivariant $\overline{Q}_\ell$-sheaf $\mathcal{F}_\theta$ over X_w. We show that $\mathcal{F}_\theta$ is of the form $\mathcal{F}^L_{\dot{w},n}$ for some (L,n) as in (iii) with $w \in Z_{L,n}$. Let n be the smallest integer ≥ 1 such that $\theta^n = 1$. Then $(n,p) = 1$ and θ is a composition $\psi \circ \overline{\theta}$ where $\psi : \mu_n \to \overline{Q}_\ell^*$ is as in 2.3 and $\overline{\theta}$ is a homomorphism $T_w \to \mu_n \subset k^*$. Since T_w is a closed subgroup of T, the homomorphism $\overline{\theta} : T_w \to k^*$ is the restriction of a homomorphism of algebraic groups $\lambda_1 : T \to k^*$. The homomorphism $\lambda_1^n : T \to k^*$ is trivial on T_w, hence it is trivial on the kernel of the étale homomorphism $t \to F'(t)\dot{w}^{-1}t^{-1}\dot{w} : T \to T$. Hence, there exists a homomorphism of algebraic groups $\lambda : T \to k^*$ such that $\lambda(F'(t)\dot{w}^{-1}t^{-1}\dot{w}) = \lambda_1^n(t)$ for all $t \in T$. Let $L \in \mathfrak{L}$ be the line bundle corresponding to $\lambda : T \to k^*$. Then $w \in Z_{L,n}$ and $\mathcal{F}_\theta = \mathcal{F}^L_{\dot{w},n}$. This implies (iii), by [DL_1, 7.7]. The statements (i), (ii), when reformulated in terms of the sheaves $\mathcal{F}_\theta$, instead of $\mathcal{F}^L_{\dot{w},n}$, are consequences of [$DL_1$, 6.8].

6.3. LEMMA. *Let* ρ *be an irreducible representation of* $G^{F'}$ *(over* $\overline{Q}_\ell$*). The following two conditions are equivalent:*

(a) *ρ is cuspidal*

(b) *for any* $L \in \mathfrak{L}$, $n \geq 1$, $(n,p) = 1$, *any* $w \in Z_{L,n}$ *such that* $<\rho, \Sigma(-1)^i H_c^i(X_w, \mathcal{F}^L_{\dot{w},n})>_{G^{F'}} \neq 0$ *and any* $x \in W$, *the element* $F'^{-1}(x)wx^{-1}$ *is not contained in any proper, standard* F'*-stable parabolic subgroup of* W.

Proof. This is a consequence of $[L_4, 2.18]$.

6.4. LEMMA. *Let* P *be an* F'*-stable parabolic subgroup of* G, *containing* B *and let* U_P *be its unipotent radical. Let* $L \in \mathfrak{L}$, $n \geq 1$, $(n,p) = 1$, *and let* $w \in Z_{L,n}$ *be such that* w *lies in the standard parabolic subgroup* $W^{(P)}$ *of* W *corresponding to* P. *Then, for any* i, *the* $G^{F'}$*-module* $H_c^i(X_w, \mathcal{F}^L_{\dot{w},n})$ *(resp.* $H^i(\overline{X}_w, \mathcal{F}^L_{\dot{w},n})$ *), is induced from the* $P^{F'}$*-module* $H_c^i(X_w^{(P)}, \mathcal{F}^{L(P)}_{\dot{w},n})$ *(resp.* $H^i(\overline{X}_w^{(P)}, \mathcal{F}^{L(P)}_{\dot{w},n})$ *) where* $X^{(P)}$, $\overline{X}_w^{(P)}$, $\mathcal{F}^{L(P)}_{\dot{w},n}$ *are defined in the same way as* X_w, $\overline{X}_w$, $\mathcal{F}^L_{\dot{w},n}$ *but with respect to* P/U_P *instead of* G. *(Note that the flag manifold of* P/U_P *is naturally embedded into that of* G, *hence* L *can be restricted to it.)*

The proof is along the lines of $[DL_1, 8.2]$. The variety $X_w^{(P)}$ (resp. $\overline{X}_w^{(P)}$) may be identified with a simultaneously open and closed subvariety of X_w (resp. of $\overline{X}_w$); it is the intersection of X_w (resp. $\overline{X}_w$) with the subvariety P/B of G/B. The variety X_w (resp. $\overline{X}_w$) is then the disjoint union of translates of $X_w^{(P)}$ (resp. $\overline{X}_w^{(P)}$) under a set of representatives for the cosets $G^{F'}/P^{F'}$ and $\mathcal{F}^{L(P)}_{\dot{w},n}$ is the restriction of $\mathcal{F}^L_{\dot{w},n}$ to $X_w^{(P)}$. From this, the lemma follows immediately.

We shall derive some consequences of the lemma. The assumption in the lemma that $Z_{L,n}$ contains an element in $W^{(P)}$ implies that the element w_1 of minimal length in $Z_{L,n}$ is in $W^{(P)}$. Indeed, let w_1' be the element of minimal length in $W^{(P)} \cap Z_{L,n}$. Then if $R_1 \subset R$ is the root system corresponding to $W^{(P)}$, we have $w_1'(R^+_{L,n} \cap R_1) \subset R^+$ (see 1.9(i)). But, since $w_1' \in W^{(P)}$, we have $w_1'(R^+ - R_1) \subset R^+$. Hence $w_1'(R^+_{L,n}) \subset R^+$ so that $w_1' = w_1$.

Now let $\tilde{V}$ be any finite dimensional $\bar{\mathbf{Q}}_\ell[\tilde{W}_{L,n}]$-module (see 3.1). Then we can define $R_{\tilde{V}} \in \mathcal{R}(G^{F'}) \otimes \bar{\mathbf{Q}}_\ell$ as in (3.7.1) with $\tilde{E}$ replaced by $\tilde{V}$. Similarly, if $\tilde{V}_1$ is a finite dimensional $\bar{\mathbf{Q}}_\ell[(W_{L,n} \cap W^{(P)})^\sim]$-module, we have an element $R_{\tilde{V}_1} \in \mathcal{R}((P/U_P)^{F'}) \otimes \bar{\mathbf{Q}}_\ell$. Let $\mathrm{Ind}(\tilde{V}_1)$ be the $\bar{\mathbf{Q}}_\ell[\tilde{W}_{L,n}]$-module induced by the $\bar{\mathbf{Q}}_\ell[(W_{L,n} \cap W^{(P)})^\sim]$-module $\tilde{V}_1$. Then $R_{\mathrm{Ind}(\tilde{V}_1)}$ is a well-defined element of $\mathcal{R}(G^{F'}) \otimes \bar{\mathbf{Q}}_\ell$, and we have

$$R_{\mathrm{Ind}(\tilde{V}_1)} = \mathrm{Ind}_{P^{F'}}^{G^{F'}}(R_{\tilde{V}_1}) \tag{6.4.1}$$

(where $R_{\tilde{V}_1}$ is regarded as an element in $\mathcal{R}(P^{F'}) \otimes \bar{\mathbf{Q}}_\ell$ in a natural way). Indeed, the left-hand side of (6.4.1) is equal to

$$(1/|W_{L,n}|) \sum_{\substack{y \in W_{L,n} \\ i \geq 0}} (-1)^i \,\mathrm{Tr}\,(\gamma y, \mathrm{Ind}(\tilde{V}_1))\, H_c^i(X_{w_1 y}, \mathcal{F}^L_{(w_1 y)^\cdot, n}) .$$

Using 6.2 (i), this can be written as

$$(1/|W_{L,n}|) \sum_{y \in W_{L,n}} (1/|W_{L,n} \cap W^{(P)}|) \sum_{\substack{z \in W_{L,n} \\ \gamma^{-1}(z) y z^{-1} \in W^{(P)}}} \mathrm{Tr}\,(z \gamma y z^{-1}, \tilde{V}_1) \cdot$$

$$\cdot \sum_{i \geq 0} (-1)^i H_c^i(X_{w_1 y'}, \mathcal{F}^L_{(w_1 y')^\cdot, n})$$

where $y' = \gamma^{-1}(z) y z^{-1}$. This can be also written as

$$(1/|W_{L,n} \cap W^{(P)}|) \sum_{\substack{y' \in W_{L,n} \cap W^{(P)} \\ i \geq 0}} (-1)^i \,\mathrm{Tr}(\gamma y', \tilde{V}_1)\, H_c^i(X_{w_1 y'}, \mathcal{F}^L_{(w_1 y')^\cdot, n})$$

which, by the lemma, is equal to the right-hand side of (6.4.1). Thus, (6.4.1) is proved. On the other hand, if we denote $\mathrm{Res}\,(\tilde{V})$ the restriction of a finite dimensional $\bar{\mathbf{Q}}_\ell[\tilde{W}_{L,n}]$-module $\tilde{V}$ to $\bar{\mathbf{Q}}_\ell[(W_{L,n} \cap W^{(P)})^\sim]$, we have

$$(6.4.2)\quad \langle (R_{\tilde{V}})^{U_P^{F'}}, \rho \rangle_{(P/U_P)^{F'}} = \langle R_{\mathrm{Res}(\tilde{V})}, \rho \rangle_{(P/U_P)^{F'}} \in \mathcal{R}((P/U_P)^{F'}) \otimes \bar{\mathbf{Q}}_\ell ,$$

where the upper-script $U_P^{F'}$ denotes the homomorphism $\mathcal{R}(G^{F'}) \otimes \bar{\mathbf{Q}}_\ell \to \mathcal{R}((P/U_P)^{F'}) \otimes \bar{\mathbf{Q}}_\ell$ defined on a $G^{F'}$-module by taking the subspace of $U_P^{F'}$-invariant vectors and ρ is any irreducible $(P/U_P)^{F'}$-module in $\mathcal{E}_{L(P),n}$.

The following statement is equivalent to (6.4.2):

$$(6.4.3)\qquad \langle R_{\tilde{V}}, \mathrm{Ind}_{P^{F'}}^{G^{F'}}(\rho) \rangle_{G^{F'}} = \langle R_{\mathrm{Res}\,\tilde{V}}, \rho \rangle_{(P/U_P)^{F'}}$$

for any irreducible $(P/U_P)^{F'}$-module ρ as in (6.4.2). Using [DL_3, (7.1)], the left-hand side of (6.4.3) can be written

$$(1/|W_{L,n}|) \sum_{y \in W_{L,n}} \mathrm{Tr}(\gamma y, \tilde{V}) \cdot$$

$$\cdot\,(1/|W_{L,n} \cap W^{(P)}|) \sum_{\substack{z \in W_{L,n} \\ \gamma^{-1}(z) y z^{-1} \in W^{(P)}}} \langle \sum_i (-1)^i H_c^i(X_{w_1 y'}^{(P)}, \mathcal{F}_{(w_1 y')^{\cdot},n}^{L(P)}), \rho \rangle_{(P/U_P)^{F'}}$$

where $y' = \gamma^{-1}(z) y z^{-1}$. This is clearly equal to the right-hand side of (6.4.3). Thus (6.4.3) and hence (6.4.2) are proved.

In the rest of this chapter, G is assumed to have connected center.

6.5. PROPOSITION. (i) *Let* $L \in \mathfrak{L}$, $n \geq 1$, $(n,p) = 1$ *be such that* $Z_{L,n}$ *is nonempty and let* w_1 *be the element of minimal length in* $Z_{L,n}$. *Let* $s \in S$ *be such that* $s \notin W_{L,n}$. *Then the map* $z \mapsto szs$ *is an isomorphism of Coxeter groups* $W_{L,n} \xrightarrow{\sim} W_{sL,n}$, *compatible with the automorphism* γ *of* $W_{L,n}$ *and* $W_{sL,n}$ *(see (2.15.1)), the map* $w \to F'(s)^{-1} ws$ *is a bijection* $Z_{L,n} \to Z_{sL,n}$ *and* $w_1' = F'(s)^{-1} w_1 s$ *is the element of minimal length in* $Z_{sL,n}$. *Moreover, for any* $w \in Z_{L,n}$ *and any integer* i, *we have*

$$\text{(6.5.1)} \qquad H^i(\bar{X}_w, \mathcal{F}^L_{\dot{w},n}) \cong H^{i+\ell(w')-\ell(w)}(\bar{X}_{w'}, \mathcal{F}^{{}^sL}_{(\dot{w}'),n})$$

(as $G^{F'}$*-modules), where* $w' = F'(s)^{-1}ws$.

(ii) *If* L, n *are as in (i) and* $x \in W$, *then* $\mathcal{E}_{L,n} = \mathcal{E}_{{}^xL,n}$.

(iii) *For any irreducible representation* ρ *of* $G^{F'}$, *there exists* $L \in \mathcal{L}$, $n \geq 1$, $(n,p) = 1$ *with* (L,n) *indivisible and* $Z_{L,n}$ *nonempty such that* $\rho \in \mathcal{E}_{L,n}$.

(iv) *For any irreducible representation* ρ *of* $G^{F'}$, *we have*

$$\rho \in \mathcal{E}_{L,n} \Longleftrightarrow \langle \rho, R_{\tilde{E}} \rangle_{G^{F'}} \neq 0 \text{ for some } E \in (W_{L,n})^\vee_{ex}.$$

(v) *Let* $L, L' \in \mathcal{L}$, $n, n' \geq 1$, $(n,p) = 1$, $(n',p) = 1$ *be such that* (L,n), (L',n') *are indivisible and* $Z_{L,n}$, $Z_{L',n'}$ *are nonempty. Then the sets* $\mathcal{E}_{L,n}$, $\mathcal{E}_{L',n'}$ *are either disjoint or coincide. They coincide if and only if* $n = n'$ *and there exists* $x \in W$ *such that* $L' = {}^xL \otimes L_0^{\otimes n}$ *for some* $L_0 \in \mathcal{L}$.

Proof. The first assertion in (i) is already contained in the proof of 1.17. The fact that w_1' has minimal length in $Zs_{L,n}$ is checked using 1.9 (i). The identity (6.5.1) is proved by rewriting both sides using 3.8 (ii) as combination of elements $R_{\tilde{E}}$, and then applying 6.2 (i). To check (ii), we write x as a product of simple reflections and are hence reduced to the case where $x = s \in S$. If $s \in W_{L,n}$, the result is obvious. If $s \notin W_{L,n}$, the result follows from (6.5.1).

We now check (iii). Given ρ, let L, n, $w \in Z_{L,n}$ be such that the conclusion of 6.2 (iii) holds. We may assume that w has minimal length among the elements of $Z_{L,n}$ for which the conclusion of 6.2 (iii) holds. Then, by (2.8.1), we have

$$\sum (-1)^i H^i_c(X_w, \mathcal{F}^L_{\dot{w},n}) = \sum (-1)^i H^i(\bar{X}_w, \mathcal{F}^L_{\dot{w},n}) \in \mathcal{R}(G^{F'})$$

and it follows that $\rho \in \mathcal{E}_{L,n,w,i}$ for some i. Hence $\rho \in \mathcal{E}_{L,n}$. The statement (iv) follows from 3.8 (ii) and 3.10 (ii). We now check (v). Assume that $\mathcal{E}_{L,n}$, $\mathcal{E}_{L',n'}$ are not disjoint. Then there exists $w \in Z_{L,n}$, $w' \in Z_{L',n'}$ and integers i, i' such that

$$<H^i(\overline{X}_w, \mathcal{F}^L_{\dot{w},n}), H^{i'}(\overline{X}_{w'}, \mathcal{F}^{L'}_{(w')^{\cdot},n'})>_{G^{F'}} \neq 0 .$$

Using 3.8 (ii) it then follows that there exist $y \in Z_{L,n}$, $y' \in Z_{L',n'}$ such that

$$<\sum (-1)^i H^i_c(X_y, \mathcal{F}^L_{\dot{y},n}), \sum (-1)^i H^i_c(X_{y'}, \mathcal{F}^{L'}_{(y')^{\cdot},n'})>_{G^{F'}} \neq 0 .$$

Using now 6.2 (ii) we see that $n = n'$ and that $L' = {}^xL \otimes L_0^{\otimes n}$ for some $x \in W$ and some $L_0 \in \mathfrak{L}$. The converse follows from (ii). The proposition is proved.

REMARK. The irreducible representations of $G^{F'}$ appearing in $\mathcal{E}_{L,n}$ for L = trivial line bundle and $n = 1$ are said to be *unipotent*. This agrees with the definition in [DL_1, 7.8].

6.6. To any irreducible representation ρ of $G^{F'}$ (over $\overline{Q}_\ell$), we associate a sign $\varepsilon_\rho \in \{\pm 1\}$ as follows. Let L, n be as in 6.5 (iii), such that $\rho \in \mathcal{E}_{L,n,w,i}$ (see 2.19), where $w \in Z_{L,n}$ and i an integer. We define $\varepsilon_\rho = (-1)^i$. To show that this is well defined, we assume that we have also $\rho \in \mathcal{E}_{L',n',w',i'}$ $(w' \in Z_{L',n'}$, (L',n') indivisible), and we must show that $i \equiv i'$ (mod 2). By 6.5 (v) we have $n = n'$, and $L' = {}^xL \otimes L_0^{\otimes n}$ for some $x \in W$ and $L_0 \in \mathfrak{L}$. By replacing L' by $L' \otimes (L_0^{\otimes n})^*$, we are reduced to the case where $L' = {}^xL$, $(x \in W)$. By decomposing x in a product of simple reflections, we are reduced to the case where x is the unit element or a simple reflection. The case where x is a simple reflection is reduced to the case where x is identity by (6.5.1). (Note that $\ell(w') \equiv \ell(w)$ (mod 2) in (6.5.1).) If $L = L'$, the congruence $i \equiv i'$ (mod 2) follows from 2.18 (i).

Thus, ε_ρ is well defined.

6.7. We now associate to an irreducible representation ρ of $G^{F'}$ (over $\overline{\mathbf{Q}}_\ell$) a sign $\hat{\varepsilon}_\rho$ in another way.

We define $\sigma(G)$ to be the largest integer $i \geq 0$ such that there exists a strictly increasing sequence of F'-stable parabolic subgroups $P_0 \underset{\neq}{\subset} P_1 \underset{\neq}{\subset} \cdots \underset{\neq}{\subset} P_i$ of G. For any F'-stable parabolic subgroup $P \subset G$, we define $\sigma(P)$ in the same way, except that P_i is required to be contained in P; equivalently, we could define $\sigma(P)$ to be $\sigma(P/U_P)$, with respect to $F' : P/U_P \to P/U_P$, where U_P is the unipotent radical of P.

Given ρ as above, there exists an F'-stable parabolic subgroup $P \subset G$ containing B such that $\rho^{U_P^{F'}}$ is a cuspidal representation of $(P/U_P)^{F'}$. (Here $\rho^{U_P^{F'}}$ is the space of $U_P^{F'}$-invariant vectors in ρ.) We then define

$$\hat{\varepsilon}_\rho = (-1)^{\sigma(P)} .$$

It is well known that $\sigma(P)$ is independent of the choice of P. Hence $\hat{\varepsilon}_\rho$ is well defined.

6.8. We recall the definition of the duality homomorphism $D : \mathcal{R}(G^{F'}) \to \mathcal{R}(G^{F'})$, (cf. [A$_1$], [Ka], [DL$_2$]). For any irreducible representation ρ of $G^{F'}$ over $\overline{\mathbf{Q}}_\ell$ we have

$$D(\rho) = \sum_P (-1)^{\sigma(P)} \operatorname{Ind}_{P^{F'}}^{G^{F'}} (\rho^{U_P^{F'}}) \in \mathcal{R}(G^{F'}) \tag{6.8.1}$$

(sum over all F'-stable parabolic subgroups P containing B), where $\rho^{U_P^{F'}}$ is regarded as a $P^{F'}$-module, in a natural way.

According to [*loc. cit.*], we have

$$\hat{\varepsilon}_\rho \cdot D(\rho) = \text{irreducible representation of } G^{F'} , \tag{6.8.2}$$

$$D^2(\rho) = \rho . \tag{6.8.3}$$

It follows from (6.8.2) and (6.8.3) that

(6.8.4) $$\langle D(\alpha),\beta\rangle_{G^{F'}} = \langle \alpha, D(\beta)\rangle_{G^{F'}}$$

for any $\alpha,\beta \in \mathcal{R}(G^{F'})$.

If $L \in \mathfrak{L}$, $n \geq 1$, $(n,p)=1$ and $w \in Z_{L,n}$, we have (cf. [DL₃]):

(6.8.5) $$D\Big(\sum_i (-1)^i H_c^i(X_w, \mathcal{F}^L_{\dot{w},n})\Big) = (-1)^{\ell(w)}\Big(\sum (-1)^i H_c^i(X_w, \mathcal{F}^L_{\dot{w},n})\Big) .$$

It follows that for any irreducible $Q[\widetilde{W}_{L,n}]$-module $\widetilde{E}$ whose restriction to $W_{L,n}$ is irreducible, we have

(6.8.6) $$D(R_{\widetilde{E}}) = (-1)^{\ell(w_1)} R_{\widetilde{E}\otimes\widetilde{\text{sign}}}$$

where w_1 has minimal length in $Z_{L,n}$, sign is the sign representation of $W_{L,n}$ and $\widetilde{\text{sign}}$ is its extension to $\widetilde{W}_{L,n}$ on which $\gamma \in \widetilde{W}_{L,n}$ acts trivially.

Using now (6.8.6), (6.8.4) and 6.5 (iv), we see that

(6.8.7) $$\rho \in \mathfrak{S}_{L,n} \Rightarrow \hat{\varepsilon}_\rho D(\rho) \in \mathfrak{S}_{L,n} .$$

6.9. PROPOSITION. *Assume that* G *has connected center. Let* $L \in \mathfrak{L}$, $n \geq 1$, $(n,p)=1$ *be such that* $Z_{L,n} \neq 0$, *and let* w_1 *be the element of minimal length in* $Z_{L,n}$. *Then, for any* $z \in W_{L,n}$ *and any integer* $i \geq 0$, *the virtual* $G^{F'}$*-module*

(6.9.1) $$D((-1)^i H^i(\overline{X}_{w_1 z}, \mathcal{F}^L_{(w_1 z)^\cdot,n}))$$

is equal to the coefficient of $u^{i/2}$ *in*

(6.9.2) $$(-1)^{\ell(w_1 z)} u^{\frac12(\ell(w_1 z)-\widetilde{\ell}(z))} \times$$

$$\times \sum_E \mathrm{Tr}\Big(\sum_{v\in W_{L,n}} (-u)^{\widetilde{\ell}(z)-\widetilde{\ell}(v)} P_{v,z}(u^{-1}) T_{\gamma v}, \widetilde{E}(u)\Big) R_{\widetilde{E}} \in \mathcal{R}(G^{F'}) \otimes Q[u^{1/2}] .$$

(E *runs over* $(W_{L,n})^{\vee}_{ex}$, *and* $\widetilde{E}$ *are chosen as in 3.6.)*

If, for $\rho \in E_{L,n}$, *we denote* $\rho' = \hat{\varepsilon}_\rho D(\rho)$ *(see 6.8.7), then we have*

$$\varepsilon_\rho = \varepsilon_{\rho'} \quad \text{(see 6.6)}. \tag{6.9.3}$$

Let us assume that (L, n) *has the following property:*

$$\varepsilon_\rho = \hat{\varepsilon}_\rho \tag{6.9.4}$$

for any $\rho \in \mathcal{E}_{L,n}$. *Then (6.9.1) is an actual* $G^{F'}$*-module.*

Proof. Note that for any $E \in (W_{L,n})^{\vee}_{ex}$, and any $v \in W_{L,n}$, we have

$$\operatorname{Tr}(T_{\gamma v}, (\widetilde{E} \otimes \widetilde{\varepsilon})(u)) = (-u)^{\tilde{\ell}(v)} \overline{\operatorname{Tr}(T_{\gamma v}, \widetilde{E}(u))} \tag{6.9.5}$$

where $\tilde{\ell}$ is the length function on $(W_{L,n}, S_{L,n})$ and $a \to \bar{a}$ is the involution of $\mathbb{Q}[u^{1/2}, u^{-1/2}]$ defined in 3.3. (We write $\widetilde{\varepsilon}$ instead of $\widetilde{\text{sign}}$, see (6.8.6).) The right-hand side of (6.9.5) is equal, by (3.3.2), to

$$(-u)^{\tilde{\ell}(v)} \operatorname{Tr}(T^{-1}_{(\gamma v)^{-1}}, \widetilde{E}(u)) . \tag{6.9.6}$$

By 3.8 (ii), the expression (6.9.1) is the coefficient of $u^{i/2}$ in

$$u^{\frac{1}{2}(\ell(w_1 z) - \tilde{\ell}(z))} \sum_E \operatorname{Tr}\Big(\sum_{v \in W_{L,n}} P_{v,z}(u) T_{\gamma v}, \widetilde{E}(u)\Big) D(R_{\widetilde{E}}) .$$

Using (6.8.6), (6.9.4), (6.9.5), this can be rewritten

$$(-1)^{\ell(w_1)} u^{\frac{1}{2}(\ell(w_1 z) - \tilde{\ell}(z))} \sum_E \operatorname{Tr}\Big(\sum_v P_{v,z}(u) T_{\gamma v}, \widetilde{E}(u)\Big) R_{\widetilde{E} \otimes \widetilde{\varepsilon}}$$

$$= (-1)^{\ell(w_1)} u^{\frac{1}{2}(\ell(w_1 z) - \tilde{\ell}(z))} \sum_E \operatorname{Tr}\Big(\sum_v P_{v,z}(u) T_{\gamma v}, (\widetilde{E} \otimes \widetilde{\varepsilon})(u)\Big) R_{\widetilde{E}}$$

$$= (-1)^{\ell(w_1)} u^{\frac{1}{2}(\ell(w_1 z) - \tilde{\ell}(z))} \sum_E \operatorname{Tr}\Big(\sum_v (-u)^{\tilde{\ell}(v)} P_{v,z}(u) T_\gamma T^{-1}_{v^{-1}}, \widetilde{E}(u)\Big) R_{\widetilde{E}} .$$

Using now (5.1.1), we see that this is equal to (6.9.2).

From (5.1.1), (5.1.2), (5.1.3), we see that for any $x \in W_{L,n}$, we have

$$C_z = \sum_{x \leq y \leq z} (-1)^{\widetilde{\ell}(z)-\widetilde{\ell}(y)} Q_{x,y}(u) P_{y,z}(u^{-1}) u^{((\ell(x)+\ell(z))/2)-\ell(y)} C'_x$$

hence $C_z = \sum_{\substack{x \in W_{L,n} \\ j}} f^{(j)}_{x,z} u^{j/2} C'_x$, where $f^{(j)}_{x,z} \in \mathbf{Z}$ and j runs over the integers such that $j \equiv \ell(x) + \ell(z) \pmod 2$. Substituting this into the equality of (6.9.1) and (6.9.2) we get that (6.9.1) is equal to the coefficient of $u^{i/2}$ in

$$\sum_{x,j} (-1)^{\ell(w_1 z)} u^{\frac{1}{2}(\ell(w_1 z))} f^{(j)}_{x,z} u^{j/2} \operatorname{Tr}(T_\gamma C'_x, \widetilde{E}(u)) R_{\widetilde{E}}$$

$$= \sum_{x,j,i'} (-1)^{\ell(w_1 z)} u^{\frac{1}{2}(\ell(w_1 z)-\ell(w_1 x))+(j/2)} f^{(j)}_{x,z} (-1)^{i'} u^{i'/2} \mathbf{H}^{i'}(\overline{X}_{w_1 x}, \mathcal{F}^L_{(w_1,x)^\cdot,n})$$

(by 3.8 (ii)). Since $j \equiv \ell(x) + \ell(z) \pmod 2$, it follows that (6.9.1) is a $\mathbf{Z}$-linear combination of $G^{F'}$-modules of form $\mathbf{H}^{i'}(\overline{X}_{w_1 x}, \mathcal{F}^L_{(w_1 x)^\cdot,n})$, $x \in W_{L,n}$, where i' has the same parity as i. Hence (6.9.3) holds.

The last statement follows from (6.8.2) and from the assumption (6.9.4). This completes the proof of the proposition.

6.10. PROPOSITION. *We associate to* (G, F') *the integer* $q > 1$ *defined by* $q^{a'} = p^a$ *where* $(F')^{a'} = F^a : G \to G$, (a, a' *integers* ≥ 1). *We also associate to* G *the integer* $h \geq 1$ *by* $h - 1$ = *maximum height of a coroot of* G *with respect to* T. *Assume that* $q \geq h$. *Then*

(6.10.1) $$\varepsilon_\rho = \hat{\varepsilon}_\rho$$

for all irreducible representations ρ *of* $G^{F'}$.

We shall first prove the following:

6.11. LEMMA. *In the setting of 3.8, for any integer* i *and any* $z \in W_{L,n}$, *the* $G^{F'}$*-modules* $H^i(\overline{X}_{w_1 z}, \mathcal{F}^L_{(w_1 z)^{\cdot},n})$ *and* $H^{2\ell(w_1 z)-i}(\overline{X}_{w_1 z}, \mathcal{F}^L_{(w_1 z)^{\cdot},n})$ *are isomorphic.*

Proof. Using 3.8 (ii), we see that it is enough to show that for any $E \in (W_{L,n})^{\vee}_{ex}$, the element

$$\mathrm{Tr}\,(u^{-\widetilde{\ell}(z)/2} \sum_{v \in W_{L,n}} P_{v,z}(u) T_{\gamma v}, \widetilde{E}(u)) \in \mathbf{Q}[u^{1/2}, u^{-1/2}]$$

is invariant under the involution $a \to \bar{a}$ of $\mathbf{Q}[u^{1/2}, u^{-1/2}]$. Using (3.3.2), we see that it is enough to check that

$$(6.11.1) \quad u^{-\widetilde{\ell}(z)/2} \sum_{v} P_{v,z}(u) T_{\gamma v} = u^{\widetilde{\ell}(z)/2} \sum_{v} P_{v,z}(u^{-1}) T^{-1}_{(\gamma v)^{-1}} \in \widetilde{H}\,.$$

This is obtained by multiplying both sides of (5.1.2) by T_γ.

6.12. *Proof of Proposition 6.10.* Let ρ be an irreducible representation of $G^{F'}$.

Assume first that ρ is not cuspidal. By 6.3 and 6.2 (i), there exists $L \in \mathfrak{L}$, $n \geq 1$, $(n,p) = 1$ and $w \in Z_{L,n}$ such that w is contained in the standard parabolic subgroup W_P of W corresponding to an F'-stable, proper, parabolic subgroup P of G, $P \supset B$, and such that $\langle \rho, \Sigma(-1)^i H^i_c(X_w, \mathcal{F}^L_{\dot{w},n}) \rangle_{G^{F'}} \neq 0$. Since all elements $w' \in W$, such that $w' \leq w$ are contained in W_P, it follows from (2.8.1) that there exists $w' \in Z_{L,n}$, $w' \in W_P$, $w' \leq w$ such that $\langle \rho, \Sigma(-1)^i H^i(\overline{X}_{w'}, \mathcal{F}^L_{(w')^{\cdot},n}) \rangle_{G^{F'}} \neq 0$. Let i_0 be such that $\langle \rho, H^{i_0}(\overline{X}_{w'}, \mathcal{F}^L_{(w')^{\cdot},n}) \rangle_{G^{F'}} \neq 0$. We have $\varepsilon_\rho = (-i)^{i_0}$. By 6.4, there exists an irreducible $(P/U_P)^{F'}$-submodule ρ_1 of $H^{i_0}(\overline{X}^{(P)}_{w'}, \mathcal{F}^{L(P)}_{(w')^{\cdot},n})$ such that $\langle \rho, \mathrm{Ind}_{P^{F'}}^{G^{F'}}(\rho_1) \rangle_{G^{F'}} \neq 0$. We then clearly have $\hat{\varepsilon}_\rho = \hat{\varepsilon}_{\rho_1}$ and $\varepsilon_{\rho_1} = (-1)^{i_0}$. We may assume that the proposition is already proved for P/U_P instead of G. Thus, we have $\hat{\varepsilon}_{\rho_1} = \varepsilon_{\rho_1}$. It follows that $\hat{\varepsilon}_\rho = (-1)^{i_0} = \varepsilon_\rho$, as required.

Next, we assume that ρ is cuspidal. Let $L \in \mathfrak{L}$, $n \geq 1$, $(n,p) = 1$ be such that there exists $w \in Z_{L,n}$ and $i_0 \geq 0$ with $\langle \rho, H_c^{i_0}(X_w, \mathcal{F}^L_{\dot{w},n}) \rangle_{G^{F'}} \neq 0$. We may assume that for all $w' \in Z_{L,n}$, $w' < w$, we have $\langle \rho, H_c^j(X_{w'}, \mathcal{F}^L_{w,n}) \rangle_{G^{F'}} = 0$, for all $j \geq 0$. In the spectral sequence considered in 2.7 we take the ρ-isotypic component. The ρ-isotypic component of the term $E_2^{i,j} = H^i(\bar{X}_w, \mathcal{H}^i(\bar{X}_w, \mathcal{F}^L_{\dot{w},n}))$ is zero if $j \neq 0$ and is the ρ-isotypic component of $H_c^i(X_w, F^L_{\dot{w},n})$ if $j = 0$. It follows that we have a canonical isomorphism between the ρ-isotypic components of $\mathbf{H}^i(\bar{X}_w, \mathcal{F}^L_{\dot{w},n})$ and $H_c^i(X_w, \mathcal{F}^L_{\dot{w},n})$, for any $i \geq 0$. Since ρ occurs in $H_c^{i_0}(X_w, \mathcal{F}^L_{\dot{w},n})$, it must also occur in $\mathbf{H}^{i_0}(\bar{X}_w, \mathcal{F}^L_{\dot{w},n})$. Hence, by 6.11, it must also occur in $\mathbf{H}^{2\ell(w)-i_0}(\bar{X}_w, \mathcal{F}^L_{\dot{w},n})$ hence also in $H_c^{2\ell(w)-i_0}(X_w, \mathcal{F}^L_{\dot{w},n})$.

The assumption $q \geq h$ implies [DL_1, 9.7] that X_w is affine hence that $H_c^j(X_w, \mathcal{F}^L_{\dot{w},n}) = 0$ for $j < \ell(w)$. Since $H_c^i(X_w, \mathcal{F}^L_{\dot{w},n})$, $H_c^{2\ell(w)-i}(X_w, \mathcal{F}^L_{\dot{w},n})$ are both nonzero, we must have $i \geq \ell(w)$, $2\ell(w) - i \geq \ell(w)$ hence $i = \ell(w)$. (Compare the proof of [DL_1, 9.9].) Thus, ρ occurs in $H_c^{\ell(w)}(X_w, \mathcal{F}^L_{\dot{w},n})$, hence it occurs in $\mathbf{H}^{\ell(w)}(\bar{X}_w, \mathcal{F}^L_{\dot{w},n})$. It follows that $\varepsilon_\rho = (-1)^{\ell(w)}$. By 6.3, for any $x \in W$, $F'^{-1}(x)wx$ is not contained in any proper, standard, F'-stable parabolic subgroup of W. Consider the Coxeter complex whose simplices are in 1-1 correspondence with the right cosets in W with respect to the various standard parabolic subgroups of W, other than W itself. It is a sphere of dimension $r(G) - 1$, where $r(G)$ is the semi-simple rank of G (over k). It has a natural action of W and of F'; the composition $F' \circ w$ acts without fixed points. It is well known that the top (reduced) homology (over $\mathbf{Q}$) of the Coxeter complex is the sign representation of W. The map F' acts on the top reduced homology as $(-1)^{r(G)-\sigma(G)}$, (see 6.7), as one can see from the Lefschetz fixed point formula. (The fixed point set of F' on the Coxeter complex is a sphere of dimension $\sigma(G)-1$.) Hence $F'\circ w$ acts on the top reduced homology as $(-1)^{\ell(w)+r(G)-\sigma(G)}$. By the Lefschetz fixed point formula, this must be equal to $(-1)^{r(G)}$ since $F'\circ w$ acts without fixed points. It follows that $\ell(w)\equiv\sigma(G) \pmod 2$. Since ρ is cuspidal, we have $\varepsilon_\rho=(-1)^{\sigma(G)}$, hence $\varepsilon_\rho=\hat{\varepsilon}_\rho$. The proposition is proved.

6.13. When γ = identity, we define, for any $\sigma\in\mathcal{R}(W_{L,n})$ an element

$$R_\sigma=|W_{L,n}|^{-1}\sum_{\substack{y\in W_{L,n}\\ i\geq 0}}(-1)^i\mathrm{Tr}(y,\sigma)H_c^i(X_{w_1y},\mathcal{F}^L_{(w_1y)^\cdot,n})\in\mathcal{R}(G^{F'})\otimes\mathbf{Q}\,.$$

In particular, R_E is defined for any $E\in(W_{L,n})^\vee$.

Without assumptions on γ, we define for any element $\sigma'\in\mathcal{R}(W_{L,n},\gamma)$, (see 4.10), an element

$$R_{\sigma'}=|W_{L,n}|^{-1}\sum_{\substack{y\in W_{L,n}\\ i\geq 0}}(-1)^i\mathrm{Tr}(\gamma y,\sigma')H_c^i(X_{w_1y},\mathcal{F}^L_{(w_1y)^\cdot,n})\in\mathcal{R}(G^{F'})\otimes\mathbf{Q}\,.$$

If γ = identity and σ,σ' correspond to each other under (5.10.1), we have $R_\sigma=R_{\sigma'}$. This definition of $R_{\sigma'}$ is compatible with the definition of $R_{\tilde{E}}$ in (3.7.1).

We can now state:

6.14. INTEGRALITY THEOREM. (i) *For any* $x\in W_{L,n}$, *the elements* $R_{a_{\gamma x}}$ *(see (5.10.2)) and* $R_{\mathfrak{A}_{\gamma x}}$ *(see (5.11.6)) are in* $\mathcal{R}(G^{F'})$. *More precisely,* $R_{a_{\gamma x}}$ *and* $R_{\mathfrak{A}_{\gamma x}}$ *are* $\mathbf{Z}$*-linear combinations of irreducible* $G^{F'}$*-modules* $\rho\in\mathcal{E}_{L,n}$ *such that* $\varepsilon_\rho=(-1)^{\ell(w_1x)-a(x)}$ *(see 6.6, (5.27.1)), where* w_1 *is the element of minimal length in* $Z_{L,n}$.

(ii) *We have* $R_{\mathfrak{a}_{\gamma x}} = (-1)^{\ell(w_1)} D(R_{a_{\gamma x}})$.

Proof. Statement (ii) follows from (6.8.6) and (5.11.7). We now prove statement (i). We shall apply (5.3.1) and (5.3.6) with $\mathbb{W} = W_{L,n}$. We see that there exist elements

$$a_{y,x} = \sum_{\substack{i \geq 0 \\ i \equiv \tilde{\ell}(x) + \tilde{\ell}(y) \,(\mathrm{mod}\, 2)}} a_{y,x}^{(i)} u^{i/2} \in \mathbf{Z}[[u^{1/2}]]$$

defined for all y, $y \underset{LR}{<} x$, such that

(6.14.1) $$a_{y,x}^{(i)} \in \mathbf{Z}, \quad a_{y,x}^{(0)} = 0,$$

(6.14.2) $$C_x - \sum_{\substack{y \\ y \underset{LR}{\sim} x}} a_{y,x} C_y \in \bigoplus_{\substack{E \\ E \underset{LR}{\sim} x}} I_{E,\mathcal{K}}.$$

From (6.14.2) it follows that the sum

(6.14.3) $$(-1)^{\tilde{\ell}(x)} \sum_{E \in (W_{L,n})^{\vee}_{ex}} \operatorname{Tr}\Big(T_\gamma C_x - \sum_{\substack{y \\ y \underset{LR}{\sim} x}} a_{y,x} T_\gamma C_y, \tilde{E}(u)_{\mathcal{K}}\Big) \tilde{E}$$

(where $\tilde{E}$ are as in 3.6) is equal to the same sum restricted to those E for which $E \underset{LR}{\sim} x$. For such E, we have $a(x) = a_E$, see (5.27.1), and

$$\operatorname{Tr}(a_{y,x} T_\gamma C_y, \tilde{E}(u)_{\mathcal{K}}) \in u^{-(a(x)/2)+(1/2)} \mathbf{Z}[[u]]$$

(cf. (5.2.1) and (6.14.1)).

Hence (6.14.3) is equal to

$$\sum_{\substack{E \in (W_{L,n})^{\vee}_{ex} \\ E \underset{LR}{\sim} x}} c_{\gamma x, \tilde{E}} u^{-(a(x)/2)} \tilde{E} + \text{terms involving only higher powers of } u^{1/2}.$$

By 5.2 (ii), this is also equal to

$$(6.14.4) \qquad \sum_{E \epsilon (W_{L,n})^{\vee}_{ex}} c_{\gamma x, \widetilde{E}} u^{-(a(x)/2)} \widetilde{E} + \text{terms involving only higher powers of } u^{1/2}.$$

Let us set for any $h \epsilon \widetilde{H}_{\mathcal{K}}$ and $\widetilde{E}(u)_{\mathcal{K}}$ as above,

$$(6.14.5) \qquad \mathrm{Tr}(h, \widetilde{E}(u)_{\mathcal{K}}) = \sum_{i \epsilon \mathbf{Z}} \mathrm{Tr}(h, \widetilde{E}(u); i/2) u^{i/2}$$

where $\mathrm{Tr}(h,\widetilde{E}(u); i/2) \epsilon \mathbf{Q}$. From the equality of (6.14.3) and (6.14.4) we have

$$R_{a_{\gamma x}} = (-1)^{\widetilde{\ell}(x)} \sum_{E} \mathrm{Tr}\Big(T_\gamma C_x - \sum_{y} a_{y,x} T_\gamma C_y, \widetilde{E}(u)_{\mathcal{K}}; -(a(x)/2)\Big) R_{\widetilde{E}}$$

$$= (-1)^{\widetilde{\ell}(x)} \Big(\sum_{E} \mathrm{Tr}(T_\gamma C_x, \widetilde{E}(u)_{\mathcal{K}}; -(a(x)/2)) R_{\widetilde{E}}$$

$$- \sum_{\substack{y \\ y \underset{LR}{<} x}} \sum_{\substack{i>0 \\ i \equiv \widetilde{\ell}(x)+\widetilde{\ell}(y)(2)}} a^{(i)}_{y,x} \sum_{E} \mathrm{Tr}(T_\gamma C_y, \widetilde{E}(u)_{\mathcal{K}}, -(a(x)/2)-(i/2)) R_{\widetilde{E}} \Big).$$

(E runs over $(W_{L,n})^{\vee}_{ex}$.)

By 6.9, this is equal to

$$(-1)^{\widetilde{\ell}(x)-a(x)} D(H^{\ell(w_1 x)-a(x)} (\overline{X}_{w_1 x}, \mathcal{F}^{L}_{(w_1 x)^{\cdot},n})$$

$$- \sum_{\substack{y \\ y \underset{LR}{<} x}} \sum_{\substack{i>0 \\ i \equiv \widetilde{\ell}(x)+\widetilde{\ell}(y)(2)}} (-1)^{\widetilde{\ell}(x)-a(x)-i} a^{(i)}_{y,x} D(H^{\ell(w_1 y)-a(x)-i} (\overline{X}_{w_1 y}, \mathcal{F}^{L}_{(w_1 y)^{\cdot},n}))$$

hence, by (6.9.3), $R_{a_{\gamma x}}$ is as required. Using (ii) and (6.9.3), we see that $R_{\mathfrak{A}_{\gamma x}}$ is also as required. The theorem is proved.

6.15. PROPOSITION. *For any* $x \in W_{L,n}$, *we have*

$$\text{(6.15.1)}\quad D((-1)^{\tilde{\ell}(x)-a(x)} H^{\ell(w_1x)-a(x)}(\bar{X}_{w_1x}, \mathcal{F}^L_{(w_1x)^{\cdot},n})) = R_{a_{\gamma x}} + \sum_{x'} n_{x',x} R_{a_{\gamma x'}} .$$

$$\text{(6.15.2)}\quad (-1)^{\ell(w_1x)-a(x)} H^{\ell(w_1x)-a(x)}(\bar{X}_{w_1x}, \mathcal{F}^L_{(w_1x)^{\cdot},n}) = R_{\mathfrak{a}_{\gamma x}} + \sum_{x'} n_{x',x} R_{\mathfrak{a}_{\gamma x'}} .$$

Both sums over x' *are taken over the set* $\{x' \in W_{L,n} | x' \underset{LR}{<} x,\ \tilde{\ell}(x)-a(x) \equiv \tilde{\ell}(x')-a(x') \pmod 2\}$ *and* $n_{x',x}$ *are certain rational numbers.*

Proof. From 6.9, we see that the left-hand side of (6.15.1) is equal to

$$\text{(6.15.3)}\qquad (-1)^{\tilde{\ell}(x)} \sum_{E} \operatorname{Tr}(T_\gamma C_x, \tilde{E}(u); -(a(x)/2)) R_{\tilde{E}}$$

(sum over all $E \in (W_{L,n})^\vee_{ex}$; $\tilde{E}$ is as in 3.6). The part of the last sum corresponding to those E for which $E \underset{LR}{\sim} x$, is equal to

$$(-1)^{\tilde{\ell}(x)} \sum_{\substack{E \\ E \underset{LR}{\sim} x}} \operatorname{Tr}(T_\gamma C_x, \tilde{E}(u); -(a_E/2)) R_{\tilde{E}} = \sum_{\substack{E \\ E \underset{LR}{\sim} x}} c_{\gamma x, \tilde{E}} R_{\tilde{E}} =$$

$$= \sum_{E} c_{\gamma x, \tilde{E}} R_{\tilde{E}} = R_{a_{\gamma x}} .$$

The part of the sum (6.15.3) corresponding to those E for which $E \underset{LR}{\not\leq} x$, is zero, (see the proof of 5.2). If, on the other hand, E satisfies $E \underset{LR}{<} x$, then R_E is a $\mathbb{Q}$-linear combination of elements $R_{a_{\gamma x'}}$ $(E \underset{LR}{\sim} x')$, cf. 5.13 (ii). Hence the left-hand side of (6.15.1) is equal to $R_{a_{\gamma x}} +$ $\sum_{\substack{x' \in W_{L,n} \\ x' \underset{LR}{<} x}} n_{x',x} R_{a_{\gamma x'}}$, where $n_{x',x}$ are rational numbers.

We now project (orthogonally) the last equality onto the subspace of $\mathcal{R}(G^{F'}) \otimes \mathbb{Q}$ spanned by the irreducible representations $\rho \in \mathcal{E}_{L,n}$ such that $\varepsilon_\rho = (-1)^{\ell(w_1 x) - a(x)}$. The left-hand side of (5.15.1) projects onto itself, (cf. (6.9.3)); $R_{a_{\gamma x}}$ projects onto itself (cf. 6.14 (i)). In the case where $x' \underset{LR}{<} x$, $R_{a_{\gamma x'}}$ projects onto itself if $\ell(w_1 x') - a(x') \equiv \ell(w_1 x) - a(x)$ (mod 2) and projects to zero otherwise, (cf. 6.14 (i)). The identity (6.15.1) follows. Applying to it D and using (6.8.3), 6.14 (ii), we get (6.15.2). The proposition is proved.

6.16. Let V be an Euclidean space with scalar product (,) and with a given orthonormal basis $\{e_i\}_{1 \le i \le n}$. For any vector $v \in V$, we define its support by $\operatorname{supp}(v) = \{i \mid 1 \le i \le n, (e_i, v) \ne 0\}$. Given $v \in V$, we shall say that $v \ge 0$ if $(e_i, v) \ge 0$, $(1 \le i \le n)$.

Let I be a finite set with a given preorder relation $x \le y$ and let $\sim$ be the associated equivalence relation on I. We write $x < y$ for "$x \le y$ and $x \not\sim y$." Assume given two families of elements $r_x \in V$, $\tilde{r}_x \in V$, $(x \in I)$, such that

(6.16.1) $$(r_x, r_{x'}) = 0 \text{ whenever } x \not\sim x'.$$

(6.16.2) When x runs through a fixed equivalence class for $\sim$, the r_x span over $\mathbb{R}$ the same subspace of V as the $\tilde{r}_x$. In particular, we have $(\tilde{r}_x, \tilde{r}_{x'}) = (r_x, \tilde{r}_{x'}) = 0$ if $x \not\sim x'$.

(6.16.3) For any $x \in I$, there exists a linear combination $r_x + \sum_{x' < x} d_{x',x} r_{x'}$ $(d_{x',x} \in \mathbb{R})$, which is ≥ 0.

(6.16.4) For any $x \in I$, there exists a linear combination $\tilde{r}_x + \sum_{x < x'} \tilde{d}_{x',x} \tilde{r}_{x'}$ $(\tilde{d}_{x',x} \in \mathbb{R})$, which is ≥ 0.

We shall show that these assumptions imply the following.

(6.16.5) *For any* $x \in I$, r_x *and* $\tilde{r}_x$ *are* ≥ 0.

For $x \in I$, we can write uniquely $r_x = r_x^I + r_x^{II}$ where

$$\operatorname{supp}(r_x^{II}) \subset \bigcup_{x' < x} \operatorname{supp}(r_{x'}), \quad \operatorname{supp}(r_x^I) \cap \Big(\bigcup_{x' < x} \operatorname{supp}(r_{x'})\Big) = \emptyset .$$

The assumption (6.16.3) implies that r_x^I is ≥ 0, for any $x \in I$.

Before proving (6.16.5) we shall prove:

(6.16.6) *Let* $y \in I$ *be a maximal element of* I *(i.e. such that there is no element* $y' \in I$, $y < y'$. *Then, for any* $x \in X$, $x \not\sim y$, *we have* $\operatorname{supp}(\tilde{r}_y) \cap \operatorname{supp}(r_x) = \emptyset$ *and* $\operatorname{supp}(\tilde{r}_y) \cap \operatorname{supp}(\tilde{r}_x) = \emptyset$.

First note that, by (6.16.4), we have $\tilde{r}_y \geq 0$. We now prove that $\operatorname{supp}(\tilde{r}_y) \cap \operatorname{supp}(r_x) = \emptyset$, (for $x \not\sim y$), by induction on the largest integer i such that there exists a chain $x = x_0 > x_1 > \cdots > x_i$ in I. If $i = 0$, then $r_x = r_x^I$ hence it is ≥ 0. Since $(r_x, \tilde{r}_y) = 0$ and $r_x \geq 0, \tilde{r}_y \geq 0$, we have $\operatorname{supp}(r_x) \cap \operatorname{supp}(\tilde{r}_y) = \emptyset$. Assume now that $i > 0$ and that (6.16.6) is proved for $i-1$. Write $r_x = r_x^I + r_x^{II}$, as above. By the induction hypothesis, $\operatorname{supp}(\tilde{r}_y)$ is disjoint from $\operatorname{supp}(r_x^{II})$, which is contained in $\bigcup_{x' < x} \operatorname{supp}(r_{x'})$. Hence $(\tilde{r}_y, r_x^{II}) = 0$, so that $(\tilde{r}_y, r_x^I) = (\tilde{r}_y, r_x) = 0$, (by (6.16.2)). But $\tilde{r}_y \geq 0$, $r_x^I \geq 0$ hence $\operatorname{supp}(\tilde{r}_y) \cap \operatorname{supp}(r_x) = \emptyset$. Since $\operatorname{supp}(r_x) = \operatorname{supp}(r_x^I) \cup \operatorname{supp}(r_x^{II})$, we have $\operatorname{supp}(\tilde{r}_y) \cap \operatorname{supp}(r_x) = \emptyset$, as required. The first conclusion of (6.16.6) is proved. The second conclusion follows from the first, since $\operatorname{supp}(\tilde{r}_x) \subset \bigcup_{\substack{x' \\ x' \sim x}} \operatorname{supp}(r_{x'})$, (cf. (6.16.2)). Thus (6.16.6) is proved.

We now prove that, under the assumptions (6.16.1)-(6.16.4), we have $\tilde{r}_x \geq 0$ for any $x \in I$. If I is a single equivalence class under $\sim$, then this follows immediately from (6.16.4). Assume now that I has at least two equivalence classes. Let y be a maximal element of I (see (6.16.6)) and let $I' = I - \{\text{equivalence class of } y\}$. Then I' inherits a preorder from I; the families of elements $r_x \in V$, $\tilde{r}_x \in V$, $(x \in I')$, still have the properties (6.16.1)-(6.16.4). Indeed, properties (6.16.1)-(6.16.3) are clearly satisfied. We now check property (6.16.4). Let $x \in I$ be such that $x \not\sim y$. If $x \not< y$, then (6.16.4) for x, I' clearly follows from the corresponding property for x, I. If $x < y$, property (6.16.4) for x, I gives

$$\tilde{r}_x + \sum_{\substack{x<x' \\ x' \not\sim y}} \tilde{d}_{x',x}\tilde{r}_{x'} + \sum_{\substack{y' \\ y' \sim y}} \tilde{d}_{y',x}\tilde{r}_{y'} \geq 0 .$$

We denote $\phi = \tilde{r}_x + \sum_{\substack{x<x' \\ x' \not\sim y}} \tilde{d}_{x',x}\tilde{r}_{x'}$. By (6.16.6), $\mathrm{supp}(\tilde{r}_{y'})$ is disjoint from $\mathrm{supp}(\phi)$, for any y', $y' \sim y$. Hence ϕ is ≥ 0, so that x, I' satisfy (6.16.4). By induction on the number of equivalence classes in I, we may assume that the property $\tilde{r}_x \geq 0$ holds for all $x \in I'$. If y' satisfies $y' \sim y$ then $\tilde{r}_y$ is ≥ 0 by (6.16.4). Thus, we have proved that $\tilde{r}_x$ is ≥ 0 for all $x \in I$.

A completely similar proof (or the symmetry between r_x, $\tilde{r}_x$) shows that r_x is also ≥ 0 for all $x \in I$. This completes the proof of (6.16.5).

6.17. DISJOINTNESS THEOREM. *Recall that* G *is assumed to have connected center. Assume that* (L, n) *has the following property:* $\varepsilon_\rho = \hat{\varepsilon}_\rho$, *(see 6.6, 6.7) for any* $\rho \in \mathcal{E}_{L,n}$. *(This assumption is automatically verified, for example, if* $q \geq h$, *see 6.10.)*

(i) *Let* $E, E' \in (W_{L,n})^\vee_{ex}$ *and let* $\tilde{E}, \tilde{E}'$ *be extensions of* E, E' *to irreducible* $\mathbf{Q}[\tilde{W}_{L,n}]$*-modules. Assume that* $E \underset{LR}{\not\sim} E'$, *(see 5.16). Then* $R_{\tilde{E}}, R_{\tilde{E}'}$ *are disjoint; in other words, if* ρ *is an irreducible* $G^{F'}$*-module then we cannot have simultaneously* $\langle\rho, R_{\tilde{E}}\rangle_{G^{F'}} \neq 0$, $\langle\rho, R_{\tilde{E}'}\rangle_{G^{F'}} \neq 0$.

(ii) *For any* $x \in W_{L,n}$, $(-1)^{\ell(w_1)} R_{\alpha_{\gamma x}}$ *and* $(-1)^{\ell(w_1 x) - a(x)} R_{\mathcal{Q}_{\gamma x}}$ *are actual* $G^{F'}$*-modules.* (w_1 *is as in 6.14 (i).)*

Proof. We shall apply the results in 6.16 in the following situation. We take V to be the real vector space with basis parametrized by the isomorphism classes of irreducible representations of $G^{F'}$, over $\overline{\mathbf{Q}}_\ell$, with the

Euclidean structure for which this basis is an orthonormal one. We take I to be $W_{L,n}$ with the preorder relation $\underset{LR}{\leq}$ of this Coxeter group. For $x \in W_{L,n}$ we take $r_x = (-1)^{\ell(w_1)} R_{a_{\gamma x}} \in V$, $\tilde{r}_x = (-1)^{\ell(w_1 w_0 x) - a(w_0 x)} R_{\mathfrak{a}_{\gamma w_0 x}} \in V$, where w_0 is the longest element of $W_{L,n}$.

We now verify that these data satisfy properties (6.16.1)-(6.16.4).

If $x_0 \in W$ is fixed, the following three subspaces of V coincide:

a) the subspace spanned by all $R_{\tilde{E}_1}$, where $E_1 \in (W_{L,n})^\vee_{ex}$ is such that $E_1 \underset{LR}{\sim} x_0$ and $\tilde{E}_1$ is an extension of E_1 to an irreducible $\mathbf{Q}[\tilde{W}_{L,n}]$-module;

b) the subspace spanned by all $R_{a_{\gamma x'}}$, $x' \underset{LR}{\sim} x_0$;

c) the subspace spanned by all $R_{\mathfrak{a}_{\gamma w_0 x'}}$, $x' \underset{LR}{\sim} x_0$.

Indeed, the space a) is contained in the space b) and also in the space c) in virtue of the formula

$$(6.17.1)\quad (E(u)_{\mathcal{K}} : [\Gamma]) f_E R_{\tilde{E}} = \sum_{x \in \Gamma} c'_{\gamma w_0 x, \tilde{E}} R_{\mathfrak{a}_{\gamma w_0 x}} = \sum_{x \in \Gamma} c_{\gamma x, \tilde{E}} R_{a_{\gamma x}}$$

which follows from 5.13 (ii); here Γ is a left cell in $W_{L,n}$ such that $(E(u)_{\mathcal{K}} : [\Gamma]) > 0$ and $E \in (W_{L,n})^\vee_{ex}$.

The space b) is contained in the space a), in virtue of the formula (see 5.2 (ii)):

$$(6.17.2)\qquad R_{a_{\gamma x}} = \sum c_{\gamma x, \tilde{E}} R_{\tilde{E}}$$

(sum over all $E \in (W_{L,n})^\vee_{ex}$ such that $E \underset{LR}{\sim} x$; for each such E, we denote by $\tilde{E}$ an extension of E to an irreducible $\mathbf{Q}[\tilde{W}_{L,n}]$-module).

The space c) is contained in the space a), in virtue of the formula

$$(6.17.3)\qquad R_{\mathfrak{a}_{\gamma w_0 x}} = \sum c'_{\gamma w_0 x, \tilde{E}} R_{\tilde{E}}$$

(sum over the same set as in (6.17.2)); see (5.11.6) and 5.14 (iii).

Using now the identity 3.9 (i), we see that (6.16.1) and (6.16.2) hold on our situation.

Property (6.16.3) follows from (6.15.1); indeed, our assumption $\varepsilon_\rho = \hat{\varepsilon}_\rho$ $(\rho \in \mathcal{E}_{L,n})$ implies that $D((-1)^i H^i(\overline{X}_{w_1x}, \mathcal{F}^L_{(w_1x)^{\cdot},n}))$ is an actual $G^{F'}$-module, for any $i \geq 0$. (See the last sentence in the statement of Proposition 6.9.) Now (6.15.2) implies

$$\tilde{r}_{w_0x} + \sum_{x' \underset{LR}{<} x} n_{x',x}\tilde{r}_{w_0x'} \geq 0, \quad (n_{x',x} \in \mathbf{Q}).$$

Replacing w_0x by x and using the fact that $x' \underset{LR}{<} x$ is equivalent to $w_0x \underset{LR}{<} w_0x'$ (see [KL_1]) we see that (6.16.4) holds in our situation. We may therefore apply the results of 6.16 and deduce that $r_x \geq 0$, $\tilde{r}_x \geq 0$ (in the sense defined in 6.16). Since $r_x, \tilde{r}_x$ are known to be virtual $G^{F'}$-modules (6.14 (i)), we see that (ii) is proved.

Now let $E, E', \tilde{E}, \tilde{E}'$ be as in (i). Assume that there exists an irreducible representation ρ of $G^{F'}$ such that $\langle \rho, R_{\tilde{E}} \rangle_{G^{F'}} \neq 0$, $\langle \rho, R_{\tilde{E}'} \rangle_{G^{F'}} \neq 0$. Using (6.17.1) for $\tilde{E}$ and $\tilde{E}'$, it follows that there exist $x, x' \in W_{L,n}$ such that $E \underset{LR}{\sim} x$, $E' \underset{LR}{\sim} x'$, $\langle \rho, (-1)^{\ell(w_1)} R_{a_{\gamma x}} \rangle_{G^{F'}} \neq 0$, $\langle \rho, (-1)^{\ell(w_1)} R_{a_{\gamma x'}} \rangle_{G^{F'}} \neq 0$. (In particular, since $E \underset{LR}{\not\sim} E'$, we have $x \underset{LR}{\not\sim} x'$.) Since $(-1)^{\ell(w_1)} R_{a_{\gamma x}}, (-1)^{\ell(w_1)} R_{a_{\gamma x'}}$ are actual $G^{F'}$-modules (ii), it follows that $\langle (-1)^{\ell(w_1)} R_{a_{\gamma x}}, (-1)^{\ell(w_1)} R_{a_{\gamma x'}} \rangle_{G^{F'}} \neq 0$. Since, by (6.17.2), $(-1)^{\ell(w_1)} R_{a_{\gamma x}}$ (resp. $(-1)^{\ell(w_1)} R_{a_{\gamma x'}}$) is a $\mathbf{Z}$-linear combination of elements $R_{\tilde{E}_1}$, with $E_1 \underset{LR}{\sim} x$ (resp. with $E_1 \underset{LR}{\sim} x'$), it follows that there exist E_1, E_1' in $(W_{L,n})^\vee_{ex}$, such that $E_1 \underset{LR}{\sim} x$, $E_1' \underset{LR}{\sim} x'$,

$\langle R_{\tilde{E}_1}, R_{\tilde{E}'_1}\rangle_{G^{F'}} \neq 0$. Using 3.9 (i), it follows that $E_1 = E'_1$, and hence $x \underset{LR}{\sim} x'$. This is a contradiction. The theorem is proved.

6.18. We preserve the notations in 6.17. We consider the partition $\mathcal{E}_{L,n} = \mathcal{E}^{I}_{L,n} \amalg \mathcal{E}^{II}_{L,n}$, where

$$\mathcal{E}^{I}_{L,n} = \{\rho \in \mathcal{E}_{L,n}, \varepsilon_\rho = (-1)^{\ell(w_1)}\}, \tag{6.18.1}$$

$$\mathcal{E}^{II}_{L,n} = \{\rho \in \mathcal{E}_{L,n}, \varepsilon_\rho = (-1)^{\ell(w_1)+1}\}.$$

We say that the representations in $\mathcal{E}^{I}_{L,n}$ are of the first kind and those in $\mathcal{E}^{II}_{L,n}$ are of the second kind. From 6.5 (iv) and 5.13 (ii) it follows that

(6.18.2) *For any* $\rho \in \mathcal{E}_{L,n}$ *there exists* $x \in W_{L,n}$ *such that* $\langle \rho, R_{a_{\gamma x}}\rangle_{G^{F'}} \neq 0$.

If ρ and x are as in (6.18.2) then

$$\rho \in \mathcal{E}^{I}_{L,n} \iff \tilde{\ell}(x) \equiv a(x) \pmod 2, \tag{6.18.3}$$

$$\rho \in \mathcal{E}^{II}_{L,n} \iff \tilde{\ell}(x) \not\equiv a(x) \pmod 2.$$

(This follows from 6.14 (i).)

In this section we shall determine the set $\mathcal{E}^{II}_{L,n}$ in several cases.

(6.18.4) *If* $E \in (W_{L,n})^\vee_{ex}$ *satisfies* $\mathrm{Tr}(T_{\gamma z}, \tilde{E}(u)) \in \mathbf{Z}[u]$ (*resp.* $\in u^{1/2}\mathbf{Z}[u]$) *for all* $z \in W_{L,n}$, ($\tilde{E}$ *as in 3.6*), *then* $R_{\tilde{E}}$ *is a* $\mathbf{Q}$*-linear combination of representations* $\rho \in \mathcal{E}^{I}_{L,n}$ (*resp.* $\rho \in \mathcal{E}^{II}_{L,n}$). *The following three conditions are equivalent*: (a) $\mathcal{E}^{II}_{L,n}$ *is empty.* (b) $H^i(\bar{X}_{w_1 z}, \mathcal{F}^L_{(w_1 z)^{\cdot},n}) = 0$ *for any* $z \in W_{L,n}$ *and any* $i \not\equiv \ell(w_1)$ (mod 2), (c) $\mathrm{Tr}(T_{\gamma z}, \tilde{E}(u)) \in \mathbf{Z}[u]$ *for any* $z \in W_{L,n}$ *and any* $E \in (W_{L,n})^\vee_{ex}$.

This follows from 3.8 (ii).

(6.18.5) *Assume that* $(W_{L,n}, S_{L,n})$ *is a product of irreducible (not necessarily* γ*-stable) Weyl groups* $(\mathbf{W}_i, \mathbf{S}_i)$, $(i \in I)$, *of type* B_n $(n \geq 2)$, C_n $(n \geq 2)$ *or* D_n $(n \geq 4)$. *Then* $\mathcal{E}^{II}_{L,n}$ *is empty.*

This is proved as follows. If I is empty, then $W_{L,n} = \{e\}$ and the result follows immediately from (6.18.3). Assume now $I \neq \emptyset$. We use the equivalence of (a) and (c) in (6.18.4) and the fact that (c) in (6.18.4) is a consequence of the following statement: $\mathrm{Tr}(T_{\gamma_i z_i}, \tilde{E}_i(u)) \in \mathbf{Z}[u]$, for any $i \in I$, any $z_i \in \mathbf{W}_i$ and any irreducible $\mathbf{Q}[\tilde{\mathbf{W}}_i]$-module $\tilde{E}_i$ whose restriction to $\mathbf{W}_i$ is irreducible, with $\tilde{\mathbf{W}}_i$ defined with respect to $\gamma_i : \mathbf{W}_i \to \mathbf{W}_i$, the restriction to $\mathbf{W}_i$ of the smallest power of $\gamma : W_{L,n} \to W_{L,n}$ leaving $\mathbf{W}_i$ stable. (This can be seen as in the proof of 3.2.) Using again the equivalence of (a) and (c) for a group possibly smaller than G, we see that we are reduced to the case where $W_{L,n}$ is irreducible.

Consider first the case where $(W_{L,n}, S_{L,n})$ is of type D_n and γ is of order 2. Let $\rho \in \mathcal{E}_{L,n}$. By 6.5 (iv), there exists $E \in (W_{L,n})^{\vee}_{ex}$ such that $\langle \rho, R_{\tilde{E}} \rangle_{G^F} \neq 0$, ($\tilde{E}$ as in 3.6). Let $\hat{E}$ be the extension $\neq \tilde{E}$ of E to a $\mathbf{Q}[\tilde{W}_{L,n}]$-module. By [$L_{12}$, 1.19], $\frac{1}{2}(\tilde{E} - \hat{E})$ is a $\mathbf{Q}$-linear combination of elements $c(Z, \Phi, \hat{\Phi})$ as in (5.18.1), with $|\hat{\Phi}|$ odd, hence of elements $a_{\gamma x} \in \mathcal{R}(W_{L,n}, \gamma)$ with $x \in W_{L,n}$ satisfying $\tilde{\ell}(x) \equiv a(x) \pmod 2$, see (5.18.1). Hence, for some such x, we have $\langle \rho, R_{a_{\gamma x}} \rangle_{G^F} \neq 0$. Using now (6.18.3), we see that $\rho \in \mathcal{E}^{I}_{L,n}$. The proof in the case where $W_{L,n}$ is of type B_n, C_n or D_n and γ is the identity is entirely similar; alternatively, in that case one can use (6.18.6). Finally, if $W_{L,n}$ is of type D_4 and γ is of order 3, it is enough to check condition (c) in (6.18.4). This follows from the fact that, in this case, each $\tilde{E}(u)$ can be realized over $\mathbf{Z}[u]$, using W-graphs as in [KL_1, §6].

(6.18.6) *Assume that* $(W_{L,n}, S_{L,n})$ *is a product of irreducible Weyl groups of type* $\neq E_7, E_8$ *and that* $\gamma : W_{L,n} \to W_{L,n}$ *is the identity. Then* $\mathcal{E}^{II}_{L,n}$ *is empty.*

By a result of Benson and Curtis [BC], the condition (c) in (6.18.4) is satisfied. Hence condition (a) is also satisfied.

(6.18.7) *Assume that* $(W_{L,n}, S_{L,n})$ *is a Weyl group of type* E_7. *Then* $\mathcal{E}^{II}_{L,n}$ *consists of two irreducible representations* ρ_1, ρ_2. *They are characterized by the equations*

$$R_{512_a} - R_{512'_a} = (-1)^{\ell(w_1)}(\rho_1+\rho_2), \langle \rho_i, R_{512_a} + R_{512'_a}\rangle_{G^{F'}} = 0, \ (i = 1,2).$$

According to [BC], for all $E \in (W_{L,n})^\vee$ other than 512_a, $512'_a$ and for all $z \in W_{L,n}$ we have $\mathrm{Tr}(T_z, E(u)) \in \mathbf{Z}[u]$, hence by (6.18.4), we have $\langle \rho, R_E\rangle_{G^{F'}} = 0$ for any $\rho \in \mathcal{E}^{II}_{L,n}$. Hence any $\rho \in \mathcal{E}^{II}_{L,n}$ satisfies either $\langle \rho, R_{512_a}\rangle_{G^{F'}} \neq 0$ or $\langle \rho, R_{512'_a}\rangle_{G^{F'}} \neq 0$. We have $512_a + 512'_a = a_x$ for some $x \in W_{L,n}$ with $\ell(x) \equiv a(x)$ (mod 2), (see (5.22.1)) hence, by (6.18.3), $\langle \rho, R_{512_a} + R_{512'_a}\rangle_{G^{F'}} = 0$ for any $\rho \in \mathcal{E}^{II}_{L,n}$. Hence any $\rho \in \mathcal{E}^{II}_{L,n}$ satisfies $\langle \rho, R_{512_a} - R_{512'_a}\rangle_{G^{F'}} \neq 0$. By (5.22.2) and 6.14 (i), $R_{512_a} - R_{512'_a}$ is a $\mathbf{Z}$-linear combination of irreducible $G^{F'}$-modules $\rho \in \mathcal{E}^{II}_{L,n}$. Its inner product with itself is 2, (see 3.9 (i)) hence it is equal to $\pm\rho_1 \pm \rho_2$ where ρ_1, ρ_2 are two distinct representations in $\mathcal{E}^{II}_{L,n}$. It follows that $\mathcal{E}^{II}_{L,n} = \{\rho_1, \rho_2\}$. It remains to show that $\pm\rho_1 \pm \rho_2$ above is, in fact, $(-1)^{\ell(w_1)}(\rho_1 + \rho_2)$.

Let us write the identity (6.15.2) for $x \in W_{L,n}$ such that $512'_a - 512_a = -a_x$ and $\tilde{\ell}(x) \not\equiv a(x)$ (mod 2). (See (5.22.2).) The sum over x' in (6.15.2) is zero. (Indeed, assume that there exists x', $x' \underset{LR}{\not\sim} x$, satisfying $\ell(x') \not\equiv a(x')$ (mod 2) and $\mathfrak{A}_{x'} \neq 0$ (hence $a_{x'} \neq 0$). Then $a_{x'}$ is a $\mathbf{Z}$-combination of representations $E \in (W_{L,n})^\vee$, $E \neq 512_a$, $512'_a$ (by 5.2 and 5.25) hence, by (6.18.4), $R_{a_{x'}}$ is a linear combination of representations $\rho \in \mathcal{E}^{I}_{L,n}$. On the other hand our assumptions on x' imply, by

(6.18.3), that $R_{\alpha_x'}$ is a nonzero linear combination of representations $\rho \in \mathcal{E}^{II}_{L,n}$. This is a contradiction.) Hence (6.15.2) shows that $(-1)^{\ell(w_1)} R_{\mathfrak{a}_x}$ is an actual representation of $G^{F'}$. We have $\mathfrak{a}_x = \alpha_x \otimes \mathrm{sign} = 512_a' - 512_a = -\alpha_x$, hence $(-1)^{\ell(w_1)} R_{\alpha_x} = (-1)^{\ell(w_1)}(R_{512_a} - R_{512_a'})$ is an actual representation. It is equal to $(-1)^{\ell(w_1)}(\pm\rho_1 \pm \rho_2)$ hence to $\rho_1 + \rho_2$ and (6.18.7) is proved.

(6.18.8) *Assume that* G *is simple of type* E_8, *that* $L = L_0$ *is the trivial line bundle and that* $n = 1$. *Then* $\mathcal{E}^{II}_{L_0,1}$ *consists of four irreducible representations* $\rho_1, \rho_2, \rho_3, \rho_4$. *They are characterized by the equations*

$$R_{4096_x} - R_{4096_z} = \rho_1 + \rho_2, \quad R_{4096_z'} - R_{4096_x'} = \rho_3 + \rho_4$$

$$\langle \rho_i, R_{4096_x} + R_{4096_z} \rangle_{G^{F'}} = 0, \quad \langle \rho_i, R_{4096_z'} + R_{4096_x'} \rangle_{G^{F'}} = 0,$$

$$(1 \le i \le 4).$$

According to [BC], for all $E \in \check{W}$ other than 4096_x, 4096_z, $4096_z'$, $4096_x'$, and for all $z \in W$ we have $\mathrm{Tr}(T_z, E(u)) \in \mathbf{Z}[u]$. From this, as in the proof of (6.18.7) we deduce that any $\rho \in \mathcal{E}^{II}_{L_0,1}$ must satisfy $\langle \rho, R_{4096_x} - R_{4096_z} \rangle_{G^{F'}} \neq 0$ or $\langle \rho, R_{4096_z'} - R_{4096_x'} \rangle_{G^{F'}} \neq 0$. We can write (as in the proof of (6.18.7)): $R_{4096_x} - R_{4096_z} = \pm\rho_1 \pm \rho_2$ where ρ_1, ρ_2 are distinct representations in $\mathcal{E}^{II}_{L_0,1}$ and $R_{4096_z'} - R_{4096_x'} = \pm\rho_3 \pm \rho_4$, where ρ_3, ρ_4 are distinct representations in $\mathcal{E}^{II}_{L_0,1}$. Hence any $\rho \in \mathcal{E}^{II}_{L_0,1}$ must be one of the ρ_i's, $(1 \le i \le 4)$.

From results in [L_2] it follows easily that there are four distinct irreducible representations of $G^{F'}$ appearing in the eigenspaces of F'

on $H_c^7(X_w, \overline{Q}_\ell)$, and also on $H^7(\overline{X}_w, \overline{Q}_\ell)$, corresponding to the eigenvalues $\pm\sqrt{-1}\, q^{7/2}$ (q as in 6.10, w is a Coxeter element of length 7 in a parabolic subgroup of type E_7 of W). Since $\overline{X}_w$ is smooth, we have $H^7(\overline{X}_w, \overline{Q}_\ell) = H^7(\overline{X}_w, \overline{Q}_\ell)$. Thus, $\mathcal{E}^{II}_{L_0,1}$ has at least four distinct elements. It follows that the ρ_i $(1 \le i \le 4)$ are distinct and that $\mathcal{E}^{II}_{L_0,1} = \{\rho_1, \rho_2, \rho_3, \rho_4\}$.

Now let $v, v' \in W$ be such that $\ell(v) \not\equiv a(v) \pmod 2$, $4096_z - 4096_x = -\alpha_v$, $\ell(v') \not\equiv a(v') \pmod 2$, $4096'_x - 4096'_z = -\alpha_{v'}$. We wish to apply (6.15.2) to v and v'.

Note that $v \underset{LR}{\not\le} v'$. (It is known from the theory of primitive ideals in enveloping algebras that $v \underset{LR}{\le} v'$ would imply $a(v) \ge a(v')$. However, $a(v) = 11$, $a(v') = 26$.) Hence the argument in the proof of (6.18.7) still applies to $R_{\mathcal{a}_{v'}}$ (but not to $R_{\mathcal{a}_v}$) and gives that $-R_{\mathcal{a}_{v'}}$ is an actual representation of $G^{F'}$. (In our case $w_1 = e$.) We have $\mathcal{a}_{v'} = \alpha_{v'} \otimes \text{sign} = 4096_z - 4096_x = -\alpha_v$, hence $R_{\alpha_v} = R_{4096_x} - R_{4096_z}$ is an actual representation of $G^{F'}$, hence it is equal to $\rho_1 + \rho_2$. Applying the operator D, (see (6.8.1)), and using (6.8.6) we get $D(\rho_1) + D(\rho_2) = R_{4096'_x} - R_{4096'_z}$. From the description of $\rho_1, \rho_2, \rho_3, \rho_4$ given in [L_2] we have $\hat{\varepsilon}_{\rho_i} = -1$, $(1 \le i \le 4)$, since each ρ_i is a component of a representation of $G^{F'}$ induced from a cuspidal representation of a parabolic subgroup of type E_7. Thus $-D(\rho_1), -D(\rho_2)$ are actual representations. (See (6.8.2).) Since $R_{4096'_z} - R_{4096'_x} = \pm\rho_3 \pm \rho_4$ we must have $\rho_3 = -D(\rho_1)$, $\rho_4 = -D(\rho_2)$, (interchanging, if necessary ρ_3 with ρ_4) and $R_{4096'_z} - R_{4096'_x} = \rho_3 + \rho_4$. The identities $\langle \rho_i, R_{4096_x} + R_{4096_z} \rangle_{G^{F'}} = 0$, $\langle \rho_i, R_{4096'_z} + R_{4096'_x} \rangle_{G^{F'}} = 0$, $(1 \le i \le 4)$, are proved exactly as in the proof of (6.18.7), using (5.23.1) and (6.18.3). This completes the proof of (6.18.8).

(6.18.9) *Assume that* G *is simple of type* E_8, *that* $(W_{L,n}, S_{L,n})$ *is a Weyl group of type* $E_7 \times A_1$ *and that* $n = 2$. *Then* $\mathcal{E}^{II}_{L,2}$ *consists of four irreducible representations* $\rho_1, \rho_2, \rho_3, \rho_4$. *They are characterized by the equations*

$$R_{512_a \boxtimes 1} - R_{512'_a \boxtimes 1} = \rho_1 + \rho_2, \quad R_{512_a \boxtimes \varepsilon} - R_{512'_a \boxtimes \varepsilon} = \rho_3 + \rho_4$$

$$\langle \rho_i, R_{512_a \boxtimes 1} + R_{512'_a \boxtimes 1} \rangle_{G^{F'}} = \langle \rho_i, R_{512_a \boxtimes \varepsilon} + R_{512'_a \boxtimes \varepsilon} \rangle_{G^{F'}} = 0,$$

$$(1 \leq i \leq 4).$$

(*Here* ε *is the sign representation of the* A_1*-factor. In our case we have* $w_1 = e$.)

First note that for all $E \in (W_{L,2})^\vee$ other than $512_a \otimes 1$, $512'_a \otimes 1$, $512_a \boxtimes \varepsilon$, $512'_a \boxtimes \varepsilon$ and for all $z \in W_{L,2}$ we have $\mathrm{Tr}(T_z, E(u)) \in \mathbf{Z}[u]$. The proof can be then carried out exactly as in the proof of (6.18.8) as soon as we know that $\mathcal{E}^{II}_{L,2}$ has at least four elements. Using 6.5 (i) and (ii) we see that we may assume (replacing if necessary L by ${}^xL$) that the E_7-factor of $W_{L,n}$ is a standard parabolic subgroup of (W, S).

Let P be the maximal parabolic subgroup of G containing B corresponding to the E_7-factor of $W_{L,2}$. Let L_1 be the restriction of the line bundle L to the flag manifold of P/U_P (naturally imbedded in G/B). Then $(L_1, 2)$ satisfies the assumption of (6.18.7) for the group P/U_P. Hence $\mathcal{E}^{II}_{L_1,2}$ (with respect to P/U_P) consists of two irreducible representations $\rho^{(1)}, \rho^{(2)}$ of $(P/U_P)^{F'}$. From the proof of 6.4, it follows that any irreducible component of $\mathrm{Ind}_{P^{F'}}^{G^{F'}}(\rho^{(i)})$, $(i = 1, 2)$ is in $\mathcal{E}^{II}_{L,2}$ (with respect to G).

Let M be the Levi subgroup of P containing T, and let $\omega : M^{F'} \to \overline{\mathbf{Q}}_\ell^*$ be the unique character of order 2 which is trivial or the F'-fixed part of the derived group of M. Then $\rho^{(1)} \otimes \omega$, $\rho^{(2)} \otimes \omega$ are unipotent representations of $M^{F'}$, of the second kind. From the results of $[L_2]$ it follows that

$\rho^{(1)} \otimes \omega$, $\rho^{(2)} \otimes \omega$ are both cuspidal and both can be extended to representations of $\mathfrak{N}(M)^{F'}$, ($\mathfrak{N}(M)$ = normalizer of M in G). Since ω clearly extends to a character of $\mathfrak{N}(M)^{F'}$, it follows that $\rho^{(1)}$ and $\rho^{(2)}$ are both cuspidal and both can be extended to representations of $\mathfrak{N}(M)^{F'}$. By a standard result on induced representations we then have

$$\langle \mathrm{Ind}_{P^{F'}}^{G^{F'}}(\rho^{(i)}), \mathrm{Ind}_{P^{F'}}^{G^{F'}}(\rho^{(j)}) \rangle_{G^F} = \begin{cases} 2 & \text{if } i = j \\ 0 & \text{if } i \neq j \end{cases}, \quad (1 \le i, j \le 2).$$

It follows that $\mathcal{E}_{L,2}^{II}$ has at least four elements, and (6.18.9) is proved.

(6.18.10) *Let* (W, S) *be an irreducible Weyl group and let* $w \in W$ *be an element such that* $a_w \in \mathcal{R}(W)$ *is nonzero. Assume that the two-sided cell in* W *containing* w *corresponds (5.25) to a family* $\mathcal{F}$ *in* $W^\vee$ *other than that of* $512_a'$ *(type* E_7 *) or that of* 4096_z *or* $4096_x'$ *(type* E_8 *). Then* $\ell(w) \equiv a(w) \pmod 2$.

Let $E \in \mathcal{F}$ be such that $c_{w,E} \neq 0$. By the definition of $c_{w,E}$ we have

$$(-1)^{\ell(w)} \mathrm{Tr}(T_w, \tilde{E}(u)) = c_{w,E} u^{(\ell(w)-a(w))/2} + \text{higher powers of } u^{1/2}.$$

By our assumption on $\mathcal{F}$, we have $\mathrm{Tr}(T_w, E(u) \in \mathbf{Z}[u]$. It follows that $\ell(w) \equiv a(w) \pmod 2$, as required.

6.19. PROPOSITION. *Let* (W, S), $\dot{R}$, γ *be as in 4.21. Let* $\bar{x} \in \bar{X}(W, \gamma)$ *and* $w \in W$ *be such that* $a_{\gamma w} \neq 0$. *Write* $a_{\gamma w} = \sum_i n_i E_{(i)}$ ($E_{(i)}$ *are in* $(\tilde{W})^\vee_{ex}$, $n_i \in \frac{1}{2}\mathbf{Z}$, $n_i \neq 0$) *and let* $x_i \in X(W, \gamma)$ *be the elements corresponding to* $E_{(i)}$ *under (4.21.14). Assume that* $\sum_i n_i \{\bar{x}, x_i\} \neq 0$. *Then*

$$(-1)^{\ell(w)-a(w)} = \Delta(\bar{x}), \quad (\text{see } 4.21).$$

Proof. We may immediately reduce ourselves to the case where the irreducible components of W are permuted transitively by γ. Let $\mathcal{F} \subset W^\vee$ be the family containing the $E_{(i)}$'s. If the condition

(6.19.1) $$\mathrm{Tr}(T_\gamma \cdot T_y, \widetilde{E}(u)) \in \mathbf{Z}[u]$$

is satisfied for all $y \in \mathbf{W}$ and all $E \in \mathcal{F}$, then, as in the proof of (6.18.10), we see that $\ell(w) \equiv a(w) \pmod 2$. On the other hand, if the condition

(6.19.2) $$\mathrm{Tr}(T_\gamma \cdot T_y, \widetilde{E}(u)) \in u^{1/2}\,\mathbf{Z}[u]$$

is satisfied for all $y \in \mathbf{W}$ and all $E \in \mathcal{F}$ then we see similarly that $\ell(w) \not\equiv a(w) \pmod 2$. If we write $\mathbf{W} = \mathbf{W}_1 \times \cdots \times \mathbf{W}_1$ (s factors) with $\gamma(w_1, w_2, \cdots, w_s) = (w_s, w_1, w_2, \cdots, w_{s-1})$ and $E = E_1 \otimes \cdots \otimes E_1$ with $\gamma(e_1 \otimes e_2 \otimes \cdots \otimes e_s) = \delta(e_s) \otimes e_1 \otimes \cdots \otimes e_{s-1}$ ($(\mathbf{W}_1, S_1)$ is an irreducible Weyl group, $\delta : (\mathbf{W}_1, S_1) \to (\mathbf{W}_1, S_1)$), then it is easy to see that for $y = (y_1, \cdots, y_s) \in \mathbf{W}$, $y_i \in \mathbf{W}_1$, we have

$$\mathrm{Tr}(T_\gamma T_y, \widetilde{E}(u)) = \pm \mathrm{Tr}(T_\delta T_{y_1} T_{y_2} \cdots T_{y_s}, \widetilde{E}_1(u)) .$$

It now follows from the results in [BC] already used in 6.18, that (6.19.1) is satisfied whenever $\delta = 1$ and E_1 is not of degree 512 (for $\mathbf{W}_1$ of type E_7) or of degree 4096 (for $\mathbf{W}_1$ of type E_8). Hence in these cases the proposition follows. From (6.18.5) and the equivalence of (a) and (c) in (6.18.4), we see that (6.19.1) also holds when $\mathbf{W}_1$ is of type D_n $(n \geq 4)$ and δ is of order 2 and when $\mathbf{W}_1$ is of type D_4 and δ is of order 3. Assume now that $\mathbf{W}_1$ is of type A_n $(n \geq 2)$ or E_6 and that δ has order 2.

Using (5.12.2), we then have, for any irreducible representation E_1 of $\mathbf{W}_1$ and any $y_1, y_2, \cdots, y_s \in \mathbf{W}_1$

$$\mathrm{Tr}(T_\delta T_{y_1} \cdots T_{y_s}, \widetilde{E}_1(u)) = u^{-\nu + (1/2)(a_{E_1} + A_{E_1})}\, \mathrm{Tr}(T_{w_0} T_{y_1} \cdots T_{y_s}, \widetilde{E}_1(u))$$

$$\in u^{(1/2)(a_{E_1} + A_{E_1})} \cdot \mathbf{Z}[u, u^{-1}] .$$

(Here w_0 is the longest element in $\mathbf{W}_1$ and ν is its length.) Thus, in these cases, (6.19.1) holds when $\Delta(\overline{x}) = 1$ and (6.19.2) holds when $\Delta(\overline{x}) = -1$. The proposition follows in these cases.

Assume now that $\mathbf{W}_1$ is of type E_7 and that $2a_{\gamma w} = n_1(\hat{E} - \tilde{E}) + n_1'(\hat{E}' - \tilde{E}')$ where E_1, E_1' are the two irreducible representations of degree 512 of $\mathbf{W}_1$, $\hat{E}$ is the lifting of $E_1 \otimes \cdots \otimes E_1$ to a $\mathbf{Q}[\tilde{\mathbf{W}}]$-module defined in 3.2, $\hat{E}'$ is the lifting of $E_1' \otimes \cdots \otimes E_1'$ to a $\mathbf{Q}[\tilde{\mathbf{W}}]$-module defined in 3.2, $\tilde{E}$ is the other lifting of $E_1 \otimes \cdots \otimes E_1$ and $\tilde{E}'$ is the other lifting of $E_1' \otimes \cdots \otimes E_1'$.

From [BC] it follows that $\mathrm{Tr}(T_{y_1}, E_1(u)) + \mathrm{Tr}(T_{y_1}, E_1'(u)) \in \mathbf{Z}[u]$, $(y_1 \in \mathbf{W}_1)$ and that for some $y_1 \in \mathbf{W}_1$, $\mathrm{Tr}(T_{y_1}, E_1(u))$ is not in $\mathbf{Z}[u]$: some odd power of $u^{1/2}$ appears in it with $\neq 0$ coefficient. Replacing $u^{1/2}$ by $-u^{1/2}$ in the polynomial $\mathrm{Tr}(T_{y_1}, E_1(u))$ gives the trace of T_{y_1} on some irreducible representation of the Hecke algebra other than E_1, and it follows also from [BC] that this representation must be $E_1'(u)$. Thus, for any $y_1 \in \mathbf{W}_1$,

$\mathrm{Tr}(T_{y_1}, E_1'(u)) \in \mathbf{Z}[u^{1/2}]$ is obtained from $\mathrm{Tr}(T_{y_1}, E_1(u)) \in \mathbf{Z}[u^{1/2}]$ by $u^{1/2} \to -u^{1/2}$, hence $\mathrm{Tr}(T_{y_1}, E_1(u)) - \mathrm{Tr}(T_{y_1}, E_1'(u)) \in u^{1/2} \cdot \mathbf{Z}[u]$, $(y_1 \in \mathbf{W}_1)$.

It follows that, for all $y \in \mathbf{W}$:

(6.19.3) $$\mathrm{Tr}(T_\gamma T_y, \hat{E} - \tilde{E}) + \mathrm{Tr}(T_\gamma T_y, \hat{E}' - \tilde{E}') \in \mathbf{Z}[u]$$

(6.19.4) $$\mathrm{Tr}(T_\gamma T_y, \hat{E} - \tilde{E}) - \mathrm{Tr}(T_\gamma T_y, \hat{E}' - \tilde{E}') \in u^{1/2}\,\mathbf{Z}[u]$$

and

(6.19.5) $\mathrm{Tr}(T_\gamma T_y, \hat{E}' - \tilde{E}') \in \mathbf{Z}[u^{1/2}]$ is obtained from

$$\mathrm{Tr}(T_\gamma T_y, \hat{E} - \tilde{E}) \in \mathbf{Z}[u^{1/2}] \text{ by } u^{1/2} \to -u^{1/2}.$$

Then (6.19.5) and the definition of $a_{\gamma w}$ show that $n_1 = \pm n_1' \neq 0$. Assume first that $n_1 = n_1'$. Our assumption on $\bar{x}$ shows then that $\Delta(\bar{x}) = 1$, and from (6.19.3) and the definition of $a_{\gamma w}$ we see that $\ell(w) \equiv a(w) \pmod 2$. Similarly, if $n_1 = -n_1'$, we have $\Delta(\bar{x}) = -1$ and, by (6.19.4), $\ell(w) \not\equiv a(w) \pmod 2$. The proposition follows in this case. The case where $\mathbb{W}_1$ is of type E_8 and $E_1 \otimes \cdots \otimes E_1 \underset{LR}{\sim} \mathbb{W}$ where E_1 is an irreducible representation of degree 4096 of $\mathbb{W}_1$ is entirely similar. This completes the proof.

6.20. PROPOSITION. *Assume that* G, F', L, n *have the following properties*:

a) *Theorem 4.23 holds for* (G, F', L, n) *as well as for* $(G, F'F^d, L, n)$ *where* $d \geq 0$ *is any integer such that* $n | p^d - 1$. *(Note that for such* d, *the set* $Z_{L,n}$ *and the map* $\gamma: W_{L,n} \to W_{L,n}$ *with respect to* F' *and with respect to* $F'F^d$ *are the same.)*

b) *Let* $\rho_1 \in \mathcal{E}_{L',n}$, *where* $\mathcal{E}_{L',n}$ *is defined in terms of* (M, F', L', n); *here* M *is the Levi subgroup containing* T *of an arbitrary* F'*-stable proper parabolic subgroup* P *of* G, *containing* B, *and* L' *is a line bundle on the flag manifold of* M *obtained by restricting an arbitrary line bundle in the* W*-orbit of* L. *Then* $\varepsilon_{\rho_1} = \hat{\varepsilon}_{\rho_1}$. *(See 6.6, 6.7.)*

With these assumptions and with notations in 4.23, we have $\varepsilon_\rho = \hat{\varepsilon}_\rho = (-1)^{\ell(w_1)}\Delta(\bar{x}_\rho)$ *for any* $\rho \in \mathcal{E}_{L,n}$ *(defined in terms of* G, F', L, n *).*

Proof. Since 4.23 is assumed to hold for (G, F', L, n), we see from 6.19 and from (6.18.3) that $\varepsilon_\rho = (-1)^{\ell(w_1)}\Delta(\bar{x}_\rho)$ for any $\rho \in \mathcal{E}_{L,n}$. If $\rho \in \mathcal{E}_{L,n}$ is not cuspidal, we can find (M, F', L', n) as in b) and $\rho' \in \mathcal{E}_{L',n}$ (a representation of $M^{F'}$) such that $\langle \rho, \mathrm{Ind}_{P^{F'}}^{G^{F'}}(\rho') \rangle_{G^{F'}} \neq 0$. It is clear that $\hat{\varepsilon}_\rho = \hat{\varepsilon}_{\rho'}$ and, using 6.4 and 6.5, we see that $\varepsilon_\rho = \varepsilon_{\rho'}$. By the assumption b), we have $\varepsilon_{\rho'} = \hat{\varepsilon}_{\rho'}$, and therefore $\varepsilon_\rho = \hat{\varepsilon}_\rho$. Let ${}^0\mathcal{E}_{L,n}$

be the set of all cuspidal representations in $\mathcal{E}_{L,n}$. It remains to show that $\varepsilon_\rho = \hat{\varepsilon}_\rho$ for all ρ in ${}^0\mathcal{E}_{L,n}$. Let d be an integer ≥ 0 such that $n | p^d - 1$. Let ${}^0\mathcal{E}_{L,n,d}$ be the set of cuspidal representations of $G^{F'F^d}$ in ${}^0\mathcal{E}_{L,n}$ (defined with respect to $F'F^d$). Let $\overline{X}(W_{L,n},\gamma)^0$ (resp. $\overline{X}(W_{L,n},\gamma)^0_d$) be the subset of $\overline{X}(W_{L,n},\gamma)$ corresponding to ${}^0\mathcal{E}_{L,n}$ (resp. ${}^0\mathcal{E}_{L,n,d}$) under the bijection 4.23, with respect to F' (resp. $F'F^d$). The two sets $\overline{X}(W_{L,n},\gamma)^0$ and $\overline{X}(W_{L,n},\gamma)^0_d$ coincide. Indeed, the condition that $\rho \in \mathcal{E}_{L,n}$ (with respect to $F'F^d$) is cuspidal is equivalent, by 6.3, to the condition that $\langle \rho, \sum_E \mathrm{Tr}(\gamma w, \tilde{E}) R_{\tilde{E}} \rangle_{G^{F'F^d}} = 0$ for any $w \in Z_{L,n}$ such that

(6.20.1) $F'^{-1}(x) w x^{-1}$ is contained in a proper standard F'-stable parabolic subgroup of W, for some $x \in W$.

Using now 4.23 applied to $(G, F'F^d, L, n)$ we see that the condition that $\overline{x} \in \overline{X}(W_{L,n},\gamma)$ is in $\overline{X}(W_{L,n},\gamma)^0_d$ is also equivalent to the condition that $\sum_E \mathrm{Tr}(\gamma w, \tilde{E}) \{\overline{x}, x_{\tilde{E}}\} = 0$ for any $w \in Z_{L,n}$ satisfying (6.20.1). (The sum is over all $E \in (W_{L,n})^\vee_{ex}$.) The last condition is independent of d, and our claim that $\overline{X}(W_{L,n},\gamma)^0 = \overline{X}(W_{L,n},\gamma)^0_d$ follows. It follows that given $\rho \in {}^0\mathcal{E}_{L,n}$, there exists $\rho' \in {}^0\mathcal{E}_{L,n,d}$ such that ρ and ρ' correspond to the same element $\overline{x}$ under the bijection 4.23 for F' and $F'F^d$. As we have seen earlier in the proof, both ε_ρ and $\varepsilon_{\rho'}$ are equal to $(-1)^{\ell(w_1)} \Delta(\overline{x})$. On the other hand $\hat{\varepsilon}_\rho$ and $\hat{\varepsilon}_{\rho'}$ are, by definition, both equal to $(-1)^{\sigma(G)}$, see 6.7; note that $\sigma(G)$ defined with respect to F' and with respect to $F'F^d$ clearly coincide. Finally, we note that, by taking d sufficiently large, we may assume by 6.10, that $\varepsilon_{\rho'} = \hat{\varepsilon}_{\rho'}$. It follows that $\varepsilon_\rho = \hat{\varepsilon}_\rho$, and the proposition is proved.

6.21. A subset R' of the set of coroots $R^\vee \subset \mathcal{L}^\vee = \mathrm{Hom}(\mathcal{L}, \mathbf{Z})$ is said to be rationally closed if it is the intersection of $R^\vee$ with a $\mathbf{Q}$-vector subspace of $\mathcal{L}^\vee \otimes_{\mathbf{Z}} \mathbf{Q}$. It is well known [Bo, VI, 1.8, 24] that a rationally closed subsystem R' of $R^\vee$ is itself a root system in $\mathcal{L}^\vee$ and that R'

admits a set of simple coroots which is conjugate under an element of W to a subset of the set of simple coroots $\Pi^\vee$ in $R^\vee$. If R' is an arbitrary subset of $R^\vee$, there is a well-defined smallest rationally closed subset $\bar{R}'$ of $R^\vee$ containing R'; we say that $\bar{R}'$ is the rational closure of R' in $R^\vee$. We identify $\mathfrak{L}$ with $\mathrm{Hom}(T, k^*)$ as in 1.3.

We shall denote $\phi' : T \to T$ the unique automorphism of finite order such that $F'(t) = \phi'(t)^q$ for all $t \in T$, where q is given by 6.10.

Now let (L,n) be as in 2.1 (such that $Z_{L,n} \neq \emptyset$) with $L = L_\lambda$. We wish to reduce the proof of Theorem 4.23 to the special case where $\bar{R}^\vee_{L,n} = R^\vee$.

(6.21.1) From 6.5 it follows easily that, if $y \in W$, then 4.23 holds for (L,n) if and only if it holds for $({}^yL,n)$.

By [loc. cit.], there exists a subset $\Pi' \subset \Pi^\vee$ and an element $y \in W$ such that $y(\bar{R}^\vee_{L,n}) = \bar{\Pi}'$ (= rational closure of Π' in $R^\vee$). We have $y(\bar{R}^\vee_{L,n}) = \overline{(R^\vee_{{}^yL,n})}$. Using the previous remark, we see that in order to prove 4.23, we may assume that (L,n) has the additional property that $\bar{R}^\vee_{L,n} = \bar{\Pi}'$ for a subset $\Pi' \subset \Pi^\vee$. Recall that w_1 is the element of minimal length in $Z_{L,n}$. Let $\check{\alpha} \in R^\vee_{L,n}$. Then $\lambda(\check{\alpha}(x)) = x^{na}$ $(x \in k^*)$ for some integer a. Hence, by (2.1.1), $\lambda(F'w_1(\check{\alpha}(x)) = \lambda(\check{\alpha}(x))\,\lambda_1^n(w_1(\check{\alpha}(x))) = x^{na} \cdot x^{nb}$ $(x \in k^*)$ for some integer b. Hence $\lambda(\phi' w_1 \check{\alpha}(x))^q = x^{n(a+b)}$, $(x \in k^*)$. Since q is relatively prime to n, we have $\lambda(\phi' w_1 \check{\alpha}(x)) = x^{nc}$, $(x \in k^*)$ for some integer c. Thus, $\phi' w_1 \check{\alpha} \in R^\vee_{L,n}$, i.e. $\phi' w_1$ maps $R^\vee_{L,n}$ into itself. (It also maps $R^\vee$ into itself.) Hence it maps $\bar{R}^\vee_{L,n}$ into itself. It follows that

$$F'(w_1 W_{\Pi'} w_1^{-1}) = W_{\Pi'}$$

where $W_{\Pi'}$ is the subgroup of W generated by the reflections with respect to the simple coroots in Π'.

Let w_2 be the unique element of minimal length in the coset $w_1 \cdot W_{\Pi'}$. We have

$$F'(w_2 W_{\Pi'} w_2^{-1}) = W_{\Pi'}$$

and, in addition, $\phi' w_2$ maps Π' into itself. Let P be the parabolic subgroup of G containing B and corresponding to the subset Π' of $\Pi^\vee$. Let M be the Levi subgroup of P containing T. We may identify the Weyl group of M with respect to T, with $W_{\Pi'}$. The map $g \to F'_M(g) = F'(\dot{w}_2 g \dot{w}_2^{-1})$ (with $\dot{w}_2$ as in 1.5) leaves M stable and is the Frobenius map for an F_q-rational structure on M, commuting with F. It also leaves T and $B_M = B \cap M$ stable. However, P is not necessarily F'_M-stable.

The flag manifold M/B_M may be canonically identified with P/B hence it is naturally imbedded in G/B; hence the line bundle L may be restricted to M/B_M and will be denoted again L. We may now consider $W_{L,n}^{(M)}$, $Z_{L,n}^{(M)}$ defined for M in the same way as $W_{L,n}$, $Z_{L,n}$ were defined for G.

We have $W_{L,n}^{(M)} = W_{L,n}$ (since $W_{L,n} \subset W_{\Pi'}$) and the map $w \to w_2 w$ is a bijection $Z_{L,n}^{(M)} \xrightarrow{\sim} Z_{L,n}$; it is easy to see that $w_2^{-1} w_1$ is the element of minimal length in $Z_{L,n}^{(M)}$. Hence the map $\gamma : W_{L,n} \to W_{L,n}$ (see (2.15.1)) and the map $\gamma^{(M)} : W_{L,n}^{(M)} \to W_{L,n}^{(M)}$ defined as γ (using F'_M, $w_2^{-1} w_1$ instead of F', w_1) coincide under the identification $W_{L,n}^{(M)} = W_{L,n}$. It follows that the sets $\overline{X}(W_{L,n},\gamma)$ and $\overline{X}(W_{L,n}^{(M)},\gamma^{(M)})$, (see (4.21.12)), are in a natural bijection. Let $\mathcal{E}_{L,n}^{(M)}$ be the set defined as $\mathcal{E}_{L,n}$ but with G replaced by M. We shall establish a bijection between $\mathcal{E}_{L,n}^{(M)}$ and $\mathcal{E}_{L,n}$, using the method in $[L_1]$.

Consider the algebraic variety $V = \{g \in G \mid g^{-1} F'(g) \in F'(\dot{w}_2 U_P)\}$ where U_P is the unipotent radical of P. There is a natural action of $G^{F'} \times M^{F'_M}$ on V defined by $(g_0, m_0) : g \to g_0 g m_0^{-1}$. Hence $G^{F'} \times M^{F'_M}$ acts on $H_c^i(V, \overline{Q}_\ell)$ by $((g_0,m_0)^*)^{-1}$. To any irreducible $\overline{Q}_\ell[M^{F'_M}]$-module ρ, we associate a virtual representation of $G^{F'}$:

$$R^G_{M\subset P}(\rho) = \sum_i (-1)^i (H^i_c(V,\bar{Q}_\ell)\otimes\rho)^{M^{F'_M}} .$$

(The space of $M^{F'_M}$-invariants is still a $G^{F'}$-module.) This extends to a homomorphism $R^G_{M\subset P}: \mathcal{R}(M^{F'_M}) \to \mathcal{R}(G^{F'})$.

Let $x \in Z^{(M)}_{L,n}$. Consider the algebraic variety $V^{(M)}_{\dot{x}} = \{m \in M | m^{-1}F'_M(m) \in \dot{x}(U\cap M)\}$. It has a natural $M^{F'_M} \times T^{(M)}_x$ action, by left and right translation (compare 2.1), where $T^{(M)}_x = \{t \in T | x(F'_M(t)) = t\}$. Consider also the algebraic variety $V_{\dot{x}F(\dot{w}_2)} = \{g' \in G | g'^{-1}F'(g') \in \dot{x}F(\dot{w}_2)U\}$. It has a natural $G^F \times T_{xF(w_2)}$ action, by left and right translation, where

$$T_{xF(w_2)} = \{t \in T | xF(w_2)(F'(t)) = t\} = T^{(M)}_x$$

We have a map

(6.21.2) $$V \times_{M^{F'_M}} V^{(M)}_{\dot{x}} \to V_{\dot{x}F(\dot{w}_2)}$$

defined by $(g,m) \to gm$.

Indeed, assume that $g^{-1}F'(g) \in F'(\dot{w}_2 U_P)$, $m^{-1}F'(\dot{w}_2 m\dot{w}_2^{-1}) \in \dot{x}(U\cap M)$. Then

$$\begin{aligned}(gm)^{-1}F'(gm) &= m^{-1}g^{-1}F'(g)F'(m) \in m^{-1}F'(\dot{w}_2 U_P)F'(m)\\ &= m^{-1}F'(\dot{w}_2)F'(m)F'(U_P)\\ &= m^{-1}F'(\dot{w}_2 m\dot{w}_2^{-1})F'(\dot{w}_2)F'(U_P) \subset \dot{x}(U\cap M)F'(\dot{w}_2)F'(U_P)\\ &= \dot{x}F'(\dot{w}_2)(U\cap F'(M))F'(U_P) = \dot{x}F'(\dot{w}_2)U .\end{aligned}$$

Hence the map (6.21.2) is well defined. It is easy to see (compare [L_1, Lemma 3]) that the map (6.21.2) is in fact a $G^{F'} \times T_{xF(w_2)}$-equivariant isomorphism.

This implies that for any $x \in Z^{(M)}_{L,n}$, the homomorphism $R^G_{M\subset P}$ carries

(6.21.3) $$\sum_i (-1)^i H_c^i(X_x^{(M)}, \mathcal{F}_{\dot{x},n}^{L(M)}) \in \mathcal{R}(M^{F'_M})$$

(objects defined as in 2.1, 2.3, with respect to M) to

(6.21.4) $$\sum_i (-1)^i H_c^i(X_{xF(w_2)}, \mathcal{F}_{\dot{x}F(\dot{w}_2),n}^{L'}) \in \mathcal{R}(G^{F'})\,;$$

here $L' = {}^{F(\dot{w}_2)^{-1}}L$. (Note that the map $x \to xF(w_2)$ is a bijection $Z_{L,n}^{(M)} \xrightarrow{\sim} Z_{L',n}$; indeed, as we have observed earlier, $x \to w_2 x$ is a bijection $Z_{L,n}^{(M)} \xrightarrow{\sim} Z_{L,n}$ and by 6.2 (i) $x' \to w_2^{-1}x'F'(w_2)$ is a bijection $Z_{L,n} \xrightarrow{\sim} Z_{L',n}$.)

Making use of [L_1, Thm. 8, Remark 11], we see that there exists a sign $\varepsilon = \pm 1$ such that $\varepsilon R_{M\subset P}^G$ defines a bijection $\mathcal{E}_{L,n}^{(M)} \xrightarrow{\sim} \mathcal{E}_{L',n}$, and such that, for any $\rho \in \mathcal{E}_{L,n}^{(M)}$ and $x \in Z_{L,n}^{(M)}$ the multiplicity of ρ in (6.19.3) is equal to ε times the multiplicity of $\varepsilon R_{M\subset P}^G(\rho)$ in (6.21.4). Moreover, by [loc. cit.] we have $\varepsilon = (-1)^{\sigma(G)-\sigma(M)}$ where $\sigma(G)$ and $\sigma(M)$ are defined in 6.7 with respect to F', F'_M respectively. It is easy to see that $(-1)^{\sigma(G)-\sigma(M)} = (-1)^{\ell(w_2)}$. On the other hand $(-1)^{\ell(w_2)} = (-1)^{\ell(w'_1)} \cdot (-1)^{\ell(w''_1)}$ where w'_1 (resp. w''_1) is the element of minimal length in $Z_{L,n}^{(M)}$ (resp. $Z_{L',n}$), indeed, $w'_1 = w_2^{-1}w_1$ and w''_1 is conjugate to w_1. It follows that 4.23 holds for (L,n) and M, F'_M if and only if it holds for (L′,n) and G, F' and (by (6.21.1)) if and only if it holds for (L,n) and G, F'. *Hence the proof of 4.23 is reduced to the special case where* $\bar{R}_{L,n}^\vee = R^\vee$. Indeed, the last equality is satisfied by our (L,n), with respect to M.

Chapter 7

SOME EXCEPTIONAL GROUPS

In this chapter we use the results in Chapter 6 to verify Theorem 4.23 in the case where $W_{L,n}$ is a not too complicated Weyl group. (Essentially, we study the case where any family in $(W_{L,n})^\vee$ has at most five elements.)

7.1. PROPOSITION. *Assume that the Weyl group* (W,S) *of* G *(see 1.2) is irreducible and that* $F':G\to G$ *is a power* F^s *of* F, *so that* F' *acts trivially on* W. *Assume also that Theorem 4.23 holds for* G, *for* $(L,n) =$ *(trivial line bundle,* 1 *) and for sufficiently large* s. *Let* Γ *be a left cell of* W, *let* $\mathcal{F} \subset W^\vee$ *be the family containing the irreducible components of* $[\Gamma]_1$ *(see 5.25) and let* $\mathcal{G} = \mathcal{G}_{\mathcal{F}}$ *be the corresponding finite group (see 4.14). Let* $E_{(x,\sigma)} \in \mathcal{F}$ *be the representation of* W *corresponding to* $(x,\sigma) \in \mathfrak{M}(\mathcal{G})$ *under the imbedding* $\mathcal{F} \hookrightarrow \mathfrak{M}(\mathcal{G})$, *(see 4.14); note that* $E_{(x,\sigma)}$ *is not defined for all* $(x,\sigma) \in \mathfrak{M}(\mathcal{G})$. *Assume that* $\Delta:\mathfrak{M}(\mathcal{G})\to \pm 1$ *is identically* 1, *(see 4.14).*

(i) *Let* $w \in \Gamma$ *be such that* $a_w \in \mathcal{R}(W)$ *is nonzero. If* $E = E_{(1,1)} \in \mathcal{F}$ *then* $c_{w,E} > 0$ *(see (5.1.22)). If* $E' \in \mathcal{F}$, $E' \neq E_{(1,1)}$ *and if* Γ' *is a left cell such that* E' *appears in* $[\Gamma']_1$ *then there exists* $w' \in \Gamma'$, *such that* $c_{w',E'} < 0$. *If* $E' \in \mathcal{F}$ *is such that* $c_{w,E'} \neq 0$ *then* E' *is a component of* $[\Gamma]_1$.

(ii) *Assume that* $[\Gamma]_1 = \sum_x E_{(x,1)}$, *sum over* $x \in \mathcal{G}_0$, *a set of representatives for the conjugacy classes in* $\mathcal{G}$.

Then there exists a unique bijection

$$\{w \in \Gamma \mid a_w \neq 0\} \leftrightarrow \{\text{irreducible characters of } \mathcal{G}\}$$

with the following property: if $w_\phi \in \Gamma$ *corresponds to the irreducible character* ϕ *of* $\mathcal{G}$, *we have* $a_{w_\phi} = \sum_{x \in \mathcal{G}_0} \phi(x) E_{(x,1)} \in \mathcal{R}(W)$.

(iii) *The elements* w_ϕ *in (ii) satisfy* $\ell(w_\phi) \equiv a(w_\phi) \pmod 2$.

Proof. (i) According to 6.17 (ii), $R_{a_w} \in R(G^{F^s})$ is an actual representation of G^{F^s} for s sufficiently large. (R_{a_w} is defined with respect to (L, n) and $F' = F^s$ as in the statement of the proposition.) Since it is nonzero, we have

$$\dim R_{a_w} > 0, \quad (\text{s large}). \tag{7.1.1}$$

By 5.2 (ii) we have

$$R_{a_w} = c_{w,E} R_E + \sum_{\substack{E' \in \mathcal{F} \\ E' \neq E}} c_{w,E'} R_{E'} \tag{7.1.2}$$

and by 3.9 (i) we have

$$c_{w,E} = \langle R_{a_w}, R_E \rangle_{G^{F^s}}. \tag{7.1.3}$$

Now let ρ be an irreducible representation of G^{F^s} such that $\langle \rho, R_{a_w} \rangle_{G^{F^s}} \neq 0$. We then have $\langle \rho, R_{E'} \rangle_{G^{F^s}} \neq 0$ for some $E' \in \mathcal{F}$. Using 4.23 in the present case (for s large), we see that to ρ corresponds an element $(x, \sigma) \in \mathfrak{M}(\mathcal{G})$ such that $\langle \rho, R_{E_{(y,\tau)}} \rangle_{G^{F^s}} = \{(x,\sigma),(y,\tau)\}$, whenever $E_{(y,\tau)}$ is defined. In particular, we have $\langle \rho, R_E \rangle_{G^{F^s}} = \{(x,\sigma),(1,1)\} = \dim \sigma / |Z_{\mathcal{G}}(x)|$. Let $b_{x,\sigma} = \langle \rho, R_{a_w} \rangle_{G^{F^s}}$. We have

$$\langle R_{a_w}, R_E \rangle_{G^{F^s}} = \sum_\rho \langle \rho, R_{a_w} \rangle_{G^{F^s}} \langle \rho, R_E \rangle_{G^{F^s}}.$$

Using now (7.1.3) we get

$$c_{w,E} = \sum_{(x,\sigma)\in\mathfrak{M}(\mathcal{G})} b_{x,\sigma}(\dim\sigma/|Z_{\mathcal{G}}(x)|) .$$

Since R_{a_w} is an actual representation ($\neq 0$), all the $b_{x,\sigma}$ are ≥ 0 integers and at least one of them is > 0. It follows that $c_{w,E} > 0$, as required.

Now let $E' \in \mathcal{F}$ be such that $c_{w,E'} \neq 0$. As in the proof of 5.2, we see that $I_{E',\mathcal{K}} \cdot C_w \neq 0$ hence (with notations in 5.1) we have $I_{E',\mathcal{K}} \cap I_w^L \neq 0$. Assume that $I_{E',\mathcal{K}} \cap \hat{I}_w^L \neq 0$. Then there exists $w' \leq_L w$, $w' \not\sim_L w$ such that E' appears in the left cell containing w'. By (4.1.13), we have $w' \not\sim_{LR} w$ hence E' does not appear in the two-sided cell containing w. Since $c_{w,E'} \neq 0$, we have a contradiction with 5.2 (ii).

Thus, we have $I_{E',\mathcal{K}} \cap \hat{I}_w^L = 0$. This, together with $I_{E',\mathcal{K}} \cap I_w^L \neq 0$ implies that E' appears in $[\Gamma]_1$; note that $[\Gamma] = I_w^L/\hat{I}_w^L$.

Now let $E' \in \mathcal{F}$, $E' \neq E$. Let Γ' be a left cell such that E' appears in $[\Gamma']_1$. Since E is a special representation in the same family as E', it must also appear in $[\Gamma']_1$. Clearly Γ' is contained in the two-sided cell which contains Γ. By 5.8, we have $\sum_{w'\in\Gamma'} c_{w',E'}c_{w',E} = 0$. Assume that $c_{w',E'} \geq 0$ for all $w' \in \Gamma'$. By the first part of the proof, we have $w' \in \Gamma'$, $c_{w',E'} \neq 0 \Rightarrow a_{w'} \neq 0 \Rightarrow c_{w',E} > 0$. It follows that for any $w' \in \Gamma'$ we have $c_{w',E'} = 0$. But this contradicts $\sum_{w'\in\Gamma'} c^2_{w',E'} \neq 0$ (5.8). It follows that $c_{w',E'} < 0$ for some $w' \in \Gamma'$, and (i) is proved.

(ii) Let $w \in \Gamma$ be such that $a_w \neq 0$. According to (i), a_w is of the form

$$a_w = \sum_{x\in\mathcal{G}_0} \phi_w(x) E_{(x,1)}$$

where ϕ_w is a class function $\mathcal{G} \to \mathbf{Z}$.

Let σ be an irreducible representation of $\mathcal{G}$ and let ρ be the irreducible representation of G^{F^s} in $\mathcal{E}_{L,n}$ corresponding under 4.23 to $(1,\sigma) \in \mathfrak{M}(\mathcal{G})$, (s large). By 4.23, we have

$$\langle R_{\alpha_w}, \rho \rangle_{G^{F^s}} = \sum_{x \in \mathcal{G}_0} \phi_w(x) \langle R_{E_{(x,1)}}, \rho \rangle_{G^{F^s}}$$

$$= \sum_{x \in \mathcal{G}_0} \phi_w(x) \{(1,\sigma),(x,1)\}$$

$$= \sum_{x \in \mathcal{G}_0} \phi_w(x) \operatorname{Tr}(x,\sigma)/|Z_{\mathcal{G}}(x)| .$$

But $\langle R_{\alpha_w}, \rho \rangle_{G^{F^s}}$ is an integer ≥ 0, (cf. 6.17 (ii)), for large s. Hence $\sum_{x \in \mathcal{G}_0} \phi_w(x) \operatorname{Tr}(x,\sigma)/|Z_{\mathcal{G}}(x)|$ is an integer ≥ 0 for any irreducible representation σ of $\mathcal{G}$. It follows that there exists a $\mathbf{Q}[\mathcal{G}]$-module $\Phi_w \neq 0$ such that $\phi_w(x) = \operatorname{Tr}(x,\Phi_w)$ for all $x \in \mathcal{G}_0$. By 5.8, we have

$$\sum_w \phi_w(x)\phi_w(y) = \begin{cases} f_{E_{(x,q)}}, & \text{if } x = y \\ 0, & \text{if } x \neq y \end{cases}, \quad (x,y \in \mathcal{G}_0),$$

(sum over all $w \in \Gamma$ such that $\alpha_w \neq 0$) and, by (4.14.2), we have

$$f_{E_{(x,1)}} = |Z(x)|, \quad x \in \mathcal{G}_0 .$$

It follows that

$$\sum_w \operatorname{Tr}(x,\Phi_w)\operatorname{Tr}(y,\Phi_w) = \begin{cases} |Z(x)|, & \text{if } x = y \\ 0, & \text{if } x \neq y \end{cases} \quad (x,y \in \mathcal{G}_0)$$

(the sum over w is as before). In other words, the $\mathcal{G}\times\mathcal{G}$-module

$\sum_w \Phi_w \otimes \Phi_w$ is isomorphic to the two-sided regular representation of $\mathcal{G}\times\mathcal{G}$. (Note that all irreducible representations of $\mathcal{G}$ are defined over $\mathbf{Q}$.) It follows immediately that each Φ_w is an irreducible $\mathcal{G}$-module, that they are mutually distinct and that they exhaust the set of irreducible $\mathcal{G}$-module; (ii) is proved.

Finally, (iii) follows from (6.18.10).

REMARK. It is easy to see that for any $\mathcal{F}$ satisfying our assumption ($\Delta : \mathfrak{M}(\mathcal{G}_{\mathcal{F}}) \to \pm 1$ is identically 1), the W-module $\sum_{x\in\mathcal{G}_0} E_{(x,1)}$ is well defined and constructible; it follows then from 5.29 that there exists a left cell $\Gamma \subset W$ such that $[\Gamma]_1 = \sum_{x\in\mathcal{G}_0} E_{(x,1)}$. We shall use this result in three special cases: $[\Gamma]_1 = Y_{1,1}$ (type F_4), $[\Gamma]_1 = 80_s + 60_s + 10_s$ (type E_6), $[\Gamma]_1 = X_{7,1}$ (type E_8). (See the last sentence in 5.29.)

7.2. PROPOSITION. *We preserve the assumptions of 7.1 and we assume in addition that* (W, S) *is of type* E_6 *and that* $\mathcal{F}$ *is the family of* 80_s. *Then each of the following elements of* $\mathcal{R}(W)$ *is of the form* a_w *for some* $w \in W$ *such that* $\ell(w) \equiv a(w) \pmod 2$:

$$80_s + 60_s + 10_s\ ,\quad 80_s - 60_s + 10_s\ ,\quad 2.80_s - 10_s$$

$$80_s + 60_s + 90_s\ ,\quad 80_s - 60_s + 90_s\ ,\quad 2.80_s - 90_s$$

$$80_s + 2.90_s + 20_s\ ,\quad 80_s - 20_s\ ,\quad 80_s - 90_s + 20_s\ .$$

Proof. For $80_s + 60_s + 10_s$, $80_s - 60_s + 10_s$, $2.80_s - 10_s$, this follows from 7.1 and the remark at its end. The W-module $80_s + 60_s + 90_s$ is constructible (5.29) hence it is of the form $[\Gamma]_1$ for some left cell $\Gamma \subset W$. Let $w_1, w_2, \cdots, w_N$ be the elements w of Γ such that $a_w \neq 0$. According to 7.1 (i), each a_{w_i} is a $\mathbf{Z}$-linear combination $c_{1,i}80_s + c_{2,i}60_s + c_{3,i}90_s$ with $c_{1,i} > 0$. By 5.8, we have $\sum_i c_{2,i}^2 = 2$ hence we may assume that $c_{2,1} = \pm 1$, $c_{2,2} = \pm 1$, $c_{2,i} = 0$ for $i > 2$. Again from 5.8, we have $\sum_i c_{1,i}c_{2,i} = 0$. Since $c_{1,i} > 0$, it follows that $c_{2,1}$, $c_{2,2}$ have opposite signs (we may assume that $c_{2,1} = 1$, $c_{2,2} = -1$) and $c_{1,1} = c_{1,2}$. Since $\sum_i c_{1,i}^2 = 6$, we must have $c_{1,1} = c_{1,2} = 1$. Since $\sum_i c_{3,i}^2 = 3$, (5.8), we have $c_{3,i} = \pm 1$ for three values of i and $c_{3,i} = 0$ for the remaining values of i. Since $\sum_i c_{1,i}c_{3,i} = 0$, we cannot have $c_{1,i} = 1$ for each i such that $c_{3,1} = \pm 1$. Hence there is an index i such that $c_{1,i} \geq 2$, $c_{3,1} = \pm 1$ and we may assume that $i = 3$. From $\sum_i c_{1,i}^2 = 6$ it follows that $c_{1,3} = 2$ and $c_{1,i} = 0$ for $i \neq 1,2,3$ hence $N = 3$. We have $c_{3,1} = \pm 1$, $c_{3,2} = \pm 1$, $c_{3,3} = \pm 1$. From $\sum_i c_{1,i}c_{3,i} = 0$, it follows that $c_{3,1} + c_{3,2} + 2c_{3,3} = 0$ hence $c_{3,1} = c_{3,2} = -c_{3,3}$. If $c_{3,1} = c_{3,2} = -c_{3,3} = -1$, then $a_{w_1} = 8\theta_s + 6\theta_s - 9\theta_s$ and we get a contradiction as follows. Let ρ be the representation in $\mathcal{E}_{(L,1)}$ corresponding under 4.23 to $(1,1) \in \mathfrak{M}(\mathcal{G}_{\mathcal{F}})$, ($s$ large); by 4.23 we have $\langle \rho, R_{a_{w_1}} \rangle_{G^{F^s}} = (1/6) + (1/2) - (1/3)$; this is not an integer, contradicting 6.17 (ii).

Hence, we have $c_{3,1} = c_{3,2} = -c_{3,3} = 1$ so that $a_{w_1}, a_{w_2}, a_{w_3}$ are $8\theta_s + 6\theta_s + 9\theta_s$, $8\theta_s - 6\theta_s + 9\theta_s$, $2.8\theta_s - 9\theta_s$ respectively, as desired, (see (6.18.10)).

The W-module $8\theta_s + 2.9\theta_s + 2\theta_s$ is constructible (5.29) hence it is of the form $[\Gamma']_1$ for some left cell $\Gamma' \subset W$. Let $w_1, \cdots, w_N$ be the elements $w \in \Gamma'$ such that $a_w \neq 0$. According to 7.1 (i), each a_{w_i} is a Z-linear combination $c_{1,i}8\theta_s + c_{2,i}9\theta_s + c_{3,i}2\theta_s$ with $c_{1,i} > 0$.

Writing the condition that $\langle R_{a_{w_i}}, \rho \rangle$ is an integer ≥ 0 for $\rho \in \mathcal{E}_{L,1}$ corresponding to $(x,1) \in \mathfrak{M}(\mathcal{G}_{\mathcal{F}})$ (s large), we see that

$$c_{1,i}\,\mathrm{Tr}(x,1) + c_{2,i}\,\mathrm{Tr}(x,r) + c_{3,i}\,\mathrm{Tr}(x,\varepsilon) \in |Z(x)| \cdot \mathbf{N}.$$

Here $1, r, \varepsilon$ denote the irreducible representations of $G = \mathfrak{S}_3$, (as in 4.3).

In other words, if $\Psi_i = c_{1,i}1 + c_{2,i}r + c_{3,i}\varepsilon$, (a virtual representation of $\mathfrak{S}_3$), then $\mathrm{Tr}(x,\Psi_i) \in |Z(x)| \cdot \mathbf{N}$ for all $x \in \mathfrak{S}_3$. Equivalently, we have $\Psi_i = \sum_{\mathcal{O}} n_{\mathcal{O},i}\Psi_{(\mathcal{O})}$ (sum over all conjugacy classes $\mathcal{O} \subset \mathfrak{S}_3$), where $n_{\mathcal{O},i} \in \mathbf{N}$ and $\Psi_{(\mathcal{O})} = \mathrm{Tr}(x,1)\cdot 1 + \mathrm{Tr}(x,r)\, r + \mathrm{Tr}(x,\varepsilon)\,\varepsilon$, for $x \in \mathcal{O}$. By 5.8, we have

$$\sum_i c_{j,i}c_{j',i} = \begin{cases} 6 = |\mathfrak{S}_3|, & \text{if } j = j' \\ 0, & \text{if } j \neq j' \end{cases}, \quad (1 \le j, j' \le 3).$$

This can be also written in the form $\sum_i \Psi_i \otimes \Psi_i = |\mathfrak{S}_3|\ (1\otimes 1 + r\otimes r + \varepsilon\otimes\varepsilon)$ (an equality of representations of $\mathfrak{S}_3 \times \mathfrak{S}_3$). The left-hand side of the last equality is $\sum_{i,\mathcal{O},\mathcal{O}'} n_{\mathcal{O},i}n_{\mathcal{O}',i}\Psi_{(\mathcal{O})}\otimes\Psi_{(\mathcal{O}')}$, and the right-hand side is $\sum |\mathcal{O}|\Psi_{(\mathcal{O})}\otimes\Psi_{(\mathcal{O})}$. Since the characters of $\Psi_{(\mathcal{O})}\otimes\Psi_{(\mathcal{O}')}$ are linearly independent, it follows that $n_{\mathcal{O},i}n_{\mathcal{O}',i} = 0$ for any i and any $\mathcal{O} \neq \mathcal{O}'$ and that $\sum_i n^2_{\mathcal{O},i} = |\mathcal{O}|$ for any $\mathcal{O}$. Since $|\mathcal{O}| = 1$, 2 or 3, we must have $n_{\mathcal{O},i} = 0$ or 1, and from $n_{\mathcal{O},i}n_{\mathcal{O}',i} = 0$ for $\mathcal{O} \neq \mathcal{O}'$, we see that each Ψ_i is equal to $\Psi_{(\mathcal{O})}$ for some $(\mathcal{O})$ depending on i. We see that the sequence $\Psi_1, \Psi_2, \cdots, \Psi_N$ is the same as the sequence of the $\Psi_{(\mathcal{O})}$ with each $\Psi_{(\mathcal{O})}$ repeated $|\mathcal{O}|$ times. Therefore, $N = 6$ and the sequence $a_{w_1}, a_{w_1}, \cdots, a_{w_6}$ consists of $80_s + 2.90_s + 20_s$ (once), $80_s - 20_s$ (three times) and $80_s - 90_s + 20_s$ (twice). The congruence $\ell(w_i) \equiv a(w_i)$ (mod 2) follows from (6.18.10). The proposition is proved.

7.3. COROLLARY. *Let* $(\mathbf{W}, S)$ *be a Weyl group of type* E_6, E_7 *or* E_8 *and let* $\mathcal{F} \subset \mathbf{W}^\vee$ *be a family such that* $\mathcal{G}_{\mathcal{F}} = \mathfrak{S}_3$. *Let* $E_{(x,\sigma)} \in \mathcal{F}$ *be the representation corresponding to* $(x,\sigma) \in \mathcal{M}(\mathcal{G}_3)$, *(see 4.11-4.13), where* (x,σ) *is one of the elements* $(1,1)$, $(g_2,1)$, $(1,r)$, $(g_3,1)$, $(1,\varepsilon)$. *Then each of the following elements of* $\mathcal{R}(\mathbf{W})$ *is of the form* a_w *for some* $w \in \mathbf{W}$ *such that* $\ell(w) \equiv a(w)$, (mod 2):

$$E_{(1,1)}+E_{(g_2,1)}+E_{(g_3,1)},\ E_{(1,1)}-E_{(g_2,1)}+E_{(g_3,1)},\ 2\cdot E_{(1,1)}-E_{(g_3,1)}$$

$$E_{(1,1)}+E_{(g_2,1)}+E_{(1,r)},\ E_{(1,1)}-E_{(g_2,1)}+E_{(1,r)},\ 2E_{(1,1)}-E_{(1,r)}$$

$$E_{(1,1)}+2E_{(1,r)}+E_{(1,\varepsilon)},\ E_{(1,1)}-E_{(1,\varepsilon)}\quad E_{(1,1)}-E_{(1,r)}+E_{(1,\varepsilon)}\,.$$

Proof. (When $(\mathbf{W}, S)$ is of type E_6, this is just 7.2. We use the fact that 4.23 holds in the situation in 7.2, for large s, see $[L_8]$.) Assume now that $(\mathbf{W}, S)$ is of type E_7. If $\mathcal{F}$ is the family containing $315'_a$, then $J_{E_6}^{\mathbf{W}}$ is a 1-1 correspondence between the family of 80_s in E_6, see (4.12.2), and $\mathcal{F}$ and the result follows from the case of E_6 and (5.10.5). If $\mathcal{F}'$ is the family containing 315_a, the desired result follows from that for $\mathcal{F}$ and from (5.12.4) and (5.12.6); we use the fact that the longest element in $\mathbf{W}$ acts on $315'_a$, 280_b, $280'_a$, $70'_a$, $35'_a$ as -1, 1, -1, -1, -1 respectively. For example the information that $315'_a + 280_b + 70'_a$ (resp. $315'_a - 280_b + 70'_a$) are of form a_w, $\ell(w)$ odd, is converted by application of (5.12.4) and (5.12.6) in the information that $315_a - 280'_b + 70_a$ (resp. $315_a + 280'_b + 70_a$) are of form $a_{w_0 w}$, $\ell(w)$ odd (i.e. $\ell(w_0 w)$ even). Assume now that $(\mathbf{W}, S)$ is of type E_8. The cases of the families $\mathcal{F}_1$ (containing 1400_z) and $\mathcal{F}_2$ (containing 1400_x) are reduced, using $J_{E_7}^{\mathbf{W}}$ and $J_{E_6\times A_1}^{\mathbf{W}}$ to already known results in E_7 and $E_6\times A_1$ (cf. (5.10.5)). The cases of the families $\mathcal{F}_3 = \mathcal{F}_1 \otimes \text{sign}$, $\mathcal{F}_4 = \mathcal{F}_2 \otimes \text{sign}$ are reduced to the already known cases of $\mathcal{F}_1$, $\mathcal{F}_2$ by (5.12.4) and (5.12.6); we use the fact that the longest element in $\mathbf{W}$ acts on 1400_z, 1344_x, 1008_z, 448_z, 56_z as -1, 1, -1, -1, -1 respectively and on 1400_x, 1050_x, 1575_x, 175_x, 350_x as identity.

7.4. PROPOSITION. *Assume that the Weyl group* (W,S) *of* G *(see 1.2) is of type* F_4 *and that* $F' : G \to G$ *is a power* F^s *of* F. *Then each of the following elements* $Y_{i,j} \in \mathcal{R}(W)$ *is of the form* a_w *for some* $w \in W$ *such that* $\ell(w) \equiv a(w) \equiv 0 \pmod 2)$:

	12_1	9_3	9_2	1_3	1_2	4_1	4_4	4_3	6_1	6_2	16_1
$Y_{1,1}$	1		1					1	1		1
$Y_{2,1}$	1	1					1		1		1
$Y_{3,1}$	1		2		1			1		1	1
$Y_{4,1}$	1	2		1			1			1	1
$Y_{5,1}$	1	1	1			1				1	2
$Y_{3,2}$	1				–1			–1		1	1
$Y_{5,2}$	1	–1	1			–1				1	
$Y_{4,2}$	1			–1			–1			1	1
$Y_{5,3}$	1	1	–1			–1				1	
$Y_{1,2}$	1		1					–1	1		–1
$Y_{2,2}$	1	1					–1		1		–1
$Y_{3,3}$	1		2		1			–1		1	–1
$Y_{4,3}$	1	2		1			–1			1	–1
$Y_{5,4}$	1	1	1			1				1	–2
$Y_{3,4}$	1				–1			1		1	–1
$Y_{4,4}$	1			–1			1			1	–1
$Y_{1,3}$	2		2						–1		
$Y_{1,4}$	3		–1					–1			1
$Y_{1,5}$	3		–1					1			–1

The elements $Y_{i,j}$ (one in each row) are linear combinations of irreducible representations of W in the family $\mathcal{F}$ containing 12_1. The representations in $\mathcal{F}$ parametrize the column in the matrix above, which describes the coefficients with which a representation in $\mathcal{F}$ appears in $Y_{i,j}$. When the coefficient is zero, it is omitted.

Proof. Our statement for $Y_{i,1}$ $(1 \leq i \leq 5)$ is contained in 5.20. For $Y_{1,i}$ $(1 \leq i \leq 5)$ it follows by applying 7.1 (ii) to a left cell $\Gamma \subset W$ such that $[\Gamma]_1 = Y_{1,1}$; such Γ exists since $Y_{1,1}$ is constructible (5.29). (In order to be able to apply 7.1 (ii), we make use of the fact that 4.23 holds for our G, $(L,n) =$ (trivial bundle, 1), and large s, see $[L_8]$.)

Then $Y_{1,i}$ $(1 \leq i \leq 5)$ are the a_{w_ϕ}, (7.1 (ii)), associated to the five irreducible characters ϕ of $\mathfrak{S}_4$.

With the notations in 4.10, we have:

$$Y_{5,3} = J_{W_1}^{W}\left(\begin{bmatrix} 0 & & 1 & & 3 \\ & 1 & & 2 & \end{bmatrix} - \begin{bmatrix} 0 & & 1 & & 2 \\ & 1 & & 3 & \end{bmatrix}\right)$$

$$Y_{5,2} = J_{W_4}^{W}\left(\begin{bmatrix} 0 & & 1 & & 3 \\ & 1 & & 2 & \end{bmatrix} - \begin{bmatrix} 0 & & 1 & & 2 \\ & 1 & & 3 & \end{bmatrix}\right)$$

$$Y_{4,2} = J_{W_1}^{W}\left(\begin{bmatrix} 0 & & 1 & & 3 \\ & 1 & & 2 & \end{bmatrix} - \begin{bmatrix} 1 & & 2 & & 3 \\ & 0 & & 1 & \end{bmatrix}\right)$$

$$Y_{3,2} = J_{W_4}^{W}\left(\begin{bmatrix} 0 & & 1 & & 3 \\ & 1 & & 2 & \end{bmatrix} - \begin{bmatrix} 1 & & 2 & & 3 \\ & 0 & & 1 & \end{bmatrix}\right)$$

hence $Y_{5,3}, Y_{5,2}, Y_{4,2}, Y_{3,2}$ are of the required form, (cf. (5.10.5) and (5.17.3)).

We now apply (5.12.4) (in the form (5.12.6)) to the elements $Y_{2,1}$, $Y_{3,1}, Y_{3,2}, Y_{4,1}, Y_{4,2}, Y_{5,1}$ which are already known to have the required form. We deduce that the elements

$$Y_{2,2}, Y_{3,3}, Y_{3,4}, Y_{4,3}, Y_{4,4}, Y_{5,4}$$

have the required form. (To apply (5.12.6) we note that the longest element w_0 in W acts as -1 on 16_1, 4_3, 4_4 and as $+1$ on the other representations in the family $\mathcal{F}$ containing 12_1; in other words, if a representation E in this family $\mathcal{F}$ corresponds to $(x, \sigma) \in \mathfrak{S}_4$, then w_0 acts on E as -1 if and only if x is an odd permutation in $\mathfrak{S}_4$.)

7.5. PROPOSITION. *Assume that the Weyl group* (W, S) *of* G, *(see 1.2) is of type* E_8 *and that* $F' : G \to G$ *is a power* F^s *of* F. *Then each of the following elements* $X_{i,j} \in \mathcal{R}(W)$ *in the table, as well as* $X_{3,2} + X_{3,3}$ *is of the form* a_w *for some* $w \in W$ *such that* $\ell(w) \equiv a(w) \pmod 2$:

	4480	3150	4200	4536	5670	420	1134	1400	2688	1680	168	70	7168	1344	2016	5600	448
$X_{1,1}$	1		1	2	2			1	1	1			2			2	
$X_{1,2}$	1		–1	2	2			1	–1	1							
$X_{1,3}$	1		1					1	–1	–1							
$X_{2,1}$	1	1	1	1	1				1				2		1	1	
$X_{2,2}$	1	1	–1	1	1				–1						1	–1	
$X_{2,3}$	2	–1		2	2								2		–1		
$X_{2,4}$	1	1	–1	1	1				–1						–1	1	
$X_{3,1}$	1			3	3			2		3		1	1			2	1
$X_{3,2}$	1			1	1					–1		–1	1				–1
$X_{3,3}$	1							–1				1	1			–1	1
$X_{4,1}$	1	1		1	2		1			1			1			2	1
$X_{4,2}$	1	1		1			–1			–1			1				–1
$X_{5,1}$	1		2	1	1			1			1		1	1		1	
$X_{5,2}$	1			1	1			1			–1		1	–1		1	
$X_{6,1}$	1	1	1		1		1						1	1		1	
$X_{7,1}$	1	1	1			1							1	1	1		
$X_{6,2}$	1	1	1		1		1						–1	–1		–1	
$X_{7,2}$	1	1	1			1							–1	–1	–1		
$X_{3,4}$	1							–1				1	–1			1	–1
$X_{2,5}$	1	1	1	1	1				1				–2		–1	–1	
$X_{5,3}$	1			1	1			1			–1		–1	1		–1	
$X_{3,5}$	1			1	1					–1		–1	–1				1
$X_{4,3}$	1	1		1			–1			–1			–1				1
$X_{5,4}$	1		2	1	1			1			1		–1	–1		–1	
$X_{4,4}$	1	1		1	2		1			1			–1			–2	–1
$X_{1,4}$	1		1	2	2			1	1	1			–2			–2	
$X_{7,3}$	4	1				–1							2		–1		
$X_{7,4}$	6		–2			1											

The elements $X_{i,j}$ (one in each row) are linear combination of irreducible representations of W in the family $\mathcal{F}$ containing 4480_y. The representations in $\mathcal{F}$ parametrize the columns in the matrix above, (we omit the index y or w when naming a representation in $\mathcal{F}$). The matrix describes the coefficient with which a representation in $\mathcal{F}$ appears in $X_{i,j}$. When the coefficient is zero, it is omitted.

Proof. Our statement for $X_{i,1}$ $(1 \le i \le 7)$, is contained in 5.23. The statement for $X_{7,1}$, $X_{7,2}$, $X_{7,3}$, $X_{7,4}$ follows by applying 7.1 (ii) to a left cell $\Gamma \subset W$ such that $[\Gamma]_1 = X_{7,1}$; such Γ exists since $X_{7,1} = J^W_{A_4 \times A_3}(\text{sign})$ is constructible (see 5.29). (In order to apply 7.1 (ii) we make use of the fact that 4.23 holds for our G, (L,n) = (trivial bundle, 1) and large s, by [L$_8$].) Then $X_{7,1}$, $X_{7,2}$, $X_{7,3}$, $X_{7,4}$ are of the form a_{w_ϕ}, (see 7.1 (ii)), for the irreducible characters $\phi = 1, \lambda^4, \lambda^1, \lambda^2$ of $\mathfrak{G}_5$, (see 4.3). With the notations in 4.13, we have

$$X_{2,2} = J^W_{E_7}(315_a - 280'_b + 70_a)$$

$$X_{2,3} = J^W_{E_7}(2.315_a - 70_a)$$

$$X_{1,2} = J^W_{E_7}(315_a - 280'_b + 280_a)$$

$$X_{3,2} = J^W_{E_7}(315_a - 35_a)$$

$$X_{3,3} = J^W_{E_7}(315_a - 280_a + 35_a)$$

$$X_{3,2} + X_{3,3} = J^W_{E_7}(2.315_a - 280_a)$$

$$X_{2,4} = J^W_{E_6 \times A_1}(30'_p \boxtimes \text{sign} - 15'_q \boxtimes \text{sign})$$

$$X_{4,2} = J^W_{E_6 \times A_1}(30'_p \boxtimes \text{sign} - 15'_p \boxtimes \text{sign})$$

$$X_{5,2} = J^W_{D_7}\left(\begin{bmatrix} 1 & 2 & 3 & 5 \\ 0 & 1 & 3 & 4 \end{bmatrix} - \begin{bmatrix} 1 & 3 & 4 & 5 \\ 0 & 1 & 2 & 3 \end{bmatrix}\right)$$

$$X_{1,3} = J^W_{D_7}\left(\begin{bmatrix} 1 & 2 & 3 & 5 \\ 0 & 1 & 3 & 4 \end{bmatrix} - \begin{bmatrix} 1 & 2 & 3 & 4 \\ 0 & 1 & 3 & 5 \end{bmatrix}\right).$$

Using (5.10.5), 7.3 (for type E_7), (5.21.1) and (5.18.1) we see that the elements $X_{2,2}, X_{2,3}, X_{1,2}, X_{3,2}, X_{3,3}, X_{3,2}+X_{3,3}, X_{2,4}, X_{4,2}, X_{5,2}, X_{1,3}$ have the required form. We now apply (5.12.4) (in the form (5.12.6)) to the elements

$$X_{6,1}, X_{3,3}, X_{2,1}, X_{5,2}, X_{3,2}, X_{4,2}, X_{5,1}, X_{4,1}, X_{1,1}, X_{3,2}+X_{3,3}$$

which are already known to have the required form. We deduce that the elements

$$X_{6,2}, X_{3,4}, X_{2,5}, X_{5,3}, X_{3,5}, X_{4,3}, X_{5,4}, X_{4,4}, X_{1,4}, X_{3,4}+X_{3,5}$$

have the required form. (To apply (5.12.6) we note that the longest element $w_0 \in W$ acts as -1 on $7168_w, 1344_w, 2016_w, 5600_w, 448_w$ and as $+1$ on the other representations in the family $\mathcal{F}$ containing 4480_y; in other words, if a representation E in this family $\mathcal{F}$ corresponds to $(x, \sigma) \in \mathfrak{G}_5$ then w_0 acts on E as -1 if and only if x is an odd permutation in $\mathfrak{G}_5$.) This completes the proof.

7.6. PROPOSITION. *Assume that all the irreducible components of the Weyl group* $(W_{L,n}, S_{L,n})$ *are of type* A, *except possibly for one component which is of type* D_m $(4 \le m \le 8)$ *or* B_m $(2 \le m \le 5)$ *or* C_m $(2 \le m \le 5)$. *Then Theorem 4.23 holds for* (L, n). *Moreover, we have* $\varepsilon_\rho = \hat{\varepsilon}_\rho$ *for all* $\rho \in \mathcal{E}_{L,n}$.

Proof. Let us write $W_{L,n} = W^{(0)} \times W^{(1)} \times W^{(2)} \times \cdots \times W^{(r)}$, $S_{L,n} = S^{(0)} \amalg S^{(1)} \amalg \cdots \amalg S^{(r)}$ where each $(W^{(i)}, S^{(i)})$, $(1 \le i \le r)$, is a Weyl group of type $A_{m_i} \times \cdots \times A_{m_i}$ (with irreducible components permuted transitively by γ) and $(W^{(0)}, S^{(0)})$ is either $\{e\}$ or is a Weyl group of type D_m, B_m, or C_m, with m bounded as above. Then $(W^{(0)}, S^{(0)})$ is automatically γ-stable. Let $\mathcal{F} = \mathcal{F}^{(0)} \times \mathcal{F}^{(1)} \times \cdots \times \mathcal{F}^{(n)}$ be a γ-stable family in $W_{L,n}^\vee = \prod_{i=0}^{r} (W^{(i)})^\vee$, where $\mathcal{F}^{(i)}$ is a γ-stable family in $(W^{(i)})^\vee$ for each i. Our bounds on m imply that $\mathcal{F}^{(0)}$ consists either of 1 or of 3 irreducible representations of $W^{(0)}$. (Families with >3 representations occur for the first time for D_9, B_6, C_6, see 4.5, 4.6.) On the other hand $\mathcal{F}^{(i)}$ $(1 \leq i \leq r)$ consists of a single representation. It follows that $\mathcal{F} = \{E\}$ or $F = \{E, M, N\}$, where E is the special representation in $\mathcal{F}$. Note that all representations in $\mathcal{F}$ are in $(W_{L,n})^{\vee}_{ex}$. This follows from the fact that in an irreducible Weyl group (in our case, $(W^{(0)}, S^{(0)})$) with an ordinary automorphism, any family stable by the automorphism is pointwise fixed by it. Write $E = E^{(0)} \otimes (\bigotimes_{i=1}^{r} E^{(i)})$; if defined, write also $M = M^{(0)} \otimes (\bigotimes_{i=1}^{r} E^{(i)})$, $N = N^{(0)} \otimes (\bigotimes_{i=1}^{r} E^{(i)})$ where $E^{(i)}$, $(0 \leq i \leq r)$ are irreducible special representations in $\mathcal{F}^{(i)}$, and $M^{(0)}$, $N^{(0)}$ (if defined) are irreducible nonspecial representations in $\mathcal{F}^{(0)}$.

We let γ act on $E^{(i)}$, $(1 \leq i \leq r)$, as follows. Write $W^{(i)} = W_1^{(i)} \times \cdots \times W_1^{(i)}$, ($s_i$ factors), where $(W_1^{(i)}, S_1^{(i)})$ is an irreducible Weyl group, and $\gamma : W^{(i)} \to W^{(i)}$ is given by $\gamma(w_1, w_2, \cdots, w_{s_i}) = (\delta_i(w_{s_i}), w_1, \cdots, w_{s_i-1})$, $(w_j \in W_1^{(i)})$ where δ_i is an automorphism of $(W_1^{(i)}, S_1^{(i)})$. Write $E^{(i)} = E_1^{(i)} \otimes \cdots \otimes E_1^{(i)}$ (external tensor product) where $E_1^{(i)}$ is an irreducible $\mathbf{Q}[W_1^{(i)}]$-module. Now the action of γ on $E^{(i)}$ is defined by

$$(7.6.1) \quad \gamma(e_1 \otimes e_2 \otimes \cdots \otimes e_{s_i}) = \delta_i(e_{s_i}) \otimes e_1 \otimes \cdots \otimes e_{s_i-1}, \quad (e_j \in E_1^{(i)}),$$

where $\delta_i(e_{s_i}) = e_{s_i}$ if $\delta_i = 1$ on $W_1^{(i)}$ and δ_i acts on $E_1^{(i)}$ as the longest element in $W_1^{(i)}$ if $\delta_i \neq 1$ on $W_1^{(i)}$. (We say that this is the *standard* extension of $E^{(i)}$ to a $\mathbf{Q}[\widetilde{W}^{(i)}]$-module, where $\widetilde{W}^{i}$ is defined in terms of γ, and we denote it $\hat{E}^{(i)}$; the other extension is said to be nonstandard and is denoted $\hat{E}^{(i)\prime}$.)

Let Γ' be a left cell in $W_1^{(i)}$ such that $[\Gamma']_1 = E_1^{(i)}$. Then $\Gamma = \Gamma' \times \cdots \times \Gamma'$ (s_i copies) is a left cell in $W^{(i)}$ and $[\Gamma]_1 = E^{(i)}$. Since $f_{E^{(i)}} = 1$ (cf. (4.1.6) and (4.4.1)) we see from (5.8.1) that

$\sum_{x\in\Gamma} c^2_{\gamma x,\hat{E}^{(i)}} = 1$ hence there is a unique element $x_i \in \Gamma$ such that $c_{\gamma x_i,\hat{E}^{(i)}} = \pm 1$; for all other elements $x' \in \Gamma$, we have $c_{\gamma x',\hat{E}^{(i)}} = 0$. It follows that

$$(7.6.2) \qquad a_{\gamma x_i} = (1/2)\,\varepsilon_i(\hat{E}^{(i)} - \hat{E}^{(i)'}) \in \mathcal{R}(W^{(i)},\gamma), \text{ where } \varepsilon_i = \pm 1 .$$

We have

$$(7.6.3) \qquad \tilde{\ell}(x_i) - a(x_i) \equiv \begin{cases} 0 & (\text{mod } 2), \quad \text{if } \delta_i = 1 \\ a_{E_1^{(i)}} + A_{E_1^{(i)}} & (\text{mod } 2), \quad \text{if } \delta_i \neq 1 \end{cases} .$$

This can be proved as follows. Since $c_{\gamma x_i,\hat{E}^{(i)}} = \pm 1$, $u^{-(1/2)(\ell(x_i) - a_{E_i})}$ appears with nonzero coefficient in $\mathrm{Tr}(T_\gamma T_{x_i}, \hat{E}^{(i)}(u)) =$ $\mathrm{Tr}(T_{\delta_i} T_{z_1} T_{z_2} \cdots T_{z_{s_i}}, E_1^{(i)}(u))$. (Here $z_1, z_2, \cdots, z_{s_i}$ are the components of x_i in $W_1^{(i)}$, $T_{z_1}, \cdots, T_{z_{s_i}}$ are the corresponding elements of the Hecke algebra of $W_1^{(i)}$, T_{x_i} is the element of the Hecke algebra of $W^{(i)}$ corresponding to x_i.) When $\delta_i = 1$, we use the fact that $\mathrm{Tr}(T_{z_1} T_{z_2} \cdots T_{z_{s_i}}, E^{(i)}(u))$ must be a polynomial in u and (7.6.3) follows. When $\delta_i \neq 1$, then, by (5.12.2), δ_i acts on $E_1^{(i)}(u)$ as $u^{-\nu^{(i)} + (1/2)(a_{E^{(i)}} + A_{E^{(i)}})} T_{y_0}$ (y_0 = longest element in $W_1^{(i)}$, $\nu^{(i)}$ = its length), and we get again (7.6.3) by the same argument as in the case $\delta_i = 1$.

The sign ε_i in (7.6.2) is given by

$$(7.6.4) \qquad \varepsilon_i = \begin{cases} 1 & , \quad \text{if } \delta_i = 1 \\ (-1)^{A_{E_1^{(i)}}} & , \quad \text{if } \delta_i \neq 1 \end{cases} .$$

This is a statement intrinsic to $(W^{(i)},\gamma)$. Hence to prove it, we may assume that $W^{(i)}$ is the Weyl group of G, γ is the action of F' on it, G is adjoint and $(L,n) = (L_0,1)$, where L_0 is the trivial bundle; moreover, we may assume that q (see 6.10) is sufficiently large so that the conclusion of 6.17 (ii) holds in the present case. It follows that $R_{\alpha_{\gamma x_i}} = \varepsilon_i R_{\hat{E}^{(i)}}$ is an actual representation of $G^{F'}$, hence $\varepsilon_i \dim(R_{\hat{E}^{(i)}}) > 0$.

Using [L_4, (3.19.1)], we see that $\dim(R_{\hat{E}^{(i)}})$ is a polynomial with integral coefficient in q^{s_i} and the highest power of q is $q^{s_i A_{E_1^{(i)}}}$, which appears with coefficient 1 if $\delta_i = 1$ and $(-1)^{A_{E^{(i)}}}$ if $\delta_i \neq 1$. For sufficiently large q, the sign of $\dim(R_{\hat{E}^{(i)}})$ is determined by the sign of the coefficient of this highest power of q, and (7.6.4) follows.

We now turn to $W^{(0)}$. Assume that it is non-trivial. We define an extension $\hat{E}^{(0)}$ of $E^{(0)}$ to a $\mathbf{Q}[\tilde{W}^{(0)}]$-module as follows: If γ acts trivially on $W^{(0)}$, we let γ act trivially on $E^{(0)}$. If γ acts as an automorphism of order 2 of $W^{(0)}$ we let γ act in such a way that, with the notations in 4.18, it has square 1 on $E^{(0)}$ and the resulting representation of $\tilde{W}_2^{(0)}$ is $[[\Lambda]]$ where Λ is of the form (4.18.1) with $\lambda_1 \geq \mu_1 \geq \lambda_2 \geq \mu_2 \geq \cdots \geq \lambda_m \geq \mu_m$. If γ acts as an automorphism of order 3 of $W^{(0)}$ we let $\hat{E}^{(0)}$ be the unique extension of $E^{(0)}$ to a $\mathbf{Q}[\tilde{W}^{(0)}]$-module on which $\gamma^3 = 1$.

Let $\hat{E}^{(0)'}$ be the extension $\neq \hat{E}^{(0)}$ of $E^{(0)}$ to a $\mathbf{Q}[\tilde{W}^{(0)}]$-module. If $M^{(0)}$, $N^{(0)}$ are defined, we choose extensions $\hat{M}^{(0)}$, $\hat{N}^{(0)}$ of M,N to $\mathbf{Q}[\tilde{W}^{(0)}]$-modules and let $\hat{M}^{(0)'}$, $\hat{N}^{(0)'}$ be the extensions other than $\hat{M}^{(0)}$, $\hat{N}^{(0)}$. Then the following holds.

(7.6.5) If $\mathcal{F}^{(0)} = \{E^{(0)}\}$, then $(1/2)(\hat{E}^{(0)} - \hat{E}^{(0)'}) \in \mathcal{R}(W^{(0)},\gamma)$ is of the form $\alpha_{\gamma x_0}$ for some $x_0 \in W^{(0)}$ such that $\tilde{\ell}(x_0) \equiv a(x_0) \pmod 2$. If $\mathcal{F}^{(0)} = \{E^{(0)}, M^{(0)}, N^{(0)}\}$, then each of the four elements of $\mathcal{R}(W^{(0)},\gamma)$: $(1/2)(\hat{E}^{(0)} - \hat{E}^{(0)'}) \pm (1/2)(\hat{M}^{(0)} - \hat{M}^{(0)'})$, $(1/2)(\hat{E}^{(0)} - \hat{E}^{(0)'}) \pm (1/2)(\hat{N}^{(0)} - \hat{N}^{(0)'})$ is of the form $\alpha_{\gamma x_0}$ for some $x_0 \in W^{(0)}$ such that $\tilde{\ell}(x_0) \equiv a(x_0) \pmod 2$.

When $\gamma^2 = 1$ on $W^{(0)}$, this follows from results in 5.17, 5.18. When $\gamma : W^{(0)} \to W^{(0)}$ has order 3 (so that $W^{(0)}$ is of type D_4), this follows easily from (5.10.5) and (5.12.4), except that when $E^{(0)} = \begin{bmatrix} 1 & 3 \\ 0 & 2 \end{bmatrix}$, $M^{(0)} = \begin{bmatrix} 2 & 3 \\ 0 & 1 \end{bmatrix}$, $N^{(0)} = \begin{bmatrix} 1 & 2 \\ 0 & 3 \end{bmatrix}$, only $(1/2)(\hat{E}^{(0)} - \hat{E}^{(0)'}) \pm (1/2)(\hat{M}^{(0)} - \hat{M}^{(0)'})$ can be handled in this way. To handle $(1/2)(\hat{E}^{(0)} - \hat{E}^{(0)'}) \pm (1/2)(\hat{N}^{(0)} - \hat{N}^{(0)'})$ we apply (5.8.1) to a left cell Γ in W^0 such that $[\Gamma]_1 = E^{(0)} + N^{(0)}$. Since $f_{E^0} = f_{N^0} = 2$, (5.8.1) reads $\sum_{x\epsilon\Gamma} c^2_{\gamma x, \hat{E}^0} = 2$, $\sum_{x\epsilon\Gamma} c^2_{\gamma x, \hat{N}^0} = 2$, $\sum_{x\epsilon\Gamma} c_{\gamma x, \hat{E}^{(0)}} c_{\gamma x, \hat{N}^{(0)}} = 0$. It follows that there exist two elements $x, x' \epsilon \Gamma$ such that $c_{\gamma x, \hat{E}^{(0)}} = c_{\gamma x, \hat{N}^{(0)}} = \varepsilon$, $c_{\gamma x', \hat{E}^{(0)}} = -c_{\gamma x', \hat{N}^{(0)}} = \varepsilon'$ where $\varepsilon, \varepsilon' \epsilon \{\pm 1\}$.

We want to prove that $\tilde{\ell}(x) \equiv a(x) \pmod 2$, $\tilde{\ell}(x') \equiv a(x') \pmod 2$, $\varepsilon = \varepsilon' = 1$. For this, we may assume that $W^{(0)}$ is the Weyl group of G (of type D_4), $(L,n) = (L_0, 1)$, γ is the action of F' on it, and that q is large enough. The congruences $\tilde{\ell}(x) \equiv a(x) \pmod 2$ $\tilde{\ell}(x') \equiv a(x') \pmod 2$ follow then from (6.18.3), (6.18.5). We have $R_{a_{\gamma x}} = \varepsilon R_{\hat{E}^{(0)}} + \varepsilon R_{\hat{N}^{(0)}}$, $R_{a_{\gamma x'}} = \varepsilon' R_{\hat{E}^{(0)}} - \varepsilon' R_{\hat{N}^{(0)}}$. By 6.17 (ii), $\dim(R_{a_{\gamma x}}) > 0$, $\dim(R_{a_{\gamma x'}}) > 0$. On the other hand, $\dim(\varepsilon R_{\hat{E}^{(0)}} + \varepsilon R_{\hat{N}^{(0)}}) = \varepsilon q^9 +$ lower powers of q, (using $[L_4, (3.19.1)]$). Since q can be taken large, it follows that $\varepsilon = 1$. Similarly, we have $\varepsilon' = 1$, and (7.6.5) is verified.

The extensions $\hat{E}^{(0)}$ of $E^{(0)}$ and $\hat{E}^{(i)}$ of $E^{(i)}$ give rise, by tensor product, to an extension $\hat{E}$ of E to a $\mathbf{Q}[\tilde{W}_{L,n}]$-module. Let $\hat{E}'$ be the other extension of E to a $\mathbf{Q}[\tilde{W}_{L,n}]$-module. If M,N are defined, let $\hat{M}, \hat{M}'$ be the two extensions of M to a $\mathbf{Q}[\tilde{W}_{L,n}]$-module and let $\hat{N}, \hat{N}'$ be the two extensions of N to a $\mathbf{Q}[\tilde{W}_{L,n}]$-module.

Let ε_E be 1 if $\sum_{\substack{1 \le i \le r \\ \delta_i \ne 1}} A_{E^{(i)}}$ is even, and -1, otherwise.

Let ε'_E be 1 if $\sum_{\substack{1\le i\le r\\ \delta_i\ne 1}} a_{E(i)}$ is even, and -1, otherwise.

The earlier part of the proof gives the following.

(7.6.6) If $\mathcal{F} = \{E\}$ then $\varepsilon_E(1/2)(\hat{\tilde{E}} - \hat{E}') \in \mathcal{R}(W_{L,n}, \gamma)$ is of the form $a_{\gamma x}$ for some $x \in W_{L,n}$ such that $(-1)^{\ell(x)-a(x)} = \varepsilon_E \varepsilon'_E$.

(7.6.7) If $\mathcal{F} = \{E,M,N\}$, then each of the four elements $\varepsilon_E((1/2)(\hat{E}-\hat{E}') \pm (1/2)(\hat{M}-\hat{M}'))$, $\varepsilon_E((1/2)(\hat{E}-\hat{E}') \pm (1/2)(\hat{N}-\hat{N}'))$ of $\mathcal{R}(W_{L,n}, \gamma)$ is of the form $a_{\gamma x}$ for some $x \in W_{L,n}$ such that $(-1)^{\ell(x)-a(x)} = \varepsilon_E \varepsilon'_E$.

Now let ρ be an irreducible representation of $G^{F'}$ in $\mathcal{E}_{L,n}$. Then there exists a family $\mathcal{F}_1 \subset (W_{L,n})^\vee$, stable under γ, and a representation $E_1 \in \mathcal{F}_1$ such that $\langle \rho, R_{\tilde{E}_1} \rangle_{G^{F'}} \ne 0$ ($\tilde{E}_1$ is an extension of E_1 to a $\mathbf{Q}[\tilde{W}_{L,n}]$-module). We want to show that $\mathcal{F}_1$ is uniquely determined by ρ. Assume this is not so, and let $\mathcal{F}_2$ be another γ-stable family such that $\langle \rho, R_{\tilde{E}_2} \rangle_{G^{F'}} \ne 0$ for some $E_2 \in \mathcal{F}_2$. Assume first that $\mathcal{F}_1 = \{E_1\}$. By (7.6.6), $\varepsilon_{E_1} R_{\hat{E}_1} = R_{a_{\gamma x}}$ is a virtual representation of $G^{F'}$ (see 6.14 (i)). Its inner product with itself is 1 (see 3.9 (i)) hence it is $\pm$ an irreducible representation. It follows that it is $\pm\rho$. Since $E_2 \ne E_1$, we cannot then have $\langle \rho, R_{\tilde{E}_2} \rangle_{G^{F'}} \ne 0$ (see 3.9 (i)). Hence $\mathcal{F}_1$ must consist of three representations. Similarly, $\mathcal{F}_2$ must consist of three representations.

Assume that $\mathcal{F}_1 = \{E,M,N\}$, with E special. From (7.6.7), using 6.14 (i), it follows that $R_{\hat{E}}+R_{\hat{M}}$, $R_{\hat{E}}-R_{\hat{M}}$, $R_{\hat{E}}+R_{\hat{N}}$, $R_{\hat{E}}-R_{\hat{N}}$ are virtual representations of $G^{F'}$. Hence $2R_{\hat{E}}$, $2R_{\hat{M}}$, $2R_{\hat{N}}$ are virtual representations of $G^{F'}$. Hence the inner products $\langle \rho, R_{\hat{E}} \rangle_{G^{F'}}$, $\langle \rho, R_{\hat{M}} \rangle_{G^{F'}}$, $\langle \rho, R_{\hat{N}} \rangle_{G^{F'}}$ take values in $\{0, \pm 1/2, \pm 1\}$. (They cannot be >1 since $\langle \rho, \rho \rangle_{G^{F'}} = 1$.) Moreover, these three inner products differ one from another by an integer. Hence, if one is $\pm 1/2$ then all three are of the form $\pm 1/2$. Since they are not all zero, it follows that $<\rho,R_{\hat{E}}>^2_{G^{F'}} + <\rho,R_{\hat{M}}>^2_{G^{F'}} + <\rho,R_{\hat{N}}>^2_{G^{F'}} \geq 3/4$. The similar sum corresponding to the family $\mathcal{F}_2$ is also $\geq 3/4$.

It follows that $\sum_{E'} <\rho,R_{\hat{E}'}>^2_{G^{F'}} \geq (3/4)+(3/4) > 1$, (E' runs over $(W_{L,n})^\vee_{ex}$). Since the $R_{\hat{E}'}$ form an orthonormal set (3.9 (i)), this contradicts the equality $<\rho,\rho>_{G^{F'}} = 1$. Thus, we have proved that $\mathcal{F}_1$ is indeed uniquely determined by ρ. This implies that, whenever $x,x' \in W_{L,n}$ are in distinct two-sided cells, the virtual representations $R_{\mathfrak{A}_{\gamma x}}$, $R_{\mathfrak{A}_{\gamma x'}}$ are necessarily disjoint (i.e. no irreducible representation of $G^{F'}$ appears in both). From this, and from (6.15.2), we deduce that $(-1)^{\ell(w_1 x)-a(x)} R_{\mathfrak{A}_{\gamma x}}$ is an actual representation of $G^{F'}$, for any $x \in W_{L,n}$. (Here w_1 is the element of minimal length in $Z_{L,n}$.)

Now let $E \in (W_{L,n})^\vee_{ex}$ be such that $\{E\}$ is a family. By (7.6.6), we have $\varepsilon_E R_{\hat{E}} = R_{a_{\gamma x}}$ for some x such that $(-1)^{\tilde{\ell}(x)-a(x)} = \varepsilon_E \varepsilon'_E$. Applying to this the operator D and using (6.8.6) and 6.14 (ii), we get $(-1)^{\ell(w_1)} \varepsilon_E R_{\hat{E}\otimes\widetilde{\text{sign}}} = (-1)^{\ell(w_1)} R_{\mathfrak{A}_{\gamma x}}$. Here $\widetilde{\text{sign}}$ is the extension of sign to a $\mathbf{Q}[\widetilde{W}_{L,n}]$-module, on which γ acts trivially. It follows that $(-1)^{\ell(w_1)} \varepsilon'_E R_{\hat{E}\otimes\widetilde{\text{sign}}}$ is an actual representation. Its inner product with itself is one, hence it is an irreducible representation. It follows easily from definition that $R_{\hat{E}\otimes\widetilde{\text{sign}}} = \varepsilon_{\text{sign}} R_{\widehat{E\otimes\text{sign}}}$ and $\varepsilon'_E \varepsilon_{\text{sign}} = \varepsilon_{E\otimes\text{sign}}$. Thus, $(-1)^{\ell(w_1)} \varepsilon_{E\otimes\text{sign}} R_{\widehat{E\otimes\text{sign}}}$ is an irreducible representation. Since $\{E\}$ is a family if and only if $\{E\otimes\text{sign}\}$ is a family, we see also that $(-1)^{\tilde{\ell}(w_1)} \varepsilon_E R_{\hat{E}}$ is an irreducible representation.

Assume now that $\mathcal{F}$ is a γ-stable family $\{E,M,N\}$, with E special. Using (6.8.7) and 6.14 (ii) we see, just as in the previous argument, that

$$(-1)^{\ell(w_1)} \varepsilon_{E\otimes\mathrm{sign}}(R_{\widehat{E\otimes\mathrm{sign}}} \pm R_{\widehat{M\otimes\mathrm{sign}}})$$

and

$$(-1)^{\ell(w_1)} \varepsilon_{E\otimes\mathrm{sign}}(R_{\widehat{E\otimes\mathrm{sign}}} \pm R_{\widehat{N\otimes\mathrm{sign}}})$$

are actual representations of $G^{F'}$.

Replacing E,M,N by $E\otimes\mathrm{sign}$, $M\otimes\mathrm{sign}$, $N\otimes\mathrm{sign}$, respectively (which form again a family), we see that $(-1)^{\ell(w_1)}\varepsilon_E(R_{\hat{E}} \pm R_{\hat{M}})$, $(-1)^{\ell(w_1)}\varepsilon_E(R_{\hat{E}} \pm R_{\hat{N}})$ are actual representations of $G^{F'}$. Their mutual inner products can be computed from 3.9 (i). By an elementary lemma (see 7.7 (i)), one can find four distinct irreducible representations $\rho_1,\rho_2,\rho_3,\rho_4 \in \mathcal{E}_{L,n}$ such that

$$(-1)^{\ell(w_1)}\varepsilon_E(R_{\hat{E}}+R_{\hat{M}}) = \rho_1+\rho_2$$

$$(-1)^{\ell(w_1)}\varepsilon_E(R_{\hat{E}}-R_{\hat{M}}) = \rho_3+\rho_4$$

$$(-1)^{\ell(w_1)}\varepsilon_E(R_{\hat{E}}+R_{\hat{N}}) = \rho_1+\rho_3$$

$$(-1)^{\ell(w_1)}\varepsilon_E(R_{\hat{E}}-R_{\hat{N}}) = \rho_2+\rho_4 .$$

Thus, we have

$$(-1)^{\tilde{\ell}(w_1)}\varepsilon_E R_{\hat{E}} = (1/2)(\rho_1+\rho_2+\rho_3+\rho_4)$$

$$(-1)^{\tilde{\ell}(w_1)}\varepsilon_E R_{\hat{M}} = (1/2)(\rho_1+\rho_2-\rho_3-\rho_4)$$

$$(-1)^{\tilde{\ell}(w_1)}\varepsilon_E R_{\hat{N}} = (1/2)(\rho_1-\rho_2+\rho_3-\rho_4)$$

so that the pattern of decomposition of any $R_{\hat{E}}$ is determined. Hence Theorem 4.23 holds for (L,n). The last sentence in the proposition follows now from 6.20, by induction on $|W_{L,n}|$. This completes the proof.

7.7. LEMMA. *Let* V *be an euclidean space with scalar product* $(\ ,\ )$ *and with a given orthonormal basis* X, *and let* V_Z^+ *be the set of vectors in* V *which are linear combinations with integral,* ≥ 0 *coefficients of elements in* X.

(i) *Let* $\{\alpha, \beta, \gamma\} \subset V$ *be an orthonormal set such that*

$$\alpha+\beta,\ \alpha-\beta,\ \alpha+\gamma,\ \alpha-\gamma \text{ are in } V_Z^+ .$$

Then there exist four distinct elements $a,b,c,d \in X$ *such that*

$$\alpha+\beta = a+b,\ \alpha-\beta = c+d,\ \alpha+\gamma = a+c,\ \alpha-\gamma = b+d .$$

Hence we have $\alpha = (1/2)(a+b+c+d)$, $\beta = (1/2)(a+b-c-d)$, $\gamma = (1/2)(a-b+c-d)$.

(ii) *Let* $\{\alpha, \beta, \gamma, \delta\} \subset V$ *be an orthonormal set such that* $\alpha+\beta+\delta$, $\alpha-\beta+\delta$, $2\alpha-\delta$, $\alpha+\beta+\gamma$, $\alpha-\beta+\gamma$, $2\alpha-\gamma$ *are in* V_Z^+. *Then there exist eight distinct elements* $a,b,c,d,e,f,g,h \in X$ *such that*

$$\alpha+\beta+\delta = a+b+d,\ \alpha-\beta+\delta = d+e+f,\ 2\alpha-\delta = b+c+f+g+h ,$$

$$\alpha+\beta+\gamma = a+b+c,\ \alpha-\beta+\gamma = c+e+f,\ 2\alpha-\gamma = b+d+f+g+h .$$

Hence we have

$$\begin{aligned}
\alpha &= (1/6)(a+3b+2c+2d+e+3f+2g+2h)\\
\beta &= (1/2)(a+b-e-f)\\
\gamma &= (1/3)(a+2c-d+e-g-h)\\
\delta &= (1/3)(a-c+2d+e-g-h)
\end{aligned}$$

(iii) *Suppose that, in addition to the vectors* $\alpha, \beta, \gamma, \delta$ *in (ii), we are given a fifth vector* ε *of length* 1, *orthogonal to* α, β, γ *and* δ *and such that* $\alpha-\varepsilon$ *is in* V_Z^+. *Then, (with the notations in (ii)), we have*

$$\alpha-\varepsilon = b+f$$

hence

$$\varepsilon = (1/6)(a-3b+2c+2d+e-3f+2g+2h) .$$

Moreover, if v *is one of the vectors* a,b,c,d,e,f, *we have* $(v,\alpha)^2+(v,\beta)^2+(v,\gamma)^2+(v,\delta)^2+(v,\varepsilon)^2 > 1/2$.

The proof is easy; it is left to the reader.

7.8. LEMMA. *Let* $V, X, (\ ,\ )$ *be as in 7.7 and let* $V_{\mathbf{Z}}$ *be the lattice in* V *generated by* X. *Let* $\alpha, \beta, \gamma, \delta, \varepsilon$ *be five vectors in* V *such that*

(7.8.1) $\alpha+\beta+\delta,\ \alpha-\beta+\delta,\ 2\alpha-\delta,\ \alpha+\beta+\gamma,\ \alpha-\beta+\delta,\ 2\alpha-\delta,\ \alpha-\varepsilon$ *are in* $V_{\mathbf{Z}}$.

Then, if $v \in V_{\mathbf{Z}}$, *the sum* $(v,\alpha)^2+(v,\beta)^2+(v,\gamma)^2+(v,\delta)^2+(v,\varepsilon)^2$ *is either* 0 *or is* $>1/4$.

Proof. We may assume that $V = \mathbf{R}$, $X = \{1\}$, $V_{\mathbf{Z}} = \mathbf{Z}$, by replacing $\alpha, \beta, \gamma, \delta, \varepsilon$ in V with (v,α), (v,β), (v,γ), (v,δ), (v,ε) in $\mathbf{R}$. We assume that $(\alpha,\beta,\delta,\gamma,\varepsilon) \neq (0,0,0,0,0)$ and we want to prove that

$$\alpha^2+\beta^2+\gamma^2+\delta^2+\varepsilon^2 > 1/4 . \tag{7.8.2}$$

From (7.8.1) it follows that $6\alpha \in \mathbf{Z}$, $2\beta \in \mathbf{Z}$, $3\gamma \in \mathbf{Z}$, $3\delta \in \mathbf{Z}$, $6\varepsilon \in \mathbf{Z}$. If $|\beta| > 1/2$ then (7.8.2) is obvious. If $\beta = \pm 1/2$ then from $\alpha+\beta+\gamma \in \mathbf{Z}$ it follows that α and δ are not both zero, hence $\alpha^2+\beta^2+\delta^2 > 1/4$ and (7.8.2) holds. Hence we may assume that $\beta = 0$. If $|\gamma| \geq 2/3$ then (7.8.2) is obvious. If $\gamma = \pm 1/3$ then, from $\alpha+\gamma \in \mathbf{Z}$, $\alpha-\varepsilon \in \mathbf{Z}$ it follows that $\alpha = \mp 1/3 + \text{integer}$, $\varepsilon = \mp 1/3 + \text{integer}$, hence $\alpha^2+\gamma^2+\varepsilon^2 \geq (1/9)+(1/9)+(1/9) > 1/4$ and (7.8.2) holds. Hence we may assume $\beta = \gamma = 0$. Similarly, we may assume that $\delta = 0$. But then α and ε must be integers, not both equal 0, so $\alpha^2+\varepsilon^2 \geq 1$ and (7.8.2) holds again. The lemma is proved.

7.9. PROPOSITION. *Assume that* $(W_{L,n}, S_{L,n})$ *is of type* G_2 *and that* $w_1 = e$. *Then Theorem 4.23 holds for* (L,n).

Proof. The automorphism γ of $(W_{L,n}, S_{L,n})$, (see (2.15.1)) is necessarily trivial, since it is ordinary (3.1). From 5.19 and 6.14 (i), we see that

$$(7.9.1)\quad R_1,\ R_V + R_{V'} + R_{\varepsilon_2},\ R_V - R_{V'} + R_{\varepsilon_2},\ 2R_V - R_{\varepsilon_2},$$

$$R_V + R_{V'} + R_{\varepsilon_1},\ R_V - R_{V'} + R_{\varepsilon_1},\ 2R_V - R_{\varepsilon_1},\ R_{\mathrm{sign}}$$

are virtual representations of $G^{F'}$. Moreover, by 3.9 (i), R_1 and R_{sign} are $\pm$ irreducible representations, orthogonal to each other and to the virtual representations other than R_1, R_{sign} in the list (7.9.1). It follows that R_E, $R_{E'}$ are disjoint (in the sense of 6.17 (i)) whenever E, E' are in distinct families. From this, and from (6.15.2), we see just as in the proof of 7.6, that the representations in the list (7.8.1) are actual representations of $G^{F'}$. Applying 7.7 (ii) to the four vectors $\alpha = R_V$, $\beta = R_{V'}$, $\gamma = R_{\varepsilon_1}$, $\delta = R_{\varepsilon_2}$ in the euclidean space $\mathcal{R}(G^{F'}) \otimes \mathbf{R}$, we see that there exist eight irreducible representations a,b,c,d,e,f,g,h of $G^{F'}$ such that $\alpha, \beta, \gamma, \delta$ are expressed in terms of them as in 7.7 (ii). These representations are necessarily in $\mathcal{E}_{L,n}$. Thus the pattern of decomposition of any R_E is determined and 4.23 is verified in our case.

7.10. PROPOSITION. *Let* $(\mathbf{W}, S)$ *be a Weyl group of type* E_6 *and let* γ *be the unique automorphism of order* 2 *of* $(\mathbf{W}, S)$. *For any irreducible* $\mathbf{Q}[\mathbf{W}]$*-module* E, *we denote* $\hat{E}$ *the extension of* E *to a* $\mathbf{Q}[\tilde{\mathbf{W}}]$*-module (see 3.1) on which* γ *acts as the longest element in* $\mathbf{W}$, *and by* $\tilde{E}$ *the other extension. Then the following elements of* $\mathcal{R}(\mathbf{W}, \gamma)$ *are of the form* $a_{\gamma x}$ $(x \in \mathbf{W},\ \ell(x) \equiv a(x) \pmod 2)$ *and also of the form* $\mathcal{A}_{\gamma y}$ $(y \in \mathbf{W},\ \ell(y) \equiv a(y) \pmod 2)$:

(a) $(-1)^{a_E}(1/2)(\hat{E} - \tilde{E})$, *where* E *is any element in* $\mathbf{W}^\vee$ *such that* $\{E\}$ *is a family and* $\dim(E) \neq 64$.

(b) $(-1)^{a_E}((1/2)(\hat{E} - \tilde{E}) \pm (1/2)(\hat{M} - \tilde{M}))$,

$(-1)^{a_E}((1/2)(\hat{E} - \tilde{E}) \pm (1/2)(\hat{N} - \tilde{N}))$, *for any family* $\{E, M, N\} \subset \mathbf{W}^\vee$ *with* E *special.*

(c) $-((1/2)(\widehat{80_s}-\widetilde{80_s}) \pm (1/2)(\widehat{60_s}-\widetilde{60_s}) + (1/2)(\widehat{10_s}-\widetilde{10_s}))$,
$-((\widehat{80_s}-\widetilde{80_s}) - (1/2)(\widehat{10_s}-\widetilde{10_s}))$,
$-((1/2)(\widehat{80_s}-\widetilde{80_s}) \pm (1/2)(\widehat{60_s}-\widetilde{60_s}) + (1/2)(\widehat{90_s}-\widetilde{90_s}))$,
$-((\widehat{80_s}-\widetilde{80_s}) - (1/2)(\widehat{90_s}-\widetilde{90_s}))$,
$-((1/2)(\widehat{80_s}-\widetilde{80_s}) - (1/2)(\widehat{20_s}-\widetilde{20_s}))$.

Moreover, $-(1/2)(\widehat{64'_p}-\widetilde{64'_p})$ *and* $(1/2)(\widehat{64_p}-\widetilde{64_p})$ *are of the form* $-a_{\gamma x}$ *for some* $x \in W$, $\ell(x) \not\equiv a(x) \pmod 2$ *and also of the form* $\mathfrak{A}_{\gamma y}$ *for some* $y \in W$, $\ell(y) \not\equiv a(y) \pmod 2$. (*Note that* 64_p, $64'_p$ *are the only representations* $E \subset W^\vee$ *such that* $a_E + A_E$ *is odd*: *we have* $a_{64_p} = 4$, $A_{64_p} = 23$, $a_{64'_p} = 13$, $A_{64'_p} = 32$.)

Proof. This follows, using (5.12.4), from the corresponding result in the case where γ is replaced by the identity map, (see 5.21 and 7.3).

7.11. PROPOSITION. *Assume that* $(W_{L,n}, S_{L,n})$ *is of type* E_6. *Then Theorem 4.23 holds for* (L,n). *Moreover, we have* $\varepsilon_\rho = \hat{\varepsilon}_\rho$ *for all* $\rho \in \mathcal{E}_{L,n}$.

Proof. Assume first that γ (see (2.15.1)) acts non-trivially on $W_{L,n}$. We want to show that $R_{\hat{E}_1}$, $R_{\hat{E}_2}$ are disjoint (in the sense of 6.17 (i)) whenever $E_1, E_2 \in (W_{L,n})^\vee$ are in distinct families $\mathcal{F}_1 \neq \mathcal{F}_2$. If $\mathcal{F}_1$ and $\mathcal{F}_2$ consist of one or three representations, this is done just as in the proof of 7.6, (using 7.10 and 6.14 (i)). Hence, we may assume that $\mathcal{F}_1$ is the family of 80_s and $\mathcal{F}_2$ is a family with one or three elements. In the same way as in the proof of 7.6, we see that for any $\rho \in \mathcal{E}_{L,n}$ such that $\langle \rho, R_{\hat{E}_2} \rangle_{G^{F'}} \neq 0$, $(E_2 \in \mathcal{F}_2)$, we have $\sum_{E \in \mathcal{F}_2} \langle \rho, R_{\hat{E}} \rangle^2_{G^{F'}} \geq 3/4$.

Suppose that we have also $\langle \rho, R_{\hat{E}_1} \rangle_{G^{F'}} \neq 0$, $(E_1 \in \mathcal{F}_1)$. We apply Lemma 7.8 to the five vectors $R_{\hat{E}}$ $(E \in \mathcal{F}_1)$ in $\mathcal{R}(G^{F'}) \otimes \mathbf{R}$; the condition

(7.8.1) follows from 7.10 and 6.14 (i). It follows that $\sum_{E\in\mathcal{F}_1} \langle\rho,R_{\hat{E}}\rangle^2_{G^{F'}} > 1/4$. But then $\sum_{E\in\mathcal{F}_1\cup\mathcal{F}_2} \langle\rho,R_{\hat{E}}\rangle^2_{G^{F'}} > (3/4)+(1/4) = 1$, contradicting the fact that $\langle\rho,\rho\rangle_{G^{F'}} = 1$. Thus, we have shown that $R_{\hat{E}_1}, R_{\hat{E}_2}$ are disjoint whenever $E_1,E_2 \in (W_{L,n})^\vee$ are in distinct families. From this, and from (6.15.2) it follows just as in the proof of 7.6, that $(-1)^{\ell(w_1x)-a(x)} R_{\mathfrak{A}_{\gamma x}}$ is actual representation of $G^{F'}$, for any $x \in W_{L,n}$. Using 7.10, we see now that

a) $(-1)^{\ell(w_1)+a_E} R_{\hat{E}}$ is an irreducible representation of $G^{F'}$ for any $E \in (W_{L,n})^\vee$ such that $\{E\}$ is a family, $\dim E \neq 64$.

b) $(-1)^{\ell(w_1)+a_E}(R_{\hat{E}} \pm R_{\hat{M}})$, $(-1)^{\ell(w_1)+a_E}(R_{\hat{E}} \pm R_{\hat{N}})$ are actual representations of $G^{F'}$ for any family $\{E,M,N\} \subset (W_{L,n})^\vee$ with E special.

c) $(-1)^{\ell(w_1)+1}(R_{\widehat{80_s}} \pm R_{\widehat{60_s}} + R_{\widehat{10_s}})$, $(-1)^{\ell(w_1)+1}(2R_{\widehat{80_s}} - R_{\widehat{10_s}})$,
$(-1)^{\ell(w_1)+1}(R_{\widehat{80_s}} \pm R_{\widehat{60_s}} + R_{\widehat{90_s}})$, $(-1)^{\ell(w_1)+1}(2R_{\widehat{80_s}} - R_{\widehat{90_s}})$,
$(-1)^{\ell(w_1)+1}(R_{\widehat{80_s}} - R_{\widehat{20_s}})$
are actual representations of $G^{F'}$.

d) $(-1)^{\ell(w_1)+1} R_{\widehat{64_p}}$, $(-1)^{\ell(w_1)} R_{\widehat{64'_p}}$ are irreducible representations of $G^{F'}$.

In case b), we apply 7.7 (i) to the three vectors $\alpha = (-1)^{\ell(w_1)+a_E} R_{\hat{E}}$, $\beta = (-1)^{\ell(w_1)+a_E} R_{\hat{M}}$, $\gamma = (-1)^{\ell(w_1)+a_E} R_{\hat{N}}$ in $\mathcal{R}(G^{F'}) \otimes \mathbf{R}$, and we see that there exist four (distinct) irreducible representations a,b,c,d of $G^{F'}$ such that $\alpha = (1/2)(a+b+c+d)$, $\beta = (1/2)(a+b-c-d)$, $\gamma = (1/2)(a-b+c-d)$.

In case c), we apply 7.7 (ii) and (iii) to the five vectors

$\alpha = (-1)^{\ell(w_1)+1} R_{\widehat{80_s}}$, $\beta = (-1)^{\ell(w_1)+1} R_{\widehat{60_s}}$, $\gamma = (-1)^{\ell(w_1)+1} R_{\widehat{90_s}}$,

$\delta = (-1)^{\ell(w_1)+1} R_{\widehat{10_s}}$, $\varepsilon = (-1)^{\ell(w_1)+1} R_{\widehat{20_s}}$ in the euclidean space

$\mathcal{R}(G^{F'}) \otimes R$ (with orthonormal basis given by the irreducible representations). We see that there exist eight (distinct) irreducible representations a,b,c,d,e,f,g,h of $G^{F'}$ such that $\alpha, \beta, \gamma, \delta, \varepsilon$ are expressed in terms of them as in 7.7 (ii) and (iii). We see that Theorem 4.23 holds for (L,n). The case where γ acts trivially on $W_{L,n}$ is entirely similar (using 5.21 and 7.3 instead of 7.10). The last sentence in the proposition now follows from 6.20; the assumption (b) in 6.20 is satisfied, due to 7.6. This completes the proof.

7.12. PROPOSITION. *Assume that* $(W_{L,n}, S_{L,n})$ *is of type* $E_6 \times A_1$ *or* E_7. *Then Theorem 4.23 holds for* (L,n). *Moreover, we have* $\varepsilon_\rho = \hat{\varepsilon}_\rho$ *for all* $\rho \in \mathcal{E}_{L,n}$.

Proof. Using 6.21, we can reduce the proof of 4.23 for (L,n) to the case where $R^\vee$ is the rational closure of $R^\vee_{L,n}$. (This will imply also the last statement of the proposition by 6.20; the assumption (b) in 6.20 is satisfied, cf. 7.6 and 7.11.) But one checks easily that, if $R^\vee$ is the rational closure of $R^\vee_{L,n}$ and $R^\vee_{L,n}$ is of type E_7 or $E_6 \times A_1$, then $R^\vee = R^\vee_{L,n}$. In this case, $\mathcal{E}_{L,n}$ contains a one-dimensional representation $\theta: G^{F'} \to \bar{Q}_\ell^*$, which is trivial on the image of the F'-fixed part of the simply-connected covering of the derived group of G, (see [DL_1, 5.11 (i)]) and tensoring by θ defines a bijection between $\mathcal{E}_{(L_0,1)}$ and $\mathcal{E}_{(L,n)}$, where L_0 is the trivial line bundle. Using [DL_1, 1.27], we see easily that 4.23 holds for (L,n) if and only if it holds for $(L_0,1)$. Thus, we are reduced to the case where $L = L_0$, $n = 1$; moreover, using [DL_1, 7.10], we see that we can further assume that G is adjoint. If W is of type $E_6 \times A_1$, then G itself will decompose in a product of two F'-stable adjoint groups, one of type E_6 and one of type A_1. Since 4.23 holds for the restriction of $(L_0,1)$ to each of the two factors (7.6, 7.11), it follows easily that it also holds for G. Assume now that G is adjoint, of type E_7. According to [L4, 3.24], $G^{F'}$ has precisely two irreducible cuspidal representations in $\mathcal{E}_{L_0,1}$; they are in $\mathcal{E}^{II}_{L_0,1}$. (See (6.18.7).) Since $\sigma(G) = 7$ (see 6.7), we have $\varepsilon_\rho = \hat{\varepsilon}_\rho = -1$ for any cuspidal representation $\rho \in \mathcal{E}_{L_0,1}$. If $\rho \in \mathcal{E}_{L_0,1}$ is not cuspidal, then the equality $\varepsilon_\rho = \hat{\varepsilon}_\rho$ still holds; it can be deduced, as in 6.12, from the similar equality corresponding to an F'-stable Levi subgroup of a proper F'-stable parabolic subgroup of G. (For that Levi subgroup, we may use 7.6 or 7.11.) Thus, the equality $\varepsilon_\rho = \hat{\varepsilon}_\rho$ holds for all $\rho \in \mathcal{E}_{L_0,1}$. Therefore, 6.17 is applicable. We now write that R_α is an actual representation of $G^{F'}$ for each of the elements $\alpha \in \mathcal{R}(W)$ listed in (5.22.1) and 7.3. We may then use Lemma 7.7, just as in the proof of 7.11, to find the explicit pattern of decomposition of any R_E where $E \in W^\vee$ is in a family with one, three or five representations. Consider the family $\{512'_a, 512_a\}$. Then $R_{512'_a} + R_{512_a}$ is an actual representation of $G^{F'}$, of length 2, hence it is of form $a + b$ where a,b are distinct irreducible representations in $\mathcal{E}^{I}_{L_0,1}$ (cf. (6.18.3)). Note also that $R_{512_a} - R_{512'_a} = \rho_1 + \rho_2$ where ρ_1, ρ_2 are the two representations in $\mathcal{E}^{II}_{L_0,1}$ (see (6.18.7)). We then have

$$R_{512_a} = a + b + \rho_1 + \rho_2, \quad R_{512'_a} = a + b - \rho_1 - \rho_2 .$$

Thus 4.23 is verified for $(L_0,1)$, for G adjoint of type E_7. The proposition follows.

7.13. PROPOSITION. *Assume that* G *is adjoint of type* E_8 *and that* $(W_{L,n}, S_{L,n})$ *is either of type* $E_7 \times A_1$ *and* $n = 2$, *or of type* $E_6 \times A_2$, *with* γ *acting trivially and* $n = 3$. *Then Theorem 4.23 holds for* (L,n).

Proof. We first prove that $\varepsilon_\rho = \hat{\varepsilon}_\rho$, for all $\rho \in \mathcal{E}_{L,n}$. As in the proof of 7.12, it is enough to prove this for ρ cuspidal. If $W_{L,n}$ is of type

$E_6 \times A_2$ (with γ acting trivially), then $\mathcal{E}^{II}_{L,n}$ is empty, see (6.18.6). If $W_{L,n}$ is of type $E_7 \times A_1$, then $\mathcal{E}^{II}_{L,n}$ contains no cuspidal representations (see (6.18.9) and its proof). In both cases, any cuspidal representation ρ in $\mathcal{E}_{L,n}$ (if any) is of the first kind, hence it satisfies $\varepsilon_\rho = (-1)^{\ell(w_1)}$. Since γ acts trivially on $W_{L,n}$, we have $w_1(\alpha^\vee) = \alpha^\vee$ for any $\alpha^\vee \in R^\vee_{L,n}$ (see (2.15.1)) hence $w_1 = e$ since $R^\vee_{L,n}$ is rationally closed in $R^\vee$. Hence $\varepsilon_\rho = 1$ for all cuspidal $\rho \in \mathcal{E}_{L,n}$. On the other hand, we have $\hat{\varepsilon}_\rho = 1$ for all cuspidal $\rho \in \mathcal{E}_{L,n}$ since $\sigma(G) = 8$, (see 6.7). Thus, we have $\varepsilon_\rho = \hat{\varepsilon}_\rho$ for all cuspidal $\rho \in \mathcal{E}_{L,n}$ hence also for all $\rho \in \mathcal{E}_{L,n}$. The rest of the proof proceeds as in 7.12.

7.14. PROPOSITION. *Assume that* G *is adjoint of type* E_8 *and that* $(W_{L,n}, S_{L,n})$ *is of type* $E_6 \times A_2$ *with* γ *acting non-trivially on* $W_{L,n}$ *and* $n = 3$. *Then Theorem 4.23 holds for* (L,n).

Proof. We can find $x \in W$ such that the E_6-factor of $W_{L',n}$ $(L' = {}^xL)$ is a standard parabolic subgroup of W. If Theorem 4.23 holds for (L',n) then it holds also for (L,n), (see 6.5), hence we may assume that $L' = L$. Since γ (2.15.1) is a Coxeter group automorphism of $(W_{L,n}, S_{L,n})$ it follows that $w_1(R^+_{L,n}) = R^+_{L,n}$. But there are only two elements in W keeping $R^+_{L,n}$ stable: $w = e$ and $w = w_0^{E_6} w_0^{E_7}$ (where $w_0^{E_6}, w_0^{E_7}$ are the longest elements in the standard parabolic subgroups of type E_6, E_7 respectively). Since γ acts non-trivially on $W_{L,n}$, we have $w_1 = w_0^{E_6} w_0^{E_7}$. In particular, we have $(-1)^{\ell(w_1)} = -1$. Note also that we have necessarily $q \equiv -1 \pmod 3$. Indeed, since n and q are relatively prime, and $n = 3$, we have $q \not\equiv 0 \pmod 3$; if we have $q \not\equiv 1 \pmod 3$ then it would follow that $e \in Z_{L,n}$ hence $w_1 = e$, a contradiction. From $w_1 = w_0^{E_6} w_0^{E_7}$, it follows that γ acts non-trivially on both irreducible factors of $W_{L,n}$.

Let P be the maximal parabolic subgroup of G, of type E_7, containing B and let M be its Levi subgroup containing T. Let $W^{(P)}$ be the Weyl group of M with respect to T; it is a standard parabolic subgroup of W. The intersection $W_{L,n}^{(P)} = W_{L,n} \cap W^{(P)}$ is the standard parabolic subgroup of type E_6 of W. Note that $w_1 \in W^{(P)}$, hence it is the element of minimal length in $Z_{L,n}^{(P)}$ ($= Z_{L,n}$ defined with respect to M, with respect to the restriction of L to the flag manifold of M, see 6.4, and n).

According to 7.11, Theorem 4.23 holds for (M,F',L,n). In particular, there exist eight (distinct) irreducible representations $a_1, b_1, c_1, d_1, e_1, f_1, g_1, h_1$ of $M^{F'}$, in $E_{L,n}^{(P)}$ ($= E_{L,n}$ defined with respect to M), such that $\alpha_1 = R_{\widehat{80_s}}$, $\beta_1 = R_{\widehat{60_s}}$, $\gamma_1 = R_{\widehat{90_s}}$, $\delta_1 = R_{\widehat{10_s}}$, $\varepsilon_1 = R_{\widehat{20_s}}$ (with notations of 7.10) are expressed in terms of $a_1, b_1, \cdots, h_1$ as in 7.7 (ii) and (iii), (adding the subscript $_1$ throughout). The representations g_1 and h_1 are cuspidal. (This can be checked using the criterion 6.3, since 4.23 holds for M. Alternatively, we can argue as follows: using 6.21, and [L_3, 7.9 and its proof] we reduce ourselves to the similar statement, where M is replaced by a group M_1 with W of type E_6, with $\gamma : W \to W$ of order 2 and L is replaced by the trivial bundle. Moreover, as in the proof of 6.20, we may assume that q is large. We then use the description of unipotent cuspidal representations of a twisted group ${}^2E_6(q)$ for large q, given in [L_8].)

Now let $\mathfrak{N}(M)$ be the normalizer of M in G and let $\phi \to \phi'$ be the involution on the set of irreducible representations of $M^{F'}$ defined by the character identity $\phi'(m) = \phi(nmn^{-1})$, $m \in M^{F'}$, $n \in \mathfrak{N}(M)^{F'} - M^{F'}$. We want to show that $g_1 = g_1'$, $h_1 = h_1'$. Let M_{ad} be M/C, where C is the center of M. If $n \in \mathfrak{N}(M)^{F'} - M^{F'}$, then $m \to nmn^{-1}$ defines an automorphism of $(M_{ad})^{F'}$ which is necessarily inner. It is therefore enough to prove that g_1, h_1 are compositions of $M^{F'} \to (M_{ad})^{F'}$ with representations of $(M_{ad})^{F'}$, or that the restriction of g_1, h_1 to $C^{F'}$ is trivial. Assume that $C^{F'}$ acts on g_1 according to a character θ. Since $n = 3$, we have $\theta^3 = 1$. But C has order q-1, hence $\theta^{q-1} = 1$. As $(q-1,3)=1$, we have $\theta = 1$, as desired. Similarly, $C^{F'}$ acts trivially on h_1. By a well-known property of induction, from $g_1 = g_1'$, $h_1 = h_1'$, it follows that there exist four (distinct) irreducible representations r_1, r_2, r_3, r_4 of $G^{F'}$ (necessarily in $\mathcal{E}_{L,n}$, see 6.4) such that $\mathrm{Ind}_{P^{F'}}^{G^{F'}}(g_1) = r_1 + r_2$, $\mathrm{Ind}_{P^{F'}}^{G^{F'}}(h_1) = r_3 + r_4$. Let $1, \varepsilon, r$ be the unit, sign, reflection representations of the A_2-factor of $W_{L,n}$. It is easy to see, using 6.4, that $\mathrm{Ind}_{P^{F'}}^{G^{F'}}(R_{\widehat{80_s}}) = R_{\widehat{80_s \boxtimes 1}} - R_{\widehat{80_s \boxtimes \varepsilon}}$ (where the last two R are defined with respect to G,L,n and $\widehat{80_s \boxtimes 1}$, $\widehat{80_s \boxtimes \varepsilon}$ denote the extensions of $80_s \boxtimes 1$, $80_s \boxtimes \varepsilon$ to $Q[\widetilde{W}_{L,n}]$-modules on which γ acts the longest elements in $W_{L,n}$). The same holds with 80_s replaced by 60_s, 90_s, 10_s or 20_s. We have

$$\langle g_1, R_{\widehat{80_s}} \rangle_{M^{F'}} = \langle h_1, R_{\widehat{80_s}} \rangle_{M^{F'}} = \langle g_1, R_{\widehat{20_s}} \rangle_{M^{F'}} = \langle h_1, R_{\widehat{20_s}} \rangle_{M^{F'}} = 1/3$$

$$\langle g_1, R_{\widehat{90_s}} \rangle_{M^{F'}} = \langle h_1, R_{\widehat{90_s}} \rangle_{M^{F'}} = \langle g_1, R_{\widehat{10_s}} \rangle_{M^{F'}} = \langle h_1, R_{\widehat{10_s}} \rangle_{M^{F'}} = -1/3$$

$$\langle g_1, R_{\widehat{60_s}} \rangle_{M^{F'}} = \langle h_1, R_{\widehat{60_s}} \rangle_{M^{F'}} = 0 .$$

Since $\mathrm{Ind}_{P^{F'}}^{G^{F'}}(g_1) = r_1 + r_2$, the subspace of r_1 (or r_2) of vectors fixed by $U_P^{F'}$ (U_P = unipotent radical of $P^{F'}$) is isomorphic as a $M^{F'}$-module to g_1. Hence

(7.14.1) $\langle r_1, R_{\widehat{80_s \boxtimes 1}} - R_{\widehat{80_s \boxtimes \varepsilon}} \rangle_{G^{F'}} = \langle g_1, R_{\widehat{80_s}} \rangle_{M^{F'}} = 1/3$. Similarly,

$$\langle r_1, R_{\widehat{60_s \boxtimes 1}} - R_{\widehat{60_s \boxtimes \varepsilon}} \rangle_{G^{F'}} = 0, \quad \langle r_1, R_{\widehat{90_s \boxtimes 1}} - R_{\widehat{90_s \boxtimes \varepsilon}} \rangle_{G^{F'}} = -1/3,$$

$$\langle r_1, R_{\widehat{10_s \boxtimes 1}} - R_{\widehat{10_s \boxtimes \varepsilon}} \rangle_{G^{F'}} = -1/3, \quad \langle r_1, R_{\widehat{20_s \boxtimes 1}} - R_{\widehat{20_s \boxtimes \varepsilon}} \rangle_{G^{F'}} = 1/3$$

and the same holds when r_1 is replaced by r_2, r_3, or r_4.

Assume that the following statement is known.

(7.14.2) *If* $\rho \in \mathcal{E}_{L,n}$ *is cuspidal, then* $\rho \in \mathcal{E}^{II}_{L,n}$.

Since a cuspidal representation ρ in $\mathcal{E}_{L,n}$ satisfies $D(\rho) = \hat{\varepsilon}_\rho \rho$ and $\hat{\varepsilon}_\rho = 1$, we see from (7.14.2) that $\varepsilon_\rho = \hat{\varepsilon}_\rho = 1$ for all cuspidal $\rho \in \mathcal{E}_{L,n}$. As in the proof of 7.12, this implies that $\varepsilon_\rho = \hat{\varepsilon}_\rho$ for all $\rho \in \mathcal{E}_{L,n}$ (using 7.6 and 7.11). Hence 6.17 is applicable, and the proof can then be carried out just as in 7.11; as in that proof, there are only three kinds of families in $W^\vee_{L,n}$: with one, three or five elements.

It remains to prove (7.14.2). Assume that there exists $\rho \in \mathcal{E}^{I}_{L,n}$ which is cuspidal. Then ρ satisfies $D(\rho) = \rho$. Let $E \in (W_{L,n})^\vee$ be such that $\langle \rho, R_{\hat{E}} \rangle_{G^{F'}} \neq 0$. ($\hat{E}$ is the extension of E to a $\mathbf{Q}[\tilde{W}_{L,n}]$-module on which γ acts as the longest element w_0 in $W_{L,n}$.) Using (5.8.4) and (5.8.6) we see that $\langle \rho, R_{\widetilde{\hat{E}\otimes \mathrm{sign}}} \rangle_{G^{F'}} = -\langle \rho, D(R_{\hat{E}}) \rangle_{G^{F'}} = -\langle D(\rho), R_{\hat{E}} \rangle_{G^{F'}} = -\langle \rho, R_{\hat{E}} \rangle_{G^{F'}}$. Hence we have $\langle \rho, R_{\widehat{E\otimes \mathrm{sign}}} \rangle_{G^{F'}} \neq 0$. Using (5.12.2), we have for any $w \in W_{L,n}$:

$$\mathrm{Tr}(T_\gamma T_w, \hat{E}(u)) = u^{(1/2)(a_E + A_E) - \nu}\, \mathrm{Tr}(T_{w_0} T_w, \hat{E}(u)) \in u^{(1/2)(a_E + A_E)} \mathbf{Z}[u, u^{-1}],$$

$$(\nu = \tilde{\ell}(w_0)).$$

Since $\langle \rho, R_{\hat{E}} \rangle_{G^{F'}} \neq 0$ and $\rho \in \mathcal{E}^{I}_{L,n}$, it follows from (6.18.4) that $(a_E + A_E)$ is even. It follows that E and $E \otimes \mathrm{sign}$ belong to distinct families in $W^\vee_{L,n}$. (There is only one family in $W^\vee_{L,n}$ stable under $\otimes$sign, and all representations E' in it satisfy $a_{E'} + A_{E'} =$ odd.) The families in $W^\vee_{L,n}$ consists of either one, three or five representations. The argument in the proof of 7.6 shows that $R_{\hat{E}_1}, R_{\hat{E}_2}$ are disjoint if E_1 and E_2 belong to distinct families with one or three elements. Since $R_{\hat{E}}, R_{\widehat{E\otimes \mathrm{sign}}}$ are not disjoint, either E or $E \otimes \mathrm{sign}$ must belong to a family with five elements. Hence both E and $E \otimes \mathrm{sign}$ belong to families with five elements. Since $a_E + A_E$ is even, this forces E to be in the family of $80_s \boxtimes 1$ or in the family of $80_s \boxtimes \varepsilon$ (1 and ε denote the unit and sign representations of the A_2-factor of $W_{L,n}$).

Let $x_1, x_2, x_3 \in W_{L,n}$ be such that $80_s \boxtimes \varepsilon \underset{LR}{\sim} x_1$, $80_s \boxtimes r \underset{LR}{\sim} x_2$, (r = reflection representation of the A_2-factor), $80_s \boxtimes 1 \underset{LR}{\sim} x_3$. It is clear that

(7.14.3) $$x_1 \underset{LR}{<} x_2 \underset{LR}{<} x_3 .$$

We now write the identity (6.15.2) for x_1. Assume that $\tilde{\ell}(x_1) \equiv a(x_1) (\mathrm{mod}\ 2)$. We see that

(7.14.4) $$(-R_{\mathcal{A}_{x_1}} + \text{a linear combination of } R_{\hat{E}_1}, E_1 \otimes \text{sign} \underset{LR}{<} x_1)$$
$$\text{is an actual representation of } G^{F'} .$$

By (7.14.3), if E_1 is such that $E_1 \otimes \text{sign} \underset{LR}{<} x_1$, then $E_1 \otimes \text{sign}$ is not in the family of $80_s \boxtimes \varepsilon$ or $80_s \boxtimes r$ or $80_s \boxtimes 1$; hence E_1 is in a family with one or three elements. By the argument in the proof of 7.11 it follows that $R_{\hat{E}_1}$ is disjoint from any $R_{\hat{E}_2}$ where $E_2 \otimes \text{sign} \underset{LR}{\sim} 80_s \otimes \varepsilon$. Hence (7.14.4) implies that

(7.14.5) $$(-R_{\mathcal{A}_{x_1}}) \text{ is an actual representation of } G^{F'}, \text{ for any } x_1 \in W_{L,n}$$
$$\text{such that } 80_s \otimes \varepsilon \underset{LR}{\sim} x_1 , \ \tilde{\ell}(x_1) \equiv a(x_1) (\mathrm{mod}\ 2) .$$

We may now apply Lemma 7.7 (ii) and (iii) to the five vectors

$$\alpha = R_{\widehat{80_s \boxtimes 1}} , \ \beta = R_{\widehat{60_s \boxtimes 1}} , \ \gamma = R_{\widehat{90_s \boxtimes 1}} , \ \delta = R_{\widehat{10_s \boxtimes 1}} , \ \varepsilon = R_{\widehat{20_s \boxtimes 1}}$$

in $\mathcal{R}(G^{F'}) \otimes \mathbf{R}$ and we see that there exist eight (distinct) irreducible representations a,b,c,d,e,f,g,h of $G^{F'}$ such that $\alpha, \beta, \gamma, \delta, \varepsilon$ are expressed in terms of them as in 7.7 (ii), (iii). (The assumptions of 7.7 (ii),

(iii) are verified using (7.14.5), 7.10 and the formula $\mathcal{A}_{\gamma s_1 s_2 s_1} = \hat{1}$ valid for a Weyl group $<s_1,s_2>$ of type A_2 with $\gamma(s_1) = s_2$, $\gamma(s_2) = s_1$; $\hat{1}$ is the unit representation with trivial action of γ.) If $\mathcal{F}_1$ is the family of $80_s \boxtimes 1$, it follows from 7.7 (iii) that $\sum_{E_1 \in \mathcal{F}_1} <a,R_{\tilde{E}_1}>^2_{G^{F'}} > 1/2$ and the same is true when a is replaced by b,c,d,e or f.

Using now (6.8.3), (6.8.4) and (6.8.6) we see that $\sum_{E_2 \in \mathcal{F}_2} <D(a),R_{\hat{E}_2}>^2_{G^{F'}} > 1/2$ (where $\mathcal{F}_2$ is the family of $80_s \boxtimes \varepsilon$) and the same is true when D(a) is replaced by D(b), D(c), D(d), D(e) or D(f). In particular, $\pm D(a)$ cannot be equal to one of the representations a,b,c,d,e,f, since this would imply $\sum_{E' \in \mathcal{F}_1 \cup \mathcal{F}_2} <D(a),R_{\hat{E}'}>^2_{G^{F'}} > 1$, contradicting $<a,a>_{G^{F'}} = 1$. Similarly, neither of $\pm D(b)$, $\pm D(c)$, $\pm D(d)$, $\pm D(e)$, $\pm D(f)$ can be in the set a,b,c,d,e,f. Since $\rho = D(\rho)$ and ρ is in the set $\{a,b,c,d,e,f,g,h\}$, it follows that ρ is g or h. We may choose notation so that $\rho = g$. Then neither of $\pm D(a)$, $\pm D(b)$, $\pm D(c)$, $\pm D(d)$, $\pm D(e)$, $\pm D(f)$ is equal to g, (since $D(g) = g$). Assume that $\pm D(a)$ is equal to h. The inner product $<\gamma,D(\beta)>_{G^{F'}}$ can be computed in two ways; on the one hand it is zero, by 3.9 (i). On the other hand it is $<(1/3)(a+2c-d+e-g-h), (1/2)(D(a)+D(b)-D(e)-D(f))>_{G^{F'}} = -(1/6)<h,D(a)>_{G^{F'}} = \pm(1/6)$, a contradiction. Hence $\pm D(a) \neq h$. The same argument shows that $\pm D(b)$, $\pm D(e)$, $\pm D(f)$ are not equal to h. Assume that $\pm D(c) = h$. The inner product $<\gamma-\delta,D(\gamma)>_{G^{F'}}$ can be computed as above in two ways; one way leads to the value 0, the other way leads to the value $\pm 1/3$, a contradiction. Hence $\pm D(c) \neq h$. The same argument shows that $\pm D(d) \neq h$. Thus, the sets of irreducible representations $\{a,b,c,d,e,f,g,h\}$, $\{\pm D(a), \pm D(b), \pm D(c), \pm D(d), \pm D(e), \pm D(f)\}$ are disjoint. We now compute the inner product $<\gamma,D(\gamma)>_{G^{F'}}$ in two different ways, as above. On the one hand it is zero, on the other hand, it is $<-g-h,-D(g)-D(h)>_{G^{F'}} = <h,D(h)>_{G^{F'}}+1$. It follows that $D(h) = -h$.

We have $<a,R_{\widehat{80_s \boxtimes 1}} - R_{\widehat{80_s \boxtimes \varepsilon}}>_{G^{F'}} = <a,R_{\widehat{80_s \boxtimes 1}}>_{G^{F'}} = 1/6$.

From (7.14.1), it follows that a cannot be equal to r_1 or to r_2, or r_3 or r_4.

We have

$$<\pm D(a),R_{\widehat{80_s \boxtimes 1}} - R_{\widehat{80_s \boxtimes \varepsilon}}>_{G^{F'}} = <\pm D(a), -R_{\widehat{80_s \boxtimes \varepsilon}}>_{G^{F'}}$$

$$= \pm <a,R_{\widehat{80_s \boxtimes 1}}>_{G^{F'}} = \pm 1/6$$

hence $\pm D(a) \notin \{r_1,r_2,r_3,r_4\}$. Similarly, we see that none of the representations b,c,d,e,f, $\pm D(b)$, $\pm D(c)$, $\pm D(d)$, $\pm D(e)$, $\pm D(f)$ can be in the set $\{r_1,r_2,r_3,r_4\}$. But $\{r_1,r_2,r_3,r_4\}$ must be contained in $\{a,b,c,d,e,f,g,h,$ $\pm D(a), \pm D(b), \pm D(c), \pm D(d), \pm D(e), \pm D(f)\}$, (see 7.14.1). Hence it must be contained in $\{g,h\}$. This is a contradiction since r_1,r_2,r_3,r_4 are distinct. This contradiction proves (7.14.2) and hence the proposition.

Chapter 8
DECOMPOSITION OF INDUCED REPRESENTATIONS

The main result of this chapter is Theorem 8.6, which gives generators and relations for the endomorphism algebra of the representation induced from a cuspidal representation of $(P/U_P)^{F'}$ where P is an F-stable parabolic subgroup of G, under the assumptions that G has connected center and that Theorem 4.23 holds for certain subgroups of G. (The last assumption will be later shown to be always satisfied.) This is a strengthening of the results in [HL] in the present case.

8.1. Let $(\mathbf{W},S)$ be a Weyl group and let γ be an ordinary automorphism of it, (see 3.1). A γ-stable family $\mathcal{F} \subset \mathbf{W}^\vee$ is said to be non-cuspidal if there exists a proper γ-stable standard parabolic subgroup $\mathbf{W}' \subsetneq \mathbf{W}$ and a γ-stable family $\mathcal{F}' \subset (\mathbf{W}')^\vee$ such that either

a) $J_{\mathbf{W}'}^{\mathbf{W}}$ establishes a γ-invariant bijection between $\mathcal{F}'$ and $\mathcal{F}$, (in particular, for any $E' \in \mathcal{F}'$, $J_{\mathbf{W}'}^{\mathbf{W}}(E')$ is irreducible and in $\mathcal{F}$) or

b) $J_{\mathbf{W}'}^{\mathbf{W}}$ establishes a γ-invariant bijection between $\mathcal{F}'$ and $\mathcal{F} \otimes \text{sign}$.

A γ-stable family $\mathcal{F} \subset \mathbf{W}^\vee$ is said to be *cuspidal* if it is not non-cuspidal. The cuspidal families can be easily classified, using the results in Chapter 4. Assume first that $(\mathbf{W},S)$ is irreducible. If $\mathbf{W}$ is of type A_n $(n \geq 1)$ and $\gamma = 1$ then no family $\mathcal{F} \subset \mathbf{W}^\vee$ is cuspidal. (However, if $\mathbf{W} = \{e\}$ i.e. it is of type A_0, then the unique family in $\mathbf{W}^\vee$ is cuspidal.)

If $\mathbf{W}^\vee$ is of type B_n or C_n $(n \geq 2)$ then there is no cuspidal family in $\mathbf{W}$ unless $n = d^2+d$ for some $d \geq 1$, in which case there is a unique cuspidal family: that of $\begin{bmatrix} 0,2,4,\cdots,2d \\ 1,3,\cdots,2d-1 \end{bmatrix}$. This is proved as follows. (See $[L_{11}, 5.10]$.) Let Λ be a symbol of rank n and defect 1. Using shift (see 4.5) we may assume that 0 does not appear in both rows of Λ. Let t_0 be the largest entry in Λ. If some number i, $0 \leq i < t_0$ does not appear in any of the rows of Λ, then $[\Lambda]$ is of the form $J_{W'}^{W}(\varepsilon_r \otimes [\Lambda'])$, where W' is of type $A_{r-1} \times B_{n-r}$ or $A_{r-1} \times C_{n-r}$ $(n \geq 1)$, ε_r is the sign representation of the A_{r-1} factor and Λ' is a symbol of rank $n-r$ and defect 1. (See 4.5.4.) Moreover, we see that $J_{W'}^{W}$ is a bijection between the family of $\varepsilon_r \otimes [\Lambda']$ and the family $\mathcal{F}$ of $[\Lambda]$. Hence $\mathcal{F}$ is non-cuspidal. Assume now that all numbers i, $0 \leq i \leq t_0$ appear in at least one of the rows of Λ. Consider the symbol Λ' defined as in (4.5.5) with respect to $t = t_0$. Then the largest entry in Λ' is t_0, 0 does not appear in both rows of Λ' and there exists j, $0 \leq j < t_0$ which does not appear in any of the rows of Λ'. By the first part of the argument, we see that the family of Λ' $(= \mathcal{F} \otimes \text{sign})$ is non-cuspidal, hence $\mathcal{F}$ is non-cuspidal. In the remaining case, Λ contains each number i, $0 \leq i \leq t_0$ exactly once. Hence $n = d^2 + d$ and F is the family of $\begin{bmatrix} 0,2,4,\cdots,2d \\ 1,3,\cdots,2d-1 \end{bmatrix}$. This family is cuspidal since it contains more representations than any family of a proper parabolic subgroup.

If W is of type D_n $(n \geq 4)$ and γ is arbitrary, then there is no cuspidal γ-stable family in $W^\vee$ unless $n = d^2$ for some $d \geq 2$, in which case there is a unique cuspidal family: that of $\begin{bmatrix} 0,2,4,\cdots,2d-2 \\ 1,3,\cdots,2d-1 \end{bmatrix}$. This is proved, just as for type B_n, using results in 4.6.

If W is of type E_6 and γ is arbitrary, or of type E_7, E_8, F_4 or G_2, then there is a unique cuspidal γ-stable family in $W^\vee$: it is the family of 80_s (type E_6), of $512_a'$ (type E_7), of 4480_y (type E_8), of 12_1 (type F_4) and of the reflection representation (type G_2). (This follows easily from results in 4.10-4.13.)

If W is of type A_n $(n \geq 2)$ and γ is of order 2, then there is no cuspidal family in $W^\vee$ unless $n = ((d^2+d)/2) - 1$ for some $d \geq 2$, in which case there is a unique cuspidal family: that of $[1,3,5,\cdots,2d-1]$, (see 4.4). The proof is similar to that for type B_n; it is left to the reader.

If $\mathbf{W}$ is not irreducible, then we write it as a product of irreducible Weyl groups; $\mathcal{F} \subset \mathbf{W}^\vee$ corresponds to a collection of families, one for each irreducible factor of $\mathbf{W}$ and the condition that $\mathcal{F}$ is cuspidal is that each of the corresponding families for an irreducible factor of $\mathbf{W}$ is cuspidal (with respect to the automorphism of that factor defined by the smallest power ≥ 1 of γ leaving that factor stable).

In particular, we see that for given $(\mathbf{W},S)$ and γ, there is at most one cuspidal γ-stable family in $\mathbf{W}^\vee$. It is necessarily stable under tensoring with the sign representation of W, (since a cuspidal family tensored by sign is again cuspidal).

8.2. Let $(\mathbf{W},S)$ and γ be as in 8.1 and let $(\mathbf{W}',S')$ be a proper, standard γ-stable parabolic subgroup of $(\mathbf{W},S)$; note that $S' \subset S$. For any γ-stable subset $I \subset S$ such that $S' \subset I$, we denote by w_0^I the longest element in the parabolic subgroup $\mathbf{W}_I$ generated by I. Assume that there exists a cuspidal family $\mathcal{F}$ in $(\mathbf{W}')^\vee$ (with respect to $\gamma: \mathbf{W}' \to \mathbf{W}'$). ($\mathcal{F}$ is then unique by 8.1.) This assumption restricts $\mathbf{W}'$ (and $\gamma: \mathbf{W}' \to \mathbf{W}'$) so drastically that it forces it to have the following property:

(8.2.1) *For any* I *as above,* $\mathbf{W}'$ *is normalized by* w_0^I. *In fact, a stronger property holds: if* $(\mathbf{W}'',S'')$ *is another* γ*-stable standard parabolic subgroup of* $(\mathbf{W},S)$ *such that there exists an isomorphism* $(\mathbf{W}',S') \approx (\mathbf{W}'',S'')$ *commuting with* γ, *then* $\mathbf{W}' = \mathbf{W}''$.

In particular we may apply the results in [L_2, §5] to the pair $\mathbf{W},\mathbf{W}'$. According to [loc. cit.], the group $\mathcal{W} = \{w \in \mathbf{W} | \gamma(w) = w, w \cdot S' \cdot w^{-1} = S'\}$ is a Coxeter group with canonical generators (of order 2) s_o (one for each orbit o of γ on $S - S'$): we have $s_o = w_0^{S' \cup o} \cdot w_0^{S'} = w_0^{S'} \cdot w_0^{S' \cup o}$; for two such orbits $o \neq o'$ the order $m(o,o')$ of the product $s_o s_{o'}$ is given by

$$m(o,o') \cdot (\ell(w_0^{S' \cup o}) + \ell(w_0^{S' \cup o'}) - 2\ell(w_0^{S'})) = 2(\ell(w_0^{S' \cup o \cup o'}) - \ell(w_0^{S'})),$$

and is in fact always one of the integers 2, 3, 4 or 6.

Let $\mathcal{S}$ be the set of γ-orbits in $S - S'$. We define a function $\lambda : \mathcal{S} \to \mathbf{Z}$ as follows. Let $\alpha \in \mathcal{S}$ and let $I = S' \cup \alpha$. One can check easily (case by case) that

(8.2.2) $J_{W'}^{W_I}$ *takes any irreducible representation in the cuspidal family* $\mathcal{F} \subset (W')^\vee$ *to an irreducible representation of* W_I.

Let E be any irreducible representation in $\mathcal{F}$ and let $E' = J_{W'}^{W_I}(E)$, $E'' = E' \otimes \text{sign}$. Then E', E'' are irreducible representations of W_I, and we set

(8.2.3) $$\lambda(\alpha) = a_{E''} - a_{E'} .$$

This is independent of the choice of E, cf. (4.14.1).

The computation of the function λ for W not irreducible can be immediately reduced to the case where W is irreducible. In the case where W is irreducible, the function λ can be easily computed in all cases; it turns out to be the same as the function described in the last column of Table II in [L_4, p. 35]. (It appears as the exponent of q in that table.)

It follows that

(8.2.4) $$\lambda(\alpha) \geq 1 \quad \textit{for any} \quad \alpha \in \mathcal{S}$$

and

(8.2.5) $$\lambda(\alpha) = \lambda(\alpha') \quad \textit{if} \quad \alpha \neq \alpha' \quad \textit{in} \quad \mathcal{S} \quad \textit{satisfy} \quad m(\alpha,\alpha') = 3 .$$

8.3. In this section, the center of G is not necessarily connected.

Let P be a proper F'-stable parabolic subgroup of G containing B and let M be its Levi subgroup containing T. Let δ be an irreducible cuspidal representation of $M^{F'}$ (over $\overline{\mathbf{Q}}_\ell$). We wish to study the induced representation $\mathrm{Ind}_{P^{F'}}^{G^{F'}}(\delta)$ where δ is regarded as a $P^{F'}$-module in a natural way. Some general results on such induced representations have been obtained by Howlett and Lehrer [HL]. (See also [L_2, §5], [L_4, 3.25] for the case where δ is unipotent and [L_3, §5], [As_1] for the case where G is a classical group.) Let T_0 be the largest F_q-split torus (q as in 6.10) in the center of M. Then the group $W(T_0) = \mathfrak{N}_G(M)^{F'}/M^{F'}$, ($\mathfrak{N}$ = normalizer), acts faithfully on T_0 by conjugation. It also acts by conjugation on the set of irreducible representations of $M^{F'}$; let W_δ be the stabilizer of δ in $W(T_0)$. By a theorem of Harish-Chandra [HC], we have

(8.3.1) $$\dim E_\delta = |W_\delta|, \text{ where } E_\delta = \mathrm{End}_{G^{F'}}(\mathrm{Ind}_{P^{F'}}^{G^{F'}}(\delta)) .$$

A closed subgroup M_1 of G is said to be admissible if it is the centralizer of a subtorus T_1 of codimension 1 in T_0 such that T_1 is the largest F_q-split torus in the center of M_1 and if $\mathfrak{N}_{M_1}(M)^{F'}/M^{F'}$ is non-trivial. If M_1 is admissible then it is F'-stable, $\mathfrak{N}_{M_1}(M)^{F'}/M^{F'}$ is necessarily of order two, and we denote s_{M_1} its non-trivial element; we regard s_{M_1} as an element of $W(T_0)$.

(8.3.2) Let W'_δ be the subgroup of $W(T_0)$ generated by all s_{M_1} (M_1 admissible) satisfying properties (a), (b) below.

(a) $s_{M_1} \in W_\delta$. (This implies, by (8.3.1), that if Q is an F'-stable parabolic subgroup in M_1 with Levi subgroup M, then $\mathrm{Ind}_{Q^{F'}}^{M^{F'}}(\delta)$ is a sum of two inequivalent irreducible representations $\delta' \oplus \delta''$.)

(b) $\dim \delta'/\dim \delta'' \neq 1$.

It is clear that W'_δ is a group generated by reflections in the real vector space $\mathrm{Hom}(k^*, T_0) \otimes \mathbf{R}$. The following result is contained in [HL].

(8.3.3) *Let ℓ' be the length function on W'_δ, regarded as a Coxeter group with respect to a set of simple reflections $s_{M_1}, \ldots, s_{M_t} \in W'_\delta$, where M_i are certain admissible subgroups. To each $w \in W'_\delta$, one can associate an element $T_w \in E_\delta$ such that T_1 = identity,*

and such that $T_w \cdot T_{w'} = T_{ww'}$ *if* $\ell'(ww') = \ell'(w) + \ell'(w')$,
$(T_{s_{M_i}} - 1)(T_{s_{M_i}} + p_i) = 0$ *if* s_{M_i} *is a simple reflection of* W'_δ.
Here p_i *is the ratio* $\dim \delta' / \dim \delta''$ *(assumed to be* > 1 *), where* δ', δ'' *are defined as in (b) above for* s_{M_i}. *Moreover, the endomorphisms* T_w, $(w \in W'_\delta)$, *are linearly independent and, for* $w \neq 1$, T_w *has trace* 0 *as an endomorphism of* $\mathrm{Ind}_{P^{F'}}^{G^{F'}}(\delta)$.

Our goal in this chapter is to show that (8.3.3) can be strengthened considerably (when the center of G is connected) if we assume the truth of Theorem 4.23.

8.4. We shall denote by G^* a group *dual* to G. Recall [DL_1, 5.21] that G^* is a connected reductive group over k with a maximal torus T^*, a Borel subgroup B^* containing T^*, and a F_q-rational structure with Frobenius map $F' : G^* \to G^*$, (q as in 6.10), such that $F'(T^*) = T^*$, $F'(B^*) = B^*$. Moreover, there is a given isomorphism

$$\mathrm{Hom}(k^*, T^*) \approx \mathrm{Hom}(T, k^*) \tag{8.4.1}$$

carrying the set of coroots of G^* with respect to T^* onto the set of roots of G with respect to T, and the set of simple coroots of G^* with respect to T^* and B^* onto the set of simple coroots of G with respect to T and B; moreover, the contragradient isomorphism

$$\mathrm{Hom}(T^*, k^*) \approx \mathrm{Hom}(k^*, T) \tag{8.4.2}$$

carries the set of roots of G^* with respect to T^* onto the set of coroots of G with respect to T. The following property is assumed to be satisfied: if $\nu : k^* \to T^*$ and $\lambda : T \to k^*$ correspond to each other under (8.4.1), then $F' \circ \nu : k^* \to T^*$ and $\lambda \circ F' : T \to k^*$ correspond to each other under (8.4.1).

Note that (G^*, T^*, B^*, F') is well defined up to isomorphism by (G, T, B, F'). Since G has connected center, the derived group DG of G

has the following property: its simply-connected covering maps bijectively onto it. It is known that this implies that the centralizer $Z(s)$ of any semisimple element $s \in G^*$ is connected.

The bijection considered above between the set of simple roots for G and G^* gives rise to a bijection between the set of parabolic groups P in G containing B and the set of parabolic groups P^* in G^* containing B^*. If P and P^* correspond, the condition that $F'P = P$ is equivalent to the condition that $F'P^* = P^*$. Let $W^* = \mathfrak{N}(T^*)/T^*$ be the Weyl group of G^* with respect to T^* and let $S^* \subset W^*$ be the set of simple reflections with respect to B^*. We may identify W and W^* by the following requirement: $w \in W$ corresponds to $w^* \in W^*$ if for any $\nu : k^* \to T^*$, $\lambda : T \to k^*$ (corresponding under (8.1.1)), the homomorphisms $x \to (\dot{w}^*)^{-1}\nu(x)\dot{w}^*$, $(k^* \to T^*)$, and $t \to \lambda(\dot{w}t\dot{w}^{-1})$, $(T \to k^*)$, correspond under (8.4.1). This isomorphism $W \approx W^*$ takes S to S^* and the action of F' on W corresponds to the action of $(F')^{-1}$ on W^*.

(8.4.3) *We shall assume chosen, once and for all, an isomorphism* ι *of* k^* *with the group* $(\mathbf{Q}/\mathbf{Z})_{p'}$ *(= subgroup of* $\mathbf{Q}/\mathbf{Z}$ *consisting of all elements of order prime to* p *).*

Let $L \in \mathfrak{L}$ and let n be an integer ≥ 1 such that $(n,p) = 1$. We associate to (L,n) an element in T^* as follows. Write $L = L_\lambda$ where $\lambda : T \to k^*$ (see 1.3.2); let $\nu : k^* \to T^*$ be the element corresponding to λ under (8.4.1), and let $s = \nu(\iota^{-1}(1/n)) \in T^*$. (We regard $1/n$ as an element of $(\mathbf{Q}/\mathbf{Z})_{p'}$; then $\iota^{-1}(1/n) \in k^*$ is given by (8.4.3).) This gives an isomorphism of $\mathfrak{L} \otimes (\mathbf{Q}/\mathbf{Z})_{p'}$ with T^*. If (L,n) and s are as above, the condition that $Z_{L,n}$ is non-empty is clearly equivalent to the condition that $F'(s) = \dot{w}^* s (\dot{w}^*)^{-1}$ for some $w^* \in W^*$, hence to the condition that the conjugacy class of s in G^* is F'-stable. Note also that if to (L,n) corresponds $s \in T^*$ then $({}^xL,n)$, $(x \in W)$, corresponds to a conjugate of s under W^*. If we now put $\mathcal{E}_{L,n} = \mathcal{E}_{(s)}$ ((s) = conjugacy class of s in G^*) we see from 6.2 and 6.5 that:

(8.4.4) $\mathcal{E}_{(s)}$ *is well defined and non-empty for any* F'*-stable semisimple class* (s) *in* G^*. *The sets* $\mathcal{E}_{(s)}$ *form a partition of the set of all irreducible representations of* $G^{F'}$.

(See also [DL_1, §10].)

If (L,n) corresponds to (s) as above, the Weyl group $(W_{L,n}, S_{L,n})$ may be identified with the Weyl group $(W_{Z(s_0)}, S_{Z(s_0)})$ of $Z(s_0)$ where s_0 is an element in (s) such that $F'(s_0) = s_0$. Since $Z(s_0)$ is F'-stable, there is a natural action of F' on the Weyl group of $Z(s_0)$; it corresponds to the action of γ^{-1} (2.15.1) on $(W_{L,n}, S_{L,n})$. We can then define $\tilde{W}_{Z(s_0)}$ to be the semidirect product of $W_{Z(s_0)}$ with the infinite cyclic group with generator γ, acting on $W_{Z(s_0)}$ as F'^{-1}. This may be identified with $\tilde{W}_{L,n}$ of 3.1. Then any $\bar{Q}_\ell[\tilde{W}_{Z(s_0)}]$-module $\tilde{E}$ (with $\tilde{W}_{Z(s_0)}$ acting through a finite quotient, as in 3.1) may be regarded as a $\bar{Q}_\ell[\tilde{W}_{L,n}]$-module and therefore $R_{\tilde{E}}$ (see 6.4) has a meaning. It is a $\bar{Q}_\ell$-linear combination of irreducible representations in $\mathcal{E}_{(s)}$. We note also that, if w_1 is the element of minimal length in $Z_{L,n}$, we have $\ell(w_1) \equiv \sigma(G) - \sigma(Z(s)) (\text{mod } 2)$, (see 6.7).

The following result is a consequence of [L_4, 2.18], (see also 6.3):

(8.4.5) *If* $\mathcal{E}_{(s)}$ *contains some cuspidal representation of* $G^{F'}$ *and if* s_0 *is an element of* (s) *such that* $F'(s_0) = s_0$, *then the largest* F_q*-split torus in the center of* $Z(s_0)$ *coincides with the largest* F_q*-split torus in the center of* G^*.

8.5. Now let P, M, δ be as in 8.3. Let P^* be the corresponding F'-stable parabolic subgroup of G^* containing B^* and let M^* be the Levi subgroup of P^* containing T^*. Then M^* is F'-stable and is a dual group to M in the same way as G^* is a dual group to G. Note also that M has connected center (since G does). Let s be a semisimple element in $(M^*)^{F'}$ such that $\delta \in \mathcal{E}_{(s),M}$ (= $\mathcal{E}_{(s)}$ defined with respect to M). From 6.4 it follows that

(8.5.1) *All irreducible components of* $\mathrm{Ind}_{P^{F'}}^{G^{F'}}(\delta)$ *are in* $\mathcal{E}_{(s)}$ *(defined with respect to* G *); note that* s *can be also regarded as a semisimple element in* $(G^*)^{F'}$.

Let T_0^* be the largest F_q-split torus in the center of M^*, and let $W^*(T_0^*)$ be the group $\mathfrak{N}_{G^*}(M^*)^{F'}/(M^*)^{F'}$. It acts faithfully on T_0^* by conjugation. Under the identification $W = W^*$, $W(T_0)$ corresponds to $W^*(T_0^*)$. Hence W_δ can be regarded as a subgroup of $W^*(T_0^*)$. We define the notion of admissible subgroup of G^*, as in 8.3. We say that a closed subgroup M_1^* of G^* is admissible if it is the centralizer of a subtorus T_1^* of codimension 1 in T_0^* such that T_1^* is the largest F_q-split torus in the center of M_1^* and if $\mathfrak{N}_{M_1^*}(M^*)^{F'}/(M^*)^{F'}$ is non-trivial (hence of order two); for such M_1^* we define $s_{M_1^*}$ to be the non-trivial element in $\mathfrak{N}_{M_1^*}(M^*)^{F'}/(M^*)^{F'}$, regarded as an element of $\mathfrak{N}_{G^*}(M^*)^{F'}/(M^*)^{F'}$. The admissible subgroups of G^* are in 1-1 correspondence with the admissible subgroups of G, and corresponding admissible subgroups are dual to each other. Hence W'_δ may be identified with the subgroup of $W^*(T_0^*)$ generated by all $s_{M_1^*}$ (M_1^* admissible in G^*) such that the corresponding s_{M_1} satisfies (a), (b) in (8.3.2).

We now show that W_δ is contained in the image of the homomorphism

(8.5.2) $(\mathfrak{N}_{G^*}(M^*)^{F'} \cap Z(s))^{F'}/(M^* \cap Z(s))^{F'} \hookrightarrow \mathfrak{N}_{G^*}(M^*)^{F'}/(M^*)^{F'}$.

An element $n \in \mathfrak{N}_G(M^*)^{F'}$ maps $\mathcal{E}_{(s),M}$ to $\mathcal{E}_{(nsn^{-1}),M}$. If the image of n in $W^*(T_0^*)$ is in W_δ, then by definition, δ is equal to its n-conjugate hence δ is both in $\mathcal{E}_{(s),M}$ and in $\mathcal{E}_{(nsn^{-1}),M}$. Since the sets $\mathcal{E}_{(s),M}$ and $\mathcal{E}_{(nsn^{-1}),M}$ are disjoint or coincide (8.4.4), it follows that $nsn^{-1} \in (s)$ and hence the class of n is in the image of (8.5.2).

We now show that

(8.5.3) $$\mathfrak{N}_{Z(s)}(Z_{M^*}(s))^{F'} = \mathfrak{N}_{G^*}(M^*)^{F'} \cap Z(s)\,.$$

The inclusion $\mathfrak{N}_{G^*}(M^*)^{F'} \cap Z(s) \subset \mathfrak{N}_{Z(s)}(Z_{M^*}(s))^{F'}$ is obvious. Conversely, let $g \in \mathfrak{N}(Z_{M^*}(s))^{F'}$. Then g normalizes the largest F_q-split torus in the center of $Z_{M^*}(s)$ which, by (8.4.5) applied to M instead of G, is equal to the largest F_q-split torus in the center of M^*, hence is equal to T_0^*. But M^* is the centralizer of T_0^* in G^*, hence g normalizes M^*. This proves (8.5.3). Using now the equality (8.5.3), we can rewrite the homomorphism (8.5.2) as

(8.5.4) $$\mathfrak{N}_{Z(s)}(M^* \cap Z(s))^{F'}/(M^* \cap Z(s)) \hookrightarrow \mathfrak{N}_{G^*}(M^*)^{F'}/(M^*)^{F'}$$

and we see that

(8.5.5) W_δ *is contained in the image of (8.5.4).*

We now prove the following

(8.5.6) *If Theorem 4.23 holds for representations in* $\mathcal{E}_{(s),M}$ *then there exists a cuspidal* F'*-stable family* $\mathcal{F} \subset (W_{Z_{M^*}(s)})^\vee_{ex}$ *(with respect to* $F': W_{Z_{M^*}(s)} \to W_{Z_{M^*}(s)}$*) and* $E \in (W_{Z_{M^*}(s)})^\vee_{ex}$, $(E \in \mathcal{F})$, *such that* $\langle\delta, R_{\tilde{E}}\rangle_{M^{F'}} \neq 0$; $\tilde{E}$ *is an extension of* E *to a* $\mathbf{Q}[\tilde{W}_{Z_{M^*}(s)}]$*-module.*

Indeed, let $E \in (W_{Z_{M^*}(s)})^\vee_{ex}$, and $\tilde{E}$ an extension of E to a $\mathbf{Q}[\tilde{W}_{Z_{M^*}(s)}]$-module be such that $\langle\delta, R_{\tilde{E}}\rangle_{M^{F'}} \neq 0$. Let $\mathcal{F} \subset (W_{Z_{M^*}(s)})^\vee$ be the family containing E. It is clearly F'-stable. Since $D(\delta) = \pm\delta$, (see (6.8.1)), and $D(R_{\tilde{E}}) = \pm R_{\tilde{E}\otimes \mathrm{sign}}$, (see (6.8.6)), it follows that we have also $\langle\delta, R_{\tilde{E}\otimes\mathrm{sign}}\rangle_{M^{F'}} \neq 0$. From Theorem 4.23 (for $\mathcal{E}_{(s),n}$) it follows then that E and $E \otimes \text{sign}$ must be in the same family. Hence $\mathcal{F} = \mathcal{F} \otimes \text{sign}$. Assume that $\mathcal{F}$ is not a cuspidal family. Then there exists a proper F'-stable standard parabolic subgroup W_1 of $(W_{Z_{M^*}(s)}, S_{Z_{M^*}(s)})$ and an irreducible representation $E_1 \in (W_1)^{\vee}_{ex}$ such that $E = J_{W_1}^{W_{Z_{M^*}(s)}}(E_1)$. This implies that $\mathrm{Ind}_{W_1}^{W_{Z_{M^*}(s)}}(E_1)$ contains E with multiplicity 1 and all its other irreducible components have the a-invariant (4.1.1) strictly bigger than a_E, hence all these other irreducible components must belong to families other than $\mathcal{F}$. Let $\tilde{E}_1$ be an extension of E_1 to a $\mathbf{Q}[\tilde{W}_1]$-module and let $\mathrm{Ind}(\tilde{E}_1)$ be the corresponding induced $\mathbf{Q}[\tilde{W}_{Z_{M^*}(s)}]$-module. Let P_1^* be an F'-stable parabolic subgroup of M^* such that $P_1^* \ni s$ and such that W_1 is the Weyl group of the reductive part of $P_1^* \cap Z_{M^*}(s)$. Using (6.4.1), we see that $R_{\mathrm{Ind}(\tilde{E}_1)} \in \mathcal{R}(M^{F'}) \otimes \mathbf{Q}$ is induced from an element of $\mathcal{R}((P_1/U_{P_1})^{F'}) \otimes \mathbf{Q}$ where P_1 is an F'-stable proper parabolic subgroup of M. Since δ is cuspidal, we must then have $\langle \delta, R_{\mathrm{Ind}(\tilde{E}_1)} \rangle_{M^{F'}} = 0$.

Let us decompose $\mathrm{Ind}(\tilde{E}_1)$ as $\bigoplus_{i=1}^{N} \tilde{E}^{(i)}$ where $\tilde{E}^{(i)}$ are irreducible $\bar{\mathbf{Q}}_\ell[\tilde{W}_{Z_{M^*}(s)}]$-modules. Then $R_{\mathrm{Ind}\,\tilde{E}_1} = \sum_{i=1}^{N} R_{\tilde{E}^{(i)}}$. We may assume that $\tilde{E}^{(1)}$ is an extension of E to a $\bar{\mathbf{Q}}_\ell[\tilde{W}_{Z_{M^*}(s)}]$-module, hence $\langle \delta, R_{\tilde{E}^{(1)}} \rangle_{M^{F'}} \neq 0$. For $i > 1$, we have either that $\tilde{E}^{(i)}$ is not irreducible as a $\bar{\mathbf{Q}}_\ell[\tilde{W}_{Z_{M^*}(s)}]$-module (in which case, clearly, $\langle \delta, R_{\tilde{E}^{(i)}} \rangle_{M^{F'}} = 0$) or that the restriction of $\tilde{E}^{(i)}$ to $W_{Z_{M^*}(s)}$ is irreducible in a family $\neq \mathcal{F}$ (in which case we have again $\langle \delta, R_{\tilde{E}^{(i)}} \rangle_{M^{F'}} = 0$, as a consequence of 4.23 for $\mathcal{E}_{(s),M}$). It follows that

$$\langle \delta, R_{\mathrm{Ind}(\tilde{E}_1)} \rangle_{M^{F'}} = \langle \delta, R_{\tilde{E}^{(1)}} \rangle_{M^{F'}} \neq 0 .$$

This contradicts the equality $<\delta, R_{\mathrm{Ind}(\tilde{E}_1)}>_{M^{F'}} = 0$. Thus, (8.5.6) is proved. We shall consider $(W_{Z_{M^*}(s)}, S_{Z_{M^*}(s)})$ as an F′-stable parabolic subgroup of $(W_{Z(s)}, S_{Z(s)})$, $Z(s) = Z_{G^*}(s)$, using the fact that $Z_{M^*}(s)$ is the Levi subgroup of the F′-stable parabolic subgroup $P^* \cap Z(s)$ of $Z(s)$. Let us denote

(8.5.7) $\mathcal{W} = \{w \in W_{Z(s)}^{F'} | w S_{Z_{M^*}(s)} w^{-1} = S_{Z_{M^*}(s)}\} = \mathcal{N}_{Z(s)}(Z_{M^*}(s))^{F'}/Z_{M^*}(s)^{F'}$.

From (8.5.6) and 8.2 it follows that

(8.5.8) $\mathcal{W}$ *is a Coxeter group with canonical generators* $s_{\mathfrak{a}}$, *one for each orbit* $\mathfrak{a}$ *of* F′ *on* $S_{Z(s)} - S_{Z_{M^*}(s)}$.

The generators $s_{\mathfrak{a}}$ are described in 8.2. An equivalent description is as follows. If $\mathfrak{a}$ is an orbit of F′ on $S_{Z(s)} - S_{Z_{M^*}(s)}$ then there is a unique parabolic subgroup $P_{\mathfrak{a}}^*$ of $Z(s)$, of type $S_{Z_{M^*}(s)} \cup \mathfrak{a}$, such that $P_{\mathfrak{a}}^*$ contains $P^* \cap Z(s)$; let $\bar{M}_{\mathfrak{a}}^*$ be the unique Levi subgroup of $P_{\mathfrak{a}}^*$ containing $Z_{M^*}(s)$. Then $\mathcal{N}_{\bar{M}_{\mathfrak{a}}^*}(Z_{M^*}(s))^{F'}/Z_{M^*}(s)^{F'}$ is a group of order 2 and $s_{\mathfrak{a}}$ is its non-trivial element. The image of $s_{\mathfrak{a}}$ under the imbedding (8.5.4) can be described as follows. Let $T_{\mathfrak{a}}^*$ be the largest F_q-split torus in the center of $\bar{M}_{\mathfrak{a}}^*$. Then $\bar{M}_{\mathfrak{a}}^*$ is the centralizer of $T_{\mathfrak{a}}^*$ in $Z(s)$, and $T_{\mathfrak{a}}^*$ is a codimension 1 torus in T_0^* (= the largest F_q-split torus in the center of $Z_{M^*}(s)$). Let $M_{\mathfrak{a}}^*$ be the centralizer of $T_{\mathfrak{a}}^*$ in G^*. Then $M_{\mathfrak{a}}^* \supset M^*$. The largest F_q-split torus in the center of $M_{\mathfrak{a}}^*$ contains $T_{\mathfrak{a}}^*$ and is contained in T_0^*; if it is not equal to $T_{\mathfrak{a}}^*$ then it must be equal to T_0^* (since $\mathrm{codim}_{T_0^*}(T_{\mathfrak{a}}^*) = 1$). This would imply that $M_0^* = M^*$ hence $\bar{M}_{\mathfrak{a}}^* = M_{\mathfrak{a}}^* \cap Z(s) = M^* \cap Z(s)$, a contradiction. Hence the largest F_q-split torus in the center of $M_{\mathfrak{a}}^*$ is T_0^*. It follows that $\mathcal{N}_{M_{\mathfrak{a}}^*}(M^*)^{F'}/(M^*)^{F'}$ is of order one or two. Since it contains the image of

s_α under (8.5.4), it must be of order two. Let s'_α be its non-trivial element (the image of s_α under (8.5.4)). We see from (8.5.8) that

(8.5.9) M^*_α *is an admissible subgroup of* G^* *and the image of (8.5.4) is generated by the elements* s'_α *just described.*

Let us fix α as above and let M_α be the admissible subgroup of G corresponding to M^*_α. (Then M^*_α is a dual group for M_α.) Let Q_α be the F'-stable parabolic subgroup $M_\alpha \cap P$. It has M as a Levi subgroup. Let $I(\delta)$ be the representation $\mathrm{Ind}_{Q_\alpha^{F'}}^{M_\alpha^{F'}}(\delta)$, where δ is regarded as $Q_\alpha^{F'}$-module via the natural projection $Q_\alpha^{F'} \to M^{F'}$. Let $\mathbf{W}' = W_{Z_{M^*}(s)}$, $\mathbf{W}''$ = subgroup of $W_{Z(s)}$ generated by $S_{Z_{M^*}(s)} \cup \alpha$. Let $E \in \mathcal{F} \subset (\mathbf{W}')^\vee$ and let E be as in (8.5.6). Thus, we have $\langle \delta, R_{\tilde{E}} \rangle_{M^{F'}} \neq 0$. Let $\mathrm{Ind}(\tilde{E})$ be the $\mathbf{Q}[\tilde{\mathbf{W}}'']$-module induced by the $\mathbf{Q}[\tilde{\mathbf{W}}']$-module $\tilde{E}$, and let $E' = J_{\mathbf{W}'}^{\mathbf{W}''}(E)$; it is an irreducible representation of $\mathbf{W}''$, (see (8.2.2)), which appears with multiplicity 1 in $\mathrm{Ind}(E)$ (= the $\mathbf{Q}[\mathbf{W}'']$-module induced by E). It follows that there is a unique irreducible $\mathbf{Q}[\tilde{\mathbf{W}}'']$-submodule $\tilde{E}'$ of $\mathrm{Ind}(\tilde{E})$ whose restriction to $\mathbf{W}''$ is E'. The restriction $\mathrm{Res}\,\tilde{E}'$ of $\tilde{E}'$ to $\tilde{\mathbf{W}}'$ can be written as a direct sum $\bigoplus_{i=1}^{N} \tilde{E}'^{(i)}$ of irreducible $\overline{\mathbf{Q}}_\ell[\tilde{\mathbf{W}}']$-submodules. We may assume that for $1 \leq i \leq N_1$, $\tilde{E}'^{(i)}$ restricted to $\mathbf{W}'$ is irreducible in $\mathcal{F}$, for $N_1 < i \leq N_2$, $\tilde{E}'^{(i)}$ restricted to $\mathbf{W}'$ is irreducible, not in $\mathcal{F}$, and for $N_2 < i \leq N$, $\tilde{E}'^{(i)}$ restricted to $\mathbf{W}'$ is not irreducible. Clearly, for $N_2 < i \leq N$, we have $R_{\tilde{E}'^{(i)}} = 0$ (in $\mathcal{R}(M^{F'}) \otimes \overline{\mathbf{Q}}_\ell$). For $N_1 < i \leq N_2$, we have $\langle \delta, R_{\tilde{E}'^{(i)}} \rangle_{M^{F'}} = 0$ (since $\langle \delta, R_{\tilde{E}} \rangle_{M^{F'}} \neq 0$ and Theorem 4.23 is assumed to be applicable to δ). For $1 \leq i \leq N_1$, if we denote $E'^{(i)}$ the restriction of $\tilde{E}'^{(i)}$ to $\mathbf{W}'$, we see that $J_{\mathbf{W}'}^{\mathbf{W}''}(E'^{(i)})$ contains E'. Using again (8.2.2), it follows that $N_1 = 1$ and $\tilde{E}'^{(1)} = \tilde{E}$. Thus, we have

$$\langle \delta, R_{\mathrm{Res}\,\tilde{E}'} \rangle_{M^{F'}} = \langle \delta, R_{\tilde{E}} \rangle_{M^{F'}} \neq 0 \, .$$

Using now (6.4.2) (with G replaced by M_α), we have $\langle \delta, R_{\mathrm{Res}\,\tilde{E}'} \rangle_{M^{F'}} = \langle I(\delta), R_{\tilde{E}'} \rangle_{M_\alpha^{F'}}$. Hence the last inner product is non-zero. From $D(\delta) = \pm\delta$ it follows that $D(I(\delta)) = \pm I(\delta)$. It follows that $\langle I(\delta), R_{\widetilde{E' \otimes \mathrm{sign}}} \rangle_{M_\alpha^{F'}} = \pm \langle I(\delta), D(R_{\tilde{E}'}) \rangle_{M_\alpha^{F'}} = \pm \langle D(I(\delta)), R_{\tilde{E}'} \rangle_{M_\alpha^{F'}} = \pm \langle I(\delta), R_{\tilde{E}'} \rangle_{M_\alpha^{F'}} \neq 0$.

It follows that there exists an irreducible component δ' of $I(\delta)$ and an irreducible component δ'' of $I(\delta)$ such that $\langle \delta', R_{\tilde{E}'} \rangle_{M_\alpha^{F'}} \neq 0$, $\langle \delta'', R_{\widetilde{E' \otimes \mathrm{sign}}} \rangle_{M_\alpha^{F'}} \neq 0$. Let $E'' = E' \otimes \mathrm{sign}$. Then $a_{E''} > a_{E'}$ by (8.2.3) and (8.2.4). In particular, E' and E'' are not in the same family. If we assume that Theorem 4.23 holds for $M_\alpha^{F'}$ and the corresponding $\mathcal{E}_{(s)}$, it follows that δ' cannot be isomorphic to δ''. Since $I(\delta)$ has one or two components (see (8.3.1)), it follows that

(8.5.10) $$I(\delta) = \delta' \oplus \delta'' \quad \text{and} \quad s'_\alpha \in W_\delta \, .$$

We now show that

(8.5.11) $\dim(\delta')/\dim(\delta'') = q^{\pm\lambda(\alpha)}$, *where* $\lambda(\alpha)$ *is defined as in (8.2.3).*

From $D(I(\delta)) = \pm I(\delta)$, it follows that either $D(\delta') = \pm\delta'$, $D(\delta'') = \pm\delta''$ or $D(\delta') = \pm\delta''$, $D(\delta'') = \pm\delta'$. If $D(\delta') = \pm\delta'$, we have $\langle \delta', R_{\widetilde{E' \otimes \mathrm{sign}}} \rangle_{M_\alpha^{F'}} = \pm \langle \delta', D(R_{\tilde{E}'}) \rangle_{M_\alpha^{F'}} = \pm \langle D(\delta'), R_{\tilde{E}'} \rangle_{M_\alpha^{F'}} = \pm \langle \delta', R_{\tilde{E}'} \rangle_{M_\alpha^{F'}} = 0$. This is impossible (by 4.23) since E', E'' are in different families. Thus, we must have $D(\delta') = \pm\delta''$. We see that (8.5.11) is a consequence of the following statement.

(8.5.12) *For any* $\rho \in \mathcal{E}_{(s)}$ *(defined with respect to* $M_{\mathfrak{a}}$ *) we have* $\dim(\rho)/\dim(\hat{\varepsilon}_\rho D(\rho)) = q^{(a_{E_1}-a_{E_2})}$ *where* E_1 *is a special representation in a family* $\mathfrak{F} \subset (W'')^{\vee}_{ex}$ *such that* $\langle \rho, R_{\tilde{E}_1} \rangle_{M^F_{\mathfrak{a}}}{}' \neq 0$, *(* $\tilde{E}_1$ *= extension of* E_1 *to a* $\mathbf{Q}[\tilde{W}'']$*-module) and* E_2 *is the special representation in* $\mathfrak{F} \otimes \text{sign}$ *(assuming that 4.23 holds for* $\mathcal{E}_{(s)}$ *and* $M_{\mathfrak{a}}$ *).*

To prove this, we first note that, with the notations in 4.21, 4.22, we have $\{\bar{x}, x_{\widetilde{E_1 \otimes \text{sign}}}\} = \pm\{\bar{x}, x_{\tilde{E}_2}\}$ for any $\bar{x} \in \bar{X}(W_{L,n}, \gamma)$. Hence, using 4.23, we have $\langle D(\rho), R_{\tilde{E}_2} \rangle_{M^F_{\mathfrak{a}}}{}' = \pm \langle D(\rho), R_{\widetilde{E_1 \otimes \text{sign}}} \rangle_{M^F_{\mathfrak{a}}}{}' = \pm \langle \rho, R_{\tilde{E}_1} \rangle_{M^F_{\mathfrak{a}}}{}' \neq 0$.

By (4.26.3), there exist two integers N, N', relatively prime to p, such that

$$\dim(\rho) = \langle \rho, R_{\tilde{E}_1} \rangle_{M^F_{\mathfrak{a}}}{}' \cdot q^{a_{E_1}} \cdot N$$

$$\dim(\hat{\varepsilon}_\rho D(\rho)) = \langle \hat{\varepsilon}_\rho D(\rho), R_{\tilde{E}_2} \rangle_{M^F_{\mathfrak{a}}}{}' \cdot q^{a_{E_2}} \cdot N' .$$

From this we get

$$\begin{aligned}\dim(\rho)/\dim(\hat{\varepsilon}_\rho D(\rho)) &= \langle \rho, R_{\tilde{E}_1} \rangle_{M^F_{\mathfrak{a}}}{}' \cdot \langle \hat{\varepsilon}_\rho D(\rho), R_{\tilde{E}_2} \rangle^{-1}_{M^F_{\mathfrak{a}}}{}' \cdot q^{a_{E_1}-a_{E_2}} \cdot N \cdot N'^{-1} \\ &= \pm q^{a_{E_1}-a_{E_2}} \cdot N \cdot N'^{-1} .\end{aligned}$$

According to [A_1, 5.5(3)], the quotient $\dim(\rho)/\dim(\hat{\varepsilon}_\rho D(\rho))$ is an integral power of p. Since N and N' are relatively prime to p, it follows that

$$\dim(\rho)/\dim(\hat{\varepsilon}_\rho D(\rho)) = q^{a_{E_1}-a_{E_2}}$$

and (8.5.12) is proved.

From (8.5.9), (8.5.10) and (8.5.11), we see that $s'_\alpha \in W'_\delta$ and therefore, the image of (8.5.4), which is generated by the s'_α is contained in W'_δ. On the other hand, that image of (8.5.4) contains W_δ (see (8.5.5)) and $W'_\delta \subset W_\delta$. It follows that

(8.5.13) *We have* $W_\delta = W'_\delta$ = *image of (8.5.4).*

Since (8.5.4) gives an isomorphism of $\mathfrak{W}$ (see (8.5.7)) onto W'_δ, we may take the s_α in (8.5.8) as set of simple reflections for W'_δ. We also see that the $T_{w'}$ $(w \in W'_\delta)$ in (8.3.3) form a basis for E_δ, (cf. (8.3.1)). Hence we have the following result which strengthens (8.3.3), for G with connected center.

8.6. THEOREM. *Assume that 4.23 holds for* $\mathcal{E}_{(s)}$ *with respect to* M *and for* $\mathcal{E}_{(s)}$ *with respect to* M_α *for each orbit* α *of* F' *on* $S_{Z(s)} - S_{Z_{M^*}(s)}$. *(Recall also that* G *has connected center.) Then there exists a basis* $T_{w'}$ $(w \in \mathfrak{W})$ *for* $\mathrm{End}_{G^{F'}} \mathrm{Ind}_{P^{F'}}^{G^{F'}}(\delta)$ *(where* $\mathfrak{W}$ *is defined by (8.5.7)) such that* T_1 = *identity and such that*

$$T_w \cdot T_{w'} = T_{ww'} \quad \text{if} \quad \bar{\ell}(ww') = \bar{\ell}(w) + \bar{\ell}(w')$$

$$(T_{s_\alpha} + 1)(T_{s_\alpha} - q^{\lambda(\alpha)}) = 0 .$$

Here $\bar{\ell}$ *is the length function on the Coxeter group* $\mathfrak{W}$ *(see (8.5.8)) and* s_α *are the simple reflections in* $\mathfrak{W}$. *The exponent* $\lambda(\alpha)$ *is an integer* ≥ 1, *defined as in (8.2.3). Moreover, if* $w \neq 1$, T_w *has trace zero as an endomorphism of* $\mathrm{Ind}_{P^{F'}}^{G^{F'}}(\delta)$.

8.7. COROLLARY. *With the assumptions of 8.6, the irreducible representations of* $G^{F'}$ *appearing in* $\mathrm{Ind}_{P^{F'}}^{G^{F'}}(\delta)$ *are in 1-1 correspondence with the irreducible representations of* $\mathfrak{W}$, *(the correspondence depends, in general, on the choice (2.19.2) of* $p^{1/2}$ *in* $\overline{\mathbf{Q}}_\ell$); *let us denote* $\delta[\chi]$ *the irreducible representation of* $G^{F'}$ *corresponding to the irreducible representation* χ *of* W. *We have*

$$\dim(\delta[\chi]) = \dim(\delta)\cdot(|G^{F'}|/|P^{F'}|)\cdot(D_\chi/\mathcal{P}) .$$

Here, $D_\chi \in \mathbf{Q}[q]$ *are the formal degrees of the Hecke algebra corresponding to* $\mathfrak{W}$, *with parameters* $q^{\lambda(\alpha)}$, *and* $\mathcal{P} = \sum_{w\in\mathfrak{W}} \phi(T_w) \in \mathbf{Z}[q]$, *where* ϕ *is the one-dimensional representation of that Hecke algebra defined by* $\phi(T_{s_\alpha}) = q^{\lambda(\alpha)}$, *for all* α.

8.8. We shall show how the verification of 4.23 for G can be reduced to the special case where the derived group D(G) of G is simply-connected and (absolutely) simple modulo its center. The arguments in this section are taken from [As$_1$, §2]. Given G, we can find a reductive group G′ over F_q, with connected center and a surjective homomorphism of reductive groups $G' \xrightarrow{\pi} G$ defined over F_q such that D(G′) is simply-connected and the kernel of π is connected and contained in the center of G′. π induces a surjective map on the groups of F_q-rational points and the irreducible representations of $G(F_q)$ may be identified with the irreducible representations of $G'(F_q)$ which are trivial on $\ker(\pi)(F_q)$. It follows easily that the truth of 4.23 for G′ implies the truth of 4.23 for G. Hence we are reduced to check 4.23 in the case where D(G) is simply-connected.

Now let $G \hookrightarrow G''$ be an imbedding (over F_q) of G into another restrictive group G″ over F_q, with connected center which induces isomorphism on derived groups. Then it is again easy to see that

(8.8.1) *the truth of 4.23 for* G *is equivalent to the truth of 4.23 for* G″ .

Since D(G) is simply-connected, we can write $D(G) = \prod_i R_{F_{q^{n_i}}/F_q}(H_i)$ where H_i are simply-connected groups over $F_{q^{n_i}}$ which are absolutely simple modulo their center, and R denotes restriction of scalars. We can imbed (over $F_{q^{n_i}}$) H_i into a reductive group G_i (over $F_{q^{n_i}}$) with connected center in such a way that $H_i \hookrightarrow G_i$ induces an isomorphism of H_i onto the derived group of G_i. Let $G'' = \prod_i R_{F_{q^{n_i}}/F_q}(G_i)$; it has connected center. Then we have a natural imbedding (over F_q) of $D(G)$ into G'' which induces an isomorphism $D(G) \approx D(G'')$. According to [As_1, 2.3.2] there exists a reductive group G''' (over F_q) with connected center and imbeddings (over F_q) $G \hookrightarrow G'''$, $G'' \hookrightarrow G'''$ which induce isomorphisms $D(G) \approx D(G''')$, $D(G'') \approx D(G''')$ (which are compatible with the earlier isomorphism $D(G) \approx D(G'')$). Using (8.8.1), we see that the truth of 4.23 for G is equivalent with the truth of 4.23 for G''' and is equivalent with the truth of 4.23 for G''. But it is easy to see that the truth of 4.23 for G'' is equivalent to the truth of 4.23 for each G_i. Thus the verification of 4.23 is reduced to the case where G has the property that $D(G)$ is simply-connected and (absolutely) simple modulo its center.

Chapter 9
CLASSICAL GROUPS

In this chapter, Theorem 4.23 will be proved in the case where G is a classical proof. For the case of unipotent representations this was done in $[L_{11}]$, $[L_{12}]$ for q large. In $[As_3]$, the case of small q was reduced to that of large q using lifting $[Ka_2]$; unlike the proof in $[L_{11}]$, $[L_{12}]$, the present proof does not rely on the formulas in $[L_3$, §6] on the number of conjugacy classes in $G^{F'}$.

9.1. Let Y be a finite dimensional vector space over the field F_2 with two elements. We assume given (a) a basis $e_1, e_2, \cdots, e_{d'}$ $(d' = \dim Y)$ of Y; (b) a symmetric F_2-bilinear form $(\ ,\ ): Y \times Y \to F_2$ such that

$$(e_i, e_j) = 0 \quad \text{if} \quad |i-j| \neq 1 .$$

Let $\mathrm{Rad}(Y)$ be the radical of $(\ ,\ )$ on Y. Given a basis element e_i of Y such that $e_i \notin \mathrm{Rad}\, Y$, we consider the vector space $Y_{e_i} = \{y \in Y | (y,e_i) = 0\}/\{0,e_i\}$ of dimension $d'-2$. Then Y_{e_i} inherits a basis from Y, consisting of the e_j $(j = i-1, i, i+1)$ and (in the case where e_{i-1}, e_{i+1} are both defined) of

$$\begin{array}{lll} e_{i-1} + e_{i+1} & \text{if} & (e_{i-1}, e_i) = (e_i, e_{i+1}) = 1 \\ e_{i+1} & \text{if} & (e_{i-1}, e_i) = 1,\ (e_i, e_{i+1}) = 0 \\ e_{i-1} & \text{if} & (e_{i-1}, e_i) = 0,\ (e_i, e_{i+1}) = 1 . \end{array}$$

Note also that Y_{e_i} inherits from Y a symmetric bilinear form, which satisfies property (b) above with respect to the basis just defined.

We now define, by induction on d', a family $\mathcal{T}(Y)$ of subspaces of Y. If $\mathrm{Rad}(Y) = Y$, then $\mathcal{T}(Y)$ consists of $\{Y\}$. Assume now that not all basis elements e_i of Y are in $\mathrm{Rad}(Y)$ and that whenever $e_i \notin \mathrm{Rad}(Y)$, $\mathcal{T}(Y_{e_i})$ is already defined. By definition, a subspace $C \subset Y$ is in $\mathcal{T}(Y)$ if and only if there exists a basis element e_i for Y, $e_i \notin \mathrm{Rad}(Y)$, such that $e_i \in C \subset \{y \in Y | (y,e_i) = 0\}$ and such that the image of C under the natural map $\{y \in Y | (y,e_i) = 0\} \to Y_{e_i}$ is in $\mathcal{T}(Y_{e_i})$.

It follows easily that any subspace in $\mathcal{T}(Y)$ is a maximal isotropic subspace of Y (with respect to $(\ ,\ )$); in particular, it contains $\mathrm{Rad}(Y)$. Let us fix a linear form $\xi_0 : \mathrm{Rad}(Y) \to F_2$.

We consider an euclidean space H with scalar product $<\ ,\ >$ and an orthonormal basis (b_α) and assume that for each $C \in \mathcal{T}(Y)$ and each linear form $\xi : C \to F_2$ such that $\xi | \mathrm{Rad}(Y) = \xi_0$, we are given an element $\rho_{C,\xi} \in H$ such that

(9.1.1) $\rho_{C,\xi}$ *is a combination with integral,* ≥ 0 *coefficients of basis elements* b_α.

(9.1.2) *For any* C, ξ *and* C', ξ' *as above, we have* $<\rho_{C,\xi}, \rho_{C',\xi'}> =$ *number of linear forms* $\eta : Y \to F_2$ *such that* $\eta | C = \xi$, $\eta | C' = \xi'$.

Let Θ be the set of basis elements b_α which appear with > 0 coefficient in some $\rho_{C,\xi}$ as above.

Now let X be the set $\{\eta : Y \to F_2, \text{ linear} : \eta | \mathrm{Rad}(Y) = \xi_0\}$.

We shall consider two element subsets $D \subset X$ for which there exists a basis element e_i of Y, ($e_i \notin \mathrm{Rad}(Y)$ such that:

(9.1.3) If $D = \{\eta', \eta''\}$ then $\eta'(y) + \eta''(y) = (e_i, y)$ for all $y \in Y$.

(9.1.4) $$\eta'(e_i) = \eta''(e_i) = 0 .$$

We shall assume that for any such D, we are given an element $\rho_D \in H$ such that

(9.1.5) ρ_D is a combination with integral, ≥ 0 coefficients of basis elements b_α in Θ.

(9.1.6) $\langle \rho_D, \rho_{C,\xi} \rangle$ = number of $\eta \in D$ such that $\eta|C = \xi$ (for any such C, ξ as above).

With these assumptions, we shall prove:

9.2. PROPOSITION. *There exists a bijection* $X \leftrightarrow \Theta$, $(\eta \in X \leftrightarrow b_\eta \in \Theta)$, *such that*

$$\rho_{C,\xi} = \sum_{\substack{\eta \in X \\ \eta|C=\xi}} b_\eta, \quad \textit{for any } C \in \mathfrak{J}(Y) \textit{ and any linear form } \xi : C \to F_2,$$

$$\rho_D = \sum_{\eta \in D} b_\eta, \quad \textit{for any 2-element subset } D \subset X \textit{ satisfying (9.1.3), (9.1.4).}$$

Proof. Let C' be the subspace of Y generated by Rad(Y) and by $e_1, e_3, e_5, \cdots$; let C'' be the subspace of Y generated by Rad(Y) and by $e_2, e_4, e_6, \cdots$. It is easy to see that $C' \in \mathfrak{J}(Y)$, $C'' \in \mathfrak{J}(Y)$ and $C' \cap C'' = \mathrm{Rad}(Y)$. It follows that the map $\eta \to \{\xi', \xi''\}$ where $\xi' = \eta|C'$, $\xi'' = \eta|C''$ is a bijection between X and the set of pairs of linear forms on C' and C'' whose restriction to Rad(Y) is ξ_0. If $\eta \in X$ corresponds in this way to $\{\xi', \xi''\}$, then from (9.1.2) we see that $\langle \rho_{C',\xi'}, \rho_{C'',\xi''} \rangle = 1$. Using now (9.1.1), it follows that there is a unique basis element b_α of H which appears with > 0 coefficient in both $\rho_{C',\xi'}$ and $\rho_{C'',\xi''}$; we denote this basis element as b_η. Then b_η appears with coefficient 1 in both $\rho_{C',\xi'}$ and $\rho_{C'',\xi''}$.

If $\eta \neq \tilde{\eta}$ (in X) then $b_\eta \neq b_{\tilde{\eta}}$. Indeed, if η corresponds to $\{\xi', \xi''\}$ as above and $\tilde{\eta}$ corresponds to $\{\tilde{\xi}', \tilde{\xi}''\}$, we have $\xi' \neq \tilde{\xi}'$ or $\xi'' \neq \tilde{\xi}''$. Assume, for example, that $\xi' \neq \tilde{\xi}'$. Then from (9.1.2) we see that $\langle \rho_{C',\xi'}, \rho_{C',\tilde{\xi}'} \rangle = 0$ hence a basis element of H cannot appear with > 0 coefficient in both $\rho_{C',\xi'}$ and $\rho_{C',\tilde{\xi}'}$ (cf. (9.1.1)); in particular, we have $b_\eta \neq b_{\eta'}$ as stated.

Now let for $C \in \mathcal{T}(Y)$, $h_C = \sum_{\xi} \rho_{C,\xi}$ (sum over all linear forms $\xi : C \to F_2$ such that $\xi | \mathrm{Rad}(Y) = \xi_0$). If $C' \in \mathcal{T}(Y)$, we have $\langle h_C, h_C \rangle = \langle h_{C'}, h_{C'} \rangle = \langle h_C, h_{C'} \rangle =$ number of linear forms $\eta : Y \to F_2$ such that $\eta | \mathrm{Rad}(Y) = \xi_0$ (cf. (9.1.2)) hence $\langle h_C - h_{C'}, h_C - h_{C'} \rangle = 0$, so that h_C is independent of the choice of C. Clearly any $b_\alpha \in \Theta$ appears with >0 coefficient in some h_C and hence in $h_{C'}$. But we have $h_{C'} = \sum_{\eta \in X} b_\eta$. (Indeed, $(h_{C'}, h_{C'}) = |X|$ by (9.1.2), $(h_{C'}, \sum_{\eta \in X} b_\eta) = |X|$, by the definition of b_η and $(\sum_{\eta \in X} b_\eta, \sum_{\eta \in X} b_\eta) = |X|$ since $b_\eta \neq b_{\eta'}$ for $\eta \neq \eta'$.) It follows that any basis element in Θ is of the form b_η for a unique $\eta \in X$. Thus $\eta \to b_\eta$ is a 1-1 correspondence $X \leftrightarrow \Theta$. We also see that each $\rho_{C,\xi}$ is a combination with coefficients 0 or 1 of elements b_η, $(\eta \in X)$, since this is so for h_C. Consider a two-element subset $D \subset X$ satisfying (9.1.3), (9.1.4) with respect to $e_i \notin \mathrm{Rad}(Y)$. We shall show that $\rho_D = \rho_{\eta'} + \rho_{\eta''}$ where $D = \{\eta', \eta''\}$. We have $e_i \in C'$ or $e_i \in C''$. Assume, for example, that $e_i \in C'$. Let ξ' be the restriction of η' to C'' and let ξ'' be the restriction of η'' to C''. Let ξ be the restriction of η' to C' or equivalently (cf. (9.1.3)) the restriction of η'' to C'. We have $\xi' \neq \xi''$ since $(e_i, C'') \neq 0$, (see (9.1.3)). Hence by (9.1.6) we have $\langle \rho_D, \rho_{C'',\xi'} \rangle = \langle \rho_D, \rho_{C'',\xi''} \rangle = 1$. It follows that there is a unique $\tilde{\eta}' \in X$ such that $b_{\tilde{\eta}'}$ appears with coefficient 1 in both ρ_D and $\rho_{C'',\xi'}$ and there is a unique $\tilde{\eta}'' \in X$ such that $b_{\tilde{\eta}''}$ appears with coefficient 1 in both ρ_D and $\rho_{C'',\xi''}$. Since $\xi' \neq \xi''$, we have $\langle \rho_{C'',\xi'}, \rho_{C'',\xi''} \rangle = 0$ hence $b_{\tilde{\eta}'}$ cannot appear with >0 coefficient in both $\rho_{C'',\xi'}$ and $\rho_{C'',\xi''}$. Thus we have $b_{\tilde{\eta}'} \neq b_{\tilde{\eta}''}$, hence $\tilde{\eta}' \neq \tilde{\eta}''$. Also note that $\langle \rho_D, h_{C'} \rangle = 2$ (cf. (9.1.6)). Since $h_{C'} = \sum_{\eta \in X} b_\eta$, it follows that $\rho_D = b_{\tilde{\eta}'} + b_{\tilde{\eta}''}$. We have $\langle \rho_D, \rho_{C',\xi} \rangle = 2$, (cf. (9.1.6)) hence $b_{\tilde{\eta}'}$ and $b_{\tilde{\eta}''}$ appear with coefficient 1 in $\rho_{C',\xi}$. Since $b_{\tilde{\eta}'}$ also appears with coefficient 1 in $\rho_{C'',\xi'}$, it follows that $b_{\tilde{\eta}'} = b_{\eta'}$. Since $b_{\tilde{\eta}''}$ also appears with coefficient 1 in $\rho_{C'',\xi''}$, it follows that $b_{\tilde{\eta}''} = b_{\eta''}$. Hence $\rho_D = b_{\eta'} + b_{\eta''}$, as asserted.

Let $\tilde{Y}$ be the union $\bigcup_{C \in \mathcal{J}(Y)} C \subset Y$. For each $y \in \tilde{Y}$ we define

$$(9.2.1)\quad \hat{b}_y = 2^{-\delta} \sum_{\xi} (-1)^{\xi(y)} \rho_{C,\xi} \in H, \quad (\delta = (1/2)\dim(Y/\mathrm{Rad}(Y))),$$

where C is a fixed subspace in $\mathcal{J}(Y)$ containing y, and the sum is over all linear forms $C \to F_2$ whose restriction to $\mathrm{Rad}(V)$ is ξ_0. To show that this independent of the choice of C, it is enough to show that, if $\tilde{C}$ is another subspace in $\mathcal{J}(Y)$ containing y, then $\langle 2^{-\delta} \sum_{\xi} (-1)^{\xi(y)} \rho_{C,\xi}, 2^{-\delta} \sum_{\tilde{\xi}} (-1)^{\tilde{\xi}(y)} \rho_{\tilde{C},\tilde{\xi}} \rangle$ is independent of $C, \tilde{C}$ ($\tilde{\xi}$ runs over all linear forms on $\tilde{C}$ whose restriction to $\mathrm{Rad}(V)$ is ξ_0). The last inner product can be computed using (9.1.2); it equals

$$2^{-2\delta} \sum_{\xi,\tilde{\xi}} (-1)^{\xi(y)+\tilde{\xi}(y)} \#\{\eta : Y \to F_2 : \eta|C = \xi, \eta|\tilde{C} = \tilde{\xi}\}$$

$$= 2^{-2\delta} \sum_{\xi,\tilde{\xi}} \sum_{\substack{\eta: Y \to F_2 \\ \eta|C = \xi \\ \eta|\tilde{C} = \tilde{\xi}}} (-1)^{\eta(y)+\eta(y)} = 2^{-2\delta} \#\{\eta : Y \to F_2 | \eta|\mathrm{Rad}(V) = \xi_0\}$$

$$= 1.$$

In particular, this is independent of $C, \tilde{C}$, so that $\hat{b}_y$ is well defined (independent of the choice of C).

For $\eta \in X$ and $y \in \tilde{Y}$ we define $[\eta,y] \in F_2$ as follows. We choose $C \in \mathcal{J}(Y)$ such that $y \in C$; let ξ be the unique linear form on C such that $\xi|\mathrm{Rad}(V) = \xi_0$ and b_η appears with coefficient 1 in $\rho_{C,\xi}$. We then set $[\eta,y] = \xi(y)$. We have $(-1)^{[\eta,y]} = 2^{\delta} \langle b_\eta, b_y \rangle$ which shows in particular that $[\eta,y]$ is independent of the choice of C. From the definition of $[\eta,y]$, we see that

(9.2.2) *For any* $C \in \mathfrak{J}(Y)$ *and any* $\eta \in X$, *the function* $C \to F_2$, $(y \to [\eta,y])$, *is* F_2*-linear.*

(9.2.3) *For any two element subset* $D = \{\eta', \eta''\} \subset X$ *satisfying (9.1.3), (9.1.4) for some* $e_i \notin \mathrm{Rad}(Y)$, *and any* $y \in \tilde{Y}$ *we have*

$$(-1)^{[\eta',y]} + (-1)^{[\eta'',y]} = (-1)^{\eta'(y)} + (-1)^{\eta''(y)}.$$

(Indeed, we have $(-1)^{[\eta',y]} + (-1)^{[\eta'',y]} = 2^{\delta} < b_{\eta'} + b_{\eta''}, \hat{b}_y > = 2^{\delta} < \rho_D, \hat{b}_y > = (-1)^{\eta'(y)} + (-1)^{\eta''(y)}$, by (9.1.6).)

(9.2.4) $\sum\limits_{\eta \in X} (-1)^{[\eta,y]} = \sum\limits_{\eta \in X} (-1)^{\eta(y)}$.

(Indeed, we have

$$\sum_{\eta \in X} (-1)^{[\eta,y]} = 2^{\delta} < \sum_{\eta \in X} b_{\eta}, \hat{b}_y >$$

$$= 2^{\delta} < \sum_{\xi} \rho_{C,\xi}, \hat{b}_y > = < \sum_{\xi} \rho_{C,\xi}, \sum_{\xi'} (-1)^{\xi'(y)} \rho_{C,\xi'} >$$

$$= \sum_{\xi,\xi'} \sum_{\substack{\eta \in X \\ \eta|C=\xi \\ \eta|C=\xi'}} (-1)^{\eta(y)} = \sum_{\eta \in X} (-1)^{\eta(y)};$$

here $y \in C$, and ξ, ξ' run over all linear forms $C \to F_2$ whose restriction to $\mathrm{Rad}(Y)$ is ξ_0.

It is then clear that the proposition is a consequence of the following result, which generalizes [L_{11}, 3.8].

3.3. LEMMA. *Let* $[\ ,\ ]: X \times \tilde{Y} \to F_2$ *be a function satisfying (9.2.2), (9.2.3), (9.2.4). Then* $[\eta,y] = \eta(y)$ *for all* $\eta \in X$ *and all* $y \in \tilde{Y}$.

Proof. We shall need the following statement

(9.3.1) *Given the basis element* $e_i \notin \mathrm{Rad}(Y)$ *and two elements* $y, y' \in \tilde{Y}$ *such that* $(y, e_i) = (y', e_i) = 1$, *there exists a sequence of elements* $y = y_1, y_2, \cdots, y_m = y' \in \tilde{Y}$ *and a sequence of subspaces* $C_1, C_2, \cdots, C_{m-1}$ *in* $\mathcal{T}(Y)$ *such that* $y_h, y_{h+1} \in C_h$ $(1 \le h \le m-1)$ *and* $(y_h, e_i) = 1$, $(1 \le h \le m)$.

Let $Y = \bigoplus_{j \in J} Y^{(j)}$ be the unique direct sum decomposition of Y, such that each e_i belongs to some $Y^{(j)}$ and e_i, e_{i+1} are in the same summand $Y^{(j)}$ if and only if $(e_i, e_{i+1}) = 1$. It is clear that each $Y^{(j)}$ is a space with the same kind of structure as Y. In particular, $\tilde{Y}^{(j)} \subset Y^{(j)}$ and $\mathcal{T}(Y^{(j)})$ are defined.

It is easy to see that a subspace $C \subset Y$ is in $\mathcal{T}(Y)$ if and only if it is of the form $\bigoplus_j C^{(j)}$ with $C^{(j)} \in \mathcal{T}(Y^{(j)})$. It follows that $\tilde{Y} = \prod_j \tilde{Y}^{(j)}$. It is clear that it suffices to prove (9.3.1) for Y replaced by the summand $Y^{(j)}$ containing e_i. But in that case, (9.3.1) is just [L_{11}, 3.4] if $\dim Y^{(j)}$ is even and [L_{12}, 2.3] if $\dim Y^{(j)}$ is odd.

The proof of the lemma will follow very closely that of [L_{11}, 3.8]. We fix $D = \{\eta', \eta''\} \subset X$ satisfying (9.1.3), (9.1.4) with respect to $e_i \notin \mathrm{Rad}(Y)$. Let $C \in \mathcal{T}(Y)$ be such that $e_i \in C$. Then by (9.1.3), $\eta'(y) = \eta''(y)$ for all $y \in C$ hence, by (9.2.3) we have $(-1)^{[\eta', y]} + (-1)^{[\eta'', y]} = 2(-1)^{\eta'(y)}$ for all $y \in C$. It follows that $(-1)^{[\eta', y]} = (-1)^{[\eta'', y]} = (-1)^{\eta'(y)} = (-1)^{\eta''(y)}$ hence $[\eta', y] = [\eta'', y] = \eta'(y) = \eta''(y)$ for all $y \in C$. Now let $C \in \mathcal{T}(Y)$ be such that $e_i \notin C$. Then, by (9.2.3) and (9.1.3), we have $(-1)^{[\eta', y] + \eta'(y)} + (-1)^{[\eta'', y] + \eta'(y)} = 1 + (-1)^{(e_i, y)}$ for all $y \in \tilde{Y}$ hence

$$\sum_{y \in C} (-1)^{[\eta', y] + \eta'(y)} + \sum_{y \in C} (-1)^{[\eta'', y] + \eta'(y)} = |C|$$

(since $y \to (e_i, y)$ is a linear function on C, not identically zero).

Using (9.2.2) it follows that exactly one of the linear functions $y \to [\eta', y] + \eta'(y)$, $y \to [\eta'', y] + \eta'(y)$ on C is identically zero. Similarly, exactly one of the linear functions $y \to [\eta',y]+\eta''(y)$, $y \to [\eta'',y]+\eta''(y)$ on C is identically zero. Thus there are two possibilities:

1) $[\eta',y] = \eta'(y)$ and $[\eta'',y] = \eta''(y)$ for all $y \in C$
   (we then say that C is of the first kind)
2) $[\eta',y] = \eta''(y)$ and $[\eta'',y] = \eta'(y)$ for all $y \in C$
   (we then say that C is of the second kind).

As in [*loc. cit.*] we see, using (9.3.1), that all $C \in \mathfrak{I}(Y)$ such that $e_j \notin C$ are of the same kind. Thus, given $D = \{\eta',\eta''\}$ as above, we have either

$$[\eta',y] = \eta'(y),\ [\eta'',y] = \eta''(y) \text{ for all } y \in \tilde{Y}$$

or

$$[\eta',y] = \eta''(y),\ [\eta'',y] = \eta'(y) \text{ for all } y \in \tilde{Y}.$$

We shall consider three cases for a vector $\eta \in X$.

Case 1. There exist $D_1 = \{\eta,\eta_1\} \subset X$ and $D_2 = \{\eta,\eta_2\} \subset X$ satisfying (9.1.3), (9.1.4) with respect to $e_{i_1} \notin \mathrm{Rad}(Y)$, $e_{i_2} \notin \mathrm{Rad}(Y)$ respectively, $(\eta \neq \eta_1 \neq \eta_2 \neq \eta)$. Then the function $y \to [\eta,y]$ on $\tilde{Y}$ must be equal to one of the functions $y \to \eta(y)$, $y \to \eta_1(y)$ and it must also be equal to one of the functions $y \to \eta(y)$, $y \to \eta_2(y)$. The functions $y \to \eta(y)$, $y \to \eta_1(y)$, $y \to \eta_2(y)$ on $\tilde{Y}$ are distinct since $\tilde{Y}$ spans Y. It follows that $y \to [\eta,y]$ is equal to $y \to \eta(y)$ on $\tilde{Y}$.

Case 2. There exists a unique $\eta_1 \in X$ such that $\eta \neq \eta_1$ and $\{\eta,\eta_1\}$ satisfies (9.1.3), (9.1.4) with respect to some $e_i \notin \mathrm{Rad}(Y)$. At least one of the equalities $(e_i,e_{i-1}) = 1$, $(e_i,e_{i+1}) = 1$ makes sense and is true (since $e_i \notin \mathrm{Rad}(Y)$). Assume, for example, that $(e_i,e_{i-1}) = 1$. Then $\eta(e_{i-1}) = 1$. (Indeed, if $\eta(e_{i-1}) = 0$, we could define a linear form $\eta' : Y \to F_2$ by $\eta'(y) = \eta(y)+(y,e_{i-1})$ and we would have that $\{\eta,\eta'\} \subset X$ satisfies (9.1.3), (9.1.4) with respect to e_{i-1}, contradicting the uniqueness of η_1.) From (9.1.3) for $\{\eta,\eta_1\}$ it follows that $\eta_1(e_{i-1}) = \eta(e_{i-1})+(e_{i-1},e_i) = 1+1 = 0$. Let $\eta_2 : Y \to F_2$ be the linear form $\eta_2(y) = \eta_1(y)+(y,e_{i-1})$. It is clear that $\eta_2 \in X$. We have $\eta_2(e_{i-1}) = \eta_1(e_{i-1}) = 0$, hence

$\{\eta_1,\eta_2\} \subset X$ satisfies (9.1.3), (9.1.4) with respect to e_{i-1}. Since η,η_1,η_2 are distinct it follows that Case 1 is applicable to η_1, so that $[\eta_1,y] = \eta_1(y)$ for all $y \in \widetilde{Y}$. Using now (9.2.3) with $\{\eta',\eta''\}$ replaced by $\{\eta,\eta_1\}$, we see that $[\eta,y] = \eta(y)$ for all $y \in \widetilde{Y}$.

Case 3. There is no $\eta_1 \in X$, $\eta_1 \neq \eta$ such that $\{\eta,\eta_1\}$ satisfies (9.1.3), (9.1.4) with respect to some $e_i \notin \mathrm{Rad}(Y)$. Then, for any $e_i \notin \mathrm{Rad}(Y)$, we must have $\eta(e_i) = 1$. (If we had $\eta(e_i) = 0$, we could define a linear form $\eta_1 : Y \to F_2$ by $\eta_1(y) = \eta(y) + (y,e_i)$ and $\{\eta,\eta_1\}$ would satisfy (9.1.3), (9.1.4) with respect to e_i.) On the other hand, for $e_i \in \mathrm{Rad}(Y)$, we have $\eta(e_i) = \xi_0(e_i)$, since $\eta \in X$. Hence η is uniquely determined (if it exists at all). Since for all $\eta' \in X$, $\eta' \neq \eta$, the identity $[\eta',y] = \eta'(y)$, $(y \in \widetilde{Y})$ is already known, it follows from (9.2.4) that $[\eta,y] = \eta(y)$, $(y \in \widetilde{Y})$. This proves the lemma, hence it proves also 9.2.

9.4. We shall also need a variant of Proposition 9.2. We keep the assumptions of Proposition 9.2 (see 9.1) except that we replace the assumption that for each two element subset $D \subset X$ satisfying (9.1.3), (9.1.4) we are given $\rho_D \in H$ satisfying (9.1.5), (9.1.6), by the following assumption:

(9.4.1) For any $\eta \in X$, we are given an element $\rho_\eta \in H$ which is a combination with integral, ≥ 0 coefficients of basis elements b_α in Θ and such that

$$<\rho_\eta,\rho_{C,\xi}> = \begin{cases} 1 & \text{if } \eta|C = \xi \\ 0 & \text{otherwise}, \end{cases}$$

for any $C \in \mathcal{T}(Y)$ and any linear form $\xi : C \to F_2$ such that $\xi\,|\,\mathrm{Rad}(Y) = \xi_0$.

With this assumption, we shall prove:

(9.4.2) *There exists a bijection* $X \leftrightarrow \Theta$ $(\eta \in X \leftrightarrow b_\eta \in \Theta)$ *such that* $\rho_{C,\xi} = \sum_{\substack{\eta \in X \\ \eta|C=\xi}} b_\eta$ *for any* C,ξ *as above and* $\rho_\eta = b_\eta$ *for any* $\eta \in X$.

For any $\eta \in X$ we define a basis element $b_\eta \in \Theta$ exactly as in the proof of 9.2. As in that proof, we see that $\eta \to b_\eta$ is a 1-1 correspondence between X and Θ and that each $\rho_{C,\xi}$ is a combination with coefficients 0 or 1 of elements b_η, $(\eta \in X)$. Consider an element $\eta \in X$. Let $h_{C'}$ be as in the proof of 9.2. Using (9.4.1), we have $<\rho_\eta, h_{C'}> = 1$, $(\eta \in X)$. Since $h_{C'} = \sum_{\eta \in X} b_\eta$ and ρ_η is a combination of the b_η with ≥ 0 integral coefficients, it follows that ρ_η is equal to one of the basis elements $b_{\eta'}$. From (9.4.1) we have $<\rho_\eta, \rho_{C',\xi'}> = <\rho_\eta, \rho_{C'',\xi''}> = 1$ where $\xi' = \eta|C'$, $\xi'' = \eta|C''$. From the definition of b_η, it now follows that $b_\eta = \rho_\eta$.

We now define $\hat{b}_y$ $(y \in \tilde{Y})$ as in 9.2; for $\eta \in X$, $y \in \tilde{Y}$, we define $[\eta,y] \in F_2$ as in 9.2. In particular we have $(-1)^{[\eta,y]} = 2^\delta <b_\eta, \hat{b}_y>$, ($\delta$ as in 9.2.1). In our case, (9.2.3) is replaced by the identity $(-1)^{[\eta,y]} = (-1)^{\eta(y)}$, $(\eta \in X, y \in \tilde{Y})$; the proof is the same as that of (9.2.3). It follows that $[\eta,y] = \eta(y)$. From the definition of $[\eta,y]$ it now follows that $\rho_{C,\xi} = \sum_{\substack{\eta \in X \\ \eta|C=\xi}} b_\eta$ and (9.4.2) is proved.

9.5. PROPOSITION. *Assume that* (L,n) *is such that any irreducible component of* $W_{L,n}$ *is of type* A, B, C, *or* D *and that* $\varepsilon_\rho = \hat{\varepsilon}_\rho$ *for all* $\rho \in \mathcal{E}_{L,n}$. *(This assumption is automatically satisfied if* q *is sufficiently large, see 6.10.) We assume also that* γ *(see (2.15.1)) has square one, it maps any irreducible component of type* $\neq A_1$ *to itself and is the identity on any irreducible component of type* A_m $(m = 2$ *or* $m \geq 4)$. *Then Theorem 4.23 holds for* (L,n).

Proof. Assume first that $W_{L,n}$ has no factors of type D and γ is the identity. Let s be a semisimple element in $(G^*)^{F'}$ corresponding to (L,n), (see 8.4). We identify $(W_{L,n}, S_{L,n})$ with $(W_{Z(s)}, S_{Z(s)})$ as in 8.4; the action of γ on $W_{L,n}$ corresponds to the action of $(F')^{-1}$ on $W_{Z(s)}$, hence $F' =$ identity on $W_{Z(s)}$.

Let I be a subset of $S_{Z(s)}$ and let $M_{I,1}$ be an F'-stable Levi subgroup of an F'-stable parabolic subgroup of $Z(s)$ of type I. Let $M^*_{I,2}$ be the centralizer in G^* of the largest F_q-split torus in the center of $M^*_{I,1}$. (Then $M^*_{I,1} = M^*_{I,2} \cap Z(s)$.) We have $M^*_{I,1} \subset M^*_{I',1}$ whenever $I \subset I'$. We may conjugate simultaneously all the subgroups $M^*_{I,1}$ by an element $g \in (G^*)^{F'}$ to subgroups M^*_I such that each M^*_I contains T^*. Then M^*_I is the Levi subgroup of an F'-stable parabolic subgroup of G^*. It corresponds to a closed F'-stable subgroup M_I of G containing T such that M_I is the Levi subgroup of an F'-stable parabolic subgroup P_I of G and such that M^*_I is the group dual to M_I. We have $M_I \subset M_{I'}$ whenever $I \subset I'$, and $s' \in M^*_I$ for all I, where s' is the conjugate of s under g. We may clearly assume that $s' = s$.

We may assume that 4.23 holds for $\mathcal{E}_{(s)}$ (with respect to M_I) whenever $M_I \neq G$. (Our assumption implies that $\varepsilon_{\rho'} = \hat{\varepsilon}_{\rho'}$ for any $\rho' \in \mathcal{E}_{(s)}$, with respect to M_I.) For such I, the Weyl group of $M^*_I \cap Z(s)$ may be identified with the parabolic group $\langle I \rangle$ of $W_{Z(s)}$ generated by I. Using (6.4.3), we see that for any irreducible $M_I^{F'}$-module $\rho' \in \mathcal{E}_{(s)}$, (with respect to M_I) and any irreducible $\mathbf{Q}[W_{Z(s)}]$-module E, we have

$$\langle R_E, \mathrm{Ind}_{P_I^{F'}}^{G^{F'}}(\rho') \rangle_{G^{F'}} = \langle R_{\mathrm{Res}\,E}, \rho' \rangle_{M_I^{F'}} . \tag{9.5.1}$$

where Res E is the restriction of E to a $\mathbf{Q}[\langle I \rangle]$-module.

Let $\mathcal{F} \subset (W_{Z(s)})^\vee$ be a family, and let E_0 be the special representation in $\mathcal{F}$. We have $\mathcal{F} = \prod_j \mathcal{F}^{(j)} \subset \prod_j W^{(j)}_{Z(s)} = W_{Z(s)}$, $E_0 = \otimes_j E_0^{(j)}$ where $W^{(j)}_{Z(s)}$ are the irreducible factors of $W_{Z(s)}$, $\mathcal{F}^{(j)}$ are families in $(W^{(j)}_{Z(s)})^\vee$ and $E_0^{(j)}$ is the special representation in $\mathcal{F}^{(j)}$. For any j such that $W^{(j)}_{Z(s)}$ is of type B or C we use the formalism in 4.5 to represent $E_0^{(j)}$ in the form $[Z^{(j)}]$ where $Z^{(j)}$ is a symbol of defect 1; let $Z_1^{(j)}$ be the set of singles in $Z^{(j)}$ and let $Y^{(j)}$ be the F_2-vector space consisting of all subsets of even cardinality of $Z^{(j)}$. Let

$Y = \bigoplus_j Y^{(j)}$, where $Y^{(j)} = 0$ if $W^{(j)}_{Z(s)}$ is of type A. As we have seen in 4.5, each $Y^{(j)}$ has a natural symplectic form and a natural basis. These give rise to a symplectic form $(\ , \ )$ on Y (for which the $Y^{(j)}$ are orthogonal to each other) and to a basis of Y. These satisfy the requirements (a) and (b) in the beginning of 9.1 and we have $\mathrm{Rad}(Y) = 0$. In 4.5 we have described a bijection $\Lambda_M \to M^{\#}$ between $\mathcal{F}^{(j)}$ and a subset of $Y^{(j)}$ ($= V_{Z_1^{(j)}}$, in the language of 4.5); here j is an index such that $W^{(j)}_{Z(s)}$ is of type B or C. Putting these together, we get a bijection between $\mathcal{F}$ and a subset of $Y = \bigoplus_j Y^{(j)}$; it is easy to see (compare [L_{11}, 5.7]) that this subset of Y is $\widetilde{Y}$, (see 9.2). Let E_y be the representation in $\mathcal{F}$ corresponding to $y \in \widetilde{Y}$. This notation is compatible with the notation E_0 for the special representation: it corresponds to $0 \in \widetilde{Y}$.

Using now (5.17.2) for each factor of type B or C of $W_{Z(s)}$ and 5.16 for each factor of type A of $W_{Z(s)}$, we see that

(9.5.2) *For any* $C \in \mathcal{S}(Y)$ *(see 9.1) and any linear form* $\xi : C \to F_2$, *there exists* $x \in W_{Z(s)}$ *such that* $\sum_{y \in C} (-1)^{\xi(y)} E_y = a_x$.

Using now 6.17 (ii), (which is applicable in our case) we see that $(-1)^{\ell(w_1)} R_{a_x}$ is an actual representation of $G^{F'}$ (where w_1 is the element of minimal length in $Z_{L,n}$), for x as in (9.5.2). Hence, (9.5.2) implies

(9.5.3) *For any* $C \in \mathcal{S}(Y)$ *and any linear form* $\xi : C \to F_2$, *the expression*

$$(-1)^{\ell(w_1)} \sum_{y \in C} (-1)^{\xi(y)} R_{E_y} \in \mathcal{R}(G^{F'}) \otimes \mathbf{Q}$$

is an actual representation of $G^{F'}$. *We denote it* $\rho_{C,\xi}$.

Thus the elements $\rho_{C,\xi}$ can be regarded as vectors in the euclidean space $H = \mathcal{R}(G^{F'}) \otimes \mathbf{R}$, which have ≥ 0 integral coordinates with respect to the orthonormal basis consisting of irreducible representations.

If C' is another element of $\mathcal{T}(Y)$ and ξ' is a linear form $\xi' : C' \to F_2$, then the inner product $\langle \rho_{C,\xi}, \rho_{C',\xi'} \rangle_{G^{F'}}$ is given by the formula (9.1.2). (This follows immediately from 3.9 (i).)

It is clear that, for any $y \in \tilde{Y}$, we have

$$(9.5.4) \quad (-1)^{\ell(w_1)} R_{E_y} = 2^{-\delta} \sum_{\xi} (-1)^{\xi(y)} \rho_{C,\xi}, \quad (\delta = (1/2)\dim Y),$$

where C is a fixed subspace in $\mathcal{T}(Y)$ containing y, and the sum is over all linear forms $C \to F_2$. (Compare (9.2.1).) It follows that the irreducible representations of $G^{F'}$ which appear with non-zero coefficient in some R_E $(E \in \mathcal{F})$ are exactly those which appear with >0 coefficient in some $\rho_{C,\xi}$. They form a subset θ of the standard basis of $H = \mathcal{R}(G^{F'}) \otimes \mathbf{R}$. Assume now that either:

(9.5.5) (a) For some j for which $W^{(j)}_{Z(s)}$ is of type A, $E^{(j)}_0$ is not the sign representation or

(b) for some j for which $W^{(j)}_{Z(s)}$ is of type B or C, the symbol $Z^{(j)}$ is such that some integer ≥ 0, does not appear in $Z^{(j)}$ and is strictly smaller than the largest entry in $Z^{(j)}$.

Using (4.4.1) and (4.5.4), we see that there exists a subset $I \subset S_{Z(s)}$, $|I| = |S_{Z(s)}| - 1$ and a family $\mathcal{F}' \subset \langle I \rangle^{\vee}$ such that $J^{W_{Z(s)}}_{\langle I \rangle}$ defines a bijection between $\mathcal{F}'$ and $\mathcal{F}$. Thus, $\mathcal{F}'$ may again be identified with $\tilde{Y} \subset Y$. Let $E'_y \in \mathcal{F}'$ be the representation of $\langle I \rangle$ corresponding to $y \in \tilde{Y}$. Let X be the set of F_2-linear forms on Y. Since 4.23 is assumed to hold for $\mathcal{E}_{(s)}$ with respect to M_I, there exist irreducible representations ρ'_η of $M_I^{F'}$ (for each $\eta \in X$) such that $\langle \rho'_\eta, R_{E'_y} \rangle_{M_I^{F'}} = (-1)^{\ell(w_1)} \cdot (-1)^{\eta(y)} \cdot 2^{-\delta}$. (Here $R_{E'_y} \in \mathcal{R}(M_I^{F'}) \otimes \mathbf{Q}$.) Let us consider $\mathrm{Ind}_{P_I^{F'}}^{G^{F'}}(\rho'_\eta)$ and let ρ_η be its orthogonal projection onto the subspace of $\mathcal{R}(G^{F'}) \otimes \mathbf{R}$ spanned by Θ. It is clear that ρ_η is a combination with ≥ 0, integral coefficients of the elements in Θ, and for any $E_y \in \mathcal{F}$ we have (by 6.17 (i)):

$$\langle R_{E_y}, \mathrm{Ind}_{P_I^{F'}}^{G'}(\rho'_\eta)\rangle_{G^{F'}} = \langle R_{E_y}, \rho_\eta\rangle_{G^{F'}}.$$

On the other hand, by (9.5.1), this is equal to $\langle R_{\mathrm{Res}\,E_y}, \rho'_\eta\rangle_{M_I^{F'}}$, and this is equal to $\langle R_{E'_y}, \rho'_\eta\rangle_{M_I^{F'}}$ since $\mathrm{Res}\,E_y = E'_y$ + irreducible representations of $\langle I\rangle$ in families other than $\mathcal{F}'$, and since 6.17 (i) holds for $M_I^{F'}$. Thus, we have

$$\langle R_{E_y}, \rho_\eta\rangle_{G^{F'}} = \langle R_{E'_y}, \rho'_\eta\rangle_{M_I^{F'}} = (-1)^{\ell(w_1)}(-1)^{\eta(y)}2^{-\delta}, \quad (\eta \in X,\ y \in \tilde{Y}).$$

It follows that

$$\langle \rho_\eta, \rho_{C,\xi}\rangle_{G^{F'}} = (-1)^{\ell(w_1)} \sum_{y \in C} (-1)^{\xi(y)} \langle \rho_\eta, R_{E_y}\rangle_{G^{F'}}$$

$$= \sum_{y \in C} (-1)^{\xi(y)+\eta(y)} 2^{-\delta} = \begin{cases} 1 & \text{if } \eta|C = \xi \\ 0 & \text{if } \eta|C \neq \xi . \end{cases}$$

Thus the ρ_η satisfy (9.4.1). Using now (9.4.2) in our case, it follows that the ρ_η are irreducible, that $\Theta = \{\rho_\eta,\ \eta \in X\}$ and that

$$\text{(9.5.6)} \qquad \langle \rho_\eta, R_{E_y}\rangle_{G^{F'}} = (-1)^{\ell(w_1)}(-1)^{\eta(y)}2^{-\delta} \quad (\eta \in X,\ y \in \tilde{Y}).$$

Hence the multiplicities in R_E $(E \in \mathcal{F})$ are as predicted by 4.23.

By (6.8.6) we have $D(R_{E_y}) = (-1)^{\ell(w_1)} R_{E_y \otimes \mathrm{sign}'}$ $(y \in \tilde{Y})$, and by (6.8.2), (6.18.6), $(-1)^{\ell(w_1)} D(\rho_\eta)$ is an irreducible representation of $G^{F'}$. By (9.5.6) we have $R_{E_y} = (-1)^{\ell(w_1)} 2^{-\delta} \sum_{\eta \in X} (-1)^{\eta(y)} \rho_\eta$. Applying to this

D, we get $R_{E_y \otimes \text{sign}} = (-1)^{\ell(w_1)} 2^{-\delta} \sum_{\eta \epsilon X} (-1)^{\eta(y)} (-1)^{\ell(w_1)} D(\rho_\eta)$. It follows easily that the multiplicities in $R_E(E \epsilon \mathcal{F} \otimes \text{sign})$, are as predicted by 4.23.

We now assume that neither $\mathcal{F}$ nor $\mathcal{F} \otimes \text{sign}$ satisfy (9.5.5). It follows that for each index j, $W^{(j)}_{Z(s)}$ is of type B or C and that $Z^{(j)} = \begin{pmatrix} 0,2,4,\cdots,2d_j \\ 1,3,\cdots,2d_j-1 \end{pmatrix}$, $d_j \geq 1$. (See 8.1.) If the set of indices j is empty, then $W_{Z(s)} = \{e\}$ and 4.23 holds, by 7.6. Hence, we may assume that the set of indices j is non-empty. Then $\dim(Y) \geq 2$.

Let $D = \{\eta', \eta''\}$ be a subset of X (= set of linear forms $Y \to F_2$) satisfying (9.1.3), (9.1.4) for some basis element $e_i \epsilon Y$. (We have automatically $e_i \not\epsilon \text{Rad}(Y)$ since $\text{Rad}(Y) = 0$.) The element e_i belongs to the summand $Y^{(j)}$ of Y. It is easy to see, using (4.5.4), that there exists a subset $I \subset S_{Z(s)}$, $|I| = |S_{Z(s)}| - 1$ and a family $\mathcal{F}' \subset \langle I \rangle^\vee$ such that $J^{W_{Z(s)}}_{\langle I \rangle}$ takes each representation in $\mathcal{F}'$ to the sum of two distinct representations in $\mathcal{F}$ of form $E_y + E_{y+e_i}$ $(y \epsilon \tilde{Y}, (y,e_i) = 0)$ and moreover, $J^{W_{Z(s)}}_{\langle I \rangle}$ defines a bijection between $\mathcal{F}'$ and the set $\{y \epsilon \tilde{Y} | (y,e_i)\}$ modulo the equivalence relation $y \sim y + e_i$. (It follows from $[L_{11}, 3.6]$ that $y \epsilon \tilde{Y}$, $(y,e_i) = 0$ implies $y + e_i \epsilon \tilde{Y}$.) We shall denote $E'_{\bar{y}}$ the representation in $\mathcal{F}'$ corresponding to the class $\bar{y}$ of y. Since 4.23 is assumed to hold for $\mathcal{E}_{(s)}$ with respect to M_I, there exist irreducible representations $\rho'_{\bar{\eta}}$ of $M_I^{F'}$ (for each $\bar{\eta} \epsilon \{\eta \epsilon X | \eta(e_i) = 0\}$ modulo the equivalence relation $\eta \sim \eta + (\ ,e_i)$) such that

$$\langle \rho'_{\bar{\eta}}, R_{E'_{\bar{y}}} \rangle_{M_I^{F'}} = (-1)^{\ell(w_1)} (-1)^{\eta(y)} 2^{-(\delta-1)} .$$

(Here $\bar{\eta}$ is the class of η.) We consider $\text{Ind}_{P_I^{F'}}^{G^{F'}}(\rho'_{\bar{\eta}})$ and we denote $\rho_{\bar{\eta}}$ its orthogonal projection onto the subspace of $\mathcal{R}(G^{F'}) \otimes \mathbf{R}$ spanned by

Θ. In the case where $\bar{\eta}$ is the class of η' and of η'', we denote $\rho_{\bar{\eta}} = \rho_D$. It is clear that ρ_D is a combination with ≥ 0, integral coefficients of elements in Θ and that for any $E_y \in \mathfrak{F}$, we have, (by 6.17 (ii)):

$$<R_{E_y}, \mathrm{Ind}_{P_I^{F'}}^{G^{F'}}(\rho'_{\bar{\eta}})>_{G^{F'}} = <R_{E_y}, \rho_D>_{G^{F'}} .$$

By (9.5.1), this is equal to $<R_{\mathrm{Res}\, E_y}, \rho'_{\bar{\eta}}>_{M_I^{F'}}$ and this is equal to 0 if $(y,e_i) \neq 0$ and to $<R_{E'_y}, \rho'_{\bar{\eta}}>_{M_I^{F'}}$ if $(y,e_i) = 0$, since

$$\mathrm{Res}\, E_y = \begin{cases} 0 & , \text{ if } (y,e_i) \neq 0 \\ E'_{\bar{y}} + \Phi, & \text{if } (y,e_i) = 0 \end{cases}$$

where Φ is a linear combination of irreducible representations of $<I>$ in families other than $\mathfrak{F}'$, and since 6.17 (i) holds for $M_I^{F'}$. Thus, we have

$$<R_{E_y}, \rho_D>_{G^{F'}} =$$

$$\begin{cases} 0 & , \text{ if } y \in \tilde{Y},\ (y,e_i) \neq 0 \\ <R_{E'_{\bar{y}}}, \rho'_{\bar{\eta}}>_{M_I^{F'}} = (-1)^{\ell(w_1)}(-1)^{\eta'(y)}2^{-(\delta-1)}, & \text{if } y \in \tilde{Y},\ (y,e_i) = 0 . \end{cases}$$

It follows that

$$<\rho_D, \rho_{C,\xi}>_{G^{F'}} = (-1)^{\ell(w_1)} \sum_{y \in C} <\rho_D, \rho_{E_y}>_{G^{F'}}$$

$$= \sum_{\substack{y \in C \\ (y,e_i)=0}} (-1)^{\xi(y)+\eta'(y)} 2^{-(\delta-1)}$$

$$= \begin{cases} 2 \text{ if } \eta'|C = \eta''|C = \xi \\ 1 \text{ if } \eta'|C = \xi,\ \eta''|C \neq \xi \\ 1 \text{ if } \eta'|C \neq \xi,\ \eta''|C = \xi \\ 0 \text{ if } \eta'|C \neq \xi,\ \eta''|C \neq \xi . \end{cases}$$

Thus, the ρ_D satisfy (9.1.5) and (9.1.6). Using now 9.2 in our case, it follows that there exists a bijection $\eta \leftrightarrow \rho_\eta$ between X and Θ, such that $\rho_{C,\xi} = \sum_{\substack{\eta \in X \\ \eta|C=\xi}} \rho_\eta$ for any $C \in \mathcal{T}(Y)$ and any linear form $\xi : C \to F_2$.

Using now (9.5.4), we have for any $\eta \in X$, $y \in \tilde{Y}$:

$$\langle \rho_\eta, R_{E_y} \rangle_{G^{F'}} = (-1)^{\ell(w_1)} 2^{-\delta} \sum_{\xi} (-1)^{\xi(y)} \langle \rho_\eta, \rho_{C,\xi} \rangle_{G^{F'}}$$

$$= (-1)^{\ell(w_1)} 2^{-\delta} (-1)^{\eta(y)} .$$

(C is a fixed subspace in $\mathcal{T}(Y)$ containing y.) This completes the proof of the proposition in the case where $W_{L,n}$ has no factors of type D and γ is the identity. In the case where $W_{L,n}$ does have factors of type D or γ is not necessarily the identity, the proof is quite similar. To each family $\mathcal{F}$ stable under γ we attach an F_2-vector space $Y = \oplus Y^{(j)}$ (j parametrizes the γ-stable irreducible factors $W_{L,n}^{(j)}$ of $W_{L,n}$, except that two factors of type A_1 which are switched by γ will be regarded as a single $W_{L,n}^{(j)}$ of type D_2.) We define $Y^{(j)}$ to be zero if $W_{L,n}^{(j)}$ is irreducible of type A; when $W_{L,n}^{(j)}$ is of type B, C, D_m $(m \geq 4)$ or D_3 (with $\gamma \neq 1$) or D_2 (with $\gamma \neq 1$) we represent the special representation of $W_{L,n}^{(j)}$ corresponding to the j-component of $\mathcal{F}$ as $[Z^{(j)}]$ where $Z^{(j)}$ is a symbol (see 4.5, 4.6) and we define $Y^{(j)}$ to be the F_2-vector space of all subsets of even cardinality of the set of singles of $Z^{(j)}$. Then $Y^{(j)}$ has a natural basis and a natural symplectic form. Hence Y will also have a natural basis and a natural symplectic form. The radical Rad(Y) of this form has the property that $\mathrm{Rad}(Y) \cap Y^{(j)}$ is 1-dimensional if $W_{L,n}^{(j)}$ is of type D_m $(m \geq 4)$ or D_3 (with $\gamma \neq 1$) or D_2 (with $\gamma \neq 1$) and is 0 otherwise. We define a linear form $\xi_0 : \mathrm{Rad}(Y) \to F_2$ by the condition that ξ_0 is non-trivial on $\mathrm{Rad}(Y) \cap Y^{(j)}$ if and only if γ acts non-trivially on $W_{L,n}^{(j)}$. The results in 9.2 and 9.4 are applicable to Y, and the proof is carried out as above.

9.6. We shall need a variation of Proposition 9.5. We again assume that (L,n) is such that any irreducible component of $W_{L,n}$ is of type A, B, C or D and that $\varepsilon_\rho = \hat{\varepsilon}_\rho$ for all $\rho \in \mathcal{E}_{L,n}$. We assume that all irreducible factors of $W_{L,n}$ except for two are of type A and that the remaining two factors are of type B, C or D and are interchanged by γ. We assume also that, for any irreducible factor of $W_{L,n}$, the smallest power of γ keeping stable that factor acts trivially on that component (if it has type A) and has order $\neq 3$ on that component (if it has type D_4). We want to prove that

(9.6.1) *With these assumptions, Theorem 4.23 holds for* $\mathcal{E}_{L,n}$.

The proof is essentially the same as that of 9.5; the only difference is that instead of (5.17.3) we must use an analogue of (5.17.3) for a product of two classical Weyl groups which are switched among them by γ. We shall give the precise formulation for Weyl groups of type $B_m \times B_m$. (The cases $C_m \times C_m$, $D_m \times D_m$ are treated in exactly the same way.)

Let $(\mathbf{W},S)$ be a Weyl group such that $\mathbf{W} = \mathbf{W}_1 \times \mathbf{W}_1$ with $\mathbf{W}_1$ of type B_m, $(m \geq 2)$, and let γ be the automorphism of $(\mathbf{W},S)$ which interchanges the two factors of $\mathbf{W}$. For each irreducible representation E of $\mathbf{W}_1$, we denote $\widetilde{E \otimes E}$ (resp. $\widehat{E \otimes E}$) the representation $E \otimes E$ of $\mathbf{W}_1 \times \mathbf{W}_1$ extended to a representation of $\tilde{\mathbf{W}}$ (with respect to γ) such that $\gamma(e \otimes e') = e' \otimes e$ (resp. $\gamma(e \otimes e') = -e' \otimes e$).

Let Z be a symbol of rank m and defect 1 such that [Z] is a special representation for $\mathbf{W}_1$. Let Y be the F_2-vector space consisting of all subsets of even cardinality of the set of singles of Z. Let $C \in \mathcal{S}(V)$ and let $\xi : C \to F_2$ be a linear form. As in the proof 9.5, the irreducible representations of $\mathbf{W}_1$ in the family of [Z] may be parametrized by the elements in a subset $\tilde{Y}$ of Y; we denote E_y the representation of $\mathbf{W}_1$ corresponding to $y \in \tilde{Y}$. Then the virtual representation

$$\sum_{y \in C} (-1)^{\xi(y)} E_y$$

of $\mathbf{W}_1$ is of the form $c(Z,\Phi,\hat{\Phi})$, see (5.17.2).

For the proof of (9.6.1) we only need the following statement.

(9.6.2) *If* $\xi : C \to F_2$ *is a linear form, then the element*

$$\sigma_{C,\xi} = \sum_{y \in C} (-1)^{\xi(y)} \frac{1}{2} (\widetilde{E_y \otimes E_y} - \widehat{E_y \otimes E_y}) \in \mathcal{R}(\mathbf{W},\gamma)$$

is of the form $a_{\gamma x}$ *for some* $x \in \mathbf{W}$.

This is proved by the method in $[L_{11}, 2.20]$ in exactly the same way as (5.17.3).

9.7. In the rest of this chapter G not only has connected center, but it is assumed to be a classical group in the following sense: all irreducible factors of the Weyl group W are of type A, B, C, or D and the smallest power of F' leaving stable such a factor is an automorphism of order 1 or 2 of that factor. We shall verify that Theorem 4.23 holds for certain pairs (L,n) satisfying $Z_{L,n} \neq \emptyset$ for G.

9.8. PROPOSITION. *Assume that* $p = 2$. *Let* $s \in (G^*)^{F'}$ *be a semi-simple element corresponding to* (L,n), *(see 8.4). Then the number of irreducible representations of* $G^{F'}$ *in* $\mathcal{E}_{L,n} = \mathcal{E}_{(s)}$ *is equal to the number of conjugacy classes of* p*-elements in the finite group* $(Z(s)/its\ center)^{F'}$.

Proof. Since $p = 2$, there exists a (not necessarily F'-stable) parabolic subgroup of G^* such that Z(s) is a Levi subgroup of it, and there exists a (not necessarily F'-stable) parabolic subgroup of G with an F'-stable Levi subgroup M such that Z(s) is the group dual to M. Using 6.21 we have a bijection between $\mathcal{E}_{(s)}$ (relative to G) and $\mathcal{E}_{(s)}$ (relative to M). By induction on dim(G), we may assume that the proposition is true if G is replaced by M $(M \subsetneq G)$. It follows that it is true whenever s is not in the center of G^*. If s is in the center of G^*, it is easy to see that $\#\mathcal{E}_{(s)} = \#\mathcal{E}_{(1)}$, hence we are reduced to the case where $s = 1$. To prove the proposition in this case, we may replace G by the corresponding adjoint group; we may, moreover, assume that G is a product $G_{(1)} \times G_{(2)}$

where $G_{(1)}$ has only factors of type A and $G_{(2)}$ has only factors of type $\neq A$. Furthermore, we can assume that we have either $G = G_{(1)}$ (in which case 4.23 is known to hold, see 7.6) or that $G = G_{(2)}$. Hence we can assume that G is adjoint without factors of type A. In this case, G^* has again trivial center (although it is not necessarily adjoint) and for any $s \in G^*$, $Z(s)$ has connected center; it follows that the numbers of classes of p-elements in $Z(s)^{F'}$ and in $(Z(s)/\text{its center})^{F'}$ coincide. It is then enough to prove the following identity

$$\sum_{(s)} \#\mathcal{E}_{(s)} = \sum_{(s)} \#\{\text{number of classes of p-elements in } Z(s)^{F'}\},$$

(both sums are over the set of semisimple classes in $(G^*)^{F'}$).

The left-hand side is the number of irreducible representations of $G^{F'}$ (see (8.4.4)) hence it is also the number of conjugacy classes in $G^{F'}$; the right-hand side is the number of conjugacy classes in $(G^*)^{F'}$. Hence it is enough to show that $G^{F'}$ and $(G^*)^{F'}$ have the same number of conjugacy classes. But it is well known that $G^{F'}$ and $(G^*)^{F'}$ are isomorphic as abstract groups (using the fact that $p = 2$ and that G has only factors of type B, C, D). The proposition is proved.

9.9. In addition to the assumption in 9.7, we assume that (L,n) is such that γ acts trivially on $W_{L,n}$. We shall prove that Theorem 4.23 holds for $\mathcal{E}_{L,n}$ without restriction on q. At the same time we shall prove (with the notations in the proof of 9.5)

(9.9.1) $\mathcal{E}_{(s)}$ *contains at most one cuspidal representation. It has one if and only if each irreducible factor of* $W_{Z(s)}$ *has even rank,* $(W_{Z(s)})^\vee$ *has a cuspidal family, and the largest* F_q*-split torus in the center of* $Z(s)$ *coincides with the largest* F_q*-split torus in the center of* G^*.

We shall use the notations in the proof of 9.5. We may assume that 4.23 and (9.9.1) hold for $\mathcal{E}_{(s)}$ (with respect to M_I) for any $I \subset S_{Z(s)}$ such that $M_I \neq G$.

We shall first prove the following result.

9.10. LEMMA. *With the assumptions in 9.9, we have* $\varepsilon_\rho = \hat{\varepsilon}_\rho$ *for any* $\rho \in \mathcal{E}_{(s)}$ *(with respect to* G *).*

As in the proof of 7.12, it is enough (using the induction hypothesis) to check that $\varepsilon_\rho = (-1)^{\sigma(G)}$ is satisfied whenever $\rho \in \mathcal{E}_{(s)}$ is cuspidal. According to (6.18.6) and 8.4, for all ρ in $\mathcal{E}_{(s)}$, we have $\varepsilon_\rho = (-1)^{\ell(w_1)} = (-1)^{\sigma(G)-\sigma(Z(s))}$. Hence we must only check that, if $\sigma(Z(s))$ is odd, then $\mathcal{E}_{(s)}$ contains no cuspidal representations. If $G \neq M_I$, for $I = S_{Z(s)}$, then, by (8.4.5), $\mathcal{E}_{(s)}$ contains no cuspidal representations, so there is nothing to prove. So, we may assume that $G = M_I$, for $I = S_{Z(s)}$. We shall say that $\rho \in \mathcal{E}_{(s)}$ is *subcuspidal* if it is contained in $\mathrm{Ind}_{P^{F'}}^{G^{F'}}(\delta)$ where ρ is a *maximal* F′-stable parabolic subgroup of G and δ is an irreducible cuspidal representation of $(P/U_P)^{F'}$. From the induction hypothesis and from 8.6, 8.7 we get a classification of all representations in $\mathcal{E}_{(s)}$ which are neither cuspidal nor subcuspidal: they are of the form $\delta_I[\chi]$ where δ_I is the cuspidal representation in $\mathcal{E}_{(s')}$ (corresponding to M_I) and $I \subset S_{Z(s)}$ is a subset such that $|I| \leq |S_{Z(s)}| - 2$ and such that the corresponding subgroup $<I> \subset W_{Z(s)}$ possesses a cuspidal family and has only factors of even rank; χ is any irreducible representation of the Coxeter group $\mathfrak{W}_I$ defined in terms of $\mathbf{W} = W_{Z(s)}$, $\mathbf{W}' = <I>$, $\gamma = 1$ (see 8.2). Note that $\delta_I[\chi] = \delta_{I'}[\chi']$ implies that there exists $w \in W_{Z(s)}$ such that $wIw^{-1} = I'$, hence that $I = I'$ (see (8.2.1)), and also that $\chi = \chi'$ (see 8.7). We shall consider separately the case where $p \neq 2$ and the case where $p = 2$.

Case 1 ($p \neq 2$). We can compute explicitly, using 8.7, the largest power of p dividing $\dim \delta_I[\chi]$, for any $\delta_I[\chi]$ as above. This is equal to $p^{n_1+n_2}$ where p^{n_1} is the largest power of p dividing $\dim(\delta_I)$ (which is determined by (4.26.3) since $\{\overline{x}, x_{\widetilde{E}_1}\}$ in that formula is $\pm 1/2^d$ and

$p \neq 2$) and p^{n_2} is the largest power of p dividing the formal degree D_χ in 8.7 (which is known from the work of Hoefsmit [Hf], [L_3, §4] and from the explicit description of the parameters $\lambda(\alpha)$ given in 8.6). We find that $p^{n_1+n_2}$ is in fact a power of q. Let $n(a)$ be the number of $\rho \in E_{(s)}$ which are not cuspidal nor subcuspidal such that the largest power of q dividing $\dim(\rho)$ is q^a. We find that

(9.10.1) *If* $\mathcal{E}_{(s)}$ *does not contain subcuspidal representations then* $\sum_{\substack{E \\ a_E = a}} f_E^2 = n(a)$ *for all* a. *If* $\mathcal{E}_{(s)}$ *contains subcuspidal representations, then*

$$\sum_{\substack{E \\ a_E = a}} f_E^2 = \begin{cases} n(a) & , \textit{for all but two values of } a \\ n(a)+1, & \textit{for two values of } a . \end{cases}$$

(E *runs over all special representations of* $W_{Z(s)}$ *such that* $a_E = a$.)

Let $\mathcal{E}_{(s),a}$ be the set of all $\rho \in \mathcal{E}_{(s)}$ such that the largest power of q dividing $\dim(\rho)$ is q^a.

We now show by induction on a that

(9.10.2) *For any irreducible representation* E *of* $W_{Z(s)}$ *with* $a_E \leq a$, $f_E R_E$ *is a combination with coefficients* ± 1 *of distinct irreducible representations* $\rho \in \mathcal{E}_{(s),a}$. *Moreover, if* E, E' *are in different families and* $a_E \leq a$, $a_{E'} \leq a$ *then* $R_E, R_{E'}$ *are disjoint.*

We first show that $f_E \cdot E$ and $(1/2) f_E(E+E_0)$ where E_0 is the special representation in the family of E are $\mathbb{Z}$-linear combinations of virtual representations of form α_x, $(x \in W_{Z(s)})$. We shall assume, for simplicity, that $W_{Z(s)}$ is irreducible of type B_n; the proof in the general case is similar. With the notations in 5.17, we can write $E_0 = [Z]$ and we can find an admissible arrangement Φ for Z such that $\pm E = \pm[\Lambda]$ is the term corresponding to $\Psi \subset \Phi$ in the sum (5.17.2) defining $c(Z,\Phi,\hat{\Phi})$, (see $[L_{11}, 2.22]$), for any $\hat{\Phi} \subset \Phi$. Then $[L_{11}, (2.22.1)]$ expresses $f_{[\Lambda]}[\Lambda]$ as a combination with coefficients ± 1 of the $c(Z,\Phi,\hat{\Phi})$ for the various $\hat{\Phi}$ in Φ. Hence $f_E \cdot E$ is as stated, (see (5.17.3)). By $[L_{11}, (2.22.1)]$ we can also express $f_{[Z]}[Z]$ as the sum (with coefficients $+1$) of the $c(Z,\Phi,\hat{\Phi})$ for the various $\hat{\Phi}$ in Φ. Since $f_{[\Lambda]} = f_{[Z]}$, we have $(1/2)f_{[\Lambda]}([\Lambda]+[Z])$ = combination (with coefficients $+1$) of the $c(Z,\Phi,\hat{\Phi})$ for those $\hat{\Phi} \subset \Phi$ such that $c(Z,\Phi,\hat{\Phi})$ appear with coefficient $+1$ in $f_{[\Lambda]}[\Lambda]$. Hence, by (5.17.3), $(1/2)f_E(E+E_0)$ is as stated.

It follows now from 6.14 (i) that $f_E R_E$ and $(1/2)f_E(R_E + R_{E_0})$ are virtual representations of $G^{F'}$.

We now go to the proof of (9.10.2). For $a < 0$, (9.10.2) is obvious since then there is no E satisfying $a_E \leq a$. Assume now that $a \geq 0$ and that (9.10.2) is proved for a replaced by $a-1$. It follows that for $\rho \in \mathcal{E}_{(s),a}$ we have $\langle \rho, R_E \rangle_{G^{F'}} = 0$ if $a_E < a$. From what it was said in 4.26, it follows that

$$\dim(\rho) = \sum_E \langle \rho, R_E \rangle_{G^{F'}} \dim(R_E), \text{ (sum over all } E \in W^{\vee}_{Z(s)}) .$$

Hence we have also $\dim(\rho) = \sum_{\substack{E \\ a_E \geq a}} \langle \rho, R_E \rangle_{G^{F'}} \dim(R_E)$. The coefficients $\langle \rho, R_E \rangle_{G^{F'}}$ are rational numbers without p in denominator. (Indeed, $f_E R_E$ is a virtual representation hence $\langle \rho, R_E \rangle_{G^{F'}}$ is an integer divided by a power of 2.) Since q^{a+1} does not divide $\dim(\rho)$, there must exist E with $a_E \geq a$, $\langle \rho, R_E \rangle_{G^{F'}} \neq 0$, such that q^{a+1} does not divide $\dim(R_E)$. For such E, we have $b_E \leq a \leq a_E$ hence E is special. Thus, there exists a special representation E such that $a_E = a$, $\langle \rho, f_E R_E \rangle_{G^{F'}} \neq 0$. This last inner product is an integer hence its square is ≥ 1. It follows that the integer $N(a) = |\mathcal{E}_{(s),a}|$ satisfies

$$N(a) \le \sum_{E} \langle f_E R_E, f_E R_E \rangle_{G^{F'}} = \sum_{E} f_E^2$$

(sum over all special E such that $a_E = a$). Moreover, we see that, if at least one of the multiplicities $\langle \rho, f_E R_E \rangle_{G^{F'}}$ above is ≥ 2 in absolute value then $N(a) \leq \sum_E f_E^2 - 3$. On the other hand from (9.10.1) we see that $\sum_E f_E^2 \leq n(a) + 1 \leq N(a) + 1$ hence $N(a) \geq \sum_E f_E^2 - 1$. It follows that the multiplicities $\langle \rho, f_E R_E \rangle_{G^{F'}}$ above must be ≤ 1 in absolute value. It also follows that the virtual representations $f_E R_E$ (E special, $a_E = a$) are disjoint, with multiplicities ± 1 and that all their components (except possibly for one) are in $\mathcal{E}_{(s),a}$. Assume that the exceptional situation actually occurs, i.e. that there exists $\rho \notin \mathcal{E}_{(s),a}$ and E special with $a_E = a$ and $\langle \rho, f_E R_E \rangle_{G^{F'}} = 1$. For all other components ρ'_i of $f_E R_E$ we have $\rho'_i \in \mathcal{E}_{(s),a}$ hence $\langle \rho'_i, R_{E'} \rangle_{G^{F'}} = 0$ if E' is irreducible and $a_{E'} < a$ (by the induction hypothesis). Hence for such E' we have $0 = \langle f_E R_E, R_{E'} \rangle_{G^{F'}} = \langle \pm\rho + \Sigma(\pm\rho'_i), R_{E'} \rangle_{G^{F'}} = \pm\langle \rho, R_{E'} \rangle_{G^{F'}}$. Hence $\langle \rho, R_{E'} \rangle_{G^{F'}} = 0$. Note also that $\langle \rho, R_{E''} \rangle_{G^{F'}} = 0$ if E'' is special, $a_{E''} = a$, $E'' \neq E$ since $R_{E''}$ and R_E are disjoint. It follows that

$$\dim(\rho) = \pm(1/f_E)\dim R_E + \sum_{\substack{E_1 \\ b_{E_1} > a}} \langle \rho, R_{E_1} \rangle_{G^{F'}} \dim R_{E_1}$$

which shows that

$$\dim(\rho) = (1/f_E)q^a + \mathbf{Z}[1/2]\text{-combination of } q^{a+1}, q^{a+2}, \ldots$$

hence $\rho \in \mathcal{E}_{(s),a}$, a contradiction. We have proved that for any special representation E with $a_E = a$, $f_E R_E$ is a combination with coefficients ± 1 of (distinct) representations in $\mathcal{E}_{(s),a}$. Now let E_1 be an irreducible representation of $W_{Z(s)}$ with $a_{E_1} = a$ and let E be the special representation in the family of E_1. Then $f_{E_1} = f_E$ and both $f_E R_E$, $f_E R_{E_1}$ are virtual representations. If some $\rho \in \mathcal{E}_{(s),a}$ is a component of $f_{E_1} R_E$ (necessarily with coefficient ± 1) then it must appear with odd coefficient in $f_{E_1} R_{E_1}$ since $(1/2) f_{E_1}(R_E + R_{E_1})$ is a virtual representation. Since $\langle f_{E_1} R_{E_1}, f_{E_1} R_{E_1} \rangle_{G^{F'}} = \langle f_{E_1} R_E, f_{E_1} R_E \rangle = f_{E_1}^2$, it follows that $f_{E_1} R_{E_1}$ is a combination with coefficients ± 1 of exactly the same representations ρ which enter in $f_{E_1} R_E$. (The signs might differ from those in $f_{E_1} R_E$.) From this it follows that (9.10.2) holds for our a, and the proof of (9.10.2) is completed.

From (9.10.2) it follows that $N(a) = \sum_{\substack{E \text{ special} \\ a_E = a}} f_E^2$. Using now (9.10.1), it follows that either A) $\sum_a N(a) = \sum_a n(a)$, or B) $\sum_a N(a) = 2 + \sum_a n(a)$ and in the case B), $\mathcal{E}_{(s)}$ contains some subcuspidal representation. In case A), it follows that $\mathcal{E}_{(s)}$ contains no cuspidal representations. In case B) it follows that either $\mathcal{E}_{(s)}$ contains two subcuspidal and no cuspidal representations or that $\mathcal{E}_{(s)}$ contains exactly one subcuspidal and one cuspidal representation. We now show that the last case is impossible. Assume that $\mathcal{E}_{(s)}$ contains exactly one subcuspidal representation. Then there exists $I \subset S_{Z(s)}$, $|I| = |S_{Z(s)}| - 1$ such that $\mathcal{E}_{(s')}$ (corresponding to M_I) has a unique cuspidal representation δ_I and such that the stabilizer of δ_I in $\mathfrak{N}_G(M_I)^{F'}/M_I^{F'}$ is trivial. Consider the imbedding (8.5.4) with M^* replaced by M_I^* and s by s′. It is an imbedding of a group of order 2 (cf. 8.2.1) into a group of order 1 or 2 (since P_I is a maximal F′-stable parabolic subgroup), hence it is an isomorphism. The non-trivial element in $\mathfrak{N}_{G^*}(M_I^*)^{F'}/(M_I^*)^{F'}$ can be represented by an element in Z(s′) hence it maps $\mathcal{E}_{(s')}$ (corresponding to M_I) into itself. It must take δ_I to a cuspidal representation, hence it takes δ_I to itself, since

δ_I is the only cuspidal representation in $\mathcal{E}_{(s')}$ (corresponding to M_I). This shows that δ_I has a non-trivial stabilizer in $\mathfrak{N}_{G^*}(M_I^*)^{F'}/(M_I^*)^{F'} = \mathfrak{N}_G(M_I)^{F'}/M_I^{F'}$. This is a contradiction. Thus, $\mathcal{E}_{(s)}$ contains no cuspidal representations and the lemma is proved for $p \neq 2$.

Case 2 $(p = 2)$. Let m be an integer ≥ 1 with the following properties: (a) $q^m >$ height of any coroot of G, (b) F'^m acts on W and on $W_{Z(s)}$ in the same way as F', (c) for any unipotent element $u \in \bar{Z}(s) = Z(s)/$its center, the order $|\Gamma_u|$ of the group of connected components of the centralizer of u in $\bar{Z}(s)$ satisfies $m \equiv 1 \pmod{|\Gamma_u|}$. Such an integer exists, since there are only finitely many unipotent classes in $\bar{Z}(s)$. We shall denote by $\mathcal{E}^m_{(s)}$, the set $\mathcal{E}_{(s)}$ defined with respect to F'^m instead of F'.

By an earlier part of the proof we see that the number of representations of $G^{F'}$ in $\mathcal{E}_{(s)}$ which are not cuspidal and not subcuspidal is the same as the number of representations of $G^{F'^m}$ in $\mathcal{E}^m_{(s)}$ which are not cuspidal and not subcuspidal. As in the case $p \neq 2$ we see that either both $\mathcal{E}_{(s)}$ and $\mathcal{E}^m_{(s)}$ contain no subcuspidal representations or both contain exactly two subcuspidal representations. If we show that $\#\mathcal{E}_{(s)} \leq \#\mathcal{E}^m_{(s)}$ it will follow that $\mathcal{E}_{(s)}$ does not contain cuspidal representations whenever $\mathcal{E}^m_{(s)}$ does not contain cuspidal representations. If $\sigma(Z(s))$ is odd (σ defined with respect to F'^m) then $\mathcal{E}^m_{(s)}$ contains no cuspidal representations, since the lemma holds for $\mathcal{E}^m_{(s)}$ by 6.10. Hence $\mathcal{E}_{(s)}$ contains no cuspidal representations. Note that $\sigma(Z(s))$ is the same when it is defined with respect to F' or F'^m since F'^m acts on $W_{Z(s)}$ in the same way as F'. Hence the lemma would follow.

It remains to check that $\#\mathcal{E}_{(s)} \leq \#\mathcal{E}^m_{(s)}$. Let C be a unipotent class in $\bar{Z}(s)$ such that $C^{F'} = \emptyset$; let $z \in C^{F'}$ and let Γ be the group of connected components of the centralizer of z in $\bar{Z}(s)$. The number of $\bar{Z}(s)^{F'}$-orbits in $C^{F'}$ is equal to $\#\{\Gamma$ modulo equivalence relation $g \sim g_0^{-1} g F(g_0), g_0 \in \Gamma\}$ and the number of $\bar{Z}(s)^{F'^m}$-orbits in $C^{F'^m}$ is equal to

$$\#\{\Gamma \text{ modulo equivalence relation } g \sim g_0^{-1} g F^m(g_0), g_0 \in \Gamma\}.$$

These two numbers coincide since, by the definition of m, we have $F^m = F$ on Γ. It follows that the number of conjugacy classes of p-elements in $\bar{Z}(s)^{F'}$ is not bigger than the analogous number for $\bar{Z}(s)^{F'^m}$. Using 9.1, it follows that $\# \mathcal{E}_{(s)} \leq \# \mathcal{E}^m_{(s)}$. The lemma is proved.

9.11. With the assumptions in 9.9, it now follows from 9.5 and 9.10 that 4.23 holds for $\mathcal{E}_{L,n}$. We now show that (9.9.1) holds for $\mathcal{E}_{L,n} = \mathcal{E}_{(s)}$. By (8.4.5) we may assume that the largest F_q-split tori in the centers of $Z(s)$ and G^* coincide. By (8.5.6) we may also assume that $(W_{Z(s)})^\vee$ contains a cuspidal family. Using 8.6, 8.7 and (9.9.1) for smaller groups we may count the number of non-cuspidal representations in $\mathcal{E}_{(s)}$. From 4.23 we know the total number of representations in $\mathcal{E}_{(s)}$. Subtracting, we find that the number of cuspidal representations in $\mathcal{E}_{(s)}$ is as stated in (9.9.1).

The proof of Lemma 9.10 remains valid (with minor changes) if instead of assuming that γ = identity, we assume that γ acts on $W_{L,n}$ as in the statement of 9.5 or as in the beginning of 9.6. It follows that in these more general cases, we have automatically $\varepsilon_\rho = \hat{\varepsilon}_\rho$ for all $\rho \in \mathcal{E}_{L,n}$ and hence the assumption $\varepsilon_\rho = \hat{\varepsilon}_\rho$ $(\rho \in \mathcal{E}_{L,n})$ can be removed from 9.5 and (9.6.1). Hence the conclusions of 9.5 and (9.6.1) are valid without restriction on q.

9.12. We now show that Theorem 4.23 holds for G as in 9.7. By 8.8, we can assume that the derived group of G is simply-connected and absolutely simple module its center. By 7.6, we may assume that $D(G)$ is not of type A. By 6.21, we can furthermore reduce ourselves to the case of $\mathcal{E}_{(s)}$ for $s \in (G^*)^{F'}$ a semisimple element such that the identity component of the center of $Z(s)$ is the same as the identity component of the center of G. In this case, $W_{Z(s)}$ is irreducible of type B, C, or D, or is a product of two such groups (which may be permuted by F'). In this case, the discussion in 9.11 is applicable, and the truth of 4.23 for our G is established.

Chapter 10

COMPLETION OF THE PROOF OF THEOREM 4.23

From the results in Chapter 7 it follows that Theorem 4.23 holds for G and any semisimple element $s \in (G^*)^{F'}$ provided that the Weyl group of G is of type G_2, D_4, E_6 or E_7, and that it holds for any non-central semisimple element $s \in (G^*)^{F'}$ provided that the Weyl group of G is of type E_8 or F_4. The case of classical groups was considered in Chapter 9. By the results in 8.8, in order to verify that 4.23 holds in general, it is enough to verify it in the case where G is simple adjoint of type E_8 or F_4 and $s = e \in (G^*)^{F'}$ is the identity element. This case will be considered in this chapter. We shall write $\mathcal{E}$, $\mathcal{E}^{I}$, $\mathcal{E}^{II}$ instead of $\mathcal{E}_{L_0,1}$, $\mathcal{E}^{I}_{L_0,1}$, $\mathcal{E}^{II}_{L_0,1}$ where L_0 is the trivial bundle. Unless otherwise indicated, in this chapter G is assumed to be simple adjoint of type E_8 or F_4.

10.1. PROPOSITION. *For any* $\rho \in \mathcal{E}$ *we have* $\varepsilon_\rho = \hat{\varepsilon}_\rho$. *Hence the conclusions of 6.17 hold for* $\mathcal{E}$.

Proof. As the similar result for smaller groups is already known (7.6, 7.11, 7.12), we see as in the proof of 7.12 that it is enough to show that $\varepsilon_\rho = \hat{\varepsilon}_\rho$ for any cuspidal $\rho \in \mathcal{E}$. In the proof of (6.18.8) we have seen that, if G is of type E_8 and $\rho \in \mathcal{E}^{II}$ then ρ cannot be cuspidal; if G is of type F_4, then from (6.18.6) we see that $\mathcal{E}^{II}$ is empty. In both cases, it follows that any cuspidal ρ in $\mathcal{E}$ must satisfy $\varepsilon_\rho = 1$; it clearly satisfies $\hat{\varepsilon}_\rho = 1$ since $\sigma(G) = 4$ or 8 (see 6.7). The proposition is proved.

10.2. For any family $\mathfrak{F} \subset W^{\vee}$ with exactly two elements (which exists for E_8 but not F_4) we see from (6.18.8) that R_E $(E \in \mathfrak{F})$ decomposes according to the pattern in 4.23. For a family $\mathcal{F} \subset W^\vee$ with exactly one element we see from 5.20, 5.23 and 6.17 that R_E, $(\mathcal{F} = \{E\})$, is an irreducible representation of $G^{F'}$ as predicted by 4.23. For a family $\mathcal{F} = \{E,E_1,E_2\} \subset W^\vee$ with E special, we see from 5.20, 5.23 and 6.17 that $R_E \pm R_{E_1}$, $R_E \pm R_{E_2}$ are actual representations of $G^{F'}$, hence, using 7.7, R_E, R_{E_1}, R_{E_2} decompose according to the pattern in 4.23. Similarly, for any family $\mathcal{F} \subset W^\vee$ with exactly five elements (which exists for E_8 but not for F_4), we see from 7.3, 7.7 and 6.17 that R_E, $(E \in \mathcal{F})$, decompose according to the pattern in 4.23. It remains to analyze the case of the cuspidal family $\mathcal{F}_0$ of W (of type E_8 or F_4).

The non-cuspidal representations in $\mathcal{E}$ can be classified as in 8.6, 8.7 (see the appendix). We say that $\rho \in \mathcal{E}$ corresponds to the family $\mathcal{F}$ (and write $\rho : \mathcal{F}$) if $\langle \rho, R_E \rangle_{G^{F'}} \neq 0$ for some $E \in \mathcal{F}$ or, equivalently, if $\langle \rho, R_{E'} \rangle_{G^{F'}} = 0$ for all $E' \in W^\vee$, $E' \notin \mathcal{F}$, (see 6.17 (i)). We want to determine explicitly how the various non-cuspidal representations in $\mathcal{E}$ correspond to families. (From the proof of (8.5.6) we see that any cuspidal representation in $\mathcal{E}$ corresponds to a family $\mathcal{F}$ such that $\mathcal{F} = \mathcal{F} \otimes \text{sign}$. In the case F_4 it follows that $\mathcal{F} = \mathcal{F}_0$; in the case E_8 one observes that the irreducible representation R_{2100_y} is not cuspidal since it appears in $\mathrm{Ind}_{B^{F'}}^{G^{F'}}(1)$ and it follows again that $\mathcal{F} = \mathcal{F}_0$.)

We first consider the irreducible representations $[\chi]$ in $\mathrm{Ind}_{B^{F'}}^{G^{F'}}(1)$, (with the notations in the appendix); thus χ is an irreducible representation of W. For each $\mathcal{F} \neq \mathcal{F}_0$ and each $\chi \in \mathcal{F}$, we define $\rho_\chi \in \mathcal{E}$, $\rho_\chi : \mathcal{F}$ to be the representation corresponding under 4.23 to the element of $\mathfrak{M}(\mathcal{G}_{\mathcal{F}})$ which is the image of χ under the imbedding $\mathcal{F} \hookrightarrow \mathfrak{M}(\mathcal{G}_{\mathcal{F}})$. For each parabolic subgroup P containing B, the inner product $\langle \rho_\chi, \mathrm{Ind}_{P^{F'}}^{G^{F'}}(1) \rangle_{G^{F'}}$ can be computed explicitly from 4.23 and is equal to $\langle [\chi], \mathrm{Ind}_{P^{F'}}^{G^{F'}}(1) \rangle_{G^{F'}} =$

$<\chi, \mathrm{Ind}_{W_P}^{W}(1)>_W$, where W_P is the parabolic subgroup of W corresponding to P. (Note that $\mathrm{Ind}_{P^{F'}}^{G^{F'}}(1) = \sum_E <E, \mathrm{Ind}_{W_P}^{W}(1)>_W R_E$ and $<\rho_\chi, R_E>_{G^{F'}} = <\rho_E, R_\chi>_{G^{F'}}$.) In particular, ρ_χ is a component of $\mathrm{Ind}_{B^{F'}}^{G^{F'}}(1)$; by a result of Benson-Curtis, [BC], χ is uniquely determined by the multiplicities $<\chi, \mathrm{Ind}_{W_P}^{W}(1)>_W$ for the various P, except for a small ambiguity for χ of degree 4096 in type E_8. It follows that $\rho_\chi = [\chi]$ for $\chi \notin \mathfrak{F}_0$, (except possibly that if $\{\chi_1, \chi_2\}$ is a family with two elements for W of type E_8 we have $\{\rho_{\chi_1} \rho_{\chi_2}\} = \{[\chi_1], [\chi_2]\}$ but ρ_{χ_i} is not necessarily equal to $[\chi_i]$). From 4.23, we see also that for $\rho : \mathfrak{F}$ which is not of form ρ_χ, $(\chi \in \mathfrak{F})$, we have $<\rho, \mathrm{Ind}_{B^{F'}}^{G^{F'}}(1)>_{G^{F'}} = 0$, hence ρ is not of form $[\chi']$, $\chi' \in W^\vee$. It follows that the representations $\rho \in \mathcal{E}$, $\rho : \mathfrak{F}_0$ which appear in $\mathrm{Ind}_{B^{F'}}^{G^{F'}}(1)$ are precisely the $[\chi]$, for $\chi \in \mathfrak{F}_0$.

Let us now consider, for G of type F_4, the unipotent representations of type $B_2[\chi]$, (with the notations in the appendix) where χ is an irreducible representation of a certain Coxeter group of type B_2. Let $\mathfrak{F}_1$ be the family of 4_2 and let $\mathfrak{F}_2 = \mathfrak{F}_1 \otimes \mathrm{sign}$. Let ρ_i be the unique representation in $\mathcal{E}$ such that $\rho_i : \mathfrak{F}_i$ and ρ_i corresponds under 4.23 to $(g_2, \varepsilon) \in \mathfrak{M}(\mathcal{G}_{\mathfrak{F}_i}) = \mathfrak{M}(\mathfrak{G}_2)$, $(i = 1$ or $2)$. Then ρ_i does not appear in $\mathrm{Ind}_{B^{F'}}^{G^{F'}}(1)$ and is not cuspidal since $\mathfrak{F} \neq \mathfrak{F}_0$ hence it must be of form $B_2[\chi]$. Its degree is known from (4.26.2). With the notations in 8.7, it follows that $\mathfrak{D}_{\chi_1} = 1$, $D_{\chi_2} = q^{12}$. This forces χ_1 to be the unit representation and χ_2 to be the sign representation of W. For the other three irreducible representations χ of W we must necessarily have $B_2[\chi] : \mathfrak{F}_0$. Hence the distribution of non-cuspidal representations of $\mathcal{E}$ in families (for G of type F_4) is as described in the appendix.

Now let G be of type E_8. For any parabolic subgroup $P \subset G$ containing the parabolic subgroup $P_1 \supset B$ of type D_4, we consider the Levi subgroups M of P, M_1 of P_1, containing T and we consider the unipotent representation $\delta[1] \subset \mathrm{Ind}_{(M \cap P_1)^{F'}}^{M^{F'}}[\delta]$, (notation 8.7), where δ is the unipotent cuspidal representation of $M_1^{F'}$. When $P = G$, then $\delta[1]$ has degree $(1/2)q^3$ times a p-adic unit (by 8.7); this forces $\delta[1]$ to be the unique representation corresponding to the family of 112_z, which is not contained in $\mathrm{Ind}_{B^{F'}}^{G^{F'}}(1)$. Hence $\langle R_E, \delta[1]\rangle_{G^{F'}}$ is known from 4.23 for any $E \in W^\vee$. Similarly, if $P_1 \subset P \subsetneq G$, $\langle R_{E_1}, \delta[1]\rangle_{M^{F'}}$ is known for any irreducible representation E_1 of the Weyl group of M. Hence, for any $E \in W^\vee$, the inner product $\langle R_E, \mathrm{Ind}_{P^{F'}}^{G^{F'}}\delta[1]\rangle_{G^{F'}} = \langle R_{\mathrm{Res}\, E}, \delta[1]\rangle_{M^{F'}}$ is known.

We now consider for each family $\mathcal{F} \subset W^\vee$ with three or five elements, the representation $\rho_1 \in \mathcal{E}$, $\rho_1 : \mathcal{F}$, corresponding under 4.23 to $(g_2, \varepsilon) \in \mathfrak{M}(\mathcal{G}_{\mathcal{F}})$. We wish to compute the inner product $A = \langle \rho_1, \mathrm{Ind}_{P^{F'}}^{G^{F'}}\delta[1]\rangle_{G^{F'}}$ for $P \supset P_1$ as above. We replace ρ_1 by $\rho_1 + \rho_2$ (if $|\mathcal{F}| = 3$) or by $\rho_1 + \rho_2 + \rho_3$ if $|\mathcal{F}| = 5$, where $\rho_2 : \mathcal{F}$ corresponds under 4.23 to $(1, \varepsilon) \in \mathfrak{M}(\mathcal{G}_{\mathcal{F}})$ and, if $|\mathcal{F}| = 5$, $\rho_3 : \mathcal{F}$ corresponds under 4.23 to $(1, r) \in \mathfrak{M}(\mathcal{G}_{\mathcal{F}})$. Note that ρ_2 or ρ_3 (if defined) is a component of $\mathrm{Ind}_{B^{F'}}^{G^{F'}}(1)$, hence the inner product A is not changed if ρ_1 is replaced by $\rho_1 + \rho_2$ or $\rho_1 + \rho_2 + \rho_3$. Note that $\rho_1 + \rho_2$ (if $|\mathcal{F}| = 3$) and $\rho_1 + \rho_2 + \rho_3$ (if $|\mathcal{F}| = 5$) is an explicit combination of R_E, $(E \in \mathcal{F})$, cf. 4.23, hence our inner product A can be evaluated explicitly. Being non-zero for $P = P_1$, it follows that $\rho_1 = D_4[\chi]$ (notation as in appendix) where χ is an irreducible representation of a certain Coxeter group $\mathfrak{W}$ of type F_4. By the result of Benson-Curtis used earlier, χ is completely determined from the knowledge of the inner products A above, for the various P, $P_1 \subset P \subset G$. Hence for each family $\mathcal{F}$ with $|\mathcal{F}| = 3, 5$ the representation $\rho_1, \rho_1 : \mathcal{F}$ is of form $D_4[\chi]$, where χ is an irreducible representation of W which can be explicitly determined in each case. (The various possibilities for χ are described in the appendix.)

We now consider for each family $\mathcal{F} \subset W^\vee$ with five elements, the representations $\rho_\theta, \rho_{\theta^2}$ corresponding under 4.23 to (g_3, θ), (g_3, θ^2) in $\mathfrak{M}(\mathcal{G}_\mathcal{F})$. Here θ is a fixed cubic root $\neq 1$ in $\bar{Q}_\ell^*$. Let M_0 be the Levi subgroup containing T of the parabolic subgroup P of type E_6 containing B. Let $\delta_\theta, \delta_{\theta^2}$ be the unipotent representations of $M_0^{F'}$ corresponding under 4.23 to (g_3, θ), (g_3, θ^2) with respect to the cuspidal family of the Weyl group of M_0. Then $\delta_\theta, \delta_{\theta^2}$ are cuspidal. We can compute the inner product $A' = \langle \rho_\theta + \rho_{\theta^2}, \mathrm{Ind}_{P_0^{F'}}^{G^{F'}}(\delta_\theta)\rangle_{G^{F'}}$ as follows. We replace $\rho_\theta + \rho_{\theta^2}$ by $\rho_\theta + \rho_{\theta^2} + \rho_3$ (where $\rho_3 : \mathcal{F}$ corresponds to $(1, r) \in \mathfrak{M}(\mathcal{G}_\mathcal{F})$); this does not change the inner product, since ρ_3 is a component of $\mathrm{Ind}_{B^{F'}}^{G^{F'}}(1)$. But $\rho_\theta + \rho_{\theta^2} + \rho_3$ is by 4.23, an explicit combination of R_E $(E \in \mathcal{F})$ hence the inner product A' can be expressed in terms of $\langle R_E, \mathrm{Ind}_{P_0^{F'}}^{G^{F'}}(\delta_\theta)\rangle_{G^{F'}} = \langle R_{\mathrm{Res}\, E}, \delta_\theta\rangle$, which is known. In this way, we see that $\langle \rho_\theta + \rho_{\theta^2}, \mathrm{Ind}_{P_0^{F'}}^{G^{F'}}(\delta_\theta)\rangle_{G^{F'}} = 1$. Hence precisely one of $\rho_\theta, \rho_{\theta^2}$ appears in $\mathrm{Ind}_{P_0^{F'}}^{G^{F'}}(\delta_\theta)$. Similarly, one of $\rho_\theta, \rho_{\theta^2}$ appears in $\mathrm{Ind}_{P_0^{F'}}^{G^{F'}}(\delta_{\theta^2})$. Since $\mathrm{Ind}_{P_0^{F'}}^{G^{F'}}(\delta_\theta)$, $\mathrm{Ind}_{P_0^{F'}}^{G^{F'}}(\delta_{\theta^2})$ are disjoint, it follows that we may assume that ρ_θ is a component of $\mathrm{Ind}_{P_0^{F'}}^{G^{F'}}(\delta_\theta)$ and ρ_{θ^2} is a component of $\mathrm{Ind}_{P_0^{F'}}^{G^{F'}}(\delta_{\theta^2})$. Hence, with the notations in the appendix, we have $\rho_\theta = E_6[\theta, \chi]$, $\rho_{\theta^2} = E_6[\theta^2, \chi']$ where χ, χ' are irreducible representations of a certain Coxeter group $\mathfrak{W}$ of type G_2. The degree of $\rho_{\theta}, \rho_{\theta^2}$ is known from (4.26.2), hence by 8.7, D_χ and $D_{\chi'}$ are also known. It is easy to see that for the Hecke algebra of type G_2 with parameters q, q^9, the integer D_χ determines completely χ. Hence χ, χ' can be determined explicitly. (They are given in the appendix.) We now see that for each family $\mathcal{F} \neq \mathcal{F}_0$, the representations $\rho, \rho : \mathcal{F}$, are as described in the appendix (for G of type E_8).

10.3. The method in 10.2 applies also to the unipotent representations of $M^{F'}$ where M is the Levi subgroup containing T of a proper parabolic subgroup of G (G of type E_8 or F_4) and allows us to describe the relation between the parametrization 4.23 and parametrization 8.7 of unipotent representations of $M^{F'}$. For M of type E_6 or E_7 this relation is given in the appendix.

10.4. Let P' be a parabolic subgroup $\neq G$ containing B, let M' be the Levi subgroup containing T; let P be a maximal parabolic subgroup $\neq G$ containing P' and let M be the Levi subgroup of P. Let δ be a unipotent cuspidal representation of $M^{F'}$. The irreducible components of $\mathrm{Ind}_{P'^{F'}}^{G^{F'}}(\delta)$ are of the form $\delta[\chi]$ (see 8.7) where χ is an irreducible representation of a certain Coxeter group $\mathfrak{W}$ associated to M and G in Chapter 8; as a group we have $\mathfrak{W} = \mathfrak{N}_G(M)^{F'}/M^{F'}$. Similarly, the irreducible components of $\mathrm{Ind}_{(M\cap P')^{F'}}^{M^{F'}}(\delta)$ are of the form $\delta[\phi]$, (see 8.7), where ϕ is an irreducible representation of a parabolic subgroup $\mathfrak{W}'$ of the Coxeter group $\mathfrak{W}$; as a group we have $\mathfrak{W}' = \mathfrak{N}_M(M\cap P')^{F'}/(M\cap P')^{F'}$. For any such ϕ, we have by a well-known deformation argument (see for example [C] or [L_2, §5]):

$$\mathrm{Ind}_{P^{F'}}^{G^{F'}}\delta[\phi] = \sum \langle \chi, \mathrm{Ind}_{\mathfrak{W}'}^{\mathfrak{W}}(\phi)\rangle_{\mathfrak{W}}\, \delta[\chi] \qquad (10.4.1)$$

sum over all irreducible representations χ of $\mathfrak{W}$.

For any element $f \in \mathcal{R}(G^{F'}) \otimes \mathbf{Q}$, we denote $pr_0(f)$ the orthogonal project of f onto the subspace spanned by all unipotent representations ρ such that $\rho : \mathcal{F}_0$. From (10.4.1), we have

$$\text{(10.4.2)} \qquad pr_0(\mathrm{Ind}_{P^{F'}}^{G^{F'}} \delta[\phi]) = \sum \langle \chi, \mathrm{Ind}_{\mathfrak{W}'}^{\mathfrak{W}}(\phi) \rangle_{\mathfrak{W}} \delta[\chi]$$

sum over all irreducible representations χ of $\mathfrak{W}$ such that $\delta[\chi] : \mathcal{F}_0$. (Using the results in 10.2, we know precisely for which χ we have $\delta[\chi] : \mathcal{F}_0$.) Now let $E \in \mathcal{F}_0$. By 6.17 and (6.4.3), we have

$$\langle R_{\mathrm{Res}\,E}, \delta[\phi] \rangle_{M^{F'}} = \langle R_E, \mathrm{Ind}_{P^{F'}}^{G^{F'}} \delta[\phi] \rangle_{G^{F'}} = \langle R_E, pr_0(\mathrm{Ind}_{P^{F'}}^{G^{F'}} \delta[\phi]) \rangle_{G^{F'}}$$

$$= \sum_{\substack{\chi \\ \delta[\chi]:\mathcal{F}_0}} \langle \chi, \mathrm{Ind}_{\mathfrak{W}'}^{\mathfrak{W}}(\phi) \rangle_{\mathfrak{W}} \langle R_E, \delta[\chi] \rangle_{G^{F'}} .$$

Hence, if X denotes one of the elements $Y_{i,j} \in \mathcal{R}(W)$ in 7.4 (for G of type F_4) or one of the elements $X_{i,j} \in \mathcal{R}(W)$ in 7.5 (for G of type F_4), we have for any ϕ:

$$\text{(10.4.3)} \qquad \langle R_X, \delta_{\phi,P} \rangle_{G^{F'}} = \langle R_{\mathrm{Res}\,X}, \delta[\phi] \rangle_{M^{F'}}$$

where

$$\text{(10.4.4)} \qquad \delta_{\phi,P} = \sum_{\substack{\chi \\ \delta[\chi]:\mathcal{F}_0}} \langle \chi, \mathrm{Ind}_{\mathfrak{W}'}^{\mathfrak{W}}(\phi) \rangle_{\mathfrak{W}} \delta[\chi] \in R(G^{F'}) .$$

(Here Res X is the restriction of X to $\mathcal{R}(W_P)$ where W_P is the subgroup of W corresponding to P.) The right-hand side of (10.4.3) is known by 4.23 applied to M and by 10.3. Hence the left-hand side of (10.4.3) is explicitly known.

Assume now that X has the property $X = J_{W_P}^{W}(X_1)$ for some $X_1 \in \mathcal{R}(W_P)$. Then $\mathrm{Ind}_{W_P}^{W}(X_1) = X +$ combination of E, $E \notin \mathcal{F}_0$, hence, by 6.17 and (6.4.1), we have

$$\mathrm{pr}_0(R_X) = \mathrm{pr}_0(R_{\mathrm{Ind}_{W_P}^{W}(X_1)}) = \mathrm{pr}_0(\mathrm{Ind}_{P^{F'}}^{G^{F'}}(R_{X_1})) .$$

By 4.23 applied to M, we can write R_{X_1} as an explicit $\mathbf{Q}$-linear combination of unipotent representations ρ_1 of $M^{F'}$ and, for each such ρ_1, we can write (by 10.2 and (10.4.1)) $\mathrm{pr}_0\mathrm{Ind}_{P^{F'}}^{G^{F'}}(\rho_1)$ as an explicit linear combination of non-cuspidal representations $\rho \in \mathcal{E}$, $\rho : \mathcal{F}_0$. Hence for our X, the multiplicities with which the various ρ, $\rho : \mathcal{F}_0$, enter in R_X are explicitly known. This applies for the X in the first 16 rows ($X_{1,1}$ up to $X_{7,1}$) in the following table for E_8 and for X in the first 9 rows ($Y_{1,1}$ up to $Y_{5,3}$) in the following table for F_4. The columns of those tables are parametrized by the $\rho \in E$, $\rho : \mathcal{F}_0$. (So far, only the columns corresponding to non-cuspidal representations have a meaning for us.) The tables describe the multiplicity with which ρ in a given column enters in R_X, in the row parametrized by X. Hence the first 16 rows (resp. 9 rows) of the table for E_8 (resp. F_4) are verified.

In the table for F_4 (not all of whose columns have been defined), $F_4[\theta]$ stands for two identical columns corresponding to θ, θ^2 (see the appendix) and $F_4[i]$ stands for two identical columns corresponding to i, $-i$ (see the appendix). Similarly, in the table for E_8, each of $E_6[\theta,r]$, $E_6[\theta,r']$, $E_8[\theta]$, $E_8[-\theta]$, $E_8[i]$ stand for two identical columns and $E_8[\zeta]$ stands for four identical columns corresponding to ζ, ζ^2, ζ^3, ζ^4 (see the appendix).

We now discuss the last row ($\mathcal{C}$) in the two tables. Let c be a Coxeter element of minimal length in W. The row $\mathcal{C}$ describes the multiplicities of the various $\rho \in \mathcal{E}$, $\rho : \mathcal{F}_0$ in

$$\sum_{E \in \mathcal{F}_0} \mathrm{Tr}(c,E) R_E = \mathrm{pr}_0\Big(\sum_E \mathrm{Tr}(c,E) R_E\Big) = \mathrm{pr}_0\Big(\sum_i (-1) H_c^i(X_c, \overline{\mathbf{Q}}_\ell)\Big) .$$

Table for E_8

	[4480]	[3150]	[4200]	[4536]	[5670]	[420]	[1134]	[1400]	[2688]	[1680]	[168]	[70]	[7168]	[1344]	[2016]	[5600]	[448]	$D_4[4_1]$	$D_4[6_1]$	$D_4[6_2]$	$D_4[12_1]$	$D_4[16_1]$	$E_6[\theta,r]$	$E_6[\theta,r']$	$E_8[-1]$	$E_8^I[1]$	$E_8^{II}[1]$	$E_8[\theta]$	$E_8[-\theta]$	$E_8[i]$	$E_8[\zeta]$
$X_{1,1}$	1		1	2	2			1	1	1			2			2															
$X_{1,2}$				1	1			1		2		1				1	1			1	1	1									
$X_{1,3}$			1								1			1				1				1									
$X_{2,1}$	1	1	1	1	1				1				2		1	1															
$X_{2,2}$		1								1		1			1		1			1	1	1									
$X_{2,3}$			1	1	1			1	1	1			1			2				1	1	1	1	1							
$X_{2,4}$					1		1			1						1	1		1		1	1									
$X_{3,1}$	1			3	3			2		3		1	1			2	1														
$X_{3,2}$			1						1				1			1				1	1	1									
$X_{3,3}$		1													1								1	1							
$X_{4,1}$	1	1		1	2		1			1			1			2	1														
$X_{4,2}$			1						1				1		1				1		1	1									
$X_{5,1}$	1		2	1	1			1			1		1	1		1															
$X_{5,2}$				1	1				1	1			1			1		1				1									
$X_{6,1}$	1	1	1		1		1						1	1		1															

$X_{7,1}$	1	1	1			1							1	1	1																
$X_{6,2}$		1					1				1	1						1		1					1		1				
$X_{7,2}$						1	1				1							1	1						1		1				
$X_{3,4}$							1												1									1	1		
$X_{2,5}$							1	1			1	1							1	1					2	1	1				
$X_{5,3}$								1		1		1		1						1		1			1	1					
$X_{3,5}$											1					1	1			1		1			1	1					
$X_{4,3}$											1				1		1		1			1			1	1					
$X_{5,4}$			1	1				1			2	1						1		1					1		1				
$X_{4,4}$		1					1	1		1		2								2	1				1		1				
$X_{1,4}$				1				2		1	1	2								2					2	1	1				
$X_{7,3}$		1	1		1		1		1				1	1	1	1		1	1		1	1	1	1						1	1
$X_{7,4}$		1				2	1		1	1				1	1	1	1	1	1	1	1	2	1	1		1		1	1	2	1
$\mathcal{C}$												1								1			–1						1		1

Table for F_4

	$[12_1]$	$[9_3]$	$[9_2]$	$[1_3]$	$[1_2]$	$[4_1]$	$[4_4]$	$[4_3]$	$[6_1]$	$[6_2]$	$[16_1]$	$B_2[r]$	$B_2[\varepsilon']$	$B_2[\varepsilon'']$	$F_4[\theta]$	$F_4[i]$	$F_4^{I}[1]$	$F_4^{II}[1]$	$F_4[-1]$
Y_{11}	1		1					1	1		1								
Y_{21}	1	1					1		1		1								
Y_{31}	1		2		1			1		1	1								
Y_{41}	1	2		1			1			1	1								
Y_{51}	1	1	1			1				1	2								
Y_{32}		1				1					1	1	1						
Y_{52}			1		1			1				1	1						
Y_{42}			1			1					1	1		1					
Y_{53}		1		1			1					1		1					
Y_{12}					1				1				1					1	1
Y_{22}				1					1					1				1	1
Y_{33}			1		2					1			1					1	1
Y_{43}		1		2						1				1				1	1
Y_{54}				1	1					1							2	1	1
Y_{34}				1				1				1					1		1
Y_{44}					1		1					1					1		1
Y_{13}			1		1			1		1	1		1		1		1		
Y_{14}		1				1	1		1		1	1	1	1	1	1			
Y_{15}				1			1	1	1			1		1	1	1	1		1
$\mathcal{C}$										1		−1			1	1			

The virtual representation $\sum_i (-1)^i H_c^i(X_c, \overline{Q}_\ell)$ has been completely decomposed in $[L_2]$ hence the last row ($\mathcal{C}$) in the two tables is verified.

At the same time this defines some columns in the tables. In the table for E_8, $E_8[\zeta^i]$, $(i=1,2,3,4)$, is the irreducible cuspidal representation of $G^{F'}$ arising as the $(\zeta^i q^4)$-eigenspace of F' on $H_c^8(X_c, \overline{Q}_\ell)$ and $E_8[-\theta^i]$, $(i=1,2)$ is the irreducible cuspidal representation of $G^{F'}$ arising as the $(-\theta^i q^4)$-eigenspace of F' on $H_c^8(X_c, \overline{Q}_\ell)$. In the table for F_4, $F_4[\theta^j]$, $(j=1,2)$, (resp. $F_4[i]$, $F_4[-i]$) denotes the irreducible cuspidal representation of $G^{F'}$ arising as the $(\theta^j q^2)$, (resp. iq^2, $-iq^2$) eigenspace of F' on $H_c^4(X_c, \overline{Q}_\ell)$.

10.5. We shall describe explicitly certain elements $\delta_{\phi,P} \in \mathcal{R}(G^{F'})$, (see (10.4.4)), in the case where G is of type E_8, in the form of a table. The columns of the table are parametrized by the non-cuspidal representation $\rho \in \mathcal{E}$, $\rho : \mathcal{F}_0$. The rows are parametrized by the various $\delta_{\phi,P}$; the table describes the multiplicity of a given ρ in $\delta_{\phi,P}$. We shall describe $\delta_{\phi,P}$ by first indicating the type of P (e.g. E_7, $E_6 \times A_1$, etc.) then indicating δ (e.g. $E_6[\theta]$, $E_6[\theta^2]$, D_4 = unipotent cuspidal representation of $M'^{F'}$ for M' of type D_4) and then indicating ϕ. When δ is the unit representation we shall omit it from the notation. When P is of type D_7 we indicate instead of δ, ϕ, the symbol corresponding to $\delta[\phi]$. In the columns, we omit writing indices y or w. For example we write [4480] instead of $[4480_y]$.

10.6. We shall consider the euclidean space H with orthonormal basis given by the elements $\rho \in E$, $\rho : \mathcal{F}_0$, (for G of type E_8). The dimension of H is not yet known. We may regard $R_{X_{6,2}}, R_{X_{7,2}}, R_{X_{3,4}}, R_{X_{2,5}}, R_{X_{5,3}}, R_{X_{3,5}}, R_{X_{4,3}}, R_{X_{5,4}}, R_{X_{4,4}}, R_{X_{1,4}}, R_{X_{7,3}}, R_{X_{7,4}}$ as 12 vectors in H with the property that *all their coordinates are integral and* ≥ 0 (cf. 6.17 (ii)), *whose mutual inner products are known.* Moreover, *we know the inner products of these 12 vectors with 19 vectors in* H whose coordinates are explicitly known. These 19 vectors are: 18 vectors given by the elements $\delta_{\phi,P}$ in 10.5 and the vector $\sum_{E \in \mathcal{F}_0} \mathrm{Tr}(c,E) R_E$ whose coordinates are described in the last row of the table for E_8 in 10.4. These properties of our 12 unknown vectors are strong enough to determine

Table for $\delta_{\phi,P}$ (type E_8)

	[4480]	[3150]	[4200]	[4536]	[5670]	[420]	[1134]	[1400]	[2688]	[1680]	[168]	[70]	[7168]	[1344]	[2016]	[5600]	[448]	$D_4[4_1]$	$D_4[6_1]$	$D_4[6_2]$	$D_4[12_1]$	$D_4[16_1]$	$E_6[\theta,r]$	$E_6[\theta^2,r]$	$E_6[\theta,r']$	$E_6[\theta^2,r']$
$E_7,315_a$	1			1	1								1													
$E_7,280'_b$			1						1				1			1										
$E_7,280_a$				1	1			1		1						1										
$E_7,70_a$		1													1											
$E_7,35_a$										1		1					1									
$E_7,D_4,r\otimes\text{sign}$																				1	1	1				
$E_7,E_6[\theta]$, sign																							1		1	
$E_7,E_6[\theta^2]$, sign																								1		1
$E_6\times A_1,30'_p\boxtimes\text{sign}$	1	1		1	1								1			1										
$E_6\times A_1,15'_q\boxtimes\text{sign}$			1						1				1		1											
$E_6\times A_1,15'_p\boxtimes\text{sign}$					1		1			1						1	1									
$E_6\times A_1,D_4$, sign																			1		1	1				

$D_7,\begin{pmatrix}1\,2\,3\,5\\0\,1\,3\,4\end{pmatrix}$	1		1	1	1			1					1			1						
$D_7,\begin{pmatrix}1\,3\,4\,5\\0\,1\,2\,3\end{pmatrix}$			1								1			1								
$D_7,\begin{pmatrix}0\,1\,3\,5\\1\,2\,3\,4\end{pmatrix}$				1	1				1	1			1			1						
$D_7,\begin{pmatrix}0\,1\,2\,3\,4\,5\\1\,3\end{pmatrix}$																		1				1
$D_5 \times A_2,\begin{pmatrix}1\,2\,3\,4\\0\,1\,2\,4\end{pmatrix} \boxtimes \mathrm{sign}$	1	1	1		1		1						1	1		1						
$A_4 \times A_3$, sign	1	1	1			1							1	1	1							

explicitly the number of elements in the basis of H and the coordinates of the 12 vectors. (We leave the verification of this fact to the reader.) The coordinates turn out to be those given in the table for E_8 in 10.4. At the same time, this defines those columns in that table which have not been already defined. (They correspond to cuspidal representations.) For example, $E_8[-1]$ is the unique cuspidal representation $\rho \in \mathcal{E}$ such that ρ appears with multiplicity 2 in $R_{X_{2,5}}$. We parametrize the representations $\rho \in \mathcal{E}$, $\rho : \mathcal{F}_0$, by the elements in $\mathfrak{M}(\mathfrak{G}_5)$, as in the appendix.

Now one checks directly that any $E \in \mathcal{F}_0$ is a $\mathbf{Q}$-linear combination of the $X_{i,j}$ (in the table for E_8, in 10.4) and of $\sum_{E \in \mathcal{F}_0} \mathrm{Tr}(c,E)E$. Hence the table in 10.4 gives also the multiplicities in R_E for $E \in \mathcal{F}_0$. This completes the verification of 4.23 in our case (E_8). The case F_4 is entirely similar (and easier). This also completes the verification of 4.23 in general.

10.7. Let us consider the first 8 rows (corresponding to P of type E_7) in the table in 10.5. Those 8 rows may be put in 1-1 correspondence with the elements of $\mathfrak{M}(\mathfrak{G}_3)$: they correspond to $(1,1)$, $(g_2,1)$, $(1,r)$, $(g_3,1)$, $(1,\varepsilon)$, (g_2,ε), (g_3,θ), (g_3,θ^2), in this order. The columns of that table correspond to certain elements in $\mathfrak{M}(\mathfrak{G}_5)$, (see the appendix). Hence the table may be interpreted as giving a linear map τ_{E_7} from the space $\bar{\mathbf{Q}}_\ell^{\mathfrak{M}(\mathfrak{G}_3)}$ of $\bar{\mathbf{Q}}_\ell$-valued functions on $\mathfrak{M}(\mathfrak{G}_3)$ to the analogous space $\bar{\mathbf{Q}}_\ell^{\mathfrak{M}(\mathfrak{G}_5)}$.

(For example, the function on $\mathfrak{M}(\mathfrak{G}_3)$ which is 1 on $(1,\varepsilon) \in M(\mathfrak{G}_3)$ and 0 on other elements is mapped to the function on $\mathfrak{M}(\mathfrak{G}_5)$ which is 1 on $(1,\lambda^2)$, $(1,\lambda^3)$, (g_2,ε) and is zero elsewhere.)

Similarly, if we put the rows in that table corresponding to P of type $E_6 \times A_1$ (resp. D_7) in 1-1 correspondence with the elements of $\mathfrak{M}(\mathfrak{G}_2)$, by associating with them $(1,1)$, $(g_2,1)$, $(1,\varepsilon)$, (g_2,ε), in this order, the table gives two linear maps $\tau_{E_6\times A_1}, \tau_{D_7} : \bar{\mathbf{Q}}^{\mathfrak{M}(\mathfrak{G}_2)} \to \bar{\mathbf{Q}}_\ell^{\mathfrak{M}(\mathfrak{G}_5)}$. The rows corresponding to $D_5 \times A_2$ and $A_4 \times A_3$ give rise to linear maps

$$\tau_{D_5\times A_2}, \tau_{A_4\times A_3} : \overline{\mathbf{Q}}_\ell^{\mathfrak{M}(\mathfrak{S}_1)} \to \overline{\mathbf{Q}}_\ell^{\mathfrak{M}(\mathfrak{S}_5)} .$$

In general, let $\mathcal{G}$ be a finite group and let $\mathcal{G}'$ be a subquotient group of $\mathcal{G}$, i.e. there exists a diagram of groups

$$\begin{array}{ccc} \mathcal{G}'' & \xrightarrow{\ i\ } & \mathcal{G} \\ {\scriptstyle \pi}\downarrow & & \\ \mathcal{G}' & & \end{array} \tag{10.7.1}$$

with i an injective homomorphism and π a surjective homomorphism. Then as we shall soon see, there is a natural linear map

$$i_*\pi^* : \overline{\mathbf{Q}}_\ell^{\mathfrak{M}(\mathcal{G}')} \to \overline{\mathbf{Q}}_\ell^{\mathfrak{M}(\mathcal{G})} . \tag{10.7.2}$$

It is remarkable that the five linear maps τ_{E_7}, $\tau_{E_6\times A_1}$, τ_{D_7}, $\tau_{D_5\times A_2}$, $\tau_{A_4\times A_3}$ are all of the type (10.7.2); they correspond respectively to the following diagrams of type (10.7.1):

$$\begin{array}{ccc} \mathfrak{S}_3\times\mathfrak{S}_2 & \hookrightarrow & \mathfrak{S}_5 , \\ \downarrow {\scriptstyle pr_1} & & \\ \mathfrak{S}_3 & & \end{array} \qquad \begin{array}{ccc} \mathfrak{S}_3\times\mathfrak{S}_2 & \hookrightarrow & \mathfrak{S}_5 , \\ \downarrow {\scriptstyle pr_2} & & \\ \mathfrak{S}_2 & & \end{array} \qquad \begin{array}{ccc} \mathrm{Dyh}(8) & \hookrightarrow & \mathfrak{S}_5 , \\ \downarrow & & \\ \mathfrak{S}_2 & & \end{array}$$

$$\begin{array}{ccc} \mathfrak{S}_4 & \hookrightarrow & \mathfrak{S}_5 , \\ \downarrow & & \\ \{e\} & & \end{array} \qquad \begin{array}{ccc} \mathfrak{S}_5 & \hookrightarrow & \mathfrak{S}_5 \\ \downarrow & & \\ \{e\} & & \end{array}$$

Here, Dyh(8) is a dyhedral subgroup of order 8 in $\mathfrak{S}_5$, which can be taken to be the centralizer of $g_2' \in \mathfrak{S}_5$; the projection $\mathrm{Dyh}(8) \to \mathfrak{S}_2$ is the unique non-trivial homomorphism whose kernel contains the two transpositions in $Z(g_2')$.

The linear map (10.7.2) is defined to be the composition of a linear map $\pi^* : \bar{\mathbf{Q}}_\ell^{\mathfrak{M}(\mathcal{G}')} \to \bar{\mathbf{Q}}_\ell^{\mathfrak{M}(\mathcal{G}'')}$ and of a linear $i_* : \bar{\mathbf{Q}}_\ell^{\mathfrak{M}(\mathcal{G}'')} \to \bar{\mathbf{Q}}_\ell^{\mathfrak{M}(\mathcal{G})}$.

We regard $\bar{\mathbf{Q}}_\ell^{\mathfrak{M}(\mathcal{G})}$ as the space of formal $\bar{\mathbf{Q}}_\ell$-linear combinations of elements (x,σ) where $x \in \mathcal{G}$ and σ is an irreducible representation of $Z(x)$, modulo the subspace generated by the differences $(x,\sigma)-(gxg^{-1},\sigma^g)$ for any x,σ and any $g \in G$. (Here σ^g is the representation of $Z(gxg^{-1})$ obtained from σ and from conjugation by g.) The spaces $\bar{\mathbf{Q}}_\ell^{\mathfrak{M}(\mathcal{G}')}, \bar{\mathbf{Q}}_\ell^{\mathfrak{M}(\mathcal{G}'')}$ are regarded in a similar way.

Let $x' \in \mathcal{G}'$, σ' an irreducible representation of $Z_{\mathcal{G}'}(x')$. We define

$$\pi^*(x',\sigma') = \sum_{y'' \in I} (y'',\pi \circ \sigma') ,$$

where I is a subset of $\pi^{-1}(x')$ with the following property: any element in $\pi^{-1}(x')$ is conjugate in $\mathcal{G}''$ to a unique element in I. (Here, $\pi \circ \sigma'$ is regarded as a representation of $Z_{\mathcal{G}''}(y'')$.) This extends by linearity to a map $\pi : \bar{\mathbf{Q}}_\ell^{\mathfrak{M}(\mathcal{G}')} \to \bar{\mathbf{Q}}_\ell^{\mathfrak{M}(\mathcal{G}'')}$.

Let $x'' \in \mathcal{G}''$, σ'' an irreducible representation of $Z_{\mathcal{G}''}(x'')$. We define

$$i_*(x'',\sigma'') = \sum_{\sigma} \langle \sigma'', \sigma|_{Z_{\mathcal{G}''}(x'')} \rangle_{Z_{\mathcal{G}''}(x'')} (x'',\sigma)$$

sum over all irreducible representations σ of $Z_{\mathcal{G}}(x')$.

This extends by linearity to a map $i_* : \bar{\mathbf{Q}}_\ell^{\mathfrak{M}(\mathcal{G}'')} \to \bar{\mathbf{Q}}_\ell^{\mathfrak{M}(\mathcal{G})}$. Let $A : \bar{\mathbf{Q}}_\ell^{\mathfrak{M}(\mathcal{G})} \to \bar{\mathbf{Q}}_\ell^{\mathfrak{M}(\mathcal{G})}$ be the linear map (Fourier transform) defined by

$$A(x,\sigma) = \sum_{(y,\tau)} \{(x,\sigma),(y,\tau)\}(y,\tau)$$

(see (4.14.3)) where the sum is taken over all pairs (y,τ) where y is running over a set of representatives for the conjugacy classes in $\mathcal{G}$ and τ is running over the set of irreducible representations of $Z_{\mathcal{G}}(y)$. Similarly, we have linear maps $A : \bar{\mathbf{Q}}_\ell^{\mathfrak{M}(\mathcal{G}')} \to \bar{\mathbf{Q}}_\ell^{\mathfrak{M}(\mathcal{G}')}$, $A : \bar{\mathbf{Q}}_\ell^{\mathfrak{M}(\mathcal{G}'')} \to \bar{\mathbf{Q}}_\ell^{\mathfrak{M}(\mathcal{G}'')}$. It is easy to check that $\pi^* A = A\pi^*$, $i_* A = A i_*$. It follows that $i_*\pi^* A = A i_* \pi^*$.

Chapter 11

EIGENVALUES OF FROBENIUS

In this chapter, we assume that G modulo its center is simple and that F' is a power of F. For each unipotent representation ρ of $G^{F'}$, there is a well-defined root of 1, $\lambda_\rho \in \overline{\mathbf{Q}}_\ell^*$ with the following property. For any $i \geq 0$ and any $w \in W$, all eigenvalues of F' on the ρ-isotypic part of $H^i(\overline{X}_w, \overline{\mathbf{Q}}_\ell)$ are equal to $\lambda_\rho q^{i/2}$. (This follows from 2.20 and 3.8 (i).) From 2.8, it follows that for any $i \geq 0$ and any $w \in W$, all eigenvalues of F' on the ρ-isotypic part of $H_c^i(X_w, \overline{\mathbf{Q}}_\ell)$ are of the form λ_ρ times an integral power of $q^{1/2}$. (Compare [L_4, 3.8].)

The purpose of this chapter is to compute explicitly λ_ρ for any ρ.

11.1. Let $\mathcal{G}$ be a finite group. Define a function $\mathfrak{M}(\mathcal{G}) \to \overline{\mathbf{Q}}_\ell^*$ by $m = (x,\sigma) \mapsto \tilde{\lambda}_m$ = scalar by which x acts on the representation σ of $Z_{\mathcal{G}}(x)$. Thus $\tilde{\lambda}_m$ is a root of 1 for any $m \in \mathfrak{M}(\mathcal{G})$.

If $\mathcal{G}$ is a vector space of dimension d over F_2 then there is a unique quadratic form f on the F_2-vector space $\mathfrak{M}(\mathcal{G})$, with values in F_2 such that $\tilde{\lambda}_m = (-1)^{f(m)}$, $(m \in \mathfrak{M}(\mathcal{G}))$ and such that

$$\{m,m'\} = 2^{-d}(-1)^{f(m+m')+f(m)+f(m')} .$$

In general, we can state the following

11.2. THEOREM. *The correspondence 4.23* $\rho \leftrightarrow \overline{x}_\rho$ *between unipotent representations of* $G^{F'}$ *and elements of* $X(W) = \overline{X}(W,\mathrm{id})$, *(see (4.21.1)), may be chosen so that the following property holds. Let* ρ *be a unipotent representation of* $G^{F'}$ *and let* $\mathcal{F}$ *be the family in* $W^\vee$ *such that* $\overline{x}_\rho \in \mathfrak{M}(\mathcal{G}_{\mathcal{F}}) \subset X(W)$. *If* $\mathcal{F}$ *is not the family of* $512_a'$ *(for* E_7*) or the* *family of* 4096_z *or* $4096'_x$ *(for* E_8*) then* $\lambda_\rho = \tilde{\lambda}_{\overline{x}_\rho}$ *(with* $\tilde{\lambda}$ *as in 11.1). If* $\mathfrak{F}$ *is the family of* $512'_a$ *(for* E_7*) or the family of* 4096_z *or* $4096'_x$ *(for* E_8*), then* $\lambda_\rho = 1$ *if* $\rho = [512'_a]$ *or* $[512_a]$ *(for* E_7*) or* $\rho = [4096_z]$, $[4096_x]$, $[4096'_x]$, *or* $[4096'_z]$ *(for* E_8*) and* $\lambda_\rho = \sqrt{-1}, -\sqrt{-1}$ *for the other two* ρ. *(The notations for* ρ *are as in the appendix.)*

11.3. If ρ is an irreducible component of $\mathrm{Ind}_{B^{F'}}^{G^{F'}}(1)$, then we have clearly $\lambda_\rho = 1$. More generally, if ρ is an irreducible component of $\mathrm{Ind}_{P^{F'}}^{G^{F'}}(\delta)$, where P is a parabolic subgroup containing B and δ is a unipotent cuspidal representation of $(P/U_P)^{F'}$, then $\lambda_\rho = \lambda_\delta$, ($\lambda_\delta$ is defined with respect to P/U_P which is simple modulo its center). This follows immediately from 6.4. Hence the computation of λ_ρ is reduced to the case where ρ is cuspidal.

When G is of type B_r and C_r, λ_ρ can be computed as in [L_{11}, 6.6]. The same method applies in the case where G is of type D_r, see [L_{12}, 3.18]. (In these cases the correspondence 4.23 is uniquely determined.)

In particular, if ρ is the unique unipotent cuspidal representation of $G^{F'}$ of type B_2 or D_4, we have $\lambda_\rho = -1$.

Assume now that G is of type E_6. In this case, we must only show that the two unipotent cuspidal representations ρ, ρ' of $G^{F'}$ satisfy $\lambda_\rho = \theta$, $\lambda_{\rho'} = \theta^2$, ($\theta$ is a primitive cubic root of 1). But this fact is known from [L_2].

Assume now that G is of type E_7. In this case, we must only show that the two unipotent cuspidal representations ρ, ρ' of $G^{F'}$ satisfy $\lambda_\rho = \sqrt{-1}$, $\lambda_{\rho'} = -\sqrt{-1}$. This is again known from [L_2]. It remains to compute λ_ρ for the unipotent cuspidal representations ρ of G of type G_2, F_4 or E_8.

11.4. LEMMA. *For any finite group* $\mathcal{G}$, *we have*

(11.4.1) $$\sum_{m \in \mathfrak{M}(\mathcal{G})} \tilde{\lambda}_m \{m, m_1\} \overline{\{m, m_2\}} = \tilde{\lambda}_{m_1} \tilde{\lambda}_{m_2} \{m_1^*, m_2\}$$

for any $m_1, m_2 \in \mathfrak{M}(\mathcal{G})$. *Here,* m_1^* *is obtained from* $m_1 = (y, \tau)$ *by keeping unchanged the* y*-component of* m_1 *and replacing the* τ*-component by the contragredient representation* τ^*.

Proof. The proof will be along the lines of the proof of the lemma in [L_5, 4]. Let $m_2 = (z, \rho)$. The left-hand side of (11.4.1) is

$$A = \sum_{x \in \mathcal{G}} |Z(x)|\, |\mathcal{G}|^{-1}$$

$$\times \sum_{\substack{g, g' \\ x \cdot gyg^{-1} = gyg^{-1} \cdot x \\ x \cdot g'zg'^{-1} = g'zg'^{-1} x}} |Z(x)|^{-2} |Z(y)|^{-1} |Z(z)|^{-1}\, \overline{\mathrm{Tr}(g^{-1}xg, \tau)}\, \mathrm{Tr}(g'^{-1}xg', \rho)$$

$$\times \sum_{\sigma \in Z(x)^\vee} \tilde{\lambda}_{(x,\sigma)} \mathrm{Tr}(gyg^{-1}, \sigma)\, \overline{\mathrm{Tr}(g'zg'^{-1}, \sigma)}\,.$$

The sum over σ can be written as

$$\sum_{\sigma \in Z(x)^\vee} \mathrm{Tr}(xgyg^{-1}, \sigma)\, \overline{\mathrm{Tr}(g'zg'^{-1}, \sigma)} = \#\{u \in Z(x) | uxgyg^{-1} = g'zg'^{-1}u\}\,.$$

We now substitute this into A. We make a change of variable $x' = g^{-1}xg$, $u' = g^{-1}ug$, $\bar{g} = g^{-1}g'$:

$$A = \sum_{\substack{x' \in Z(y) \\ u' \in Z(x') \\ g, \bar{g} \\ u'y = x'^{-1} \bar{g} z \bar{g}^{-1} u'}} \overline{\mathrm{Tr}(x', \tau)}\, \mathrm{Tr}(\bar{g}^{-1} x' \bar{g}, \rho) \cdot |\mathcal{G}|^{-1} |Z(x')|^{-1} |Z(y)|^{-1} |Z(z)|^{-1}\,.$$

We now set $\tilde{g} = u'^{-1}\bar{g}$:

$$A = \sum_{\substack{x' \epsilon Z(y) \\ \tilde{g} \\ y = x'^{-1}\tilde{g} z \tilde{g}^{-1}}} |Z(y)|^{-1} |Z(z)|^{-1} \overline{\mathrm{Tr}(x',\tau)}\, \mathrm{Tr}(\tilde{g}^{-1} x' \tilde{g}, \rho)$$

$$= \sum_{\substack{g \\ gzg^{-1}y = ygzg^{-1}}} |Z(y)|^{-1} |Z(z)|^{-1} \overline{\mathrm{Tr}(gzg^{-1}y^{-1},\tau)}\, \mathrm{Tr}(zg^{-1}y^{-1}g, \rho)$$

$$= \sum_{\substack{g \\ gzg^{-1}y = ygzg^{-1}}} \tilde{\lambda}_{(y,\tau)} \tilde{\lambda}_{(z,\rho)} |Z(y)|^{-1} |Z(z)|^{-1} \overline{\mathrm{Tr}(gzg^{-1},\tau)}\, \mathrm{Tr}(g^{-1}y^{-1}g, \rho)$$

$$= \tilde{\lambda}_{m_1} \tilde{\lambda}_{m_2} \{m_1^*, m_2\}$$

as required.

11.5. For any $w \epsilon W$, we have $F' : \overline{X}_w \to \overline{X}_w$ since F' is a power of F. Note that F' is the identity on X_e and has no fixed points on X_y, $y \neq e$. Hence, by the Lefschetz fixed point formula for $F' : \overline{X}_w \to \overline{X}_w$ and for the intersection complex of $\overline{X}_w$, we have

$$\text{(11.5.1)} \qquad \mathrm{Tr}\left(F', \sum_i (-1)^i H^i(\overline{X}_w, \overline{Q}_\ell)\right) = P_{e,w}(q) \cdot |X_e| .$$

On the other hand, if $E(q)$ is the $H(q)$-module corresponding to $E \epsilon W^\vee$, then by 3.8, we have

$$\text{(11.5.2)} \quad \mathrm{Tr}\left(F', \sum_i (-1)^i H^i(\overline{X}_w, \overline{Q}_\ell)\right)$$

$$= \sum_\rho \lambda_\rho \dim(\rho) \sum (-1)^i \langle \rho, H^i(\overline{X}_w, \overline{Q}_\ell) \rangle_{G^{F'}} q^{i/2}$$

$$= \sum_\rho \lambda_\rho \dim(\rho) \sum_{E \epsilon W^\vee} \langle \rho, R_E \rangle_{G^{F'}} \mathrm{Tr}(C'_w, E(q))\, q^{\ell(w)/2} .$$

By (11.5.1), this is also equal to

$$P_{e,w}(q)|X_e| = \mathrm{Tr}(q^{\ell(w)/2}C'_w, \mathrm{Ind}_{B^{F'}}^{G^{F'}}(1))$$

where $\mathrm{Ind}_{B^{F'}}^{G^{F'}}(1)$ is regarded as a module over the Hecke algebra $H(q)$ in a natural way. (Note that $\mathrm{Tr}(T_y, \mathrm{Ind}_{B^{F'}}^{G^{F'}}(1)) = 0$ for $y \neq e$ and it is equal to $|X_e|$ for $y = e$.)

By decomposing $\mathrm{Ind}_{B^{F'}}^{G^{F'}}(1)$ with respect to the action of $H(q)$, we have therefore

$$P_{e,w}(q)|X_e| = \sum_{E \in W^\vee} \mathrm{Tr}(q^{\ell(w)/2}C'_w, E(q)) \dim \rho_E \tag{11.5.3}$$

where ρ_E is the irreducible $G^{F'}$-module in $\mathrm{Ind}_{B^{F'}}^{G^{F'}}(1)$ corresponding to $E(q)$. Thus, the expressions (11.5.2) and (11.5.3) are equal for all $w \in W$. Since the elements $q^{\ell(w)/2}C'_w$ form a basis for $H(q)$, it follows that

$$\sum_\rho \lambda_\rho \dim(\rho) \langle \rho, R_E \rangle_{G^{F'}} = \dim(\rho_E) \tag{11.5.4}$$

for any $E \in W^\vee$.

From (4.26.2) and 4.23 we have

$$\dim(\rho_E) = \sum_{\substack{E' \in W^\vee \\ E' \underset{LR}{\sim} E}} \langle \rho_E, R_{E'} \rangle_{G^{F'}} \dim R_{E'} . \tag{11.5.5}$$

Let us set for any unipotent representation ρ of $G^{F'}$, $\tilde{\lambda}_\rho = \tilde{\lambda}_m$ where $m = \bar{x}_\rho \in X(W)$ and $\tilde{\lambda}_m$ is defined in 11.1. Using now Lemma 11.4, we see that for $E, E' \in W^\vee$, $E' \underset{LR}{\sim} E$, we have

$$\langle \rho_E, R_{E'} \rangle_{G^{F'}} = \sum_\rho \tilde{\lambda}_\rho \langle \rho, R_E \rangle_{G^{F'}} \langle \rho, R_{E'} \rangle_{G^{F'}} . \tag{11.5.6}$$

(Note that if ρ_E corresponds to $m = \mathfrak{M}(\mathcal{G}_{\mathcal{F}}) \subset X(W)$ then $m = m^*$ and $\lambda_m = 1$.) Substituting this into (11.5.5) we find

$$\dim(\rho_E) = \sum_{E' \in W^\vee} \sum_{\rho} \tilde{\lambda}_\rho \langle \rho, R_E \rangle_{G^{F'}} \langle \rho, R_{E'} \rangle_{G^{F'}} \dim R_{E'} .$$

Using once more (4.26.2) and 4.23 we may replace in the previous sum $\sum_{E' \in W^\vee} \langle \rho, R_{E'} \rangle \dim R_{E'}$ by $\Delta(\bar{x}_\rho) \dim(\rho)$ so that we have

$$\dim(\rho_E) = \sum_{\rho} \tilde{\lambda}_\rho \Delta(\bar{x}_\rho) \dim(\rho) \langle \rho, R_E \rangle_{G^{F'}} . \tag{11.5.7}$$

Comparing this with (11.5.4), we get for any $E \in W^\vee$:

$$\sum_{\rho} \lambda_\rho \dim(\rho) \langle \rho, R_E \rangle = \sum_{\rho} \tilde{\lambda}_\rho \Delta(\bar{x}_\rho) \dim(\rho) \langle \rho, R_E \rangle_{G^{F'}} . \tag{11.5.8}$$

When ρ is a non-cuspidal representation other than $E_7[\pm\xi]$, $E_7[\pm\xi,1]$, $E_7[\pm\xi,\text{sign}]$, (see appendix) we know already that $\lambda_\rho = \tilde{\lambda}_\rho$ and $\Delta(\bar{x}_\rho) = 1$ hence the corresponding terms in the two sides of (11.5.8) cancel each other. In type E_7, if $\rho = E_7[\xi]$, $\rho' = E_7[-\xi]$, we have $\langle \rho, R_E \rangle = \langle \rho', R_E \rangle$, $\dim(\rho) = \dim(\rho')$, $\Delta(\bar{x}_\rho) = \Delta(\bar{x}_{\rho'}) = -1$, $\lambda_\rho + \lambda_{\rho'} = \sqrt{-1} - \sqrt{-1} = 0$ and $\tilde{\lambda}_\rho + \tilde{\lambda}_{\rho'} = 1 - 1 = 0$ hence the terms corresponding to ρ, ρ' can be cancelled in the two sides of (11.5.8). The same applies to $E_7[\pm\xi,1]$, $E_7[\pm\xi,\text{sign}]$, in type E_8. Hence from (11.5.8) we get

$$\sum_{\substack{\rho \\ \text{cuspidal} \\ \text{unipotent}}} \lambda_\rho \dim(\rho) \langle \rho, R_\alpha \rangle_{G^{F'}} = \sum_{\substack{\rho \\ \text{cuspidal} \\ \text{unipotent}}} \tilde{\lambda}_\rho \dim(\rho) \langle \rho, R_\alpha \rangle_{G^{F'}} , \tag{11.5.9}$$

for any $\alpha \in \mathfrak{R}(W)$.

Assume now that G is of type E_8. When $\rho = E_8[\zeta^i]$, $i = 1,2,3,4$, or $\rho = E_8[-\theta^j]$, $j = 1,2$, we can assume, by $[L_2]$ that $\lambda_\rho = \tilde{\lambda}_\rho$. Applying now (11.5.9) for $\alpha = X_{3,4}$ (see Table for E_8 in 10.4) we get for

$\rho_1 = E_8[\theta]$, $\rho_2 = E_8[\theta^2]$ the identity $\lambda_{\rho_1} + \lambda_{\rho_2} = \tilde{\lambda}_{\rho_1} + \tilde{\lambda}_{\rho_2} = -1$. Since $\lambda_{\rho_1}, \lambda_{\rho_2}$ are roots of 1, this implies that $\lambda_{\rho_1} = \theta$, $\lambda_{\rho_2} = \theta^2$ (by interchanging ρ_1 and ρ_2, if necessary). Thus, $\lambda_{\rho_1} = \tilde{\lambda}_{\rho_1}$, $\lambda_{\rho_2} = \tilde{\lambda}_{\rho_2}$.

Applying (11.5.9) for $\alpha = X_{7,3}$ (see Table for E_8 in 10.4) we get for $\rho_3 = E_8[i]$, $\rho_4 = E_8[-i]$ the identity $\lambda_{\rho_3} + \lambda_{\rho_4} = 0$. This is not sufficient to determine $\lambda_{\rho_3}, \lambda_{\rho_4}$. Applying (11.5.9) for $\alpha = X_{7,4}$ we get for $\rho_5 = E_8^{I}[1]$, the identity: $\lambda_{\rho_5} = \tilde{\lambda}_{\rho_5} = 1$. Applying (11.5.9) for $\alpha = X_{1,4} - X_{4,4}$, we get for $\rho_6 = E_8^{I}[-1]$, the identity $\lambda_{\rho_6} = \tilde{\lambda}_{\rho_6} = -1$. Applying (11.5.9) for $\alpha = X_{4,3}$, we get for $\rho_7 = E_8^{II}[1]$, the identity $\lambda_{\rho_7} = \tilde{\lambda}_{\rho_7} = 1$. Assume now that G is of type F_4. When $\rho = F_4[\pm i]$ or $\rho = F_4[\theta^j]$, $j = 1,2$, we can assume by $[L_2]$ that $\lambda_\rho = \tilde{\lambda}_\rho$. Applying (11.5.9) for $\alpha = Y_{1,3}$ (see Table for F_4 in 10.4) we get for $\rho_1 = F_4^{I}[1]$ the identity $\lambda_{\rho_1} = \tilde{\lambda}_{\rho_1} = 1$. Applying (11.5.9) for $\alpha = Y_{4,4}$ (see Table for F_4 in 10.4) we get for $\rho_2 = F_4[-1]$ the identity $\lambda_{\rho_2} = \tilde{\lambda}_{\rho_2} = -1$. Applying (11.5.9) for $\alpha = Y_{1,2}$, we get for $\rho_3 = F_4^{II}[1]$ the identity $\lambda_{\rho_3} = \tilde{\lambda}_{\rho_3} = 1$. A similar method applies in the case where G is of type G_2. The previous method fails only in one case: when G is of type E_8, from the identity $\lambda_{\rho_3} + \lambda_{\rho_4} = 0$ we cannot determine $\lambda_{\rho_3}, \lambda_{\rho_4}$. Note, however that the set of all roots of 1 of form λ_ρ must be stable under $\mathrm{Gal}(\overline{\mathbf{Q}}/\mathbf{Q})$. It follows that the set $\{\lambda_{\rho_3}, \lambda_{\rho_4}>$ is stable under $\mathrm{Gal}(\overline{\mathbf{Q}}/\mathbf{Q})$, hence there are two possibilities: it is either $\{\sqrt{-1}, -\sqrt{-1}\}$ or $\{1, -1\}$. To complete the proof of Theorem 11.2 it is enough to show that $\{\lambda_{\rho_3}, \lambda_{\rho_4}\}$ cannot be equal to $\{1, -1\}$. This will be done in the next section.

11.6. In this section, G is of type E_8. There is a unique conjugacy class in W with the following property: the characteristic polynomial of an element of this class on the reflection representation of W, is the cyclotomic polynomial ϕ_{24}, see [F_2], [Ca]. According to Springer [Sp_1], the elements of minimal length in this class have length 10. One can show that the elements of minimal length in this class are exactly the elements of W which admit a reduced expression of length 10, with the following property: each $s \in S-\{s_4\}$ appears exactly once in the reduced expression, s_4 appears exactly three times and between the first and second (resp. second and third) apparition of s_4 there is a unique apparition of a simple reflection in $\{s_2,s_3,s_5\}$; here, s_i are as in 4.13. (This result will not be used here.)

We shall fix an element w in our conjugacy class. We shall consider the subvariety X_w of G/B. Let $H_c^i(X_w)^{(j)}$ be the largest subspace of $H_c^i(X_w)$ on which the eigenvalues of F' are of form $q^{j/2}$ times a root of 1, and let $H_c(X_w)^{(j)} = \sum_i (-1)^i H_c^i(X_w)^{(j)}$, (a virtual representation of $G^{F'}$).

According to [DM_2], [As_2], (or from 2.8 and 3.8) we have

$$\sum_{j\geq 0} H_c(X_w)^{(j)} u^{j/2} = \sum_{E\in W^\vee} \mathrm{Tr}(T_w,E(u))R_E \in \mathcal{R}(G^{F'})\otimes \mathbf{Q}[u^{1/2}]. \tag{11.6.1}$$

We now compute $\mathrm{Tr}(T_w,E(u))$. It is clear that $w^{12} = -1$ on the reflection representation of W. Hence $w^{12} = w_0$, the longest element in W. We have $\ell(w_0) = 120 = 12\,\ell(w)$. It follows that $T_w^{12} = T_{w_0}$ in the Hecke algebra. According to (5.12.2), for each $E \in W^\vee$, T_{w_0} acts on E(u) as $u^{\phi(E)}$ times an involution, where $\phi(E) = 120 - ((a_E+A_E)/2)$. It follows that T_w acts on E(u) as $u^{\phi(E)/12}$ times a transformation of order dividing 24 hence, by specializing $u^{1/2}$ to 1, we see that $\mathrm{Tr}(T_w,E(u)) = u^{\phi(E)/12}\,\mathrm{Tr}(w,E)$. The function $E \to \phi(E)$ is constant on families of W, hence we may define $\phi(\rho)$ for any unipotent representation ρ of $G^{F'}$ to be $\phi(E)$ where $E \in W^\vee$ is such that $\langle\rho,R_E\rangle_{G^{F'}} \neq 0$. We can rewrite (11.6.1) in the form

$$(11.6.2)\quad \sum_{j\geq 0} H_c(X_w)^{(j)} u^{j/2} = \sum_\rho u^{\phi(\rho)/12} \sum_E \langle \rho, R_E\rangle_{G^{F'}} \mathrm{Tr}(w,E)\rho$$

$$= \sum u^{\phi(\rho)/12} \langle \rho, H_c(X_w)\rangle_{G^{F'}} \rho$$

(ρ runs over the unipotent representations of $G^{F'}$, and $H_c(X_w) = \sum_i (-1)^i H_c^i(X_w)$).

It follows that for any integer $s \geq 1$, we have

$$\mathrm{Tr}(F'^s, H_c(X_w)) = \sum_\rho \lambda_\rho^s\, q^{s\phi(\rho)/12} \langle \rho, H_c(X_w)\rangle_{G^{F'}} \dim(\rho)\,.$$

(One can check from 4.23 and the character table of W that for all ρ such that $\langle \rho, H_c(X_w)\rangle_{G^{F'}} \neq 0$, we have $\phi(\rho)/12 \in \mathbf{N}$.)

We now set

$$\omega_\rho = \lambda_\rho\, q^{\phi(\rho)/12}$$

and, for each $\omega \in \bar{\mathbf{Q}}_\ell^*$, we set

$$d_\omega = \sum_{\substack{\rho \\ \omega_\rho = \omega}} \langle \rho, H_c(X_w)\rangle_{G^{F'}} \dim(\rho)\,.$$

We then have

$$\mathrm{Tr}(F'^s, H_c(X_w)) = \sum_{\omega\in\bar{\mathbf{Q}}_\ell^*} d_\omega\, \omega^s$$

hence, by the Lefschetz fixed point formula,

$$(11.6.3)\qquad |X_w^{F'^s}| = \sum_{\omega\in\bar{\mathbf{Q}}_\ell^*} d_\omega\, \omega^s\,, \quad (s=1,2,3,\cdots)\,.$$

We now show that $X_w^{F'^s}$ is empty for $1 \le s \le 23$. Let $B' \epsilon X_w^{F'^s}$. Then $(B',F'B')$, $(F'B',F'^2B'), \cdots, (F'^{11}B',F'^{12}B')$ are in relative position w. Since $\ell(w^2) = 2\ell(w)$, $\ell(w^3) = 3\ell(w), \cdots, \ell(w^{12}) = 12\ell(w)$, it follows that (B',F'^iB') are in relative position w^i for $i = 1,2,\cdots,12$. Since $w^i \neq e$ for $1 \le i \le 12$, it follows that $F'^iB' \neq B'$ for $1 \le i \le 12$ so that $s \ge 13$. Assume that $13 \le s \le 23$. Then $(F'^{12}B', F^{13}B'), (F^{13}B', F^{14}B'), \cdots, (F^{s-1}(B'), F^s(B'))$ are in relative position w and $\ell(w^{s-12}) = (s-12)\ell(w)$ hence $(F^{12}B', F^sB')$ are in relation position w^{s-12}. On the other hand $F^sB' = B'$ and $(F^{12}B', B')$ are in relative position w^{12}. It follows that $w^{12} = w^{s-12}$ a contradiction.

Thus, from (11.6.3) we see that

(11.6.4) $$\sum_{\omega} d_{\omega} \omega^s = 0, \text{ for } s = 1,2,3,\cdots,23 .$$

Clearly not all d_ω are $= 0$. (We have $\sum_{\omega} d_\omega = \dim H_c(X_w) \equiv 1 \pmod p$.) Also, any 23 distinct vectors in $(\bar{Q}_\ell)^{23}$ of form $(\omega, \omega^2, \cdots, \omega^{23})$ are linearly independent. Hence, from (11.6.4), we deduce that the number of $\omega \epsilon \bar{Q}_\ell^*$ such that $d_\omega \neq 0$ is ≥ 24.

On the other hand, we have $\langle H_c(X_w), H_c(X_w)\rangle_{G^{F'}} = |$centralizer of w in $W| = 24$, hence there are at most 24 different ρ such that $\langle \rho, H_c(X_w)\rangle_{G^{F'}} \neq 0$. From the definition of d_ω it now follows that:

a) there are exactly 24 elements $\omega \epsilon \bar{Q}_\ell^*$ such that $d_\omega \neq 0$

b) there are exactly 24 ρ's such that $\langle \rho, H_c(X_w)\rangle_{G^{F'}} \neq 0$

c) the sets in a), b) are in 1-1 correspondence: for each ω in a) there is a unique ρ in b) such that $\omega_\rho = \omega$. (We then have $d_\omega = \langle \rho, H_c(X_w)\rangle_{G^{F'}} \dim(\rho)$.)

(11.6.5) In particular, if $\rho \neq \rho'$ both appear with non-zero multiplicity in $H_c(X_w)$, then $\lambda_\rho q^{\phi(\rho)/12} \neq \lambda_{\rho'} q^{\phi(\rho')/12}$.

Let us apply (11.6.5) to $\rho = \rho_3 = E_8[i]$, $\rho' = [448_w]$. (From 4.23 one can check that both ρ, ρ' appear with multiplicity 1 in $H_c(X_w)$.) We have $\phi(\rho) = \phi(\rho') = 0$, $\lambda_{\rho'} = 1$. Hence (11.6.5) shows that $\lambda_{\rho_3} \neq 1$. Similarly, we see that $\lambda_{\rho_4} \neq 1$. Hence $\{\lambda_{\rho_3}, \lambda_{\rho_4}\}$ cannot be equal to the set $\{1, -1\}$. This completes the proof of Theorem 11.2.

Chapter 12
ON THE STRUCTURE OF LEFT CELLS

The main result of this chapter is that the $\mathbf{W}$-module $[\Gamma]_1$ carried by a left cell Γ of a Weyl group $\mathbf{W}$ gives rise to a "lagrangian subspace" of the corresponding set $\mathfrak{M}(\mathcal{G}_{\mathcal{F}})$. This implies, in particular, that when $\mathbf{W}$ is of classical type, the $\mathbf{W}$-module $[\Gamma]_1$ is multiplicity free and the number of its irreducible components is a power of 2. In this chapter we take $\mathbf{W} = W$, the Weyl group of G, assumed to be adjoint simple; we assume that $F' : G \to G$ is a power of F.

12.1. Let Γ be a left cell in $\mathbf{W}$. The irreducible components of the W-module $[\Gamma]_1$ are in a fixed family $\mathcal{F}$ (see 5.25). Let $\widetilde{\mathfrak{M}} = \widetilde{\mathfrak{M}(\mathcal{G}_{\mathcal{F}})}$ be the image of the imbedding $\mathcal{F} \hookrightarrow \mathfrak{M}(\mathcal{G}_{\mathcal{F}}) = \mathfrak{M}$ (see 4.4-4.13). For $m \in \widetilde{\mathfrak{M}}$ we write E_m for the corresponding element of $\mathcal{F}$, (an irreducible representation of W). We can write

$$[\Gamma]_1 = \sum_{m \in \widetilde{\mathfrak{M}}} \phi(m) E_m, \quad (\phi(m) \in \mathbf{N}).$$

We extend ϕ to a function $\phi : \mathfrak{M} \to \mathbf{N}$ by setting $\phi(m) = 0$ for $m \in \mathfrak{M} - \widetilde{\mathfrak{M}}$.

For any function $f : \mathfrak{M} \to \overline{\mathbf{Q}}_\ell$, its Fourier transform is the function $\hat{f} : \mathfrak{M} \to \overline{\mathbf{Q}}_\ell$ defined by

$$\hat{f}(m) = \sum_{m' \in \mathfrak{M}} \{m,m'\} f(m')$$

where $\{m,m'\}$ is given by (4.14.3).

12.2. THEOREM. (i) *We have* $\hat{\phi} = \phi$.

(ii) *There is a unique element* $x_0 \in \Gamma$ *such that* $[\Gamma]_1 = a_{x_0}$ *(see (5.10.3)). In particular, (6.17),* $R_{[\Gamma]_1}$ *is an actual* $G^{F'}$*-module.*

12.3. Let $\mathrm{supp}(\phi)$ be the set of all $m \in \mathfrak{M}$ such that $\phi(m) \neq 0$. Then $\mathrm{supp}\,\phi \subset \tilde{\mathfrak{M}}$.

Let Γ^* be the set of all elements x in Γ such that $a_x \neq 0$, (see (5.10.3)). For each $x \in \Gamma^*$, let $\phi_x : \mathfrak{M} \to \mathbf{Z}$ be the function defined by

$$\phi_x(m) = \begin{cases} c_{x,E_m}, & \text{if} \quad m \in \tilde{\mathfrak{M}} \\ 0, & \text{if} \quad m \in \mathfrak{M} - \tilde{\mathfrak{M}} \end{cases}$$

(see (5.1.22)). According to 6.17, R_{a_x} is an actual representation of $G^{F'}$. Hence, if ρ_m is the unipotent representation of $G^{F'}$ corresponding to $m \in \mathfrak{M}$ under 4.23, then

$$\langle \rho_m, R_{a_x} \rangle \in \mathbf{N},$$

hence

$$\langle \rho_{m'}, \sum_{m' \in \tilde{\mathfrak{M}}} \phi_x(m') R_{E_{m'}} \rangle \in \mathbf{N}.$$

By 4.23, this is equal to

$$\sum_{m' \in \mathfrak{M}} \Delta(m)\{m,m'\}\phi_x(m') = \Delta(m) \sum_{m' \in \mathfrak{M}} \{m,m'\}\phi_x(m') = \Delta(m)\hat{\phi}_x(m).$$

Thus, we have

$$\Delta(m)\hat{\phi}_x(m) \in \mathbf{N}, \quad \text{for all } m \in \mathfrak{M}. \tag{12.3.1}$$

Next we note that, by (4.14.2), we have

(12.3.2) $$f_{E_m}\{m_0, m\} = 1, \quad (m \in \widetilde{\mathfrak{M}}),$$

where $m_0 = (1,1) \in \mathfrak{M} = \mathfrak{M}(\mathcal{G}_{\mathcal{F}})$.

We can now rewrite the identities (5.8.1) in the following form

(12.3.3) $$\sum_{x \in \Gamma^*} \phi_x(m)\phi_x(m') = \begin{cases} \phi(m)/\{m_0, m\}, & \text{if } m = m' \\ 0, & \text{if } m \neq m' \end{cases}.$$

Actually, (5.8.1) states that this holds for $m, m' \in \operatorname{supp} \phi$. However, the proof of 7.1 (i) shows that the support of ϕ_x is contained in the support of ϕ. Hence (12.3.3) is valid for any $m, m' \in \mathfrak{M}$.

We now compute

(12.3.4) $$\sum_{x \in \Gamma^*} \hat{\phi}_x(m)\hat{\phi}_x(m_0) = \sum_{\substack{x \in \Gamma^* \\ m_1, m_2 \in \mathfrak{M}}} \{m, m_1\}\{m_0, m_2\}\phi_x(m_1)\phi_x(m_2)$$

$$= \sum_{m_1 \in \mathfrak{M}} \{m, m_1\}\{m_0, m_1\}(\phi(m_1)/\{m_0, m_1\}), \text{ by (12.3.3)}$$

$$= \sum_{m_1 \in \mathfrak{M}} \{m, m_1\}\phi(m_1)$$

$$= \hat{\phi}(m).$$

Using now (12.3.1) and $\Delta(m_0) = 1$ we see that

(12.3.5) $$\Delta(m)\hat{\phi}(m) = \sum_{x \in \Gamma^*} (\Delta(m)\hat{\phi}_x(m)) \cdot (\Delta(m_0)\hat{\phi}_x(m_0))$$

hence

(12.3.6) $$\Delta(m)\hat{\phi}(m) \in \mathbf{N}, \quad (m \in \mathfrak{M}).$$

Next we note that, by a theorem of Joseph [J], we have

(12.3.7) $$\phi(m_0) = 1 .$$

It follows that

$$\hat{\phi}(m_0) = \sum_{m \in \mathfrak{M}} \{m_0,m\}\phi(m) \geq \{m_0,m_0\}\phi(m_0) > 0$$

(since $\{m_0,m\} > 0$ for all m, $\phi(m) \geq 0$ for all m) hence

(12.3.8) $$\hat{\phi}(m_0) \geq 1 .$$

We shall show that $\hat{\phi}(m_0) = 1$. To prove this, we shall have to vary Γ, hence we shall temporarily denote ϕ as ϕ_Γ to show that it depends on Γ. Let Ω be the two-sided cell in W containing Γ. We shall compute $\sum_{\Gamma'} \phi_{\Gamma'}(m_0)$ (sum over all left cells Γ' contained in Ω). From 4.23, we see that

$$\langle \rho_{m_0}, R_{[\Gamma']_1} \rangle_{G^{F'}} = \langle \rho_{m_0}, \sum_{m \in \widetilde{\mathfrak{M}}} \phi_{\Gamma'}(m) R_{E_m} \rangle_{G^{F'}} = \sum_{m \in \widetilde{\mathfrak{M}}} \{m_0,m\}\phi_{\Gamma'}(m)$$

$$= \sum_{m \in \mathfrak{M}} \{m_0,m\}\phi_{\Gamma'}(m) = \hat{\phi}_{\Gamma'}(m_0) .$$

Hence

$$\sum_{\Gamma'} \hat{\phi}_{\Gamma'}(m_0) = \langle \rho_{m_0}, \sum_{\Gamma'} R_{[\Gamma']_1} \rangle_{G^{F'}}$$

$$= \langle \rho_{m_0}, R_{[\Omega]_1} \rangle_{G^{F'}} .$$

Here $[\Omega]_1$ is the W-module $\sum_{\Gamma'} [\Gamma']_1$; it is clear that $[\Omega]_1 = \sum_{E \in \mathcal{F}} \dim(E) \cdot E$. If $E \notin \mathcal{F}$, we have $\langle \rho_{m_0}, R_{E'} \rangle_{G^{F'}} = 0$, hence

(12.3.9)
$$\sum_{\Gamma'} \hat{\phi}_{\Gamma'}(m_0) = \langle \rho_{m_0}, R \sum_{E \in W^\vee} \dim(E) \cdot E \rangle_{G^{F'}}$$
$$= \langle \rho_{m_0}, \operatorname{Ind}_{B^{F'}}^{G^{F'}}(1) \rangle_{G^{F'}}$$
$$= \dim(E_{m_0}) .$$

On the other hand, by (12.3.8), we have

$$\sum_{\Gamma'} \phi_{\Gamma'}(m_0) \geq \text{number of left cells } \Gamma', \ \Gamma' \subset \Omega ;$$

the last number is known to be equal to $\dim(E_{m_0})$. (This statement follows clearly from (12.3.7).) Comparing this with (12.3.9), we see that each of the inequalities $\hat{\phi}_{\Gamma'}(m_0) \geq 1$ must in fact be an equality. Returning to our fixed Γ, we see that

(12.3.10)
$$\hat{\phi}(m_0) = 1 .$$

Let us combine (12.3.10) with the identity (12.3.5) for $m = m_0$; we see that

$$1 = \sum_{x \in \Gamma^*} \hat{\phi}_x(m_0)^2 .$$

Since $\hat{\phi}_x(m_0) \in \mathbf{N}$, (12.3.1), it follows that there is a unique $x \in \Gamma^*$ such that $\hat{\phi}_x(m_0) = 1$ (we call it x_0); for all $x \in \Gamma^*$, $x \neq x_0$, we have $\hat{\phi}_x(m_0) = 0$. Hence (12.3.5) becomes

$$\hat{\phi}(m) = \hat{\phi}_{x_0}(m), \text{ for all } m \in \mathfrak{M} .$$

This implies that $\phi = \phi_{x_0}$. Note that for $x \in \Gamma^*$, $x \neq x_0$, we have $\hat{\phi}_x(m_0) = 0$, $\hat{\phi}(m_0) = 1$ hence $\phi \neq \phi_x$. Thus, part (ii) of the theorem is proved.

12.4. We shall now prove part (i) of the theorem in the case where $\mathcal{G}_{\mathcal{F}}$ is a vector space of dimension d over F_2 and $\Delta \equiv 1$ on $\mathcal{F}$. Then $\mathfrak{M}$ is a vector space of dimension 2d over F_2 with a natural non-degenerate quadratic form $f: \mathfrak{M} \to F_2$ such that $\tilde{\lambda}_m = (-1)^{f(m)}$, $(m \in \mathfrak{M})$, see 11.1. Let (,) be the symplectic form associated to f; in other words $(m,m') = f(m+m') + f(m) + f(m')$. We then have $\{m,m'\} = 2^{-d}(-1)^{(m,m')}$. We have the inclusion:

$$\tilde{\mathfrak{M}} \subset \{m \in \mathfrak{M} | f(m) = 0\} . \tag{12.4.1}$$

If $d = 1$, this is obvious; the inclusion is then an equality. Assume now that we are in the situation of 4.5: $\mathfrak{M} = V_{Z_1}$. With the notations in 4.5, f is defined by

$$f(M^{\#}) = (1/2)(|(M^{\#} \cup M_0) - (M^{\#} \cap M_0)| - d) (\text{mod } 2), \quad (M^{\#} \in V_{Z_1}) .$$

On the other hand, $\mathfrak{M}$ is the set of all $M^{\#} \in V_{Z_1}$ such that $|(M^{\#} \cup M_0) - (M^{\#} \cap M_0)| = d$. Hence (12.4.1) follows in this case. A similar argument applies if we are in the situation of 4.6: $\mathfrak{M} = (V'_{Z_1})^+$.

12.5. Now let V be the F_2-subspace of M spanned by the support of $\hat{\phi}$. We show that

$$\phi(m) = \begin{cases} 1, & \text{if } (m,V) = 0 \\ 0, & \text{if } (m,V) \neq 0 \end{cases} . \tag{12.5.1}$$

Assume first that $(m,V) = 0$. Then $2^d \phi(m) = \sum_{m' \in \mathfrak{M}} (-1)^{(m,m')} \hat{\phi}(m') =$

$$\sum_{m' \in V} (-1)^{(m,m')} \hat{\phi}(m') = \sum_{m' \in V} \hat{\phi}(m') = \sum_{m' \in \mathfrak{M}} \hat{\phi}(m') = 2^d \phi(m_0) = 2^d, \text{ (see}$$

(12.3.7)) hence $\phi(m) = 1$.

Assume next that $(m,V) \neq 0$. Then there exist m' in the support of $\hat{\phi}$, such that $(m,m') \neq 0$. Thus,

$$2^d\phi(m) = \sum_{m'\in \operatorname{supp}\hat{\phi}} (-1)^{(m,m')}\hat{\phi}(m') < \sum_{m'\in V} \hat{\phi}(m'),$$

(since $\hat{\phi}(m') > 0$ for $m' \in \operatorname{supp} \hat{\phi}$) hence $2^d\phi(m) < 2^d\phi(m_0) = 2^d$. Thus, $\phi(m) < 1$. On the other hand, $\phi(m) \in \mathbf{N}$, hence $\phi(m) = 0$ and (12.5.1) is proved. Thus ϕ is the characteristic function of the F_2-subspace $V' = \{m \in \mathfrak{M} | (m,V) = 0\}$. It follows that $\hat{\phi}$ is $2^{\dim(V')-d}$ times the characteristic function of the F_2-subspace V. Since $\hat{\phi}$ has integral values, we must have $\dim(V') \geq d$. On the other hand, $V' \subset \widetilde{\mathfrak{M}} \subset f^{-1}(0)$, (see (12.4.1)). But an F_2-subspace of $\mathfrak{M}$ contained in $f^{-1}(0)$ must have dimension $\leq d$. It follows that V' is a Lagrangian subspace of M (a d-dimensional subspace on which the quadratic form f is identically zero). It follows that the symplectic form is also identically zero on V', so that $V = V'$, hence $\phi = \hat{\phi}$. Thus the theorem is proved in the case where $\mathcal{G}_{\mathcal{F}}$ is an F_2-vector space of dimension d and $\Delta \equiv 1$ on $\mathcal{F}$. Before continuing the proof of the theorem we shall prove the following.

12.6. PROPOSITION. *Let* ρ_E *be the irreducible component of the* $G^{F'}$*-module* $\operatorname{Ind}_{B^{F'}}^{G^{F'}}(1)$ *corresponding to the irreducible representation* $E(q)$ *of the Hecke algebra* $H(q) \approx \operatorname{End}_{G^{F'}}(\operatorname{Ind}_{B^{F'}}^{G^{F'}}(1))$; *here,* E *is an irreducible representation of* W. *The bijection* $\rho \leftrightarrow \overline{x}_\rho$ *given by 4.23 (between unipotent representations of* $G^{F'}$ *and elements of* $X(W)$, *see (4.21.1)) has the following property. Let* x_E *be the image of* $E \in W^\vee$ *under (4.21.3) and assume that* $\dim(E) \neq 512$ *(for* E_7*),* $\dim(E) \neq 4096$ *(for* E_8*). Then*

(12.6.1) $$\overline{x}_{\rho_E} = x_E .$$

Proof. Let $\hat{\rho}$ be the unipotent representation of $G^{F'}$ such that $\overline{x}_{\hat{\rho}} = x_E$. Let I be a subset of S, let W_I be the subgroup of W it generates and let $P_I \supset B$ be the corresponding parabolic subgroup. We shall prove that

(12.6.2) $$<\hat{\rho}, \mathrm{Ind}_{P_I^{F'}}^{G^{F'}}(1)>_{G^{F'}} = <\rho_E, \mathrm{Ind}_{P_I^{F'}}^{G^{F'}}(1)>_{G^{F'}} :$$

We have

$$<\hat{\rho}, \mathrm{Ind}_{P_I^{F'}}^{G^{F'}}(1)>_{G^{F'}} = <\hat{\rho}, \sum_{E'\epsilon W^\vee} <E', \mathrm{Ind}_{W_I}^{W}(1)>_W \cdot R_{E'}>_{G^{F'}}$$

$$= \sum_{E'\epsilon W^\vee} <E', \mathrm{Ind}_{W_I}^{W}(1)>_W \{x_E, x_{E'}\} ,$$

$$<\rho_E, \mathrm{Ind}_{P_I^{F'}}^{G^{F'}}(1)>_{G^{F'}} = <E, \mathrm{Ind}_{W_I}^{W}(1)>_W .$$

Hence (12.6.2) is a consequence of

(12.6.3) $$<E - \sum_{\substack{E'\epsilon W^\vee \\ E' \underset{LR}{\sim} E}} \{x_E, x_{E'}\} E', \mathrm{Ind}_{W_I}^{W}(1)>_W = 0 .$$

If G is of low rank, (in particular, if G is an exceptional group) this can be checked by direct computation. We shall now show how to verify (12.6.3) in the case where G is a classical group. Since $\mathrm{Ind}_{W_I}^{W}(1)$ is a sum of W-modules of form $[\Gamma]_1$ where Γ are left cells of W, it is enough to show that (12.6.3) is true when $\mathrm{Ind}_{W_I}^{W}(1)$ is replaced by $[\Gamma]_1$; we can also assume that Γ is in the two-sided cell corresponding to E. If $\mathfrak{M}$, $\widetilde{\mathfrak{M}}$, ϕ, are defined as in 12.1, (with respect to Γ), the identity to be proved is

$$<E_m - \sum_{m'\epsilon\widetilde{\mathfrak{M}}} \{m,m'\} E_{m'}, \sum_{m''\epsilon\widetilde{\mathfrak{M}}} \phi(m'') E_{m''}>_W = 0$$

which is the same as

$$\phi(m) = \hat{\phi}(m) \quad (m \in \widetilde{\mathfrak{M}}) .$$

But this is known since 2.2 (i) is already proved for classical groups. Thus, (12.6.3), and hence (12.6.2) are proved. By a result of [BC], (12.6.2) implies that $\hat{\rho} = \rho_E$ except possibly when G is of type G_2 and $\dim(E) = 2$. Let us consider separately this exceptional case. If V is the reflection representation of the Weyl group W of type G_2, and V′ is the other two-dimensional irreducible representation of W, and if $E = V$ or V' then from (12.6.2) it only follows that $\hat{\rho} \in \{\rho_V, \rho_{V'}\}$. On the other hand, from $\bar{x}_{\hat{\rho}} = x_E$ and from (4.26.2), we can compute explicitly $\dim(\hat{\rho})$ and we find that it is equal to $\dim(\rho_E)$. To show that $\hat{\rho} = \rho_E$ it is therefore sufficient to check that $\dim(\rho_V) \neq \dim(\rho_{V'})$. We have ([CR]): $\dim(\rho_V) = (q/6)(q+1)^2(q^2+q+1)$, $\dim(\rho_{V'}) = (q/2)(q+1)^2(q^2-q+1)$, hence $\dim(\rho_{V'}) - \dim(\rho_V) = (q/3)(q^2-1)^2 > 0$.

Thus, we have $\hat{\rho} = \rho_E$ in all cases and the proposition is proved.

12.7. We shall now prove 12.2 (i) in the case where $\mathcal{G}_{\mathcal{F}} = \mathfrak{G}_2$ and $\Delta \neq 1$ on $\mathcal{F}$. In this case $\widetilde{\mathfrak{M}}$ consists of two elements m_0, m_1. By (12.3.7), we have $\phi(m_0) = 1$. Let $c = \phi(m_1) \in \mathbf{N}$. We have $\hat{\phi}(m_0) = (c+1)/2$ and this equals 1, by (12.3.10). Thus, $c = 1$, hence $\phi = \hat{\phi} =$ characteristic function of $\widetilde{\mathfrak{M}}$.

12.8. We now assume that $\mathcal{G}_{\mathcal{F}} = \mathfrak{G}_3, \mathfrak{G}_4$, or $\mathfrak{G}_5$. We first show that

$$\hat{\phi}(m) = 0 \quad \text{for all} \quad m \notin \widetilde{\mathfrak{M}}. \tag{12.8.1}$$

As in the proof of (12.3.8), we shall let Γ vary through the set of left cells in Ω and we shall write ϕ_Γ for the function ϕ corresponding to Γ. If $m \notin \mathfrak{M}$, it follows from 12.6 that ρ_m is not a component of $\mathrm{Ind}_{B^{F'}}^{G^{F'}}(1)$. Hence

$$0 = \langle \rho, \mathrm{Ind}_{B^{F'}}^{G^{F'}}(1) \rangle_{G^{F'}} = \langle \rho, R \sum_{E \in W^\vee} \dim(E) \cdot E \rangle_{G^{F'}}$$

$$= \langle \rho, R_{[\Omega]_1} \rangle_{G^{F'}} = \langle \rho, \sum_{\substack{\Gamma' \\ \Gamma' \subset \Omega}} R_{[\Gamma']_1} \rangle_{G^{F'}} = \sum_{\Gamma'} \langle \rho, R_{[\Gamma']_1} \rangle_{G^{F'}}$$

(compare the proof of (12.3.8)). We have

$$\langle \rho, R_{[\Gamma']_1} \rangle_{G^{F'}} = \sum_{m' \in \mathfrak{M}} \{m,m'\} \phi_{\Gamma'}(m') = \hat{\phi}_{\Gamma'}(m)$$

hence

$$0 = \sum_{\Gamma'} \hat{\phi}_{\Gamma'}(m) .$$

Since each term $\hat{\phi}_{\Gamma'}(m)$ is ≥ 0, (see (12.3.6)), it follows that $\hat{\phi}_{\Gamma'}(m) = 0$ for all $\Gamma' \subset \Omega$ and (12.8.1) is proved.

Taking into account (12.3.6), (12.3.7), (12.3.10), (12.8.1), we see that to complete the proof of 12.2 (i), it is enough to prove the following.

12.9. LEMMA. *Assume that* $\mathcal{G}_{\mathcal{F}} = \mathfrak{G}_3$, $\mathfrak{G}_4$ *or* $\mathfrak{G}_5$ *and let* $\phi : \mathfrak{M} \to \mathbf{N}$ *be a function such that*

(a) $\operatorname{supp} \phi \subset \widetilde{\mathfrak{M}}$, $\operatorname{supp} \hat{\phi} \subset \widetilde{\mathfrak{M}}$,

(b) $\hat{\phi}(m) \in \mathbf{N}$ *for all* $m \in \mathfrak{M}$,

(c) $\phi(m_0) = \hat{\phi}(m_0) = 1$.

Then $\phi = \hat{\phi}$.

Proof. It is not difficult to determine all functions $\phi : \mathfrak{M} \to \mathbf{N}$ satisfying conditions (a), (b), (c) above. We find the following complete list of possibilities for ϕ.

Table for $\mathfrak{G}_3$, $|\widetilde{\mathfrak{M}}| = 4$

(1,1)	1	1
$(g_2,1)$	1	1
(1,r)	1	0
$(g_3,1)$	0	1

Table for $\mathfrak{G}_3$, $|\widetilde{\mathfrak{M}}| = 5$

(1,1)	1	1	1	1	1
$(g_2,1)$	1	1	0	0	0
(1,r)	1	0	2	0	1
$(g_3,1)$	0	1	0	2	1
$(1,\varepsilon)$	0	0	1	1	1

Table for $\mathfrak{G}_4$

$(1,1)$	1	1	1	1	1	1	1
$(1,\lambda^1)$	1	0	1	0	2	1	1
$(1,\lambda^2)$	0	0	0	0	1	1	0
$(1,\sigma)$	0	0	1	1	1	1	1
$(g_2,1)$	1	1	2	1	1	1	1
(g_2,ε'')	1	0	0	0	1	0	1
$(g_2',1)$	0	1	1	2	0	1	1
(g_2',ε'')	0	0	1	0	0	0	0
(g_2',ε')	0	0	0	1	0	0	1
$(g_3,1)$	1	1	0	0	0	0	0
$(g_4,1)$	0	1	0	1	0	1	0

Table for $\mathfrak{G}_5$

$(1,1)$	1	1	1	1	1	1	1	1	1	1	1	1	1	1	1	1
$(g_3,1)$	1	1	0	0	1	0	1	0	1	0	2	1	0	1	0	1
$(g_2',1)$	1	1	2	0	1	1	0	1	1	1	0	0	0	0	1	1
$(1,\nu)$	0	0	1	3	1	2	1	1	1	2	1	2	2	1	2	1
$(1,\lambda^1)$	0	1	1	3	1	2	2	1	1	2	1	2	2	1	2	1
$(g_5,1)$	1	0	0	0	0	0	0	1	0	0	0	0	1	1	0	0
(g_3,ε)	0	1	0	0	0	0	1	0	0	0	0	0	0	0	0	0
$(1,\nu')$	0	0	1	2	0	1	0	1	1	2	0	1	1	0	1	0
(g_2',ε'')	0	0	0	0	1	1	0	0	0	0	0	0	0	0	0	0
$(1,\lambda^2)$	0	0	0	3	0	1	1	1	1	2	1	2	2	1	1	0
(g_2',ε')	0	0	1	0	0	0	0	0	0	0	0	0	0	0	1	1
$(1,\lambda^3)$	0	0	0	1	0	0	0	0	1	1	1	1	0	0	0	0
$(g_2,1)$	1	1	1	1	2	2	1	1	1	1	1	1	1	1	1	1
$(g_4,1)$	1	1	1	0	0	0	0	1	1	1	0	0	0	0	0	0
$(g_6,1)$	1	0	0	0	1	0	0	0	1	0	2	1	0	1	0	1
(g_2,r)	0	1	1	2	1	2	2	1	0	1	0	1	2	1	2	1
(g_2,ε)	0	0	0	1	0	0	1	0	0	0	1	1	1	1	1	1

In each of these tables, the rows correspond to the various elements of $\widetilde{\mathfrak{M}}$. Each column represents a function ϕ supported on $\widetilde{\mathfrak{M}}$, satisfying conditions a), b), c); the tables give the values of ϕ on the elements of $\widetilde{\mathfrak{M}}$. One checks by direct computation that each function ϕ in the tables satisfies $\phi = \hat{\phi}$. This completes the proof of the lemma and hence that of Theorem 12.2.

12.10. For any finite group $\mathcal{G}$, we define a "*lagrangian subspace*" of $\mathfrak{M}(\mathcal{G})$ to be a function $\phi : \mathfrak{M}(\mathcal{G}) \to \mathbf{N}$ such that

(i) $\operatorname{supp} \phi \subset \{m \in \mathfrak{M}(\mathcal{G}) | \tilde{\lambda}_m = 1\}$, (see 11.1),

(ii) $\phi = \hat{\phi}$,

(iii) $\phi(m_0) = 1$.

In the case where $\mathcal{G}$ is a d-dimensional vector space over F_2, then such ϕ is the characteristic function of a d-dimensional subspace of $\mathfrak{M}$ on which the quadratic form $f : \mathfrak{M} \to F_2$ defined by $\tilde{\lambda}_m$ (see 11.1) is identically zero. (See the proof in 12.5.) This justifies the name "lagrangian subspace." In general, ϕ is not the characteristic function of a subset of $\mathfrak{M}(\mathcal{G})$; it can have values ≥ 2 (see the tables in 12.9). Since for $G = \mathfrak{S}_3, \mathfrak{S}_4, \mathfrak{S}_5$, we have $\widetilde{\mathfrak{M}} \subset \{m \in \mathfrak{M}(\mathcal{G}) | \tilde{\lambda}_m = 1\}$, (by direct verification), we see that:

12.11. COROLLARY. *With the assumptions of 12.2, the function* ϕ *corresponding to a left cell* Γ *is a "lagrangian subspace" of* $\mathfrak{M}$.

12.12. Not all functions ϕ in the tables in 12.10 arise from left cells. The functions ϕ corresponding to the first two (resp. 3, 5, 7) columns in the tables for $\mathfrak{S}_3$, $|\widetilde{M}| = 4$, (resp. $\mathfrak{S}_3$, $|\widetilde{M}| = 5$, $\mathfrak{S}_4, \mathfrak{S}_5$) do arise from left cells, and it would follow from a conjecture in $[L_7]$ that the functions ϕ corresponding to the other columns do not arise from left cells. This is true at least in the table for $\mathfrak{S}_4$.

Indeed, let a_i, $(1 \le i \le 7)$, be the number of left cells in the two-sided cell of W of type F_4 associated to the family of 12_1, for which the corresponding function $\phi : M \to N$ is given by the i-th column in that table. Call ϕ_i this function ϕ. For any $m \in \mathfrak{M}$, the sum $\sum_{i=1}^{7} a_i \phi_i(m)$ is equal to $\dim(E_m)$. In particular, for $m = (g_2', \varepsilon')$ and $m = (1, \lambda^2)$, we have $\dim(E_m) = 1$, hence $\sum_{i=1}^{7} a_i \phi_i(m) = 1$. Since $\phi_1, \phi_2, \phi_3, \phi_4, \phi_5$ do arise from left cells, we have $a_i > 0$ for $1 \le i \le 5$. For $m = (g_2', \varepsilon')$, we have $\phi_4(m) = \phi_7(m) = 1$, $\phi_i(m) = 0$ for $i \ne 4,7$ hence $a_4 + a_7 = 1$. Since $a_4 > 0$, it follows that $a_4 = 1$ and $a_7 = 0$. Similarly, using $m = (1, \lambda^2)$, we see that $a_5 = 1$ and $a_6 = 0$. In particular, we see that ϕ_6 and ϕ_7 do not arise from left cells.

We shall now prove

12.13. THEOREM. *Assume that* $\mathcal{G}_{\mathcal{F}}$ *is an* F_2*-vector space of dimension* d, *and* $\Delta \equiv 1$ *on* $\mathcal{F}$, *(the family corresponding to the left cell* Γ *). Then*

(i) *The function* $\phi : \mathfrak{M} \to N$ *corresponding to* Γ *is the characteristic function of a* d*-dimensional vector space* $V \subset \mathfrak{M}$ *which is isotropic for the canonical quadratic form* $f : \mathfrak{M} \to F_2$. *In particular,* $[\Gamma]_1$ *is multiplicity free and the number of its irreducible components is* 2^d.

(ii) *There is a unique bijection* $\mathrm{Hom}_{F_2}(V, F_2) \xrightarrow{\sim} \Gamma^*$ *(see 12.3) with the following property. If* $x_\xi \in \Gamma^*$ *corresponds to the linear form* $\xi : V \to F_2$, *then*

$$\phi_{x_\xi}(m) = \begin{cases} (-1)^{\xi(m)}, & \text{for } m \in V \\ 0, & \text{for } m \in \mathfrak{M} - V \end{cases}.$$

Proof. (i) has been proved already, (see 12.5). We now prove (ii). As we have stated in 12.3 we have $\mathrm{supp}\, \phi_x \subset V$, $\forall x \in \Gamma^*$. It follows that $\hat{\phi}_x$ is constant on the cosets of $\mathfrak{M}$ with respect to V. Let us choose, for each linear form $\xi : V \to F_2$, an element $\dot{\xi} \in \mathfrak{M}$ such that $(\dot{\xi}, m') = \xi(m')$ for all $m' \in V$. Then $\hat{\phi}_x(\dot{\xi})$ depends only on ξ and we have

$$(12.13.1) \qquad \phi_x(m) = \sum_{\xi} \hat{\phi}_x(\dot{\xi})(-1)^{\xi(m)}, \quad m \in V.$$

Since $\hat{\phi}_x(\dot{\xi}) \in \mathbf{N}$, (12.3.1), it follows from (12.13.1) that $m \to \phi_x(m)$ is the character of a $\mathbf{Q}[V]$-module ε_x. By definition of Γ^*, we have $\phi_x \neq 0$ on V, hence $\varepsilon_x \neq 0$, $x \in \Gamma^*$. Since $\phi \equiv 1$ on V, by (i), it follows from (12.3.4) that the $\mathbf{Q}[V\times V]$-module $\sum_{x\in\Gamma^*} \varepsilon_x \boxtimes \varepsilon_x$ is isomorphic to the two-sided regular representation of V. This forces each ε_x to be an irreducible $\mathbf{Q}[V]$-module (hence of dimension 1); it also follows that $x \mapsto \varepsilon_x$ is a bijection between Γ^* and the set of irreducible $\mathbf{Q}[V]$-modules (up to isomorphism). From this, (ii) follows immediately.

12.14. The space $\mathcal{P}$ of functions $f:(G/B)^{F'} \to \overline{\mathbf{Q}}_\ell$ is in a natural way a $G^{F'}$-module and a module over the Hecke algebra $H(q)$, (see (3.4.2)). The left ideals $I^L_x, \hat{I}^L_x$ of H, $(x \in W)$, defined in 5.1 give rise by extension of scalars $(\mathbf{Z}[u^{1/2},u^{-1/2}] \to \overline{\mathbf{Q}}_\ell$, $u^{1/2} \to q^{1/2})$ to left ideals $I^L_{x,q}, \hat{I}^L_{x,q}$ of $H(q)$.

Let $\mathcal{P}^L_x = \{f \in \mathcal{P} | hf = 0 \text{ for all } h \in I^L_{x,q}\}$. We define similarly $\hat{\mathcal{P}}^L_x$ by replacing $I^L_{x,q}$ in the previous definition by $\hat{I}^L_{x,q}$. Then $\mathcal{P}^L_x$ $(x \in W)$ form a filtration of $\mathcal{P}$ by $G^{F'}$-submodules: we have $\mathcal{P}^L_x \subset \mathcal{P}^L_{x'} \Longleftrightarrow I^L_{x,q} \supset I^L_{x',q} \Longleftrightarrow x' \underset{L}{\leq} x$. Moreover, $\hat{\mathcal{P}}^L_x \supset \mathcal{P}^L_x$ and the $G^{F'}$-modules $\mathcal{P}^L_x, \hat{\mathcal{P}}^L_x/\mathcal{P}^L_x$ are disjoint (as one can see from (5.1.13)). It follows that there is a unique $G^{F'}$-stable complement $\overline{\mathcal{P}}^L_x$ for $\mathcal{P}^L_x$ in $\hat{\mathcal{P}}^L_x$. We have $\overline{\mathcal{P}}^L_x = \overline{\mathcal{P}}^L_{x'}$ if and only if x,x' are in the same left cell $\Gamma \subset W$, hence we may define for any left cell $\Gamma \subset W$, $\mathcal{P}_\Gamma = \overline{\mathcal{P}}^L_x$ where x is any element of Γ. Thus, we have defined a direct sum decomposition (as a $G^{F'}$-module)

$$(12.14.1) \qquad \mathcal{P} = \bigoplus_{\Gamma} \mathcal{P}_\Gamma$$

where Γ runs over the left cells of W. From the definition it follows easily that, as a $G^{F'}$-module, we have $\mathcal{P}_\Gamma = \sum_E n_E \rho_E$ (ρ_E as in 12.6), where for any irreducible representation E of W, n_E denotes the multiplicity of E in $[\Gamma]_1$. We can now state:

(12.14.2) *For any left cell* $\Gamma \subset W$, *we have* $\mathcal{P}_\Gamma \cong R_{[\Gamma]_1}$.

With notations in 12.1, we have $\mathcal{P}_\Gamma = \sum_{m \in \widetilde{\mathfrak{M}}} \phi(m)\rho_{E_m}$, $R_{[\Gamma]_1} = \sum_{m \in \mathfrak{M}} \hat{\phi}(m)\rho_m$. It remains to use 12.6 and 12.2 (i).

12.15. PROPOSITION. *Let* Γ, Γ' *be two left cells of* W *contained in the same two-sided cell. We have*

$$\dim \mathrm{Hom}_W([\Gamma]_1, [\Gamma']_1) = |\Gamma \cap (\Gamma')^{-1}|. \tag{12.15.1}$$

Proof. With the notations in Lemma 5.4, we have $[\Gamma] = I^L_{x,\mathcal{K}}/\hat{I}^L_{x,\mathcal{K}}$ (a left $H_{\mathcal{K}}$-module) where x is any element of Γ. We shall give an alternative description of $[\Gamma]$. Let $J^L_{x,\mathcal{K}}$ (resp. $\hat{J}^L_{x,\mathcal{K}}$) be the $\mathcal{K}$-subspace of $H_{\mathcal{K}}$ spanned by the elements D_u, $u \underset{L}{\geq} x$ (resp. by the elements D_u, $u \underset{L}{\geq} x$, $u \underset{L}{\not\sim} x$), see (5.1.6). Then, by (5.1.16), $J^L_{x,\mathcal{K}}$ and $\hat{J}^L_{x,\mathcal{K}}$ are left ideals in $H_{\mathcal{K}}$ and, clearly, $J^L_{x,\mathcal{K}} \supset \hat{J}^L_{x,\mathcal{K}}$. Let $\{\Gamma\}$ be the $H_{\mathcal{K}}$-module $J^L_{x,\mathcal{K}}/\hat{J}^L_{x,\mathcal{K}}$. We want to show that $[\Gamma]$ is isomorphic to $\{\Gamma\}$.

From (5.1.14) and (5.1.10) it follows that $J^L_{x,\mathcal{K}} = (\hat{I}^L_{x,\mathcal{K}})^\perp$, $\hat{J}^L_{x,\mathcal{K}} = (I^L_{x,\mathcal{K}})^\perp$ where $\perp$ is defined with respect to the non-degenerate symmetric bilinear form $h,h' \to \tau(h\cdot\widetilde{h'})$ on $H_{\mathcal{K}}$, (for $h' = \Sigma a_w T_w$, we set $\widetilde{h'} = \Sigma a_w T_{w^{-1}}$). Moreover, we have $\tau(T_w h\widetilde{h'}) = \tau(\widetilde{hh'T_{w^{-1}}})$ for $w \in W$. It follows that $\{\Gamma\}$ is in a natural way the dual space of $[\Gamma]$ and that the action of T_w on $\{\Gamma\}$ corresponds to the transpose of the action of $T_{w^{-1}}$ on $[\Gamma]$. Hence $\mathrm{Tr}(T_w, \{\Gamma\}) = \mathrm{Tr}(T_{w^{-1}}, [\Gamma]) = \mathrm{Tr}(T_w, [\Gamma])$, (see (3.3.3)), for all $w \in W$. Hence $\{\Gamma\}$, $[\Gamma]$ are isomorphic as $H_{\mathcal{K}}$-modules.

Now let $y \in \Gamma \cap (\Gamma')^{-1}$. By (5.1.14), the map $h \to h\cdot D_{y^{-1}}$ $(I^L_{y,\mathcal{K}} \to J^L_{y^{-1},\mathcal{K}})$ is zero on $\hat{I}^L_{y,\mathcal{K}}$, hence it defines an $H_{\mathcal{K}}$-linear map $[\Gamma] \to J^L_{y^{-1},\mathcal{K}}$. Composing with the canonical projection $J^L_{y^{-1},\mathcal{K}} \to \{\Gamma'\}$, (note that $y^{-1} \in \Gamma'$), we get an $H_{\mathcal{K}}$-linear map $\phi_y : [\Gamma] \to \{\Gamma'\}$. We now show that, as y runs over the set $\Gamma \cap (\Gamma')^{-1}$, the maps ϕ_y are linearly independent over $\mathcal{K}$. Assume that $a_y \in \mathcal{K}$, $(y \in \Gamma \cap (\Gamma')^{-1})$, are such that $\sum_y a_y \phi_y = 0$. Then, for any $z \in \Gamma$, we have

$$\text{(12.15.2)} \qquad \sum_{y \in \Gamma \cap (\Gamma')^{-1}} a_y C_z D_{y^{-1}} \in \hat{J}^L_{z_0',\mathcal{K}}$$

where z_0' is a fixed element of Γ'. We have $\hat{J}^L_{z_0',\mathcal{K}} \subset \hat{J}^{LR}_{z_0',\mathcal{K}} = \bigoplus_{\substack{E \\ E \underset{LR}{>} z_0'^{-1}}} I_{E,\mathcal{K}}$, (see (5.1.19)). On the other hand, the element (12.15.2) is in $I^{LR}_{z_0,\mathcal{K}} = \bigoplus_{\substack{E \\ E \underset{LR}{\leq} z_0}} I_{E,\mathcal{K}}$, see (5.1.17); here z_0 is a fixed element of Γ. We have $z_0 \underset{LR}{\sim} z_0'^{-1}$, (see 5.2), hence $I^{LR}_{z_0,\mathcal{K}} \cap \hat{J}^{LR}_{z_0',\mathcal{K}} = 0$. It follows that

$$\sum_{y \in \Gamma \cap (\Gamma')^{-1}} a_y C_z D_{y^{-1}} = 0$$

for all $z \in \Gamma$, hence, by (5.1.14), $h \cdot \sum_y a_y D_{y^{-1}} = 0$ for all $h \in I^L_{z_0,\mathcal{K}}$. In other words, $\sum_{y \in \Gamma \cap (\Gamma')^{-1}} a_y D_{y^{-1}} \in (I^L_{z_0,\mathcal{K}})^{\perp} = \hat{J}^L_{z_0,\mathcal{K}}$ hence $\sum_{y \in \Gamma \cap (\Gamma')^{-1}} a_y D_{y^{-1}}$ is a $\mathcal{K}$-linear combination of elements D_u, $u \underset{L}{\geq} z_0$, $u \underset{L}{\not\sim} z_0$, (in particular, $u \notin \Gamma$). It follows that $a_y = 0$ for all $y \in \Gamma \cap (\Gamma')^{-1}$. We have proved that the linear maps ϕ_y are linearly independent, hence

$$\text{(12.15.3)} \quad \dim \mathrm{Hom}_W([\Gamma]_1, [\Gamma']_1) = \dim \mathrm{Hom}_{H_{\mathcal{K}}}([\Gamma], \{\Gamma'\}) \geq |\Gamma \cap (\Gamma')^{-1}| .$$

If we sum term by term these inequalities over all pairs Γ, Γ' (with Γ, Γ'

in the same two-sided cell), and as a result, we get an equality, it will follow automatically that each of the inequalities (12.15.3) is in fact an equality, and (12.15.1) will follow. We have

$$\sum_{\Gamma,\Gamma'} \dim \mathrm{Hom}_W([\Gamma]_1, [\Gamma']_1) = \dim \mathrm{Hom}_W\Big(\sum_{\Gamma}[\Gamma]_1, \sum_{\Gamma'}[\Gamma']_1\Big)$$

$$= \dim \mathrm{Hom}_W(\mathrm{Reg}_W, \mathrm{Reg}_W)$$

$$= |W|$$

where Reg_W is the left regular representation of W. (Note that $\mathrm{Hom}_W([\Gamma]_1, [\Gamma']_1) = 0$ whenever Γ, Γ' are not in the same two-sided cell.) On the other hand,

$$\sum_{\Gamma,\Gamma'} |\Gamma \cap (\Gamma')^{-1}| = |W|$$

since any element of W belongs to a unique left cell and to a unique right cell. The proposition is proved.

12.16. COROLLARY. *If Γ is a left cell of W and Δ is a right cell of W contained in the same two-sided cell as Γ, then $\Gamma \cap \Delta \neq \emptyset$.*

Indeed, let $\Gamma' = \Delta^{-1}$. Then $[\Gamma]_1$, $[\Gamma']_1$ contain the same special representation with multiplicity one, hence $\dim \mathrm{Hom}_W([\Gamma]_1, [\Gamma']_1) > 0$. It follows that $\Gamma \cap (\Gamma')^{-1} \neq \emptyset$, as required.

12.17. COROLLARY. *Let Γ be a left cell in W and assume that W is of classical type. Let $\mathfrak{F}$ be as in 12.1 and let d be defined by the condition that $|\mathcal{G}_{\mathfrak{F}}| = 2^d$. Then $\Gamma \cap \Gamma^{-1}$ is exactly the set of involutions in Γ; it consists of 2^d elements.*

Proof. Let $\Gamma_{(2)}$ be the set of involutions in Γ. It is clear that $\Gamma_{(2)} \subset \Gamma \cap \Gamma^{-1}$. By 12.15, $|\Gamma \cap \Gamma^{-1}|$ is the number v_Γ of irreducible components of $[\Gamma]_1$ (which is multiplicity free by 12.13(i)). Hence

$|\Gamma_{(2)}| \leq v_\Gamma$. If we sum these inequalities over all Γ in W, we get $\sum_\Gamma |\Gamma_{(2)}| \leq \sum_\Gamma v_\Gamma$. Now $\sum_\Gamma \Gamma_{(2)}$ is the number of involutions in Γ, which is well known to be equal to the sum of dimensions of the irreducible representations of W, hence it is equal to number of irreducible components of the regular representation of W, and hence to $\sum_\Gamma v_\Gamma$. It follows that each inequality $|\Gamma_{(2)}| \leq v_\Gamma$ is an equality. It remains to use the identity $v_\Gamma = 2^d$, (see 12.13(i)). The fact that $\Gamma \cap \Gamma^{-1}$ is the set of involutions in Γ can be also deduced from [BV_1].

12.18. It seems likely that in general, we have

$$\Gamma^* = \Gamma \cap \Gamma^{-1} \tag{12.18.1}$$

for any left cell $\Gamma \subset W$. (Γ^* is defined in 12.3.) One can show that this is true when W is of classical type. However, for exceptional W, it may happen that $\Gamma \cap \Gamma^{-1}$ contains elements of order ≥ 3.

Chapter 13
RELATIONS WITH CONJUGACY CLASSES

In this chapter we shall express the bijection in 4.23 in a more geometric form, in terms of "special conjugacy classes" in the dual group.

13.1. In this section we shall review some results on the Springer representations of Weyl groups.

Let G_1 be a connected reductive algebraic group over k, and let W_1 be the Weyl group of G_1. For each unipotent element $u \in G_1$ we denote by $\mathcal{B}_u$ the variety of Borel subgroups of G_1 containing u, and by $A(u)$ the group of connected components of the centralizer of u in G_1 modulo the center of G_1. Then $A(u)$ acts naturally on the set of irreducible components of $\mathcal{B}_u$. (According to Spaltenstein, all irreducible components of $\mathcal{B}_u$ have the same dimension, say $e(u)$.) Hence $A(u)$ acts naturally on $H^{2e(u)}(\mathcal{B}_u, \overline{Q}_\ell)$. Springer [$Sp_2$] has defined a natural representation of W_1 on $H^{2e(u)}(\mathcal{B}_u, \overline{Q}_\ell)$ which commutes with the action of $A(u)$. (Actually, Springer's definition is valid only in large enough characteristic; a definition valid in arbitrary characteristic was given in [L_{10}].) For any irreducible representation ϕ of $A(u)$, the vector space $E_{u,\phi} = \mathrm{Hom}_{A(u)}(\phi, H^{2e(u)}(\mathcal{B}_u, \overline{Q}_\ell))$ is a W_1-module. The main result on Springer representations is that $E_{u,\phi}$ is either 0 or an irreducible W_1-module, that $E_{u,\phi} \approx E_{u',\phi'}$ if and only if (u,ϕ), (u',ϕ') are conjugate under G_1 and that all irreducible representations of W_1 arise in this way. In particular, $E_{u,1}$ is always an irreducible representation of W_1. (It is non-zero.) The correspondence between pairs (u,ϕ) as above and irreducible representations of W_1 has been explicitly described (assuming that p is a good prime for G_1) by Springer [Sp_2] for type A and G_2, by Shoji

$[\mathrm{Sh}_1]$, $[\mathrm{Sh}_2]$ for classical groups and F_4 and in [AL], [S] for groups of type E_n. The results in this section are consequences of this explicit description of the Springer correspondence.

In the following discussion we assume that p is a good prime for G_1. A case by case verification gives:

(13.1.1) *Any special irreducible representation of* W_1 *is of the form* $E_{u,1}$ *for a well-defined unipotent element* u *(up to conjugacy) in* G_1. *A unipotent element* u *(or its class in* G_1*) is said to be special if* $E_{u,1}$ *is a special representation of* W_1, *or, equivalently, if* $a_{E_{u,1}} = \dim \mathcal{B}_u$.

(13.1.2) *If* $u \in G_1$ *is a special unipotent element then for any irreducible representation* ϕ *of* $A(u)$ *such that* $E_{u,\phi} \neq 0$ *we have* $a_{E_{u,\phi}} \leq \dim \mathcal{B}_u$; *we set* $A(u)_0^\vee = \{\phi \in A(u)^\vee | E_{u,\phi} \neq 0$ *and* $a_{E_{u,\phi}} = \dim \mathcal{B}_u\}$.

This defines a canonical quotient group $\bar{A}(u)$ of $A(u)$, for a special unipotent element $u \in G_1$. By definition, $\bar{A}(u)$ is the largest quotient of $A(u)$ through which one can factor all $\phi \in A(u)_0^\vee$. (Hence the kernel of $A(u) \to \bar{A}(u)$ acts trivially on all representations ϕ in $A(u)_0^\vee$.)

The group $\bar{A}(u)$ has an alternative description, which we now sketch. (This description will not be used here.) Given u (special), we consider a parabolic subgroup $P \subset G_1$ which is minimal subject to the following property: there exists a special unipotent element $u' \in M$ = Levi subgroup of P, such that $E_{u,1} = J_{W_M}^{W_1}(E_{u',1})$, where $E_{u',1} \in W_M$ is defined with respect to M. It turns out that the pair (u',M) is uniquely defined up to G_1-conjugacy by u and that the centralizer of u in G_1 is contained in P. Hence $A(u')$ is in a natural way a quotient of $A(u)$. We have $\bar{A}(u) = A(u')$ modulo the image of the center of M. (Here $A(u')$ is defined with respect to M.)

(13.1.3) *If* u *is a special unipotent element in* G_1, *then* $\bar{A}(u)$ *is isomorphic to* $\mathcal{G}_{\mathcal{F}}$, *(see 4.14), where* $\mathcal{F}$ *is the family in* $W_1^\vee$ *containing* $E_{u,1} \otimes \text{sign}$.

Assume now that G_1 is defined over F_q and let $F': G_1 \to G_1$ be the corresponding Frobenius map. Let $\mathcal{F} \subset W_1^\vee$ be an F'-stable family. Giving $\mathcal{F}$ is the same as giving an F'-stable special unipotent class C in G_1. We choose an F'-stable element u_1 in this conjugacy class. Then $E_{u_1,1} \otimes \text{sign} \in \mathcal{F}$. Note that F' acts on $A(u_1)$ and on $\bar{A}(u_1)$ in a natural way, hence we can form the semidirect product $\tilde{A}(u_1)$ of $\bar{A}(u_1)$ with the infinite cyclic group generated by F' such that $F' \cdot \bar{a} \cdot F'^{-1} = F'(\bar{a})$, $\bar{a} \in \bar{A}(u_1)$. The pair $\bar{A}(u_1) \subset \tilde{A}(u_1)$ and the image of F' in $\tilde{A}(u_1)/\bar{A}(u_1)$ depend up to isomorphism only on the class C, not on the choice of $u_1 \in C(F_q)$; the triple $\bar{A}(u_1) \subset \tilde{A}(u_1) \supset \bar{A}(u_1) \cdot F'$ is isomorphic to $\mathcal{G}_{\mathcal{F}} \subset \tilde{\mathcal{G}}_{\mathcal{F}} \supset \mathcal{G}'_{\mathcal{F}}$, (see 4.21), defined with respect to the automorphism F' of W_1.

13.2. Now let G, F' be as in 3.4 and let G^* be a group dual to G (see 8.4) with the corresponding Frobenius map $F': G^* \to G^*$. Let $g \in G^*$ be an element whose Jordan decomposition is $g = su = us$ (s semisimple, u unipotent). We say that g (or its conjugacy class is G^*) is special if u is a special unipotent element (see (13.1.1)) of $Z_{G^*}(s)$, the centralizer of s in G^*. (This centralizer is connected, since G has connected center.) If $g \in G^*$ is a special element, we set $\bar{A}(g) = \bar{A}(u)$, the last group being defined with respect to $Z_{G^*}(s)$, (see 13.1). If, in addition, we have $F'(g) = g$, we define $\tilde{A}(g)$ to be $\tilde{A}(u)$ (defined with respect to $Z_{G^*}(s)$, see 13.1). Note that $F': Z_{G^*}(s) \to Z_{G^*}(s)$ and $F'(u) = u$. Hence $\bar{A}(g)$ is naturally imbedded in $\tilde{A}(g)$ and $\tilde{A}(g)/\bar{A}(g)$ is a cyclic group with a distinguished generator, the image of F'; the triple $\bar{A}(g) \subset \tilde{A}(g) \supset \bar{A}(g) \cdot F'$ depends (up to isomorphism) only on the conjugacy class of g in G. Using now (8.4.4), we see that the bijection 4.23 may be also interpreted as a bijection

$$(13.2.1)\quad \{\text{irreducible } G^{F'}\text{-modules over } \overline{\mathbb{Q}}_\ell, \text{ up to isomorphism}\}$$
$$\updownarrow$$
$$\coprod_C \overline{\mathfrak{M}}(\bar{A}(g) \subset \tilde{A}(g))$$

where C runs over the set of F'-stable special conjugacy classes in G^*, and g is an element in $C^{F'}$. (See $[M_1]$ for the case of GL_n.)

Under this bijection, the set $\mathcal{E}_{(s)}$ of (8.4.4) corresponds to the union over those F'-stable special conjugacy classes C which contain elements with semisimple part s.

13.3. The bijection (13.2.1) involves the set of special conjugacy classes in G^*. *We shall now construct (assuming that* p *is a good prime for* G *) a canonical surjective map* Φ *from the set of special conjugacy classes in* G^* *to the set of unipotent conjugacy classes in* G.

Let g be a special element in G^*. Let $g = su = us$ be the Jordan decomposition of g with s semisimple and u special unipotent in $Z(s) = Z_{G^*}(s)$. Replacing g by a conjugate, we may assume that $s \in T^*$. Then $W_{Z(s)}$ may be identified with the normalizer of T^* in $Z(s)$ modulo T^*; hence with a subgroup of $W = N_{G^*}(T^*)/T^*$. We consider the Springer representation $E_{u,1}$ of $W_{Z(s)}$ (see 13.1) defined for u with respect to the group $Z(s)$. It is a special representation of $W_{Z(s)}$.

The induced representation $\mathrm{Ind}_{W_{Z(s)}}^{W}(E_{u,1})$ contains a unique irreducible W-submodule $\hat{E}$ such that $b_{\hat{E}} = b_{E_{u,1}}$ (see $[L_6]$, $[L_7]$). Let us now regard $\hat{E}$ as a representation of the Weyl group of G, rather than that of G^*. Then there is a well-defined unipotent element $\hat{u} \in G$ (up to conjugacy) so that $\hat{E}$ is the Springer representation $E_{\hat{u},1}$. This fact, which was conjectured in $[L_6]$ can be checked using $[Sh_1]$, $[Sh_2]$, [AL]. Our map Φ then associates the class of $\hat{u}$ to the class of g. Note that *the dimension of the class of* g is equal to the dimension of the class of $\hat{u}$, hence Φ preserves dimensions. This follows from the equality $b_{\hat{E}_{u,1}} = b_{E_{u,1}}$ and the known identities dim(class of $\hat{u}$) $= 2(\nu - b_{E_{\hat{u},1}})$, dim(class of u in Z(s)) $= 2(\nu' - b_{E_{u,1}})$, (see [BM]), where ν (resp. ν') is the number of positive roots of G (resp. of Z(s)). The surjectivity of Φ can be also checked using [Sh_1], [Sh_2], [AL]. Using [loc. cit.], I have checked that, for any unipotent element $u \in G$, we have

$$|A(u)| = \sup |\bar{A}(g)|$$

where the supremum is taken over all special classes (g) in G^* which are mapped by Φ to the class of u. (This was conjectured in [L_6].)

The restriction of Φ to the set $\mathfrak{U}_{G^*}$ of special unipotent classes in G is a bijection $\mathfrak{U}_{G^*} \xrightarrow{\approx} \mathfrak{U}_G$. This bijection is also defined when G does not have connected center. In particular we have a natural bijection $\Psi_s : \mathfrak{U}_{Z(s)} \xrightarrow{\approx} \mathfrak{U}_{Z(s)^*}$ for any semisimple element $s \in G^*$. Composing Ψ_s^{-1} with the restriction of Φ to $\mathfrak{U}_{Z(s)}$, we get a natural map (generalizing induction): $\Phi\Psi_s^{-1}$: {special unipotent classes in $Z(s)^*$} $\to$ {unipotent classes in G} for any semisimple element $s \in G^*$. This map preserves the codimension of a class.

13.4. With the notations in 13.2, (assuming p to be a good prime for G), we consider an irreducible representation ρ of $G^{F'}$, we attach to ρ a special conjugacy class $C = C_\rho$ in G^*, ($F'C_\rho = C_\rho$), by (13.2.1) and we then consider the unipotent class $\Phi(C_\rho)$ in G, (see 13.3). It is easy to see that $\Phi(C_\rho)$ is F'-stable. We consider the map

(13.4.1) ξ : {irreducible $G^{F'}$-modules up to isomorphism}
$\to$ {F'-stable unipotent classes in G}

defined by $\xi(\rho) = \Phi(C_\rho)$. Here are some properties of the map ξ which follow from the results in 13.3 and 4.23.

(13.4.2) ξ is surjective.

(13.4.3) If $\xi(\rho) = C'$, then $\dim(\rho) = (1/n_\rho)q^{\dim(C')/2}$ + lower powers of q, and n_ρ divides $|A(g')|$, $(g' \in C')$.

(13.4.4) If C' is an F'-stable unipotent class in G and $g' \in C'$, then $|A(g')| = \sup_\rho(n_\rho)$ where ρ runs over all irreducible $G^{F'}$-modules such that $\xi(\rho) = C'$.

(13.4.5) $\xi(\rho)$ is the regular unipotent class if and only if ρ is a component of the Gelfand-Graev representation, (see [DL_1, §10]).

13.5. Let $\widetilde{X}_{\dot{w}_0}$ be defined as in 2.1, where w_0 is the longest element of W. One could conjecture that the following property holds:

(13.5.1) If ρ is an irreducible representation of $G^{F'}$ appearing in $H^i_c(\widetilde{X}_{\dot{w}_0}, \overline{Q}_\ell)$ then $i = 2\nu - (1/2)\dim(C')$ where $\nu = \ell(w_0)$ and $C' = \xi(\rho)$. In particular ρ can appear in $H^i_c(\widetilde{X}_{w_0}, \overline{Q}_\ell)$ for at most one i.

13.6. In the setting of 13.2 we consider the set

(13.6.1) $$\coprod_{C} \mathfrak{M}(\overline{A}(g) \subset \widetilde{A}(g))$$

(union over all special F'-stable conjugacy classes $C \subset G^*$; g is an element in $C^{F'}$. This set has a natural free action of M, the group of roots of 1 in $\overline{Q}_\ell$, see 4.21. For each x in the set (13.6.1), we have a corresponding element $R_x \in \mathcal{R}(G^{F'}) \otimes \overline{Q}_\ell$ defined by (4.24.1). (The pair (L,n) in (4.24.1) corresponds as in 8.4 to the semisimple part s of g, where $x \in \mathfrak{M}(\overline{A}(g) \subset \widetilde{A}(g)) \subset X(W_{L,n}, \gamma)$.) We shall say that R_x is an *almost-character* of $G^{F'}$. When x is replaced by an element in the same M-orbit, the almost-character R_x is changed into a scalar multiple of R_x, (the scalar being in M). When x runs through a set of representatives for the orbits of M on the set (13.6.1), the almost-characters R_x from a $\overline{\mathbf{Q}}_\ell$-basis of $\mathcal{R}(G^{F'}) \otimes \overline{\mathbf{Q}}_\ell$. (This follows immediately from (4.24.1) and 4.25.)

Thus, $\mathcal{R}(G^{F'}) \otimes \overline{\mathbf{Q}}_\ell$ has a natural basis (defined up to multiplication of its elements by roots of 1 in $\overline{\mathbf{Q}}_\ell$) given by the almost-characters R_x. This is, in general, different from the basis of $\mathcal{R}(G^{F'}) \otimes \overline{\mathbf{Q}}_\ell$ which consists of all irreducible representations of $G^{F'}$. Note that both basis are orthonormal, (see 4.25).

13.7. It follows from (4.24.1) and 4.25 that the knowledge of the character of any irreducible representation of $G^{F'}$ is equivalent to the knowledge of $\mathrm{Tr}(g_1, R_x)$ for any $g_1 \in G^{F'}$ and any almost-character R_x.

I conjecture that to any x in the set (13.6.1) one can associate an irreducible "perverse $\overline{\mathbf{Q}}_\ell$-sheaf" K^x (in the sense of [BBD]) on G, which is G-equivariant (for the action of G on G given by conjugation) and which has a definite F_q-rational structure compatible with the G-action such that the following holds. For any $g_1 \in G^{F'}$, the alternating sum of traces of the Frobenius map on the stalks at g_1 of the cohomology sheaves of K^x is equal to $\mathrm{Tr}(g_1, R_x)$.

Consider the natural map $\mathfrak{M}(\bar{A}(g) \subset \tilde{A}(g)) \to \mathfrak{M}(\bar{A}(g))$ obtained by sending a pair (y, τ) in $\mathfrak{M}(\bar{A}(g) \subset \tilde{A}(g))$ to (y , restriction of τ to $\bar{A}(g)$). This gives rise to a natural map $x \to \zeta(x)$ of the set (13.6.1) into the set

$$\coprod_{C} \mathfrak{M}(\bar{A}(g)) \tag{13.7.1}$$

(union over all special conjugacy classes $C \subset G^*$; g is an element of C).

I also conjecture that to any m in the set (13.7.1) one can associate an irreducible G-equivariant perverse $\overline{\mathbf{Q}}_\ell$-sheaf K^m on G (which does not depend on the F_q-rational structure of G) such that the following holds. If an F_q-rational structure on G is given and if x is an element of the set (13.6.1) defined in terms of that rational structure and if m is the image of x under the map $x \to \zeta(x)$ defined above, then K^m is isomorphic to K^x with its F_q-rational structure forgotten. Note that any m in the set (13.7.1) is in the image of the map $x \to \zeta(x)$ for a suitable F_q-rational structure (with large q) on G.

The irreducible G-equivariant perverse sheaves K^m (whose existence is conjectural) form a set $\mathrm{Irr}(G)$ in 1-1 correspondence with the set (13.7.1). This set does not depend on an F_q-rational structure on G.

For example, when G is a torus, $\mathrm{Irr}(G)$ consists of all locally constant tame $\overline{Q}_\ell$-sheaves of rank 1 over G. They are in 1-1 correspondence with the elements of G^* (once the isomorphism (8.4.3) and the imbedding $\psi: k^* \to \overline{Q}_\ell^*$, see 2.3, are chosen).

The conjectures in this section can be verified for the case of GL_n and SO_5 using the results in [G], [Sr], [L_{10}]. Let us describe the set $\mathrm{Irr}(G)$ in the case where G is GL_n. Consider the diagram .

$$G \xleftarrow{\pi} \tilde{G} \xrightarrow{\sigma} T$$

where $G = \{(g_1, g'B) \in G \times (G/B) | g'^{-1} g_1 g' \in B\}$ and $\pi(g_1, g'B) = g_1$, $\sigma(g_1, g'B) = t \in T$, where $g'^{-1} g_1 g' \in t \cdot U$. Let s be an element of T^* and let K^s be the corresponding locally constant sheaf of rank 1 over T; then $\pi_* \sigma^*(K^s)$ decomposes in a direct sum of irreducible G-equivariant perverse sheaves over G in the same way as the left regular representation of the Weyl group of $Z_{G^*}(s)$ decomposes into irreducible representations. The elements of $\mathrm{Irr}(K)$ are exactly the various direct summands of $\pi_* \sigma^*(K^s)$ for the various $s \in T^*$.

When G is not GL_n, the previous construction gives most but, in general, not all elements of $\mathrm{Irr}(G)$.

13.8. We now propose a definition of the perverse sheaves in $\mathrm{Irr}(G)$, assuming that G is a reductive connected group over $\mathbb{C}$, the complex numbers. (This is suggested by the work of Harish-Chandra [HC_1, Thm. 5].)

For each semisimple class $\mathcal{O} \subset G$, there exists an open G-invariant neighborhood $\mathcal{U}_\mathcal{O}$ of $\mathcal{O}$ (in the classical topology) in G and an analytic

G-equivariant isomorphism $\phi_{\mathcal{O}}$ of $\mathcal{U}_{\mathcal{O}}$ onto an open G-invariant neighborhood $\mathcal{V}_{\mathcal{O}}$ of the zero section in the normal bundle $N_{\mathcal{O}}$ of $\mathcal{O}$ in G such that for any G-orbit in $\mathcal{U}_{\mathcal{O}}$ (resp. $\mathcal{V}_{\mathcal{O}}$), the closure of that G-orbit in G (resp. in $N_{\mathcal{O}}$) is entirely contained in $\mathcal{U}_{\mathcal{O}}$ (resp. $\mathcal{V}_{\mathcal{O}}$).

For each orbit $\mathcal{O}'$ of G on $N_{\mathcal{O}}$ such that the closure $\bar{\mathcal{O}}'$ of $\mathcal{O}'$ in $N_{\mathcal{O}}$ contains the zero section of $N_{\mathcal{O}}$ and for any irreducible locally constant G-equivariant C-sheaf $\mathcal{S}'$ on $\mathcal{O}'$, we consider the intersection cohomology complex $IC(\bar{\mathcal{O}}',\mathcal{S}')$ as a perverse sheaf on $N_{\mathcal{O}}$ and we let $\widehat{IC}(\bar{\mathcal{O}}',\mathcal{S}')$ be its Fourier transform, which by a theorem of Deligne is a perverse sheaf on the dual vector bundle $N_{\mathcal{O}}^*$. Since the vector bundle $N_{\mathcal{O}}$ is self dual, we may regard $\widehat{IC}(\bar{\mathcal{O}}',S')$ as a G-equivariant irreducible perverse sheaf on $N_{\mathcal{O}}$; its inverse image under $\phi_{\mathcal{O}}$ is then a G-equivariant perverse sheaf on $\mathcal{U}_{\mathcal{O}}$.

The proposed definition for Irr(G) is the following. An irreducible G-equivariant perverse C-sheaf K on G is in Irr(G) if for each semisimple class $\mathcal{O} \subset G$, one can choose $\mathcal{U}_{\mathcal{O}}, \mathcal{V}_{\mathcal{O}}, \phi_{\mathcal{O}}$ as above, in such a way that the restriction of K to $\mathcal{U}_{\mathcal{O}}$ is isomorphic to a finite direct sum of perverse sheaves of the form $\phi_{\mathcal{O}}^*(\widehat{IC}(\bar{\mathcal{O}}',\mathcal{S}'))$, where $\mathcal{O}' \subset N_{\mathcal{O}}$ and $\mathcal{S}'$ are as described above.

This implies that K restricted to the set of regular semisimple elements in G is either zero or is the locally constant sheaf associated to an irreducible representation of the affine Weyl group. If $G = PGL_n(C)$, this restriction is always non-zero hence Irr(G) is in 1-1 correspondence with the irreducible representations of the affine Weyl group.

Chapter 14
CONCLUDING REMARKS

In this chapter we discuss the possibility of extending our results to the case of groups with disconnected center. We also discuss the unipotent representations of the Ree groups.

14.1. *On groups with disconnected center*

Let G, G^*, F' be as in 13.2; we shall now drop the assumption that the center of G is connected. We may imbed G into a connected reductive group G_1 (over F_q) such that the center of G_1 is connected and G, G_1 have the same derived group. The Frobenius map $F': G_1 \to G_1$ extends the Frobenius map $F': G \to G$. Correspondingly, we have a surjective homomorphism $\pi: G_1^* \to G^*$ defined over F_q, whose kernel is a torus. For each semisimple element $\tilde{s} \in (G_1^*)^{F'}$ we have defined in 8.4 a set $\mathcal{E}_{(\tilde{s})}$ of irreducible representations of $G_1^{F'}$; it depends only on the $(G_1^*)^{F'}$-conjugacy class of $\tilde{s}$. If $\tilde{s}$ is replaced by $\tilde{s}z$, $(z \in \ker((G_1^*)^{F'} \to (G^*)^{F'}))$, then $\mathcal{E}_{(\tilde{s})}$ becomes $\mathcal{E}_{(\tilde{s}z)} = \{\rho \otimes \theta_z | \rho \in \mathcal{E}_{(\tilde{s})}\}$ where θ_z is a one-dimensional representation of $G_1^{F'}$, trivial on $G^{F'}$, determined by z. For each semisimple element $s \in (G^*)^{F'}$ we define $\mathcal{E}_{(s)}$ to be the set of irreducible representations of $G^{F'}$ which are components of restrictions to $G^{F'}$ of representations of $G_1^{F'}$ in $\mathcal{E}_{(\tilde{s})}$ where $\tilde{s} \in (G_1^*)^{F'}$ is a semisimple element such that $\pi(\tilde{s}) = s$. This set is independent of the choice of $\tilde{s}$. The set $\mathcal{E}_{(s)}$ depends only on the $(G^*)^{F'}$-conjugacy class (s) of s. It is easy to see that the sets $\mathcal{E}_{(s)}$, $\mathcal{E}_{(s')}$ are disjoint whenever the semisimple elements $s, s' \in (G^*)^{F'}$ are not $(G^*)^{F'}$-conjugate, (see (8.4.4)). In the rest of this section we shall assume that the center of G is a finite cyclic group and that G modulo its center is simple. Then the imbedding

$G \subset G_1$ can be chosen so that G_1/G is a one-dimensional torus. It follows that $G_1^{F'}/G^{F'}$ is a cyclic group. This implies that the restriction of any irreducible representation ρ_1 of $G_1^{F'}$ to $G^{F'}$ is multiplicity free and the number of its irreducible components is equal to the number of 1-dimensional representations $\theta: G_1^{F'} \to \overline{\mathbf{Q}}_\ell^*$ (trivial on $G^{F'}$) such that $\rho_1 \otimes \theta$ is isomorphic to ρ_1. Moreover, if ρ_2 is another irreducible representation of $G_1^{F'}$, then either ρ_2 is isomorphic to $\rho_1 \otimes \theta$ for some $\theta: G_1^{F'} \to \overline{\mathbf{Q}}_\ell^*$ (trivial on $G^{F'}$), so that ρ_1 and ρ_2 have the same restriction to $G^{F'}$, or else, the restrictions of ρ_1 and ρ_2 to $G^{F'}$ are disjoint.

We wish to classify the irreducible representations in $\mathcal{E}_{(s)}$ (for s a semisimple element in $(G^*)^{F'}$) and to find their multiplicities in $R_{\widetilde{E}} \in \mathcal{R}(G^{F'}) \otimes \mathbf{Q}$. (Here, $\widetilde{E}$ is an irreducible $\mathbf{Q}[\widetilde{W}_{Z^0(s)}]$-module whose restriction to $W_{Z^0(s)}$ is irreducible; it corresponds to an irreducible $\mathbf{Q}[\widetilde{W}_{Z(\tilde{s})}]$-module E_1 whose restriction to $W_{Z(\tilde{s})}$ is irreducible (see 8.4), under the natural isomorphisms $W_{Z^0(s)} \cong \widetilde{W}_{Z(\tilde{s})}, \widetilde{W}_{Z^0(s)} \cong W_{Z(\tilde{s})}$, where $\tilde{s} \in (G_1^*)^{F'}$ is such that $\pi(\tilde{s}) = s$.) We shall do this by reducing to the case of G_1 which has connected center, and for which therefore 4.23 is applicable. We fix $\tilde{s} \in (G_1^*)^{F'}$, such that $\pi(\tilde{s}) = s$.

Let Θ be the group of homomorphisms $\theta: G_1^{F'} \to \mathbf{Q}_\ell^*$ which are trivial on $G^{F'}$ and are such that $\tilde{\rho} \in \mathcal{E}_{(\tilde{s})} \Rightarrow \tilde{\rho} \otimes \theta \in \mathcal{E}_{(\tilde{s})}$. Then Θ has as many elements as the group $(Z(s)/Z^0(s))^{F'}$.

The irreducible representations of $G_1^{F'}$ in $\mathcal{E}_{(\tilde{s})}$ are classified in 4.23. We want to know explicitly the action of the group Θ (by tensor product) on the set $\mathcal{E}_{(\tilde{s})}$. The action of Θ (by tensor product) on the $R_{\widetilde{E}_1}$, ($\widetilde{E}_1$ as above), is easy to determine. In most cases, the representations in $\mathcal{E}_{(\tilde{s})}$ are completely determined by their multiplicities in the various $R_{\widetilde{E}}$ (see 4.23); in those cases, the action of Θ on $\mathcal{E}_{(\tilde{s})}$ is also determined by 4.23.

The only case which requires a special argument is the case where G is simply-connected of type E_7, p is odd and $s \in (G^*)^{F'}$ is a semisimple element whose centralizer has two connected components and the identity component $Z^0(s)$ modulo its center is of type E_6. In this case, two representations of $G_1^{F'}$ in $\mathcal{E}_{(\tilde{s})}$ cannot be distinguished by their multiplicities in the various $R_{\tilde{E}_1}$, and we must show directly that they become reducible when restricted to $G^{F'}$. An equivalent problem is the following. Let P be the parabolic subgroup of type E_6 in G containing B, let M be the Levi subgroup of P containing T, and let $\omega : M^{F'} \to \overline{\mathbf{Q}}_\ell^*$ be the unique character of order 2. Let ρ_1, ρ_2 be the two unipotent cuspidal representations of $M^{F'}$. We must show that $\mathrm{Ind}_{P^{F'}}^{G^{F'}}(\rho_i \otimes \omega)$ are reducible for $i = 1, 2$. It is enough to show that the isomorphism class of the $M^{F'}$-module $\rho_i \otimes \omega$ is stable under the action (by conjugation) on $\mathfrak{N}(M)^{F'}$, where $\mathfrak{N}(M)$ is the normalizer of M in G. The fact that each ρ_i can be extended to a $N(M)^{F'}$-module is known from $[L_2]$. The fact that ω is stable under the action of $\mathfrak{N}(M)^{F'}$ is obvious.

Hence the action of Θ on $\mathcal{E}_{(\tilde{s})}$ is explicitly known in all cases.

We can now classify the irreducible representations of $G^{F'}$ in $\mathcal{E}_{(s)}$ as follows. Consider the equivalence relation on $\mathcal{E}_{(\tilde{s})}$ defined by the orbits of the Θ-action; write $\tilde{\rho} \sim \tilde{\rho}'$ for this equivalence relation. Let $\overline{\mathcal{E}}_{(\tilde{s})}$ be a set of representatives for the equivalence classes. For each $\tilde{\rho} \in \overline{\mathcal{E}}_{(\tilde{s})}$, let $n_{\tilde{\rho}}$ be the number of $\theta \in \Theta$ such that $\tilde{\rho}$ is isomorphic to $\tilde{\rho} \otimes \theta$. Then $\mathrm{Res}(\tilde{\rho})$ (= restriction of $\tilde{\rho}$ to $G^{F'}$) decomposes as $\oplus \tilde{\rho}_i$ where $\tilde{\rho}_i$ $(1 \le i \le n_{\tilde{\rho}})$ are distinct irreducible $G^{F'}$-modules; the $\tilde{\rho}_i$, $(\tilde{\rho} \in \overline{\mathcal{E}}_{(\tilde{s})}, 1 \le i \le n_{\tilde{\rho}})$, exhaust $\mathcal{E}_{(s)}$ and form a list without repetition.

It remains to describe the inner products $\langle \tilde{\rho}_i, R_{\tilde{E}} \rangle_{G^{F'}}$, where $\tilde{E}$ corresponds to $\tilde{E}_1$ as above. Now $R_{\tilde{E}}$ is the restriction of $R_{\tilde{E}_1}$ from $\mathcal{R}(G_1^{F'}) \otimes \mathbf{Q}$ to $\mathcal{R}(G^{F'}) \otimes \mathbf{Q}$, hence

$$\begin{aligned}
<\tilde{\rho}_i, R_{\tilde{E}}>_{G^{F'}} &= <\tilde{\rho}_i, \mathrm{Res}\, R_{\tilde{E}_1}>_{G^{F'}} \\
&= <\mathrm{Ind}_{G^{F'}}^{G_1^{F'}}(\tilde{\rho}_i), R_{\tilde{E}_1}>_{G^{F'}} \\
&= \sum_{\substack{\tilde{\rho}' \in \mathcal{E}_{(\tilde{s})} \\ \tilde{\rho}' \sim \tilde{\rho}}} <\tilde{\rho}', R_{\tilde{E}_1}>_{G_1^{F'}} .
\end{aligned}$$

This last sum is computable from 4.23 for G_1. It is independent of i, $(1 \leq i \leq n_{\tilde{\rho}})$. It follows that $\dim(\tilde{\rho}_i)$ is also independent of i; it is equal to $(n_{\tilde{\rho}})^{-1} \dim(\rho)$.

We now see that

(14.1.1) *If* G *has a finite cyclic center and it is simple modulo its center, then the question of classification of the irreducible representations of* $G^{F'}$ *and of computing their multiplicities in the* $R_{\tilde{E}}$ *(as above) can be reduced to the analogous question for a group with connected center and hence is solved by 4.23.*

This applies to all almost-simple G except for the simply-connected groups of type D_{2n} in odd characteristic. (See [Le] for the case of SL_n.) For a statement valid for any connected reductive group over F_q (without any assumption on the center), the reader is referred to my article to appear in the proceedings of the International Congress of Mathematicians, Warsaw, 1983.

14.2. *On the Ree groups* ${}^2F_4$

In this section, G will denote an adjoint group of type F_4 over k and p is assumed to be 2. Let $W, T \subset B$, $F: G \to G$ be as in 1.2, and let m be an odd integer ≥ 1. Chevalley [Ch] has shown that there exists an endomorphism $F': G \to G$ leaving T and B stable, such that $F'^2 = F^m$. The fixed point set $G^{F'}$ is clearly a finite group (a subgroup of G^{F^m}); it is known as the Ree group of type ${}^2F_4$ corresponding to $q = 2^{m+\frac{1}{2}}$. The methods in the previous chapters can be applied with only minor modifications to the study of irreducible representations of $G^{F'}$. We shall state the results in the most complicated case: the case of unipotent representations of $G^{F'}$. Let γ be the unique automorphism of order 2 of the Coxeter group (W,S). It is not an ordinary automorphism (3.1). It induces an involution $\gamma : W^\vee \to W^\vee$ which maps families to families. There is a unique involution $\gamma : X(W) \to X(W)$, (see (4.21.1)) such that the diagram

$$\begin{array}{ccc} W^\vee & \xrightarrow{(4.21.3)} & X(W) \\ \Big\downarrow \gamma & & \Big\downarrow \gamma \\ W^\vee & \xrightarrow{(4.21.3)} & X(W) \end{array}$$

is commutative and such that the restriction of γ to $\mathfrak{M}(\mathfrak{G}_4) \subset X(W)$ is given by: $(1,\lambda^2) \overset{\gamma}{\leftrightarrow} (g_2',\varepsilon')$, $(1,\lambda^1) \overset{\gamma}{\leftrightarrow} (g_2',1)$, $(g_4,1) \overset{\gamma}{\leftrightarrow} (g_2,\varepsilon''')$, $(g_4,-1) \overset{\gamma}{\leftrightarrow} (g_2,\varepsilon')$, $\gamma(x,\sigma) = (x,\sigma)$ for all other $(x,\sigma) \in \mathfrak{M}(\mathfrak{G}_4)$. (This is an involution on $\mathfrak{M}(\mathfrak{G}_4)$ preserving the form $\{\ ,\ \}$ on $\mathfrak{M}(\mathfrak{G}_4)$.) It turns out that the unipotent representations of $G^{F'}$ are in 1-1 correspondence with the set $X(W)^\gamma$ of fixed points of γ on $X(W)$, and hence they can be grouped together according to the γ-stable families in $W^\vee$. The degrees of unipotent representations are polynomials in $q = 2^{m+\frac{1}{2}}$ with coefficients in $\mathbf{Q}[\sqrt{2}]$.

The set of γ-stable elements of $W^\vee$ is

$$W^\vee_{ex} = \{\dot{1}_1, 1_4, 4_2, 4_5, 9_1, 9_4, 12_1, 4_1, 6_1, 6_2, 16_1\}.$$

Each representation in this set is an irreducible $\mathbf{Q}[W]$-module E such that $E \otimes K$ ($K = \mathbf{Q}(\sqrt{2})$) can be extended in exactly two different ways to a $K[\widetilde{W}]$-module $\widetilde{E}$, where $\widetilde{W}$ is the semidirect product of W with the infinite cyclic group with generator γ, acting on W as $\gamma : W \to W$. (We assume that γ acts on E as a transformation of finite order.) We define $R_{\widetilde{E}} \in \mathcal{R}(G^{F'}) \otimes K$ just as in (3.7.1) (with $W_{L,n} = W$, $w_1 = e$, $L = L_0$, $n = 1$).

If ρ is a unipotent representation, the multiplicities $\langle \rho, R_{\tilde{E}} \rangle_{G^{F'}} \in K$ are not in general rational numbers. Knowing them is equivalent to knowing the multiplicities $\langle \rho, R_w \rangle_{G^{F'}} \in \mathbf{Z}$, where $R_w = \sum_i (-1)^i H_c^i(X_w, \overline{\mathbf{Q}}_\ell) \in \mathcal{R}(G^{F'})$, ($X_w$ is defined as in 2.1, using F'). In the appendix we list these multiplicities $\langle \rho, R_w \rangle_{G^{F'}}$ in a table. The rows of the table are parametrized by the various ρ, which are indicated by their degree. We also indicate how the unipotent representations are grouped together according to families. The columns of the table are parametrized by the elements $w \in W$ up to the equivalence relation $W \sim \gamma(w_1) x w_1^{-1}$, (for any $w_1 \in W$); note that R_w depends only on the equivalence class of w. We indicate w by giving the roots of the characteristic polynomial of γw on the reflection representation of W. (These polynomials have been determined by P. Gager (Ph.D. Thesis, Warwick); see also [Sp$_1$].) We shall use the following notation: $\phi'_{24} = q^4 + q^3\sqrt{2} + q^2 + q\sqrt{2} + 1$, $\phi''_{24} = q^4 - q^3\sqrt{2} + q^2 - q\sqrt{2} + 1$, $\phi'_8 = q^2 + q\sqrt{2} + 1$, $\phi''_8 = q^2 - q\sqrt{2} + 1$, so that $\phi'_{24}\phi''_{24}$ is the cyclotomic polynomial ϕ_{24} and $\phi'_8\phi''_8$ is the cyclotomic polynomial ϕ_8. We denote by $\zeta'_{24}, \zeta''_{24}, \zeta'_8, \zeta''_8$ a root of $\phi'_{24}, \phi''_{24}, \phi'_8, \phi''_8$ respectively.

The last cuspidal representation ρ listed in the table for ${}^2F_4$ has the following property: it has even multiplicity in any R_w. Note that this does not happen for any unipotent representation of any other $G^{F'}$, (G simple). In the case where $q = \sqrt{2}$, (i.e., $m = 1$), the character table of $G^{F'}$ has been determined by Frame and Rudvalis.

In the appendix, we have also given tables for the unipotent representations of the finite groups ${}^2B_2(q)$, $q = 2^{m+\frac{1}{2}}$ and ${}^2G_2(q)$, $q = 3^{m+\frac{1}{2}}$. The conventions used are the same as for ${}^2F_4$; in the case of ${}^2G_2$, we denote $\phi'_{12} = q^2 - q\sqrt{3} + 1$, $\phi''_{12} = q^2 + q\sqrt{3} + 1$, so that $\phi'_{12}\phi''_{12}$ is the cyclotomic polynomial ϕ_{12}. We denote by ζ'_{12} a root of ϕ'_{12} and by ζ''_{12} a root of ϕ''_{12}. The character tables of ${}^2B_2(q)$, ${}^2G_2(q)$ are known [Sz], [Wa].

14.3. *The function* $x \mapsto \hat{a}(x)$

In 5.27, we have defined a function $x \mapsto a(x)$ on a Weyl group $\mathbb{W}$, which has played an essential role in this work. Unfortunately, the definition of $a(x)$ depends on the theory of primitive ideals, (via 5.25). We shall propose another definition for such a function.

Let us write for $x', x \in \mathbb{W}$

$$C_{x'} \cdot C_x = \sum_{x' \in \mathbb{W}} d_{x',x,x''} C_{x''} \in H$$

where $d_{x',x,x''} \in \mathbf{Z}[u^{1/2}, u^{-1/2}]$. It follows easily from (5.1.12) that

$$u^{\ell(x)/2} d_{x',x,x''} \in \mathbf{Z}[u^{1/2}]. \tag{14.3.1}$$

We now fix $x \in \mathbb{W}$.

(14.3.2) DEFINITION. $\hat{a}(x)$ *is the smallest integer* ≥ 0 *such that* $u^{\hat{a}(x)/2} d_{x',x,x} \in \mathbf{Z}[u^{1/2}]$, *for all* $x' \in \mathbb{W}$.

This definition has a meaning even for infinite Coxeter groups; (14.3.1) shows that $0 \leq \hat{a}(x) \leq \ell(x)$. One can show that for Weyl groups we have $\hat{a}(x) = a(x)$ for all $x \in \mathbb{W}$.

APPENDIX

Tables of Unipotent Representations

In this appendix, we assume that G is absolutely simple modulo its center.

If G is of type A_{r-1} with F' acting trivially on W, the unipotent representations of $G^{F'}$ are in 1-1 correspondence with the irreducible representations $E = [\lambda_1, \cdots, \lambda_m]$ of W; here $\lambda_1 < \lambda_2 < \cdots < \lambda_m$ are integers ≥ 0 such that $\sum_{i=1}^{m} \lambda_i = r + \binom{m}{2}$, (see 4.4). The degree of the representation of $G^{F'}$ corresponding to E is

$$\frac{\prod_{1 \leq j < i \leq m} (q^{\lambda_i} - q^{\lambda_j}) \cdot |G^{F'}|_{p'} \cdot (q-1)}{\prod_{i=1}^{m} \left(\prod_{h=1}^{\lambda_i} (q^h - 1) \right) \cdot q^{\binom{m-1}{2} + \binom{m-2}{2} + \cdots}}$$

where $|G^F|_{p'}$ is the part prime to p of $|G^F|$, (cf. Steinberg [S]); this representation is never cuspidal unless $r = 1$. The same parametrization is valid when G is of type A_{r-1} $(r \geq 3)$ with F' acting non-trivially on W. The corresponding degree is obtained (up to sign) by changing q by $-q$ in the previous formula, [LS]. The representation corresponding to $E = [1,3,5,\cdots,2m-1]$ is cuspidal (in that case $r = \binom{m+1}{2}$); to any other E there corresponds a non-cuspidal representation.

Assume now that G is of type B_r, C_r, $(r \geq 2)$, D_r or ${}^2D_r$, $(r \geq 4)$. The unipotent representations of $G^{F'}$ are parametrized by symbols (with permutable rows)

$$\begin{pmatrix} \lambda_1 < \lambda_2 < \cdots & & < \lambda_b \\ & \mu_1 < \mu_2 < \cdots < \mu_{b'} & \end{pmatrix}$$

where $\lambda_1 \geq 0$, $\mu_1 \geq 0$, $b-b'$ is odd (for type B_r, C_r), divisible by 4 (for D_r) and of form $4k+2$ (for ${}^2D_r$) and

$$\sum_{i=1}^{b} \lambda_i + \sum_{i=1}^{b'} \mu_i = \begin{cases} r + ((b+b'-1)/2)^2 & , \text{ for } B_r, C_r \\ r + ((b+b')(b+b'-2)/4), & \text{for } D_r, {}^2D_r \end{cases}.$$

These symbols are taken up to shift (see 4.5); moreover, for type D_r each symbol which is degenerate (i.e. $b = b'$, $\lambda_1 = \mu_1$, $\lambda_2 = \mu_2, \cdots, \lambda_b = \lambda_{b'}$) is repeated twice.

The unipotent representation corresponding to such a symbol has degree

$$\frac{\prod\limits_{1\leq j<i\leq b} (q^{\lambda_i}-q^{\lambda_j}) \prod\limits_{1\leq j<i\leq b'} (q^{\mu_i}-q^{\mu_j}) \prod\limits_{\substack{1\leq i\leq b \\ 1\leq j\leq b'}} (q^{\lambda_i}+q^{\mu_j})|G^{F'}|_{p'}}{2^c \prod\limits_{1\leq i\leq b} \Big(\prod\limits_{h=1}^{\lambda_i} (q^{2h}-1)\Big) \prod\limits_{1\leq j\leq b'} \Big(\prod\limits_{h=1}^{\mu_j} (q^{2h}-1)\Big)\, q^{\binom{b+b'-2}{2}+\binom{b+b'-4}{2}+\cdots}}$$

where $c = \frac{b+b'-1}{2}$ for type B_r, C_r, $c = \frac{b+b'-2}{2}$ for type D_r, ${}^2D_r$ except when the symbol is degenerate, in which case $c = b = b'$.

The condition that a unipotent representation of $G^{F'}$ is cuspidal is that the corresponding symbol is

$$\binom{0,1,2,\cdots,2d}{-} \text{ for type } B_r, C_r, \text{ (hence } r = d^2+d\text{)}$$

$$\binom{0,1,2,\cdots,2d-1}{-}, \begin{Bmatrix} d \text{ even for type } D_r \\ d \text{ odd for type } {}^2D_r \end{Bmatrix}, \text{ (hence } r = d^2\text{)}.$$

For exceptional groups we shall describe the degrees of unipotent representations in tables. (I wish to thank R. W. Carter and K. Shinoda for spotting some misprints in these tables in the preprint.) In the case where G is of type F_4, E_6, E_7, E_8 (with $F'=1$ on W), we parametrize the unipotent representations of $G^{F'}$ in the form $\delta[\chi]$ as in 8.6, 8.7. Here, δ is a unipotent cuspidal representation of $M^{F'}$, where M is the Levi subgroup of a parabolic subgroup P of G, $P \supset B$; χ denotes an irreducible representation of the Coxeter group $\mathcal{W} = \mathfrak{N}(M)/M$. When P is a Borel subgroup, we have $\mathcal{W} = W$ and we omit δ from the notation: thus, $[\chi]$ denotes the irreducible component of $\mathrm{Ind}_{B^{F'}}^{G^{F'}}(1)$ corresponding to the irreducible representation χ of W. The formulas for $\dim[\chi]$ are due to Benson-Grove-Surówski (for F_4), Surowski (for E_6, E_7), Benson (for E_8), see [BGSu], [Su], [B]. When G is of type F_4, M could be also of type B_2; then δ is uniquely determined and $\mathcal{W}$ is a Coxeter group of type B_2. We denote the irreducible representations of $\mathcal{W}$ by 1 (unit), r (reflection), ε', ε'' (one dimensional $\neq 1$, sign), ε (= sign); we write $B_2[\chi]$ instead of $\delta[\chi]$. When $M = G$, then δ is a cuspidal representation of $G^{F'}$; we denote it by one of the symbols $F_4[\theta]$, $F_4[\theta^2]$; $F_4[i]$, $F_4[-i]$, $F_4^{I}[1]$, $F_4^{II}[1]$, $F_4[-1]$. Here θ (resp. i) is a fixed primitive cubic (resp. fourth) root of 1 in $\bar{\mathbf{Q}}_\ell^*$.

If G is of type E_6 ($F'=1$ on W), M could be of type D_4; then δ is uniquely determined and $\mathcal{W}$ is of type A_2. We denote the irreducible representations of $\mathcal{W}$ by 1 (unit), r (reflection), ε (sign); we write $D_4[\chi]$ instead of $\delta[\chi]$. When $M = G$, then δ is a cuspidal representation of $G^{F'}$; we denote it by one of the symbols $E_6[\theta]$, $E_6[\theta^2]$, (θ as above).

If G is of type E_7, M could be of type D_4; then δ is uniquely determined and $\mathcal{W}$ is of type B_3. We denote 1 its unit representation, r its reflection representation, ε_1 its one-dimensional representation $\neq 1$ which is trivial on the standard parabolic subgroup of type A_2, $\sigma_3 = \rho \otimes \varepsilon_1$, σ_2 its two-dimensional irreducible character whose restriction to the standard parabolic subgroup of type B_2 contains the unit representation;

we write $D_4[\chi]$ instead of $\delta[\chi]$. We could also have M of type E_6; then $\delta = E_6[\theta]$ or $E_6[\theta^2]$ and $\mathfrak{W}$ is of type A_1. We denote the unit representation of $\mathfrak{W}$ by 1 and we write $\delta[1]$ as $E_6[\theta,1]$ or $E_6[\theta^2,1]$. When $M = G$, then δ is a cuspidal representation of $G^{F'}$; we denote it by $E_7[\xi]$ or $E_7[-\xi]$ where ξ is a fixed square root of $-q$ in $\overline{\mathbf{Q}}_\ell^*$.

If G is of type E_8, M could be of type D_4; then δ is uniquely determined and $\mathfrak{W}$ is of type F_4. The simple reflections in $\mathfrak{W}$ are in 1-1 correspondence with the simple reflections in W which are not in the standard parabolic subgroup W' of type D_4 (see 8.2); we denote them s_1, s_2, s_3, s_4 in such a way that s_3 and s_4 correspond to simple reflections in W which commute with the elements in W' and such that s_1, s_4 correspond to end points of the Coxeter graph of W. We may then parametrize the irreducible representations of $\mathfrak{W}$ as in 4.10; we write $D_4[\chi]$ instead of $\delta[\chi]$.

We could also have M of type E_6; then $\delta = E_6[\theta]$ or $E_6[\theta^2]$ and $\mathfrak{W}$ is of type G_2. We denote 1 its unit representation, $\varepsilon', \varepsilon''$ its one-dimensional representations $\neq 1$, sign; ε the sign representation, r the reflection representation, $r' = r \otimes \varepsilon'$. (We arrange notation so that ε' is non-trivial on the simple reflection in W corresponding, as in 8.2, to the simple reflection in W commuting with the elements in the parabolic subgroup of type E_6.) We write $E_6[\theta,\chi]$ or $E_6[\theta^2,\chi]$ instead of $\delta[\chi]$.

We could also have M of type E_7; then $\delta = E_7[\xi]$ or $E_7[-\xi]$ and $\mathfrak{W}$ is of type A_1. We denote the unit representation of $\mathfrak{W}$ by 1 and we write $E_7[\xi,1]$ or $E_7[-\xi,1]$ instead of $\delta[1]$.

When $M = G$, then δ is a cuspidal representation of $G^{F'}$; we denote it by one of the symbols $E_8[\zeta^j]$, $(j = 1,2,3,4)$, $E_8[-\theta^j]$, $(j=1,2)$, $E_8[\theta^j]$, $(j=1,2)$, $E_8[i]$, $E_8[-i]$, $E_8^{I}[1]$, $E_8^{II}[1]$, $E_8[-1]$. Here ζ is a primitive 5-th root in $\overline{\mathbf{Q}}_\ell^*$ and θ, i are as above.

In the tables which follow, the unipotent representations of $G^{F'}$ are grouped according to the F'-stable families in $W^\vee$. (For each such family, we group together the unipotent representations such that

$<\rho,R_E>_{G^{F'}} \neq 0$ where E is the special representation in the family.) When a family $\mathfrak{F}$ is such that $\mathfrak{F} \otimes \text{sign} \neq \mathfrak{F}$, we only indicate the representations of $G^{F'}$ corresponding to $\mathfrak{F}$. The representations of $G^{F'}$ corresponding to $\mathfrak{F} \otimes$ sign are obtained by application of the duality operator $\pm D$; if we replace the power of q in the formula for the degree by the power of q given in the column "Remarks", we get the degree of the corresponding representation obtained by application of $\pm D$.

In the case where F' acts trivially on W, the representations of $G^{F'}$ corresponding to a family $\mathfrak{F} \subset W^{\vee}$ are in 1-1 correspondence with the set $\mathfrak{M}(\mathcal{G}_{\mathfrak{F}})$. The correspondence is indicated in the tables (column "Remarks") in the case where $\mathcal{G}_{\mathfrak{F}} = \mathfrak{G}_4$ or $\mathfrak{G}_5$ and in the case where G is of type E_6 and $\mathcal{G}_{\mathfrak{F}} = \mathfrak{G}_3$. If G is of type E_7 or E_8 and $\mathcal{G}_{\mathfrak{F}} = \mathfrak{G}_3$ the representations of $G^{F'}$ corresponding to $\mathfrak{F}$ (in the order listed) correspond to the elements in $\mathfrak{M}(\mathfrak{G}_3)$, in the order $(1,1)$, $(g_2,1)$, $(1,r)$, $(g_3,1)$, $(1,\varepsilon)$, (g_2,ε), (g_3,θ), (g_3,θ^2).

If $\mathfrak{F}$ is such that $\mathcal{G}_{\mathfrak{F}} = \mathfrak{G}_2$, and $\mathfrak{F}$ is not the family of $512_a'$ (for E_7) or 4096_z (for E_8) the representations of $G^{F'}$ corresponding to $\mathfrak{F}$ (in the order listed) correspond to the elements in $\mathfrak{M}(\mathfrak{G}_2)$ in the order $(1,1)$, $(g_2,1)$, $(1,\varepsilon)$, (g_2,ε). If $\mathfrak{F}$ is the family of $512_a'$ (for E_7) or 4096_z (for E_8) the representations of $G^{F'}$ corresponding to $\mathfrak{F}$ (in the order listed) correspond to the elements in $\mathfrak{M}(\mathfrak{G}_2)$ in the order $(1,1)$, $(1,\varepsilon)$, $(g_2,1)$, (g_2,ε).

The degrees of the unipotent representations of ${}^2E_6(q)$ are obtained (up to sign) by replacing q by $-q$ in the degrees for $E_6(q)$. The unipotent representations of ${}^2E_6(q)$ corresponding to $E_6[\theta]$, $E_6[\theta^2]$ and $[80_s]$ are cuspidal and those corresponding to $[64_p]$ and $[64_p']$ are subcuspidal.

We denote the d-th cyclotomic polynomial by ϕ_d; for example, $\phi_3 = q^2+q+1$.

Degrees of Unipotent Representations of $E_6(q)$

Representation	*Degree*	*Remarks*
$[1_p]$	1	(q^{36})
$[6_p]$	$q\phi_8\phi_9$	(q^{25})
$[20_p]$	$q^2\phi_4\phi_5\phi_8\phi_{12}$	(q^{20})
$[30_p]$	$\frac{1}{2}q^3\phi_4^2\phi_5\phi_9\phi_{12}$	(q^{15})
$[15_q]$	$\frac{1}{2}q^3\phi_5\phi_8\phi_9\phi_{12}$	
$[15_p]$	$\frac{1}{2}q^3\phi_5\phi_6^2\phi_8\phi_9$	
$D_4[1]$	$\frac{1}{2}q^3\phi_1^4\phi_3^2\phi_5\phi_9$	
$[64_p]$	$q^4\phi_2^3\phi_4^2\phi_6^2\phi_8\phi_{12}$	(q^{13})
$[60_p]$	$q^5\phi_4\phi_5\phi_8\phi_9\phi_{12}$	(q^{11})
$[81_p]$	$q^6\phi_3^3\phi_6^2\phi_9\phi_{12}$	(q^{10})
$[24_p]$	$q^6\phi_4^2\phi_8\phi_9\phi_{12}$	(q^{12})
$[80_s]$	$\frac{1}{6}q^7\phi_2^4\phi_5\phi_8\phi_9\phi_{12}$	$(1,1)$
$[60_s]$	$\frac{1}{2}q^7\phi_4^2\phi_5\phi_8\phi_9\phi_{12}$	$(g_2,1)$
$[90_s]$	$\frac{1}{3}q^7\phi_3^3\phi_5\phi_6^2\phi_8\phi_{12}$	$(1,\rho)$
$[10_s]$	$\frac{1}{3}q^7\phi_5\phi_6^2\phi_8\phi_9\phi_{12}$	$(g_3,1)$
$[20_s]$	$\frac{1}{6}q^7\phi_4^2\phi_5\phi_6^2\phi_8\phi_9$	$(1,\varepsilon)$
$D_4[r]$	$\frac{1}{2}q^7\phi_1^4\phi_3^2\phi_5\phi_8\phi_9$	(g_2,ε)
$E_6[\theta]$	$\frac{1}{3}q^7\phi_1^6\phi_2^4\phi_4^2\phi_5\phi_8$	(g_3,θ)
$E_6[\theta^2]$	same	(g_3,θ^2)

Degrees of Unipotent Representations of $E_7(q)$

Representation	*Degree*	*Remarks*
$[1_a]$	1	(q^{63})
$[7'_a]$	$q\phi_7\phi_{12}\phi_{14}$	(q^{46})
$[27_a]$	$q^2\phi_3^2\phi_6^2\phi_9\phi_{12}\phi_{18}$	(q^{37})
$[56'_a]$	$\frac{1}{2}q^3\phi_2^4\phi_6^2\phi_7\phi_{10}\phi_{14}\phi_{18}$	(q^{30})
$[35_b]$	$\frac{1}{2}q^3\phi_5\phi_7\phi_8\phi_{12}\phi_{14}\phi_{18}$	
$[21_a]$	$\frac{1}{2}q^3\phi_7\phi_8\phi_9\phi_{10}\phi_{12}\phi_{14}$	
$D_4[1]$	$\frac{1}{2}q^3\phi_1^4\phi_3^2\phi_5\phi_7\phi_9\phi_{14}$	
$[21'_b]$	$q^3\phi_7\phi_9\phi_{14}\phi_{18}$	(q^{36})
$[120_a]$	$\frac{1}{2}q^4\phi_2^4\phi_5\phi_6^2\phi_9\phi_{10}\phi_{14}\phi_{18}$	(q^{25})
$[15'_a]$	$\frac{1}{2}q^4\phi_5\phi_8\phi_9\phi_{10}\phi_{12}\phi_{14}\phi_{18}$	
$[105'_a]$	$\frac{1}{2}q^4\phi_5\phi_7\phi_8\phi_9\phi_{10}\phi_{12}\phi_{18}$	
$D_4[\varepsilon_1]$	$\frac{1}{2}q^4\phi_1^4\phi_3^2\phi_5\phi_7\phi_9\phi_{10}\phi_{18}$	
$[189'_b]$	$q^5\phi_3^2\phi_6^2\phi_7\phi_9\phi_{12}\phi_{14}\phi_{18}$	(q^{22})
$[210_a]$	$q^6\phi_5\phi_7\phi_8\phi_9\phi_{10}\phi_{14}\phi_{18}$	(q^{21})
$[105_b]$	$q^6\phi_5\phi_7\phi_9\phi_{10}\phi_{12}\phi_{14}\phi_{18}$	(q^{21})
$[168_a]$	$q^6\phi_4^2\phi_7\phi_8\phi_9\phi_{12}\phi_{14}\phi_{18}$	(q^{21})
$[189_c]$	$q^7\phi_3^2\phi_6^2\phi_7\phi_9\phi_{12}\phi_{14}\phi_{18}$	(q^{20})
$[315'_a]$	$\frac{1}{6}q^7\phi_3^3\phi_5\phi_7\phi_8\phi_{10}\phi_{12}\phi_{14}\phi_{18}$	(q^{16})
$[280_b]$	$\frac{1}{2}q^7\phi_2^4\phi_5\phi_6^3\phi_7\phi_{10}\phi_{12}\phi_{14}\phi_{18}$	
$[280'_a]$	$\frac{1}{3}q^7\phi_4^2\phi_5\phi_7\phi_8\phi_9\phi_{10}\phi_{14}\phi_{18}$	
$[70'_a]$	$\frac{1}{3}q^7\phi_5\phi_7\phi_8\phi_9\phi_{10}\phi_{12}\phi_{14}\phi_{18}$	
$[35'_a]$	$\frac{1}{6}q^7\phi_5\phi_6^3\phi_7\phi_8\phi_9\phi_{10}\phi_{12}\phi_{14}$	

Representation	*Degree*	*Remarks*
$D_4[r]$	$\frac{1}{2}q^7\phi_1^4\phi_3^3\phi_5\phi_7\phi_9\phi_{10}\phi_{12}\phi_{14}$	
$E_6[\theta,1]$	$\frac{1}{3}q^7\phi_1^6\phi_2^6\phi_4^2\phi_5\phi_7\phi_8\phi_{10}\phi_{14}$	
$E_6[\theta^2,1]$	same	
$[405_a]$	$\frac{1}{2}q^8\phi_3^3\phi_5\phi_6^2\phi_8\phi_9\phi_{12}\phi_{14}\phi_{18}$	(q^{15})
$[216'_a]$	$\frac{1}{2}q^8\phi_2^4\phi_3^2\phi_6^3\phi_9\phi_{10}\phi_{12}\phi_{14}\phi_{18}$	
$[189_a]$	$\frac{1}{2}q^8\phi_3^2\phi_6^3\phi_7\phi_8\phi_9\phi_{10}\phi_{12}\phi_{18}$	
$D_4[\sigma_3]$	$\frac{1}{2}q^8\phi_1^4\phi_3^3\phi_5\phi_6^2\phi_7\phi_9\phi_{12}\phi_{18}$	
$[378'_a]$	$q^9\phi_3^2\phi_6^2\phi_7\phi_8\phi_9\phi_{12}\phi_{14}\phi_{18}$	(q^{14})
$[210_b]$	$q^{10}\phi_5\phi_7\phi_8\phi_9\phi_{10}\phi_{12}\phi_{14}\phi_{18}$	(q^{13})
$[420_a]$	$\frac{1}{2}q^{10}\phi_4^2\phi_5\phi_7\phi_8\phi_9\phi_{12}\phi_{14}\phi_{18}$	(q^{13})
$[84_a]$	$\frac{1}{2}q^{10}\phi_4^2\phi_7\phi_8\phi_9\phi_{10}\phi_{12}\phi_{14}\phi_{18}$	
$[336'_a]$	$\frac{1}{2}q^{10}\phi_2^4\phi_6^2\phi_7\phi_8\phi_9\phi_{10}\phi_{14}\phi_{18}$	
$D_4[\sigma_2]$	$\frac{1}{2}q^{10}\phi_1^4\phi_3^2\phi_5\phi_7\phi_8\phi_9\phi_{14}\phi_{18}$	
$[105_c]$	$q^{12}\phi_5\phi_7\phi_9\phi_{10}\phi_{12}\phi_{14}\phi_{18}$	(q^{15})
$[512'_a]$	$\frac{1}{2}q^{11}\phi_2^7\phi_4^2\phi_6^3\phi_8\phi_{10}\phi_{12}\phi_{14}\phi_{18}$	
$[512_a]$	same	
$E_7[\xi]$	$\frac{1}{2}q^{11}\phi_1^7\phi_3^3\phi_4^2\phi_5\phi_7\phi_8\phi_9\phi_{12}$	
$E_7[-\xi]$	same	

Degrees of Unipotent Representations of $E_8(q)$

Representation	*Degree*	*Remarks*
$[1_x]$	1	(q^{120})
$[8_z]$	$q\phi_4^2\phi_8\phi_{12}\phi_{20}\phi_{24}$	(q^{91})
$[35_x]$	$q^2\phi_5\phi_7\phi_{10}\phi_{14}\phi_{15}\phi_{20}\phi_{30}$	(q^{74})
$[112_z]$	$\frac{1}{2}q^3\phi_2^4\phi_6^2\phi_7\phi_8\phi_{10}^2\phi_{14}\phi_{18}\phi_{24}\phi_{30}$	(q^{63})
$[84_x]$	$\frac{1}{2}q^3\phi_4^2\phi_7\phi_8\phi_9\phi_{12}\phi_{14}\phi_{20}\phi_{24}\phi_{30}$	
$[28_x]$	$\frac{1}{2}q^3\phi_4^2\phi_7\phi_8\phi_{12}\phi_{14}\phi_{15}\phi_{18}\phi_{20}\phi_{24}$	
$D_4[1_1]$	$\frac{1}{2}q^3\phi_1^4\phi_3^2\phi_5^2\phi_7\phi_8\phi_9\phi_{14}\phi_{15}\phi_{24}$	
$[210_x]$	$\frac{1}{2}q^4\phi_5\phi_7\phi_8^2\phi_9\phi_{10}^2\phi_{15}\phi_{20}\phi_{24}\phi_{30}$	(q^{52})
$[50_x]$	$\frac{1}{2}q^4\phi_5^2\phi_8^2\phi_{10}\phi_{14}\phi_{15}\phi_{18}\phi_{20}\phi_{24}\phi_{30}$	
$[160_z]$	$\frac{1}{2}q^4\phi_2^4\phi_4^2\phi_5\phi_6^2\phi_{10}^2\phi_{12}\phi_{14}\phi_{15}\phi_{18}\phi_{20}\phi_{30}$	
$D_4[2_3]$	$\frac{1}{2}q^4\phi_1^4\phi_3^2\phi_4^2\phi_5^2\phi_7\phi_9\phi_{10}\phi_{12}\phi_{15}\phi_{20}\phi_{30}$	
$[560_z]$	$q^5\phi_4^2\phi_5\phi_7\phi_8^2\phi_{10}\phi_{12}\phi_{14}\phi_{15}\phi_{20}\phi_{24}\phi_{30}$	(q^{47})
$[567_x]$	$q^6\phi_3^3\phi_6^3\phi_7\phi_9\phi_{12}^2\phi_{14}\phi_{15}\phi_{18}\phi_{24}\phi_{30}$	(q^{46})
$[700_x]$	$\frac{1}{2}q^6\phi_4^2\phi_5^2\phi_7\phi_8\phi_{10}^2\phi_{12}\phi_{15}\phi_{18}\phi_{20}\phi_{24}\phi_{30}$	(q^{42})
$[400_z]$	$\frac{1}{2}q^6\phi_2^4\phi_5^2\phi_6^2\phi_8\phi_{10}^2\phi_{14}\phi_{15}\phi_{18}\phi_{20}\phi_{24}\phi_{30}$	
$[300_x]$	$\frac{1}{2}q^6\phi_4^2\phi_5^2\phi_8\phi_9\phi_{10}^2\phi_{12}\phi_{14}\phi_{15}\phi_{20}\phi_{24}\phi_{30}$	
$D_4[1_3]$	$\frac{1}{2}q^6\phi_1^4\phi_3^2\phi_5^2\phi_7\phi_8\phi_9\phi_{10}^2\phi_{15}\phi_{20}\phi_{24}\phi_{30}$	
$[1400_z]$	$\frac{1}{6}q^7\phi_4^2\phi_5^2\phi_7\phi_8^2\phi_9\phi_{12}^2\phi_{14}\phi_{18}\phi_{20}\phi_{24}\phi_{30}$	(q^{37})
$[1344_x]$	$\frac{1}{2}q^7\phi_2^4\phi_4^2\phi_6^2\phi_7\phi_8\phi_9\phi_{10}^2\phi_{12}\phi_{14}\phi_{18}\phi_{20}\phi_{24}\phi_{30}$	
$[1008_z]$	$\frac{1}{3}q^7\phi_3^3\phi_4^2\phi_6^3\phi_7\phi_8^2\phi_{12}^2\phi_{14}\phi_{15}\phi_{20}\phi_{24}\phi_{30}$	
$[448_z]$	$\frac{1}{3}q^7\phi_4^4\phi_7\phi_8^2\phi_9\phi_{12}\phi_{14}\phi_{15}\phi_{18}\phi_{20}\phi_{24}\phi_{30}$	

Representation	*Degree*	*Remarks*
$[56_z]$	$\frac{1}{6}q^7\phi_4^2\phi_7\phi_8^2\phi_9\phi_{10}^2\phi_{12}^2\phi_{14}\phi_{15}\phi_{18}\phi_{20}\phi_{24}$	
$D_4[4_2]$	$\frac{1}{2}q^7\phi_1^4\phi_3^2\phi_4^2\phi_5^2\phi_7\phi_8\phi_9\phi_{12}\phi_{14}\phi_{15}\phi_{18}\phi_{20}\phi_{24}$	
$E_6[\theta,1]$	$\frac{1}{3}q^7\phi_1^6\phi_2^6\phi_4^4\phi_5^2\phi_7\phi_8^2\phi_{10}^2\phi_{12}\phi_{14}\phi_{20}\phi_{24}$	
$E_6[\theta^2,1]$	same	
$[1400_x]$	$\frac{1}{6}q^8\phi_4^4\phi_5^2\phi_7\phi_9\phi_{10}^2\phi_{14}\phi_{15}\phi_{18}\phi_{20}\phi_{24}\phi_{30}$	(q^{32})
$[1050_x]$	$\frac{1}{2}q^8\phi_5^2\phi_7\phi_8^2\phi_9\phi_{10}^2\phi_{14}\phi_{15}\phi_{18}\phi_{20}\phi_{24}\phi_{30}$	
$[1575_x]$	$\frac{1}{3}q^8\phi_3^3\phi_5^2\phi_6^3\phi_7\phi_{10}^2\phi_{12}^2\phi_{14}\phi_{15}\phi_{20}\phi_{24}\phi_{30}$	
$[175_x]$	$\frac{1}{3}q^8\phi_5^2\phi_7\phi_9\phi_{10}^2\phi_{12}^2\phi_{14}\phi_{15}\phi_{18}\phi_{20}\phi_{24}\phi_{30}$	
$[350_x]$	$\frac{1}{6}q^8\phi_5^2\phi_7\phi_8^2\phi_9\phi_{10}^2\phi_{12}^2\phi_{14}\phi_{15}\phi_{18}\phi_{20}\phi_{30}$	
$D_4[8_3]$	$\frac{1}{2}q^8\phi_1^4\phi_2^4\phi_3^2\phi_5^2\phi_6^2\phi_7\phi_9\phi_{10}^2\phi_{14}\phi_{15}\phi_{18}\phi_{20}\phi_{30}$	
$E_6[\theta,\varepsilon']$	$\frac{1}{3}q^8\phi_1^6\phi_2^6\phi_4^4\phi_5^2\phi_7\phi_8^2\phi_{10}^2\phi_{14}\phi_{15}\phi_{20}\phi_{30}$	
$E_6[\theta^2,\varepsilon']$	same	
$[3240_z]$	$q^9\phi_3^3\phi_4^2\phi_5\phi_6^3\phi_8\phi_9\phi_{10}\phi_{12}^2\phi_{15}\phi_{18}\phi_{20}\phi_{24}\phi_{30}$	(q^{31})
$[2268_x]$	$\frac{1}{2}q^{10}\phi_3^3\phi_4^2\phi_6^4\phi_7\phi_8\phi_9\phi_{12}^2\phi_{15}\phi_{18}\phi_{20}\phi_{24}\phi_{30}$	(q^{30})
$[972_x]$	$\frac{1}{2}q^{10}\phi_3^4\phi_4^2\phi_6^3\phi_8\phi_9\phi_{12}^2\phi_{14}\phi_{15}\phi_{18}\phi_{20}\phi_{24}\phi_{30}$	
$[1296_z]$	$\frac{1}{2}q^{10}\phi_2^4\phi_3^3\phi_6^4\phi_8\phi_9\phi_{10}^2\phi_{12}^2\phi_{14}\phi_{15}\phi_{18}\phi_{24}\phi_{30}$	
$D_4[9_1]$	$\frac{1}{2}q^{10}\phi_1^4\phi_3^4\phi_5^2\phi_6^3\phi_7\phi_8\phi_9\phi_{12}^2\phi_{15}\phi_{18}\phi_{24}\phi_{30}$	
$[2240_x]$	$\frac{1}{2}q^{10}\phi_2^4\phi_4^2\phi_5\phi_6^2\phi_7\phi_8\phi_{10}^2\phi_{12}\phi_{14}\phi_{15}\phi_{18}\phi_{20}\phi_{24}\phi_{30}$	(q^{28})
$[1400_{zz}]$	$\frac{1}{2}q^{10}\phi_4^2\phi_5^2\phi_7\phi_8^2\phi_{10}\phi_{12}^2\phi_{14}\phi_{15}\phi_{18}\phi_{20}\phi_{24}\phi_{30}$	
$[840_z]$	$\frac{1}{2}q^{10}\phi_4^2\phi_5\phi_7\phi_8^2\phi_9\phi_{10}^2\phi_{12}^2\phi_{14}\phi_{15}\phi_{20}\phi_{24}\phi_{30}$	
$D_4[4_4]$	$\frac{1}{2}q^{10}\phi_1^4\phi_3^2\phi_4^2\phi_5^2\phi_7\phi_8\phi_9\phi_{10}\phi_{12}\phi_{14}\phi_{15}\phi_{20}\phi_{24}\phi_{30}$	

Degrees of Unipotent Representations of $E_8(q)$ (Continued)

Representation	*Degree*	*Remarks*
$[4096_z]$	$\frac{1}{2}q^{11}\phi_2^7\phi_4^4\phi_6^4\phi_8^2\phi_{10}^2\phi_{12}^2\phi_{14}\phi_{18}\phi_{20}\phi_{24}\phi_{30}$	(q^{26})
$[4096_x]$	same	
$E_7[\xi,1]$	$\frac{1}{2}q^{11}\phi_1^7\phi_3^4\phi_4^4\phi_5^2\phi_7\phi_8^2\phi_9\phi_{12}^2\phi_{15}\phi_{20}\phi_{24}$	
$E_7[-\xi,1]$	same	
$[525_x]$	$q^{12}\phi_5^2\phi_7\phi_9\phi_{10}^2\phi_{14}\phi_{15}\phi_{18}\phi_{20}\phi_{24}\phi_{30}$	(q^{36})
$[4200_x]$	$\frac{1}{2}q^{12}\phi_4^2\phi_5^2\phi_7\phi_8^2\phi_9\phi_{10}\phi_{12}\phi_{14}\phi_{15}\phi_{18}\phi_{20}\phi_{24}\phi_{30}$	(q^{24})
$[840_x]$	$\frac{1}{2}q^{12}\phi_4^2\phi_5\phi_7\phi_8^2\phi_9\phi_{10}^2\phi_{12}\phi_{14}\phi_{15}\phi_{18}\phi_{20}\phi_{24}\phi_{30}$	
$[3360_z]$	$\frac{1}{2}q^{12}\phi_2^4\phi_5\phi_6^2\phi_7\phi_8^2\phi_9\phi_{10}^2\phi_{14}\phi_{15}\phi_{18}\phi_{20}\phi_{24}\phi_{30}$	
$D_4[2_1]$	$\frac{1}{2}q^{12}\phi_1^4\phi_3^2\phi_5^2\phi_7\phi_8^2\phi_9\phi_{10}\phi_{14}\phi_{15}\phi_{18}\phi_{20}\phi_{24}\phi_{30}$	
$[2800_z]$	$\frac{1}{2}q^{13}\phi_2^4\phi_5^2\phi_6^4\phi_7\phi_8\phi_{10}^2\phi_{14}\phi_{15}\phi_{18}\phi_{20}\phi_{24}\phi_{30}$	(q^{25})
$[700_{xx}]$	$\frac{1}{2}q^{13}\phi_4^2\phi_5^2\phi_7\phi_8\phi_{10}^2\phi_{12}^2\phi_{14}\phi_{15}\phi_{18}\phi_{20}\phi_{24}\phi_{30}$	
$[2100_x]$	$\frac{1}{2}q^{13}\phi_4^2\phi_5^2\phi_7\phi_8\phi_9\phi_{10}^2\phi_{12}^2\phi_{14}\phi_{15}\phi_{20}\phi_{24}\phi_{30}$	
$D_4[9_3]$	$\frac{1}{2}q^{13}\phi_1^4\phi_3^4\phi_5^2\phi_7\phi_8\phi_9\phi_{10}^2\phi_{14}\phi_{15}\phi_{20}\phi_{24}\phi_{30}$	
$[4536_z]$	$q^{13}\phi_3^3\phi_4^2\phi_6^3\phi_7\phi_8\phi_9\phi_{12}^2\phi_{14}\phi_{15}\phi_{18}\phi_{20}\phi_{24}\phi_{30}$	(q^{23})
$[2835_x]$	$q^{14}\phi_3^3\phi_5\phi_6^3\phi_7\phi_9\phi_{10}\phi_{12}^2\phi_{14}\phi_{15}\phi_{18}\phi_{20}\phi_{24}\phi_{30}$	(q^{22})
$[6075_x]$	$q^{14}\phi_3^4\phi_5^2\phi_6^4\phi_9\phi_{10}^2\phi_{12}^2\phi_{15}\phi_{18}\phi_{20}\phi_{24}\phi_{30}$	(q^{22})
$[4200_z]$	$q^{15}\phi_4^2\phi_5^2\phi_7\phi_8\phi_9\phi_{10}^2\phi_{12}\phi_{14}\phi_{15}\phi_{18}\phi_{20}\phi_{24}\phi_{30}$	(q^{21})
$[5600_z]$	$\frac{1}{2}q^{15}\phi_4^4\phi_5^2\phi_7\phi_8^2\phi_{10}^2\phi_{12}^2\phi_{15}\phi_{18}\phi_{20}\phi_{24}\phi_{30}$	(q^{21})
$[3200_x]$	$\frac{1}{2}q^{15}\phi_2^4\phi_4^2\phi_5^2\phi_6^2\phi_8^2\phi_{10}^2\phi_{12}\phi_{14}\phi_{15}\phi_{18}\phi_{20}\phi_{24}\phi_{30}$	
$[2400_z]$	$\frac{1}{2}q^{15}\phi_4^4\phi_5^2\phi_8^2\phi_9\phi_{10}^2\phi_{12}^2\phi_{14}\phi_{15}\phi_{20}\phi_{24}\phi_{30}$	
$D_4[8_1]$	$\frac{1}{2}q^{15}\phi_1^4\phi_3^2\phi_4^2\phi_5^2\phi_7\phi_8^2\phi_9\phi_{10}^2\phi_{12}\phi_{15}\phi_{20}\phi_{24}\phi_{30}$	

Representation	*Degree*	*Remarks*
$[2100_y]$	$q^{20}\phi_5^2\phi_7\phi_8^2\phi_9\phi_{10}^2\phi_{14}\phi_{15}\phi_{18}\phi_{20}\phi_{24}\phi_{30}$	
$[4480_y]$	$\frac{1}{120}q^{16}\phi_2^8\phi_5^2\phi_7\phi_8^2\phi_9\phi_{12}^2\phi_{14}\phi_{15}\phi_{18}\phi_{20}\phi_{24}\phi_{30}$	$(1,1)$
$[3150_y]$	$\frac{1}{6}q^{16}\phi_3^3\phi_5^2\phi_6^4\phi_7\phi_8^2\phi_{12}^2\phi_{14}\phi_{15}\phi_{18}\phi_{20}\phi_{24}\phi_{30}$	$(g_3,1)$
$[4200_y]$	$\frac{1}{8}q^{16}\phi_4^4\phi_5^2\phi_7\phi_8^2\phi_9\phi_{12}^2\phi_{14}\phi_{15}\phi_{18}\phi_{20}\phi_{24}\phi_{30}$	$(g_2',1)$
$[4536_y]$	$\frac{1}{24}q^{16}\phi_3^4\phi_4^4\phi_7\phi_8^2\phi_9\phi_{10}^2\phi_{14}\phi_{15}\phi_{18}\phi_{20}\phi_{24}\phi_{30}$	$(1,\nu)$
$[5670_y]$	$\frac{1}{30}q^{16}\phi_3^4\phi_5^2\phi_7\phi_8^2\phi_9\phi_{10}^2\phi_{12}^2\phi_{14}\phi_{18}\phi_{20}\phi_{24}\phi_{30}$	$(1,\lambda^1)$
$[420_y]$	$\frac{1}{5}q^{16}\phi_5^2\phi_7\phi_8^2\phi_9\phi_{10}^2\phi_{12}^2\phi_{14}\phi_{15}\phi_{18}\phi_{20}\phi_{24}\phi_{30}$	$(g_5,1)$
$[1134_y]$	$\frac{1}{6}q^{16}\phi_3^4\phi_6^3\phi_7\phi_8^2\phi_9\phi_{10}^2\phi_{12}^2\phi_{14}\phi_{15}\phi_{20}\phi_{24}\phi_{30}$	(g_3,ε)
$[1400_y]$	$\frac{1}{24}q^{16}\phi_4^4\phi_5^2\phi_6^4\phi_7\phi_8^2\phi_9\phi_{14}\phi_{15}\phi_{18}\phi_{20}\phi_{24}\phi_{30}$	$(1,\nu')$
$[2688_y]$	$\frac{1}{8}q^{16}\phi_2^8\phi_6^4\phi_7\phi_8^2\phi_9\phi_{10}^2\phi_{14}\phi_{15}\phi_{18}\phi_{20}\phi_{24}\phi_{30}$	(g_2',ε'')
$[1680_y]$	$\frac{1}{20}q^{16}\phi_4^4\phi_5^2\phi_7\phi_8^2\phi_9\phi_{10}^2\phi_{12}^2\phi_{14}\phi_{15}\phi_{18}\phi_{24}\phi_{30}$	$(1,\lambda^2)$
$[168_y]$	$\frac{1}{8}q^{16}\phi_4^4\phi_7\phi_8^2\phi_9\phi_{10}^2\phi_{12}^2\phi_{14}\phi_{15}\phi_{18}\phi_{20}\phi_{24}\phi_{30}$	(g_2',ε')
$[70_y]$	$\frac{1}{30}q^{16}\phi_5^2\phi_6^4\phi_7\phi_8^2\phi_9\phi_{10}^2\phi_{12}^2\phi_{14}\phi_{15}\phi_{18}\phi_{20}\phi_{24}$	$(1,\lambda^3)$
$[7168_w]$	$\frac{1}{12}q^{16}\phi_2^8\phi_4^4\phi_6^2\phi_7\phi_9\phi_{10}^2\phi_{12}\phi_{14}\phi_{15}\phi_{18}\phi_{20}\phi_{24}\phi_{30}$	$(g_2,1)$
$[1344_w]$	$\frac{1}{4}q^{16}\phi_2^4\phi_4^2\phi_6^2\phi_7\phi_8^2\phi_9\phi_{10}^2\phi_{12}\phi_{14}\phi_{15}\phi_{18}\phi_{20}\phi_{24}\phi_{30}$	$(g_4,1)$
$[2016_w]$	$\frac{1}{6}q^{16}\phi_2^4\phi_3^3\phi_4^2\phi_6^4\phi_7\phi_{10}^2\phi_{12}^2\phi_{14}\phi_{15}\phi_{18}\phi_{20}\phi_{24}\phi_{30}$	$(g_6,1)$
$[5600_w]$	$\frac{1}{6}q^{16}\phi_2^4\phi_4^2\phi_5^2\phi_6^4\phi_7\phi_9\phi_{10}^2\phi_{12}^2\phi_{14}\phi_{18}\phi_{20}\phi_{24}\phi_{30}$	(g_2,r)
$[448_w]$	$\frac{1}{12}q^{16}\phi_2^4\phi_4^2\phi_6^4\phi_7\phi_8^2\phi_9\phi_{10}^2\phi_{12}^2\phi_{14}\phi_{15}\phi_{18}\phi_{20}\phi_{30}$	(g_2,ε)
$D_4[4_1]$	$\frac{1}{4}q^{16}\phi_1^4\phi_3^2\phi_4^2\phi_5^2\phi_7\phi_8^2\phi_9\phi_{12}\phi_{14}\phi_{15}\phi_{18}\phi_{20}\phi_{24}\phi_{30}$	$(g_4,-1)$
$D_4[6_1]$	$\frac{1}{6}q^{16}\phi_1^4\phi_3^4\phi_4^2\phi_5^2\phi_6^3\phi_7\phi_9\phi_{12}^2\phi_{14}\phi_{15}\phi_{20}\phi_{24}\phi_{30}$	$(g_6,-1)$
$D_4[6_2]$	$\frac{1}{6}q^{16}\phi_1^4\phi_3^4\phi_4^2\phi_5^2\phi_7\phi_9\phi_{10}^2\phi_{12}^2\phi_{14}\phi_{15}\phi_{18}\phi_{20}\phi_{24}$	$(g_2,-r)$
$D_4[12_1]$	$\frac{1}{12}q^{16}\phi_1^4\phi_3^4\phi_4^2\phi_5^2\phi_7\phi_8^2\phi_9\phi_{12}^2\phi_{14}\phi_{15}\phi_{18}\phi_{20}\phi_{30}$	$(g_2,-1)$

Degrees of Unipotent Representations of $E_8(q)$ (Continued)

Representation	*Degree*	*Remarks*
$D_4[16_1]$	$\frac{1}{4}q^{16}\phi_1^4\phi_2^4\phi_3^2\phi_5^2\phi_6^2\phi_7\phi_8^2\phi_9\phi_{10}^2\phi_{14}\phi_{15}\phi_{18}\phi_{24}\phi_{30}$	(g_2',r)
$E_6[\theta,r]$	$\frac{1}{6}q^{16}\phi_1^6\phi_2^8\phi_4^4\phi_5^2\phi_7\phi_8^2\phi_{10}^2\phi_{14}\phi_{15}\phi_{18}\phi_{20}\phi_{24}$	(g_3,θ)
$E_6[\theta^2,r]$	same	(g_3,θ^2)
$E_6[\theta,r']$	$\frac{1}{6}q^{16}\phi_1^6\phi_2^8\phi_4^4\phi_5^2\phi_6^2\phi_7\phi_8^2\phi_{10}^2\phi_{12}\phi_{14}\phi_{18}\phi_{20}\phi_{30}$	(g_6,θ)
$E_6[\theta^2,r']$	same	(g_6,θ^2)
$E_8[\zeta]$	$\frac{1}{5}q^{16}\phi_1^8\phi_2^8\phi_3^4\phi_4^4\phi_6^4\phi_7\phi_8^2\phi_9\phi_{12}^2\phi_{14}\phi_{18}\phi_{24}$	(g_5,ζ)
$E_8[\zeta^2]$	same	(g_5,ζ^2)
$E_8[\zeta^3]$	same	(g_5,ζ^3)
$E_8[\zeta^4]$	same	(g_5,ζ^4)
$E_8[-\theta]$	$\frac{1}{6}q^{16}\phi_1^8\phi_2^6\phi_3^2\phi_4^4\phi_5^2\phi_7\phi_8^2\phi_9\phi_{10}^2\phi_{12}\phi_{14}\phi_{15}\phi_{20}$	$(g_6,-\theta)$
$E_8[-\theta^2]$	same	$(g_6,-\theta^2)$
$E_8[\theta]$	$\frac{1}{6}q^{16}\phi_1^8\phi_2^6\phi_4^4\phi_5^2\phi_7\phi_8^2\phi_9\phi_{10}^2\phi_{14}\phi_{20}\phi_{24}\phi_{30}$	$(g_3,\varepsilon\theta)$
$E_8[\theta^2]$	same	$(g_3,\varepsilon\theta^2)$
$E_8[i]$	$\frac{1}{4}q^{16}\phi_1^8\phi_2^8\phi_3^4\phi_5^2\phi_6^4\phi_7\phi_9\phi_{10}^2\phi_{14}\phi_{15}\phi_{18}\phi_{30}$	(g_4,i)
$E_8[-i]$	same	$(g_4,-i)$
$E_8^{II}[1]$	$\frac{1}{120}q^{16}\phi_1^8\phi_7\phi_8^2\phi_9\phi_{10}^2\phi_{12}^2\phi_{14}\phi_{15}\phi_{18}\phi_{20}\phi_{24}\phi_{30}$	$(1,\lambda^4)$
$E_8[-1]$	$\frac{1}{12}q^{16}\phi_1^8\phi_3^2\phi_4^4\phi_5^2\phi_7\phi_9\phi_{12}\phi_{14}\phi_{15}\phi_{18}\phi_{20}\phi_{24}\phi_{30}$	$(g_2,-\varepsilon)$
$E_8^{I}[1]$	$\frac{1}{8}q^{16}\phi_1^8\phi_3^4\phi_5^2\phi_7\phi_8^2\phi_9\phi_{14}\phi_{15}\phi_{18}\phi_{20}\phi_{24}\phi_{30}$	(g_2',ε)

Degrees of Unipotent Representations of $F_4(q)$

Representation	*Degree*	*Remarks*
$[1_1]$	1	(q^{24})
$[4_2]$	$\frac{1}{2}q\phi_2^2\phi_6^2\phi_8$	(q^{13})
$[2_1]$	$\frac{1}{2}q\phi_4\phi_8\phi_{12}$	
$[2_3]$	$\frac{1}{2}q\phi_4\phi_8\phi_{12}$	
$B_2[1]$	$\frac{1}{2}q\phi_1^2\phi_3^2\phi_8$	
$[9_1]$	$q^2\phi_3^2\phi_6^2\phi_{12}$	(q^{10})
$[8_1]$	$q^3\phi_4^2\phi_8\phi_{12}$	(q^9)
$[8_3]$	$q^3\phi_4^2\phi_8\phi_{12}$	(q^9)
$[12_1]$	$\frac{1}{24}q^4\phi_2^4\phi_3^2\phi_8\phi_{12}$	$(1,1)$
$[9_2]$	$\frac{1}{8}q^4\phi_3^2\phi_4^2\phi_8\phi_{12}$	$(g_2',1)$
$[9_3]$	$\frac{1}{8}q^4\phi_3^2\phi_4^2\phi_8\phi_{12}$	$(1,\lambda^1)$
$[1_2]$	$\frac{1}{8}q^4\phi_4^2\phi_6^2\phi_8\phi_{12}$	(g_2',ε')
$[1_3]$	$\frac{1}{8}q^4\phi_4^2\phi_6^2\phi_8\phi_{12}$	$(1,\lambda^2)$
$[4_1]$	$\frac{1}{8}q^4\phi_2^4\phi_6^2\phi_8\phi_{12}$	(g_2',ε'')
$[4_3]$	$\frac{1}{4}q^4\phi_2^2\phi_4\phi_6^2\phi_8\phi_{12}$	$(g_4,1)$
$[4_4]$	$\frac{1}{4}q^4\phi_2^2\phi_4\phi_6^2\phi_8\phi_{12}$	(g_2,ε'')
$[6_1]$	$\frac{1}{3}q^4\phi_3^2\phi_6^2\phi_8\phi_{12}$	$(g_3,1)$
$[6_2]$	$\frac{1}{12}q^4\phi_3^2\phi_4^2\phi_6^2\phi_8$	$(1,\sigma)$
$[16_1]$	$\frac{1}{4}q^4\phi_2^4\phi_4^2\phi_6^2\phi_{12}$	$(g_2,1)$
$B_2[r]$	$\frac{1}{4}q^4\phi_1^2\phi_2^2\phi_3^2\phi_6^2\phi_8$	(g_2',r)
$B_2[\varepsilon']$	$\frac{1}{4}q^4\phi_1^2\phi_3^2\phi_4\phi_8\phi_{12}$	$(g_4,-1)$

Degrees of Unipotent Representations of $F_4(q)$ (Continued)

Representation	*Degree*	*Remarks*
$B_2[\varepsilon'']$	$\frac{1}{4}q^4\phi_1^2\phi_3^2\phi_4\phi_8\phi_{12}$	(g_2,ε')
$F_4[\theta]$	$\frac{1}{3}q^4\phi_1^4\phi_2^4\phi_4^2\phi_8$	(g_3,θ)
$F_4[\theta^2]$	same	(g_3,θ^2)
$F_4[i]$	$\frac{1}{4}q^4\phi_1^4\phi_2^4\phi_3^2\phi_6^2$	(g_4,i)
$F_4[-i]$	same	$(g_4,-i)$
$F_4^{I}[1]$	$\frac{1}{8}q^4\phi_1^4\phi_3^2\phi_8\phi_{12}$	(g_2',ε)
$F_4^{II}[1]$	$\frac{1}{24}q^4\phi_1^4\phi_6^2\phi_8\phi_{12}$	$(1,\lambda^3)$
$F_4[-1]$	$\frac{1}{4}q^4\phi_1^4\phi_3^2\phi_4^2\phi_{12}$	(g_2,ε)

Degrees of Unipotent Representations of $G_2(q)$

Degree	*Remarks*
1	(q^6)
$\frac{1}{6}q\phi_2^2\phi_3$	$(1,1)$
$\frac{1}{2}q\phi_2^2\phi_6$	$(g_2,1)$
$\frac{1}{3}q\phi_3\phi_6$	$(1,r)$
$\frac{1}{3}q\phi_3\phi_6$	$(g_3,1)$
$\frac{1}{6}q\phi_1^2\phi_6$	$(1,\varepsilon)$, cuspidal
$\frac{1}{2}q\phi_1^2\phi_3$	(g_2,ε) , cuspidal
$\frac{1}{3}q\phi_1^2\phi_2^2$	(g_3,θ), cuspidal
$\frac{1}{3}q\phi_1^2\phi_2^2$	(g_3,θ^2), cuspidal

Degrees of Unipotent Cuspidal Representations of $^3D_4(q)$

Degree	*Remarks*
1	(q^{12})
$q\phi_{12}$	(q^7)
$\frac{1}{2}q^3\phi_2^2\phi_{12}$	
$\frac{1}{2}q^3\phi_2^2\phi_6^2$	
$\frac{1}{2}q^3\phi_1^2\phi_3^2$	cuspidal
$\frac{1}{2}q^3\phi_1^2\phi_{12}$	cuspidal

The notations in the following three tables are explained in 14.2.

Table of Unipotent Representations of $^2B_2(q)$, $q = 2^{m+\frac{1}{2}}$

Degree	(± 1)	(ζ''_8)	(ζ'_8)	*Remarks*
1	1	1	1	
q^4	1	−1	−1	
$\frac{1}{\sqrt{2}}q\phi_1\phi_2$	0	−1	1	cuspidal
$\frac{1}{\sqrt{2}}q\phi_1\phi_2$	0	−1	1	cuspidal

Table of Unipotent Representations of ${}^2F_4(q)$, $q = 2^{m+\frac{1}{2}}$

Degree	(± 1)	$(\zeta''_8, \pm 1)$	$(\zeta'_8, \pm 1)$	$(i, \pm 1)$	(ζ''_8)	(ζ'_8)	(ζ'_8, ζ''_8)	(i)	(ζ_{12})	(ζ''_{24})	(ζ'_{24})	*Remarks*
1	1	1	1	1	1	1	1	1	1	1	1	
q^{24}	1	–1	–1	–1	1	1	1	1	1	1	1	
$\frac{1}{\sqrt{2}} q\phi_1\phi_2\phi_4^2\phi_{12}$	0	–1	1	0	–2	2	0	0	0	–1	1	
$\frac{1}{\sqrt{2}} q\phi_1\phi_2\phi_4^2\phi_{12}$	0	–1	1	0	–2	2	0	0	0	–1	1	
$\frac{1}{\sqrt{2}} q^{13}\phi_1\phi_2\phi_4^2\phi_{12}$	0	–1	1	0	2	–2	0	0	0	1	–1	
$\frac{1}{\sqrt{2}} q^{13}\phi_1\phi_2\phi_4^2\phi_{12}$	0	–1	1	0	2	–2	0	0	0	1	–1	
$q^2\phi_{12}\phi_{24}$	1	1	1	–1	3	3	–1	–3	0	0	0	
$q^{10}\phi_{12}\phi_{24}$	1	–1	–1	1	3	3	–1	–3	0	0	0	
$\frac{1}{4} q^4\phi_4^2\phi''^2_8\phi_{12}\phi'_{24}$	2	0	0	0	0	–4	0	0	0	–1	0	
$\frac{1}{4} q^4\phi_4^2\phi'^2_8\phi_{12}\phi''_{24}$	2	0	0	0	–4	0	0	0	0	0	–1	
$\frac{1}{2} q^4\phi_8^2\phi_{24}$	2	0	0	0	0	0	0	2	–1	0	0	
$\frac{1}{12} q^4\phi_1^2\phi_2^2\phi'''^2_8\phi_{12}\phi'_{24}$	0	0	0	0	–4	0	0	–2	0	1	0	cuspidal

$\frac{1}{12}q^4\phi_1^2\phi_2^2\phi_8''^2\phi_{12}\phi_{24}''$	0	0	0	0	0	–4	0	–2	0	0	1	cuspidal
$\frac{1}{6}q^4\phi_1^2\phi_2^2\phi_4^2\phi_{24}$	0	0	0	0	–2	–2	–2	0	1	0	0	cuspidal
$\frac{1}{4}q^4\phi_1^2\phi_2^2\phi_4^2\phi_{12}\phi_{24}''$	0	0	0	0	1	–3	–1	0	0	0	1	cuspidal
$\frac{1}{4}q^4\phi_1^2\phi_2^2\phi_4^2\phi_{12}\phi_{24}''$	0	0	0	0	1	–3	–1	0	0	0	1	cuspidal
$\frac{1}{4}q^4\phi_1^2\phi_2^2\phi_4^2\phi_{12}\phi_{24}'$	0	0	0	0	–3	1	–1	0	0	1	0	cuspidal
$\frac{1}{4}q^4\phi_1^2\phi_2^2\phi_4^2\phi_{12}\phi_{24}'$	0	0	0	0	–3	1	–1	0	0	1	0	cuspidal
$\frac{1}{3}q^4\phi_1^2\phi_2^2\phi_4^2\phi_8^2$	0	0	0	0	0	0	0	0	–1	1	1	cuspidal
$\frac{1}{3}q^4\phi_1^2\phi_2^2\phi_4^2\phi_8^2$	0	0	0	0	0	0	0	0	–1	1	1	cuspidal
$\frac{1}{3}q^4\phi_1^2\phi_2^2\phi_{12}\phi_{24}$	0	0	0	0	2	2	–2	4	0	0	0	cuspidal

Table of Unipotent Representations of ${}^2G_2(q)$, $q = 3^{m+\frac{1}{2}}$

Degree	(± 1)	(ζ''_{12})	(ζ'_{12})	(i)	*Remarks*
1	1	1	1	1	
q^6	1	–1	–1	–1	
$\frac{1}{2\sqrt{3}} q\phi_1\phi_2\phi'_{12}$	0	–1	0	1	cuspidal
$\frac{1}{2\sqrt{3}} q\phi_1\phi_2\phi'_{12}$	0	–1	0	1	cuspidal
$\frac{1}{2\sqrt{3}} q\phi_1\phi_2\phi''_{12}$	0	0	1	–1	cuspidal
$\frac{1}{2\sqrt{3}} q\phi_1\phi_2\phi''_{12}$	0	0	1	–1	cuspidal
$\frac{1}{\sqrt{3}} q\phi_1\phi_2\phi_4$	0	–1	1	0	cuspidal
$\frac{1}{\sqrt{3}} q\phi_1\phi_2\phi_4$	0	–1	1	0	cuspidal

REFERENCES

[A_1] D. Alvis, The duality operation in the character ring of a finite Chevalley group, Bull. Amer. Math. Soc. 1 (1979), 907-911.

[A_2] ———, Induce/restrict matrices for exceptional Weyl groups, preprint, M.I.T., 1981.

[AL] D. Alvis and G. Lusztig, On Springer's correspondence for simple groups of type E_n $(n = 6,7,8)$, Math. Proc. Camb. Phil. Soc. 92 (1982), 65-72.

[As_1] T. Asai, Endomorphism algebras of the reductive groups over F_q of classical type, preprint.

[As_2] ———, On the zeta functions of the varieties $X(w)$ of the split classical groups and the unitary groups, Osaka J. Math. 20 (1983), 21-32.

[As_3] ———, Unipotent characters of $SO^{\pm}_{2n}$, Sp_{2n} and SO_{2n+1} over F_q with small q, to appear in Osaka J. Math.

[As_4] ———, Unipotent characters of exceptional groups over finite fields F_q with small q, preprint.

[BV_1] D. Barbasch and D. Vogan, Primitive ideals and orbital integrals in complex classical groups, Math. Ann. 259 (1982), 153-199.

[BV_2] ———, Primitive ideals and orbital integrals in complex exceptional groups, J. of Alg., 80 (1983), 350-382.

[B] C. T. Benson, The generic degrees of the irreducible characters of E_8, Comm. in Alg. 7 (1979), 1199-1209.

[BB] A. A. Beilinson and J. Bernstein, Localisation des g-modules, C. R. Acad. Sci. Paris 292 (1981), serie A, 15-18.

[BBD] A. A. Beilinson, J. Bernstein and P. Deligne, Faisceaux pervers, Astérisque 100 (1982).

[BGSu] C. T. Benson, L. C. Grove and D. B. Surowski, Semilinear automorphisms and dimension functions for certain characters of finite Chevalley groups, Math. Z., 144 (1977), 149-159.

[BC] C. T. Benson and C. W. Curtis, On the degrees and rationality of certain characters of finite Chevalley groups, Trans. Amer. Math. Soc. 165 (1972), 251-273, 202 (1975), 405-406.

[BL] W. M. Beynon and G. Lusztig, Some numerical results on the characters of exceptional Weyl groups, Math. Proc. Camb. Phil. Soc. 84 (1978), 417-426.

[BK] J. L. Brylinski and M. Kashiwara, Kazhdan-Lusztig conjecture and holonomic systems, Invent. Math. 64 (1981), 387-410.

[BM] W. Borho and R. MacPherson, Représentations des groupes de Weyl et homologie d'intersection pour les variétés nilpotentes, C. R. Acad. Sci. Paris t. 292 (1981), série A, 707-710.

[Bo] N. Bourbaki, Groupes et algèbres de Lie, Chap. IV, V, VI, Hermann, Paris, 1968.

[Ca] R. W. Carter, Conjugacy classes in the Weyl group, Comp. Math. 25 (1972), 1-59.

[CR] B. Chang and R. Ree, The characters of $G_2(q)$, Ist. Naz. di Alta Mat. Sympos. Math. vol XIII, 1974, 385-413.

[Ch] C. Chevalley, Classification des groupes de Lie algébriques, Séminaire, Paris 1956-58.

[C] C. W. Curtis, Reduction theorems for characters of finite groups of Lie type, J. Math. Soc. Japan 27 (1975), 666-688.

[CIK] C. W. Curtis, N. Iwahori, R. Kilmoyer, Hecke algebras and characters of parabolic type of finite groups with (BN) pairs, Publ. Math. IHES, 40 (1972), 81-116.

[D] P. Deligne, La conjecture de Weil II, Publ. Math., IHES 52 (1979), 138-252.

[DL_1] P. Deligne and G. Lusztig, Representations of reductive groups over finite fields, Ann. of Math. 103 (1976), 103-161.

[DL_2] ———, Duality for representations of a reductive group over a finite field, J. of Alg. 74 (1982), 284-291.

[DL_3] ———, Duality for representations of a reductive group over a finite field II, J. of Alg. 81 (1983), 540-549.

[DM_1] F. Digne and J. Michel, Nombre de points rationnels des variétés de Deligne-Lusztig et caractères de l'algèbre de Hecke, C. R. Acad. Sci. Paris, t. 287 (1978), série A, 811-814.

[DM_2] ———, Descente de Shintani des caractères d'un groupe de Chevalley fini, C. R. Acad. Sci. Paris, t. 291 (1980), série A, 571-574.

[F_1] J. S. Frame, The classes and representations of the groups of 27 lines and 28 bitangents, Ann. Mat. Pura Appl. 32 (1951), 83-119.

[F_2] J. S. Frame, The characters of the Weyl group E_8, Computational problems in abstract algebra, (Oxford conference 1967), J. Leech ed., 111-130.

[G] J. A. Green, The characters of the finite general linear groups, Trans. Amer. Math. Soc. 80(1955), 402-447.

[HC_1] Harish Chandra, The characters of reductive p-adic groups, in Contributions to Algebra, 175-182, Academic Press, New York (1977).

[HC_2] ———, Eisenstein series over finite fields, Functional Analysis and Related Fields, edited by F. E. Browder, Springer (1970).

[Hf] N. Hoefsmit, Representations of Hecke algebras of finite groups with BN pairs of classical type. Ph.D. dissertation, Vancouver, 1974.

[H] R. Howlett, Some irreducible characters of groups with BN pairs, J. of Alg. 39(1974), 571-592.

[HL] R. Howlett and G. Lehrer, Induced cuspidal representations and generalized Hecke rings, Invent. Math. 58(1980), 37-64.

[J] A. Joseph, Goldie rank in the enveloping algebra of a semisimple Lie algebra, J. of Algebra 66(1980), 269-283, 66(1980), 284-306, 73(1981), 295-326.

[Ka_1] N. Kawanaka, Fourier transforms of nilpotently supported invariant functions on a simple Lie algebra over a finite field, Invent Math. 69(1982), 411-435.

[Ka_2] ———, Liftings of irreducible characters of finite classical groups I, J. Fac. Sci. Univ. Tokyo, 28(1982), 851-861.

[K] D. Kazhdan, Proof of Springer's hypothesis, Israel J. Math. 28 (1977), 272-286.

[KL_1] D. Kazhdan and G. Lusztig, Representations of Coxeter groups and Hecke algebras, Invent. Math. 53(1979), 165-184.

[KL_2] ———, Schubert varieties and Poincaré duality, Proc. Symp. Pure Math. vol. 36(1980), Amer. Math. Soc., 185-203.

[Ki] R. Kilmoyer, Principal series representations of finite Chevalley groups, J. of Alg. 51 (1978), 300-319.

[Ko] T. Kondo, The characters of the Weyl group of type F_4, J. Fac. Sci. Univ. Tokyo 11(1965), 145-153.

[Le] G. I. Lehrer, The characters of the finite special linear groups, J. of Alg. 26(1973), 564-583.

[L_1] G. Lusztig, On the finiteness of the number of unipotent classes, Invent. Math. 34(1976), 201-213.

[L_2] G. Lusztig, Coxeter orbits and eigenspaces of Frobenius, Invent. Math. 28 (1976), 101-159.

[L_3] ———, Irreducible representations of finite classical groups, Invent. Math. 43 (1977), 125-175.

[L_4] ———, Representations of finite Chevalley groups, C.B.M.S. Regional Conference series in Math. 39, Amer. Math. Soc., 1978.

[L_5] ———, Unipotent representations of a finite Chevalley group of type E_8, Quart. J. Math. Oxford 30 (1979), 315-338.

[L_6] ———, A class of irreducible representations of a Weyl group, Proc. Kon. Nederl. Akad., series A, 82 (1979), 323-335.

[L_7] ———, A class of irreducible representations of a Weyl group II, Proc. Kon. Nederl. Akad., series A, 85 (1982), 219-226.

[L_8] ———, On the unipotent characters of exceptional groups over finite fields, Invent. Math. 60 (1980), 173-192.

[L_9] ———, On a theorem of Benson and Curtis, J. of Alg. 71 (1981), 490-498.

[L_{10}] ———, Green polynomials and singularities of unipotent classes, Adv. in Math. 42 (1981), 169-178.

[L_{11}] ———, Unipotent characters of the symplectic and odd orthogonal groups over a finite field, Invent. Math. 64 (1981), 263-296.

[L_{12}] ———, Unipotent characters of the even orthogonal groups over a finite field, Trans. Amer. Math. Soc. 272 (1982), 733-751.

[LS] G. Lusztig and B. Srinivasan, The characters of the finite unitary groups, J. of Alg. 49 (1977), 167-171.

[LV] G. Lusztig and D. Vogan, Singularities of closures of K-orbits on a flag manifold, Invent. Math. 71 (1983), 365-379.

[M_1] I. G. Macdonald, Zeta functions attached to finite general linear groups, Math. Ann. 249 (1980), 1-15.

[M_2] ———, Symmetric functions and Hall polynomials, Oxford Univ. Press (Clarendon), Oxford, 1979.

[Sh_1] T. Shoji, On the Springer representations of Chevalley groups of type F_4, Comm. in Algebra 8 (1980), 409-440.

[Sh_2] ———, On the Springer representations of the Weyl groups of classical algebraic groups, Comm. in Alg. 7 (16) (1979), 1713-1745, 7 (18) (1979), 2027-2033.

[S] N. Spaltenstein, Appendix, Math. Proc. Camb. Phil Soc. 92 (1982), 73-78.

[Sp_1] T. A. Springer, Regular elements of finite reflection groups, Invent. Math., 25 (1974), 159-198.

[Sp_2] T. A. Springer, Trigonometric sums, Green functions of finite groups and representations of Weyl groups, Inv. Math. 36 (1976), 173-207.

[Sr] B. Srinivasan, The characters of the finite symplectic group Sp(4, q), Trans. Amer. Math. Soc. 131 (1968), 488-525.

[St] R. Steinberg, A geometric approach to the representations of the full linear group over a Galois field, Trans. Amer. Math. Soc., 71 (1951), 274-282.

[Su] D. Surowski, Degrees of irreducible characters of BN pairs of types E_6 and E_7, Trans. Amer. Math. Soc., 243 (1978), 235-249.

[Sz] M. Suzuki, On a class of doubly transitive groups, Ann. of Math. 75 (1962), 105-145.

[Ve] J. L. Verdier, Dualité dans la cohomologie des espaces localement compacts, Sém. Bourbaki (1965), N° 300.

[Vo] D. Vogan, Ordering of the primitive spectrum of a semisimple Lie algebra, Math. Ann. 248 (1980), 195-203.

[Wa] H. N. Ward, On Ree's series of simple groups, Trans. Amer. Math. Soc. 121 (1966), 62-89.

SUBJECT INDEX

admissible arrangement, 162, 164
almost-character, 347
automorphism (ordinary-), 58

coherent lifting, 27
Coxeter complex, 192

duality homomorphism, 187

family, 78
  cuspidal-, 251
formal dimension, 61
Fourier transform, 324
Frobenius map, 5
ϕ-structure, 4

Gelfand-Graev representation, 347

Hecke algebra, xi, 60

intersection cohomology complex, 10
Iwahori algebra, xi

lagrangian subspace, 334
left cell, 137

perverse sheaf, 348

rational closure, 213

special
  -conjugacy class, 344
  -representation, 77
  -unipotent element, 343
Springer representation, 342
symbol, 83, 89

two-sided cell, 137

unipotent representation, 186

NOTATION INDEX

$P_{y,w}$, 3, 134

$k = \overline{F}_p, F_q$, 4

$f^*, R^i f^*, R^i f_!$, 5

$G, B, T, U, F : G \to G, \mathfrak{N}(T), W, \ell, \mathcal{B}$, 5

$\mathcal{B}_w, \mathfrak{L}$, 6

$\alpha_s, \check{\alpha}_s$, 7

$\mu_n, \mathcal{X}^L_{w,\psi}, \overline{\mathcal{X}}^L_{w,\psi}$, 8

$\overline{\mathcal{X}}^L_{\dot{w},\psi}$, $IC(X,\mathcal{F})$, $\mathcal{H}^i(X,\mathcal{F})$, $\mathbf{H}^i(X,\mathcal{F})$, 10

$\widetilde{\mathcal{X}}^L_{w,\psi}$, 11

$W_{L,n}, R^\vee_{L,n}$, 12

$R_{L,n}, S_{L,n}, \tilde{\ell}$, 13

$\hat{\mathcal{X}}^L_{w,\psi}$, 14

$\mathfrak{N}_{L',n}$, 23

$F' : G \to G, G^{F'}, X_w$, 30

$\widetilde{X}_{\dot{w}}, T_w, Z_{L,n}$, 31

$\psi, \mathcal{F}^L_{\dot{w},n}$, 34

$d, \overline{X}_w$, 35

γ, 47

$\mathcal{E}_{L,n}, \lambda'_\rho$, 54

$\widetilde{\mathbb{W}}, \mathbb{W}^\vee, \mathbb{W}^\vee_{ex}$, 58

$\widetilde{\mathbb{W}}_c$, 59

$H, \widetilde{H}, E(u)$, 60

$\widetilde{E}(u), P_{\mathbb{W}}(u), D_E(u)$, 61

$\mathcal{R}(G^{F'}), R_{\widetilde{E}}$, 67

a_E, b_E, A_E, f_E, 76

$J^{\mathbb{W}}_{\mathbb{W}'}, \mathcal{R}(\mathbb{W})$, 77

$j^{\mathbb{W}}_{\mathbb{W}'}$, 78

$\mathfrak{M}(\mathcal{G})$, 79

W_n, 82

$[\Lambda]$, 83, 89

Δ, 109, 124, 126

$\{\ ,\ \}$, 109, 124, 128, 130

$\mathfrak{M}(\mathcal{G} \subset \widetilde{\mathcal{G}}), \mathfrak{M}'(\mathcal{G} \subset \widetilde{\mathcal{G}})$, 114

$\overline{\mathfrak{M}}(\mathcal{G} \subset \widetilde{\mathcal{G}})$, M, M_c, 115

$[[\Lambda]]$, 117

$X(\mathbb{W})$, 125

$X(\mathbb{W}, \gamma), \overline{X}(\mathbb{W}, \gamma)$, 129

$(W_{L,n})^\vee_{\mathbb{Q}}$, 130

C_x, C'_x, 134

D_x, D'_x, τ, 135

$\underset{L}{\leq}, \underset{L}{\sim}$, 136

$\underset{LR}{\leq}$, $\underset{LR}{\sim}$, 137

$\mathcal{K}$, 139

$c_{\gamma x, \widetilde{E}}$, $c'_{\gamma x, \widetilde{E}}$, 141

$\mathcal{R}(\mathbf{W}, \gamma)$, $a_{\gamma x}$, a_x, 154

$\mathcal{A}_{\gamma x}$, $\mathcal{A}_x$, 156

$c(Z, \Phi, \hat{\Phi})$, 163, 164

$a(x)$, 178

ε_ρ, 186

$\hat{\varepsilon}_\rho$, $D(\rho)$, 187

$R_{\sigma'}$, 193

$\mathcal{E}^{I}_{L,n}$, $\mathcal{E}^{II}_{L,n}$, 202

$\mathcal{E}_{(s)}$, 257

$\delta[\chi]$, 267

$\widetilde{\lambda}_m$, 313

$\mathcal{B}_u$, $E_{u,\phi}$, $A(u)$, 342

$\bar{A}(u)$, 343

$\bar{A}(g) \subset \widetilde{A}(g) \supset \bar{A}(g) \cdot F'$, 344

$\Phi(C_\rho)$, 346

Irr (G), 349

Library of Congress Cataloging in Publication Data

Lusztig, George, 1946-
Characters of reductive groups over a finite field.

(Annals of mathematics studies ; no. 107)
Includes bibliographical references and indexes.
1. Finite groups. 2. Finite fields (Algebra)
3. Characters of groups. I. Title. II. Series.
QA171.L847 1984 512′.2 83-43083
ISBN 0-691-08350-9
ISBN 0-691-08351-7 (pbk.)

George Lusztig is Professor of Mathematics at the Massachusetts Institute of Technology.